Introduction to Statistics and Probability

(Through Sports)

Reza Noubary and Dong Zhang

Cover image © Shutterstock.com

Kendall Hunt
publishing company

www.kendallhunt.com
Send all inquiries to:
4050 Westmark Drive
Dubuque, IA 52004-1840

Copyright © 2020 by Kendall Hunt Publishing Company

ISBN: 978-1-7924-1112-0

All rights reserved. No part of this publication may be reproduced, stored in a retrieval system, or transmitted, in any form or by any means, electronic, mechanical, photocopying, recording, or otherwise, without the prior written permission of the copyright owner.

Published in the United States of America

Contents

Preface		*xi*
1 ELEMENTS OF STATISTICS		**1**
1.1	NUMERICAL SUMMARIES	4
1.2	FREQUENCY TABLE AND HISTOGRAM	17
1.3	POPULATION AND SAMPLING	25
1.4	INTERPRETING THE STANDARD DEVIATION	27
1.5	MEASURES OF RELATIVE STANDING	30
	1.5.1 Percentile Ranking	30
	1.5.2 z-score	30
	1.5.3 Interpretation of z-Score for Bell-Shaped Distribution	33
1.6	CASE STUDY: COMPARISON OF PROFESSIONAL PLAYERS AND TEAMS	33
	1.6.1 Performance Measures	33
	1.6.2 How Good Was Michael Jordan?	36
1.7	PAIRED DATA AND LINEAR CORRELATION COEFFICIENT	38
	1.7.1 Correlation Coefficient	38
	1.7.2 Using Correlation for Prediction	44
	1.7.3 Interpretation of Correlation Coefficient	50
	1.7.4 Regression-to-the-Mean	51
	1.7.5 Inference and Data Mining	51
1.8	USING R	53
2 ELEMENTS OF PROBABILITY		**57**
2.1	CHANCE EXPERIMENTS	58
2.2	EVENTS AND THEIR ALGEBRA	60
2.3	PROBABILITY	65
	2.3.1 Introduction and Definitions	65
	2.3.2 Permutations and Combinations	76
	2.3.3 Misleading Use of Probability and Statistics	83

2.4	CONDITIONAL PROBABILITY AND INDEPENDENCE	84
	2.4.1 Conditional Probability, Addition, and Multiplication Rules	84
	2.4.2 Independent Events	90
	2.4.3 More on The Multiplication Rule	93
	2.4.4 The Law of Total Probability	94
	2.4.5 Bayes' Theorem	96
	2.4.6 Independence of More Than Two Events	100
2.5	ANALYSIS OF A TENNIS MATCH	102
	2.5.1 Table Tennis	107
2.6	RANDOM VARIABLES AND THEIR DISTRIBUTION	108
2.7	EXPECTATION AND VARIANCE	116
	2.7.1 Lifetimes and Life Expectancies	124
2.8	SIMPSON'S PARADOX AND HOT HAND IN SPORT	126
	2.8.1 Introduction and Examples	126
	2.8.2 Weighted Averages	129
	2.8.3 Application to "Hot Hand" in Sport	130
2.9	COINCIDENCES	135
	2.9.1 Streaks	139
2.10	USING R	142
	2.10.1 Probability of random variables	142

3 SOME SPECIAL DISTRIBUTIONS — 145

3.1	BINOMIAL DISTRIBUTION (DISCRETE)	147
	3.1.1 Binomial Distribution	147
	3.1.2 Bernoulli Distribution and System Reliability	151
3.2	GEOMETRIC AND NEGATIVE BINOMIAL DISTRIBUTIONS (DISCRETE)	154
	3.2.1 Geometric Distribution	154
	3.2.2 Negative Binomial (Pascal) Distribution	157
3.3	HYPERGEOMETRIC DISTRIBUTION (DISCRETE)	159
	3.3.1 Hypergeometric Distribution	159
3.4	POISSON DISTRIBUTION (DISCRETE)	162
	3.4.1 Binomial-Poisson Hierarchy	165
3.5	UNIFORM DISTRIBUTION (CONTINUOUS)	166
3.6	NORMAL DISTRIBUTION (CONTINUOUS)	168
3.7	RELATIVES OF THE NORMAL DISTRIBUTION: CHI-SQUARE (χ^2), t AND F-DISTRIBUTIONS (ALL CONTINUOUS)	176
	3.7.1 Chi-Square Distribution	176
	3.7.2 t-Distribution	177
	3.7.3 F-Distribution	178

Contents　　　　　　　　　　　　　　　　　　　　　　　　　　　　　　　　v

　　3.8　EXPONENTIAL, GAMMA, BETA, LOGNORMAL, GUMBEL, WEIBULL AND FRÉCHET
　　　　　DISTRIBUTIONS (ALL CONTINUOUS)　　　　　　　　　　　　　　179
　　　　　3.8.1　Exponential Distribution　　　　　　　　　　　　　　179
　　　　　3.8.2　Gamma Distribution　　　　　　　　　　　　　　　　182
　　　　　3.8.3　Beta Distribution　　　　　　　　　　　　　　　　　184
　　　　　3.8.4　Lognormal Distribution (Another Relative of the Normal Distribution)　185
　　　　　3.8.5　Gumbel Distribution　　　　　　　　　　　　　　　　187
　　　　　3.8.6　Weibull Distribution　　　　　　　　　　　　　　　　189
　　　　　3.8.7　Fréchet Distribution　　　　　　　　　　　　　　　　192
　　　　　3.8.8　Maxwell Distribution　　　　　　　　　　　　　　　　192
　　　　　3.8.9　Pareto Distribution　　　　　　　　　　　　　　　　193
　　　　　3.8.10　Relationships Between Some Distributions　　　　　　194
　　3.9　THE POISSON PROCESS　　　　　　　　　　　　　　　　　　195
　　　　　3.9.1　Facts and Applications　　　　　　　　　　　　　　198
　　3.10 USING R　　　　　　　　　　　　　　　　　　　　　　　　　200

4　LIMIT PROPERTIES AND MODELS　　　　　　　　　　　　　　　203
　　4.1　THE MODEL OF SUMS: NORMAL DISTRIBUTION　　　　　　　　204
　　　　　4.1.1　The Central Limit Theorem　　　　　　　　　　　　　204
　　　　　4.1.2　The Normal Approximation to the Binomial Distribution　　206
　　4.2　POISSON LIMIT THEOREM　　　　　　　　　　　　　　　　　207
　　4.3　NORMAL APPROXIMATION: A SUMMARY　　　　　　　　　　　209
　　4.4　THE MODEL OF PRODUCTS: THE LOGNORMAL DISTRIBUTION　　211
　　4.5　THE MODEL OF EXTREMES: THE EXTREME VALUE DISTRIBUTIONS　212
　　4.6　EXCEEDANCES　　　　　　　　　　　　　　　　　　　　　215
　　　　　4.6.1　Return Periods　　　　　　　　　　　　　　　　　　217
　　　　　4.6.2　Exceedances and English Premier League　　　　　　219
　　　　　4.6.3　Characteristic Values　　　　　　　　　　　　　　　219
　　4.7　USING R　　　　　　　　　　　　　　　　　　　　　　　　220

5　ESTIMATION　　　　　　　　　　　　　　　　　　　　　　　　223
　　5.1　THE PROBLEM DESCRIPTION　　　　　　　　　　　　　　　224
　　5.2　SAMPLING DISTRIBUTION　　　　　　　　　　　　　　　　225
　　5.3　POINT ESTIMATOR　　　　　　　　　　　　　　　　　　　227
　　5.4　MAXIMUM LIKELIHOOD PRINCIPLE　　　　　　　　　　　　229
　　　　　5.4.1　Method of Moments　　　　　　　　　　　　　　　235
　　5.5　INTERVAL ESTIMATION AND CONFIDENCE INTERVALS　　　　236
　　5.6　BOOTSTRAP CONFIDENCE INTERVAL　　　　　　　　　　　249
　　　　　5.6.1　An Example of Bootstrapping　　　　　　　　　　　250
　　5.7　USING R　　　　　　　　　　　　　　　　　　　　　　　　252
　　　　　5.7.1　Using Known Distributions　　　　　　　　　　　　252

6 STATISTICAL TESTING — 255

- 6.1 STATISTICAL HYPOTHESIS — 257
- 6.2 TESTS — 258
- 6.3 Z-TEST — 260
  - 6.3.1 Testing Using Confidence Interval — 263
- 6.4 TWO-SAMPLE Z-TEST — 264
- 6.5 OBSERVED SIGNIFICANCE LEVEL, p-VALUES — 266
- 6.6 t-TEST — 268
  - 6.6.1 Test for a Population Mean — 268
  - 6.6.2 Test for a Population Correlation Coefficient — 270
- 6.7 LARGE-SAMPLE TEST OF HYPOTHESIS FOR POPULATION PROPORTION — 271
  - 6.7.1 Determination of Sample Size — 276
- 6.8 χ^2 TEST FOR VARIANCE (STANDARD DEVIATION) — 277
- 6.9 F-TEST — 279
- 6.10 OTHER ALTERNATIVE HYPOTHESIS — 282
- 6.11 MORE ON INFERENCE ABOUT TWO POPULATIONS — 284
  - 6.11.1 Large-Sample Inference About the Difference Between Two Population Means — 285
  - 6.11.2 Small-Sample Inference About the Difference Between Two Population Means (Normal Populations) — 287
  - 6.11.3 Inference About the Difference Between Two Population Means (Unequal Variances) — 290
  - 6.11.4 Inference About the Difference Between Two Population Means: Paired Difference Experiment — 292
  - 6.11.5 Fitness Test — 295
  - 6.11.6 Inference About the Difference Between Two Population Proportions — 297
  - 6.11.7 Case Study: 2018-2019 NBA MVP — 300
- 6.12 DISTRIBUTION FREE χ^2-TEST — 303
- 6.13 TEST OF INDEPENDENCE, CONTINGENCY TABLES — 306
  - 6.13.1 Further Discussions on "Hot Hand" in Sports — 309
  - 6.13.2 Case study: Comparison of Men and Women Professional Basketball Players — 311
- 6.14 KOLMOGOROV-SMIRNOV TEST — 315
- 6.15 SIGN TEST — 319
- 6.16 FINAL WORDS — 320
- 6.17 USING R — 320
  - 6.17.1 Hypothesis test of mean values — 320
  - 6.17.2 Hypothesis test of variances — 322
  - 6.17.3 χ^2-Test — 322
  - 6.17.4 Kolmogorov-Smirnov test — 323

7 ANALYSIS OF VARIANCE AND EXPERIMENTAL DESIGN — 325

- 7.1 COMPONENTS OF VARIANCE — 325
- 7.2 ONE-WAY CLASSIFICATION — 328
- 7.3 TWO-WAY CLASSIFICATIONS — 334
- 7.4 THE EXPERIMENTAL DESIGN — 338
  - 7.4.1 The Completely Randomized Design — 339
  - 7.4.2 The Randomized Block Design — 342
  - 7.4.3 Factorial Experiments — 346
- 7.5 FINAL WORDS — 349
- 7.6 USING R — 350

8 REGRESSION ANALYSIS — 353

- 8.1 INTRODUCTION — 353
- 8.2 SIMPLE LINEAR REGRESSION — 357
- 8.3 STATISTICAL INFERENCE FOR LEAST SQUARES ESTIMATORS — 362
- 8.4 ANOVA FOR SIMPLE LINEAR REGRESSION — 365
- 8.5 THE COEFFICIENT OF DETERMINATION: A MEASURE OF THE USEFULNESS OF THE MODEL — 367
- 8.6 APPLICATION OF SIMPLE LINEAR REGRESSION TO TRACK AND FIELD — 371
  - 8.6.1 Modeling and Prediction — 372
  - 8.6.2 More on Ultimate Records — 375
  - 8.6.3 Some Examples — 378
  - 8.6.4 Olympic Trends — 383
  - 8.6.5 An Alternative Regression Model — 384
  - 8.6.6 Estimation of Ultimate Record, An Example — 385
  - 8.6.7 Least Squares Using Matrices — 386
- 8.7 MULTIPLE LINEAR REGRESSION — 388
- 8.8 BEST SUBSET SELECTION AND STEPWISE REGRESSION — 392
- 8.9 APPLICATION — 394
  - 8.9.1 Why NBA Teams Win — 394
  - 8.9.2 Prediction of Medal Totals for Olympic Games — 402
- 8.10 DIFFICULTIES OF USING MULTIPLE REGRESSION — 405
  - 8.10.1 Exclusion of a Relevant Variable — 405
  - 8.10.2 Inclusion of an Irrelevant Variable — 406
  - 8.10.3 Incorrect Functional Form — 406
  - 8.10.4 Stepwise Regression — 406
  - 8.10.5 Proxy Variables and Measurement Error — 407
  - 8.10.6 Selection Bias — 408
  - 8.10.7 Multicollinearity and Singularity — 408
  - 8.10.8 Autocorrelation — 409
  - 8.10.9 Heteroskedasticity — 410
  - 8.10.10 Outliers — 410

	8.10.11 Influential Observations	410
	8.10.12 Misconception	411
	8.10.13 Ethical Issues	411
	8.10.14 Cross Validation	411
	8.10.15 Some Remedies	411
	8.10.16 Final Word	414
8.11	THE LOGISTIC MODEL	415
	8.11.1 Effect of the Star Player	421
8.12	USING R	423

9 TIME SERIES ANALYSIS — 427

9.1	STOCHASTIC PROCESSES	427
9.2	STATIONARY PROCESS	429
9.3	AUTOREGRESSIVE PROCESSES	433
9.4	FIRST-ORDER AR PROCESS	434
9.5	GENERAL-ORDER AR PROCESS	435
9.6	FORECASTING USING TIME SERIES	439
9.7	COMPONENTS OF TIME SERIES	441
9.8	SMOOTHING TECHNIQUES	443
9.9	TREND ANALYSIS (TREND PROJECTION)	456
9.10	ANALYSIS OF DATA FOR 100, 400, AND 800-METER RUNS	462
	9.10.1 Advanced Analysis (400 m and 800 m)	463
	9.10.2 Elementary Analysis Using Minitab (400 m)	471
	9.10.3 Time Series Analysis of the Men's 100-m Run	475
9.11	USING R	477

10 NONPARAMETRIC STATISTICS — 483

10.1	ORDER STATISTICS, RANKING THE BEST	483
10.2	DISTRIBUTION OF THE i-TH ORDER STATISTICS	484
10.3	JOINT DISTRIBUTION OF THE FIRST r-ORDER STATISTICS	486
10.4	THE PROBABILITY INTEGRAL TRANSFORMATION AND UNIFORM ORDER STATISTICS	487
10.5	DISTRIBUTION-FREE CONFIDENCE INTERVALS AND TESTS	489
	10.5.1 Single Sample Sign Test	491
	10.5.2 Run Test	495
	10.5.3 Wilcoxon (Mann-Whitney) Rank Sum Test	497
	10.5.4 Wilcoxon Matched-Pairs Signed Rank Test	501
	10.5.5 Paired-Sample Sign Test	503
	10.5.6 Kruskal-Wallis and Friedman Tests	505
	10.5.7 Spearman Rank Correlation	512

10.6 EXAMPLES OF APPLICATIONS — 514
- 10.6.1 Wilcoxon Rank Sum Test for Comparing Two Populations, Independent Samples — 515
- 10.6.2 Wilcoxon Signed Rank Test for Comparing Two Populations, Paired Difference Experiment — 516
- 10.6.3 Kruskal–Wallis H Test for a Completely Randomized Design — 518
- 10.6.4 The Friedman F_r Test for a Randomized Block Design — 519

10.7 USING R — 521
- 10.7.1 Sign Test — 521
- 10.7.2 Wilcoxon Test — 522
- 10.7.3 Friedman Test — 522
- 10.7.4 Spearman's correlation — 524

Preface

Data analysis and modeling are very fertile fields in modern scientific investigations. Their goals are to analyze and translate real-life situations into scientific terms; to use defined operations to arrive at solutions; and to convert the solutions into the real-life situations. No one can analyze the real world (by definition of analyze). One can only analyze a picture (conceptual model) of the world. The main requirements on such models are that they be accurate enough for the purposes at hand and that they be tractable enough for the needed accuracy.

Approaches to modeling and data analysis have changed a great deal in recent decades. Probability and statistics are important concepts used in modern approaches to mathematical modeling. By using probability and statistics data is decomposed into a smooth part and a rough part and models are developed for each part separately. The basic decomposition is the following:

$$\text{Datum} = \text{Systematic Part} + \text{Random Part}$$

or

$$\text{Datum} = \text{Deterministic Part} + \text{Stochastic Part}$$

or

$$\text{Message} = \text{Signal} + \text{Noise}$$

It is the attention given to the second component that perhaps most distinguishes probabilistic or statistical approaches to the modeling. In fact, mathematics mostly deals with the analysis and modeling of the systematic part. In the modern approach, however, models are judged based on their closeness to reality/signal-to-noise ratio. This ratio measures the contribution of the systematic part relative to the random part. Often a large value of this ratio occurs if the context of data is well understood. When this happens, usually only the systematic part is used as a model to describe the situation and more importantly to make predictions. The random part is then used to evaluate the model and to produce bounds on prediction errors. This is where probability and statistics demonstrate their power and value.

Today, the ideas of randomness are central to much of the modern scientific disciplines. Clearly, the real world can only be analyzed and explained scientifically through physical sciences. However, many scientists expressed their frustration when attempting to do this utilizing classical method based on determinism. Historically, it was natural for physicists to be interested, at first, in the macroscopic world that surrounds us. To make quantitative predictions about it, they devised deterministic models, which perform impeccably. Such are the origins of mechanics, of thermodynamics, of optics, of electromagnetism, and of relativity. These theories remain valid in the domains for which they were designed, and they continue in a state of vigorous development. But as regard fundamentals, physics today has its cutting edge on the microscopic level, where progress is achieved by means of probabilistic models, models that allow precise quantitative predictions for random phenomena. In short, chance is inherent in the basic nature of microscopic processes, reducing determinism to a mere consequence of chance regarding mean values that is on the macroscopic level.

In general, modeling could take a different course depending on the context of data. This is because the modeling process should incorporate the rules of a particular field of application. In this book, the field of application chosen is sports and subjects related to it. Our goal, of course, is not to study sports in depth but merely to use them as a motivating theme.

WHY SPORTS?

1. Students often ask why they should study mathematics and statistics. Our answer each time is a variation on the theme, to understand rules governing the real world, to model relationships between the variables, and to learn how to make sensible decisions in the face of uncertainty. Many textbooks try to motivate students by introducing varied applications. This address both students apparent desire to see the relevance of their studies to the outside world and also their skepticism about whether mathematics and statistics have any value. In recent years, several good textbooks have been published, which are based on the belief that applications to students experience or to their chosen career fields are the key to motivation. This usually works with students who are deeply committed to a particular academic or career field. However, even for some committed students, applied examples may fail to motivate because they are not usually of immediate concern to them. So, is the prospect of motivating students entirely bleak? Despite many voids in student interests and despite many ideas that fail to motivate them, students have some common interests that we can build on as we try to teach them mathematics and statistics. Based on our experiences, connecting their studies to a familiar subject with which they have interest or concern almost always works better.

    This book utilizes a motivational strategy different from those that may be found in popular textbooks. It suggests that motivating students to study mathematics and statistics may be accomplished by linking their study to familiar subjects such as games and sports. With this focus, we can help students to build their studies on a foundation an understanding of a sport already possessed by most of them. This approach is adaptable to all educational levels from junior high to graduate schools. Many students, whose eyes would normally glaze over after 10 min of mathematics or statistics, will happily spend hours analyzing their favorite sport.

2. Sports and the data related to them provide a unique opportunity to teach probability and statistics and to test the methods they offer. I believe that there is no teaching and research setting other than sports where we could collect reliable data with highest precision possible. In sports, in addition to the quality measurements, we have access to the names,

faces, and life history of every participant and their coaches, trainers, and everybody involved in the industry. Almost all other data-producing disciplines are susceptible to data mining and error, since unlike sports they are not watched by millions of the followers and also media representatives. Clearly a neat theoretical result can only work and be tested when data are reliable and satisfies the required conditions. If only one or few people produce data and its validity cannot be confirmed, we may end up disbelieving the statistical methodology used. In fact, historically this has been the case with statistics. When results do not make sense, people blame statistics and not the user. This is unlike medical sciences where doctors get the blame, not the medical sciences. In short, teaching and learning statistics through sports eliminates the possibility of data being the cause of drawing false conclusions. If through statistical modeling one predicts the next record or outcome of a tournament, we can all find out if this prediction was accurate or else. Research by Schultz and Liu Statistical Modeling in Track and Field published as a Chapter (9) in a book tilted *Statistics a Sports* pointed out the following regarding the track and field (athletics).

The nature and general availability of these data have resulted in their extensive use by researchers, teachers, and sports enthusiasts. The data are unique in that they (1) possess a meaning that is apparent to most people; (2) are collected under very constant and controlled conditions (carefully monitored by the International Association of Athletics Federations [IAAF]), and thus are very accurate and reliable; (3) are recorded with great precision (e.g., to the hundredth of a second in races), and thus permit very fine differentiation of change and/or differences; (4) are both longitudinal (almost 100 years for men records) and cross-sectional (over different distances and across gender); and (5) are publicly available at no cost (e.g., a full listing of the progression of men and women world records is available at http://www.uta/csmipe/sport). Thus, they provide wonderful data sets to test statistical models of change as well as being a rich source of information for studying the limits of human locomotion.

Since our classroom experience led to student excitement and success, we are very hopeful that this book will prove useful to both students and the instructors teaching probability and statistics. Further reasons for choosing sports as the theme of the book are

   a. Sports have a general appeal and it is an area to which modern scientific methods are increasingly applicable.
   b. Sports have become a part of everyday life, especially for young people.
   c. Students usually enjoy sports and show a great deal of interest in mathematics and statistics applied to the problems, related to them.
   d. A major part of the analysis and statistics sequence offered at college level can be taught using a chosen sport.
   e. Most students can relate to sports and can understand the meanings of the different statistics presented to them.
   f. For students who plan to become mathematics teachers both at the secondary level and in higher education, this subject provides a valuable source of material, which can enrich their teaching of many mathematical topics.

g. The American Statistical Association has started a separate section on Statistics and Sports. Also, sections are devoted to mathematics and sports in several recent Joint Meeting of American Mathematical Society and Mathematical Association of America.

WHAT IS COVERED?

The following is a brief description of the topics covered in this book:

- Descriptive statistics, probability, random variable, classical distributions, limiting distributions, sampling distributions, estimation, testing, analysis of variance, regression, and time series analysis including examples describing the applications of these concepts and methods to sports.
- Use of R to perform data analysis including regression and time series analysis.

CHAPTER 1

Elements of Statistics

We, as human beings, are equipped with many amazing tools for communicating with the world around us and for collecting and analyzing data. Of course, some people have better senses, such as better vision, that enables them to collect, let us say, "better" data and observe or understand the details. Data collection and analysis is a field that touches almost all aspects of human life. Every day hundreds, possibly even thousands of pieces of information are either given to us directly or are slipped by cleverly in research materials, news, advertisements, and product description or rating. Each of us weigh statistics in our minds when making decisions on everyday issues, such as which detergent to buy, what car to drive, and most importantly, how to take care of our health and well-being. With the onset of the twenty-first century and the rapid increase in health problems such as heart attack, cancer, AIDS, and so on, and the arrival of home or self-tests, we need to have a more thorough understanding of how to evaluate medical information (data). More generally and in fact, more importantly, we need to learn how to know ourselves through collecting data related to our responses to the situations that arise in our lives. Needless to say that people who know themselves are less likely to make decisions or take actions that lead to failure and disappointment.

But what happens after data are collected? Most of us use our brain to classify them as important, unimportant, useful, relevant, exciting, surprising, and so on. We do this type of analysis all the time, not because we are statisticians or because we have learned statistics, but because this is something inherent; we do it subconsciously. It is part of our survival mechanism. We need to analyze data and draw conclusions. Are we equally skillful when it comes to data collection and analysis? The answer is, of course, no. However, all of us can learn from each other and improve our skills. We can try other people's approaches and become smart data handlers. But where can we find information about such skills or approaches? Well, by talking to each other, by reading, listening, seeing, watching TV, and so on. For scientific and systematic approaches, we need to study statistics and learn techniques developed for doing inference when partial, but not complete, information is available. In short, we need to learn statistics in order to develop better skills for formulating informed judgments on what data are really telling us. We also need to learn statistics to understand research studies and the research methodologies that are often used or presented to us in newspapers, television, and scientific publications.

Remember that, as human beings, we also have certain limitations regarding both the data collection and their analysis. For instance, we cannot be in two different places or look at two separate objects far from each other, or listen to two people and pay attention to details at the same time. So we have to rely on data collected by others. This means the data being analyzed may be influenced by several factors outside our control or by factors that we do not have any knowledge about.

Are all data presented to us equally reliable? The answer is no and this may be a problem. In fact, more often than not we have no way to check the validity of the data, never mind its reliability. This is one important reason for using sports data in this book. Fortunately, sports are watched by millions of people and the data collected are usually very precise.

In summary, inference (learning from data or drawing conclusions about something based on partial information) is an integral part of our lives. We do that with or without knowledge of statistics. What statistics will teach us is a more logical and systematic way of doing inference. Let us now start from scratch and talk about statistics in a more scientific manner.

Statistics, usually referred to as a science, is indeed both a science and an art. In fact, in short, it may be defined as the art of learning from data. Data carry a message but by itself neither present a complete picture nor provides sufficient information. Therefore, learning from data and drawing conclusions from it requires more than mere science. To learn from data, we first need to describe it in a form that is easy to understand and helpful in noticing any regularities or patterns. The procedure of describing data may involve numbers, graphs, tables, and so on, is collectively referred to as descriptive statistics. Descriptive statistics are an important part of learning from data since they provide a starting point. Statistical inference is the major component of the science of statistics, and involves items such as estimation, prediction, testing a hypothesis, and drawing a conclusion or generalizing from the sample. For example, if we collect data on the heights and standing long jump scores for a group of boys and girls in a certain grade in middle school, we could use them to learn about these attributes individually or about their relation to each other for the population of boys and girls in that grade. We could, for instance, estimate the average height for girls, or test a hypothesis that taller boys can jump farther. We can also test for possible differences in performance between boys and girls. So putting these together, statistics, in a narrow sense, is a branch of science that deals with making inferences about populations based on samples. In a broader sense, it encompasses collection, organization, and summarization of data, presentation of data in tabular, graphical, and numerical form, developing models for the purpose of understanding random and nonrandom phenomena, use of models for prediction, mathematical approaches to decision-making, evaluation of risks, and so forth.

Statistical inference involves learning about the whole from a part of the whole. It uses inductive reasoning and, hence, has an uncertain component that requires quantification. The systematic assessment and quantification of uncertainty involves study of the probability theory. It is necessary in measuring the reliability (or margin of error) of the inference. The probability theory applies deductive reasoning and in this sense has more similarities with mathematics. In summary, probability and statistics are simply tools to help in making decisions. The former uses deductive inference, that is, reasoning from the general to the particular, while the latter uses inductive inference, that is, reasoning from the particular to the general. With deductive inference, the concern is with the validity of the argument. With inductive inference, the concern is with the plausibility of the conclusion given that the premises are true.

Statistical inference or inferential statistics may, in general, include the following components:

Chapter 1: Elements of Statistics

1. A population of interest made up of the entire collection of individuals or objects about which information is desired, together with a procedure describing a scheme to draw a sample from it. Ordinarily, members of a population have a common property.
2. Methods for analyzing information contained in the sample (data).
3. Logic and reasonings for drawing conclusions and methods for generalizing from the sample to the population.
4. Methods and procedures for providing measures for reliability of the inferences, or equivalently, for quantitative assessment of the margins of errors.

Statistical inference is often made based on the available information in the sample. Samples (data) may be considered as a message or evidence. One goal of statistical inference is to provide communication tools to understand the messages and, if possible quantify the evidence. Samples carry information about the population or its "parameter." Data could be of different types. The main classifications are

1. Numerical (quantitative): when data are observations that are measured on a numerical scale. That is, they can be given a numerical value and respondents can be ordered according to those values.
2. Categorical (qualitative): when observations are categorical responses. That is, they can be placed into categories, but that they may not have any logical ordering.

For example, suppose that at the beginning of a school year all the fifth graders in an elementary school were asked how many days (or hours) they had spent in a swimming pool during the summer break. The results would be a group of numbers or, as defined above, numerical data. However, if they were asked whether they had spent any time swimming at all, their answers would be either "yes" or "no." The results would then form categorical data containing "yes"s and "no"s. As another example, we could take a sample of college students and record their ages. This would yield numerical data, whereas recording their majors would result in categorical data. Note that it is important to identify the type of data collected or to be collected since different data analysis methods have been developed for each type of data.

Most of the methods discussed in this book are applicable to numerical data. Numerical data are classified according to whether variables observed or measured are "discrete" or "continuous." This classification scheme is known as gappiness and to apply that, one needs to determine whether gaps exist between successively observed values of the variable of interest. If gaps exist between observations or between the possible values that could be observed, the variable is said to be discrete; if no gaps exist, the variable is said to be continuous. More precisely, a variable is discrete if, between any two potentially observable values, a value exists that is not observable. A variable is continuous if, between any two potentially observable values, another potentially observable value exists. In short, a discrete variable is one for which we could actually count its' possible values. Conversely, a continuous variable can take any value within a given interval. In layman's terms, think about the difference between the number of something and the amount of something.

Some examples of continuous variables are age, blood pressure, cholesterol level, height, weight, and time to the first goal in a soccer game, and so on. Some examples of discrete variables are number of goals scored in a soccer game, number of days a randomly selected student visits the recreation center or library, number of heads observed when flipping a coin 50 times, sex (e.g., 0 if male and 1 if female), number of injuries, group identification (e.g., 1 if member of team A and 2 if member of team B), and state of disease (e.g., 1 if a coronary heart disease case and 0 if not a coronary heart disease case).

We could also be more precise and define a discrete variable as the one for which possible observable values are either finite or infinite, but countable. Otherwise, the variable is continuous.

A PRACTICE PROBLEM Before leaving a particular restaurant, customers are asked to respond to the questions listed below. For each question, determine whether the possible response are numerical or categorical. If numerical, specify whether discrete or continuous.

(a) What is the approximate distance of the restaurant from your residence.
(b) Have you eaten at the restaurant previously?
(c) If your answer to part (b) was yes, on how many occasions?
(d) Would your overall rating of the restaurant be excellent, good, fair, or poor?
(e) What is the area code of the place you live?

Numerical data can be summarized by listing the number of observations that fall in each of the possible ranges. Such numbers are called frequency. For example, to study the sample containing hours spent in a swimming pool, we may consider five ranges: less than 20 hours, 20 to 40 hours, 40 to 60 hours, 60 to 80 hours, and more than 80 hours. We may then summarize the data by counting the number of observations in each range and providing a frequency distribution. Categorical data can be summarized similarly by counting the number of, for example, "yes"s and "no"s, or by counting the number of students in different majors attending a certain university. This type of summarization will be discussed in Section 1.2. We note that this is one way for describing and summarizing the data. Many methods have been developed for this purpose, especially for numerical data. The following sections present a few that are used frequently.

1.1 NUMERICAL SUMMARIES

Let $x_1, \ldots, x_N$ represent a set of data (sample), for example, winning times in N successive Men's 100-m run in Olympic competitions, or number of the goals scored in N successive soccer matches in Europe, or the College Board scores for N Notre Dame football players. Here N is called the sample size. We write $x'_1, \ldots, x'_N$ to represent data when rearranged according to their magnitude. The following is an example:

EXAMPLE 1.1 Consider the data presenting the winning times in the Men's 100-m Freestyle Olympic swimming competitions from 1980 to 2016.

Year	1980	1984	1988	1992	1996	2000	2004	2008	2012	2016
Time (s)	50.4	49.8	48.6	49.0	48.7	48.3	48.2	47.2	47.5	47.6

Chapter 1: Elements of Statistics

For this data, $N = 10$, $x_1 = 50.4$, $x_2 = 49.8$, and so on. The last value in the sample is $x_{10} = 47.6$. If we rearrange the list so that the values are written in ascending order from the smallest to the largest, we obtain $x'_1, \ldots, x'_{10}$ respectively as

$$47.2, 47.5, 47.6, 48.2, 48.3, 48.6, 48.7, 49.0, 49.8, 50.4$$

That is, $x'_1 = 47.2, x'_2 = 47.5$, and so on. Note that rearranging the data are useful for sport-related measurements, since here we may be interested in records that usually appear either in the beginning or the end of the list. As an example, consider the data representing times to replace a Winston Cup car's four tires in the world Championships from 1967 to 2000. The end of the list for rearranged data provide the fastest times. In fact, as of year 2002 they are; Bill Elliot's crew: 20.87 s, Mike Skinner's crew: 20.322, Jeff Gordon's crew: 19.363, Bobby Labonte's crew: 19.166, and Jeff Burton's crew: 18.355.

Mean

In practice, it is useful to summarize the data using certain numerical measures. For example, we may use the mean value (i.e., the average) of our data, denoted by $\bar{x}$ to describe its central tendency or just "center." One such a measure is the sample mean defined as

$$\bar{x} = \frac{1}{N}(x_1 + \ldots + x_N) = \frac{1}{N}\sum_{i=1}^{N} x_i, \tag{1.1}$$

where the symbol Σ means add or find the sum. Applying this definition, the mean value of the data in Example 1.1 is

$$\bar{x} = \frac{50.4 + 49.8 + \ldots + 47.6}{10} = 48.53.$$

Thus, the sample mean can be found by adding all the measurements and dividing that by the total number of measurements (sample size). Note that the mean is the point at which measurements to the left of the mean balance those to the right (see illustration). If the largest measurement in the sample in Example 1.1 shifts to the right, then the mean (center of gravity or the balance point) must shift with it.

Applying this to different professional sports in the United States, one finds the following average daily salaries for the players of the different leagues for the year 2016–2017: National Basketball Association (NBA): \$16,986; Major League Baseball (MLB); \$12,055; National Hockey League (NHL): \$7,945; and National Football League (NFL): \$5,753.

Looking at the samples, one finds that some measurements or values are very "close" to the sample mean and some are "far" from it. For each value x_i in the sample, the difference $x_i - \bar{x}$ is called the deviation of this value from the mean. For instance, in Example 1.1, the winning time for the year 1980 is 50.4 (s). The deviation of this value from the mean 48.53 is $50.4 - 48.53 = 1.87$. Note that this deviation is positive representing the fact that the value 50.4 is "above average." Deviations of other values can be computed similarly. Some of these deviations are positive and some are negative, showing that some values are above average and others are below it. In general, a deviation could be positive, negative, or even zero if the corresponding value is exactly equal to $\bar{x}$. It is important to note that the sum of all deviations from the mean is always zero, that is,

$\sum_{i=1}^{N}(x_i - \bar{x}) = 0$. We hope that students can, by using the definition of the sample mean, figure out why this is always true.

PRACTICE PROBLEMS

(1) The following data represent a number of FIFA World Cup championships by soccer superpowers from start to the 2018 World Cup.

Country	Brazil	Germany	Italy	Argentina	England	France
Title	5	4	4	2	1	2

Find the mean of this data. Find also the mean number of titles won by four European superpowers.

(2) According to the National Weather Service, the number of deaths in the United States as a result of lightning during the time period of 2008–2017 were

Year (20..)	08	09	10	11	12	13	14	15	16	17
Number of Deaths	29	35	29	26	29	23	26	28	40	16

Find the mean number of deaths in this period. As an example of statistical inference, see if you can predict the number of deaths for the year 2004. Find the exact value from their website and compare that with your prediction.

Standard Deviation

Average is a useful numerical summary. It could, for example, be used to compare two players to decide who had a better overall performance (like comparing two students based on their grade point averages [GPA]). However, the urge to average can be seductive. Recall the story about the man who reports that, though his head is in the oven and his feet are in the refrigerator, he's pretty comfortable on average. Or a soldier who shot the enemy above the head and below the feet, claiming that on average the enemy should be dead. Sometimes a reliance on averages can have serious consequences. Suppose a doctor tells you that you have a dreaded disease, the average victim of which lives for 5 years. If this is all you know, there may be some reason for hope. Perhaps two-thirds of the people who have this disease die within a year of its onset, and one-third of its victims live from 10 to 40 years. The point is, if you know only the average survival time and nothing about the distribution of the survival times, it is difficult to plan intelligently. So we need measures revealing other characteristics of the distribution from which sample is selected.

The fact that the average value of some quantity is 100 might mean that all values of this quantity are between 95 and 105, or that half of them are around 50 and half around 150, or that a fourth of them are zero, half of them are near 50, and a fourth of them are approximately 300, or any number of other distributions that have the same average. Figure 1.1 presents three samples with the same mean and median but different amount of variability. Recall the example of comparing two players. As pointed out, mean may be used as a measure of overall performance, but how do we decide who is more consistent? In fact, some coaches think that consistency is a better measure for performance than many other statistics often used.

Chapter 1: Elements of Statistics

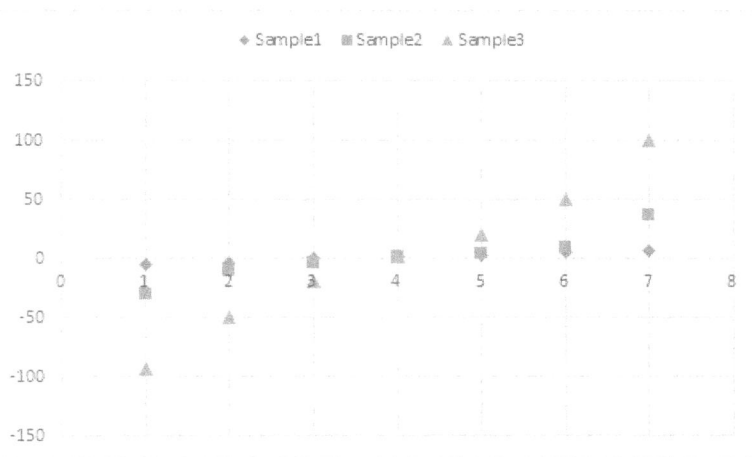

FIGURE 1.1 Three Samples with the Same Mean (9/7) and Median (2) but Different Amounts of Variability

The "average distance" of the numerical values from their "center" is described by variance and standard deviation, which are measures for variation or variability. More precisely, the variance of a sample is defined as

$$s^2 = \frac{1}{N-1} \sum_{i=1}^{N} (x_i - \bar{x})^2 \qquad (1.2)$$

and the standard deviation of a sample is defined as the positive square root s (>0) of the variance s^2, that is,

$$s = \sqrt{\frac{1}{N-1} \sum_{i=1}^{N} (x_i - \bar{x})^2} \qquad (1.3)$$

You may wonder why the sum of squares of deviations is divided by $N-1$ rather than N. Since $\sum_{i=1}^{N}(x_i - \bar{x}) = 0$ the sum $\sum_{i=1}^{N}(x_i - \bar{x})^2$ has $N-1$ "degrees of freedom." That is, once $N-1$ deviations are specified the N-th deviation can be calculated. This is one reason for using $N-1$ rather than N in the expression for s^2. It will also become clear, in later chapters, that when estimating a population variance based on sample variance, dividing by $N-1$ will yield an estimate with certain desirable statistical property. Note also that

$$\sum_{i=1}^{N}(x_i - \bar{x})^2 = \sum_{i=1}^{N} x_i^2 - N\bar{x}^2 = \sum_{i=1}^{N} x_i^2 - \frac{1}{N}(\sum_{i=1}^{N} x_i)^2.$$

Hence the variance s^2 and the standard deviation s can also be computed from

$$s^2 = \frac{1}{N-1} (\sum_{i=1}^{N} x_i^2 - N\bar{x}^2) \qquad (1.4)$$

and

$$s = \sqrt{\frac{1}{N-1} (\sum_{i=1}^{N} x_i^2 - N\bar{x}^2)} \qquad (1.5)$$

or

$$s^2 = \frac{1}{N-1}\left(\sum_{i=1}^{N} x_i^2 - \frac{1}{N}\left(\sum_{i=1}^{N} x_i\right)^2\right)$$

and

$$s = \sqrt{\frac{1}{N-1}\left(\sum_{i=1}^{N} x_i^2 - \frac{1}{N}\left(\sum_{i=1}^{N} x_i\right)^2\right)}.$$

Using the definitions, the variance and standard deviation of the data in Example 1.1 are, respectively,

$$s^2 = 1.025, \quad s = 1.012.$$

EXAMPLE Find the standard deviation of the following data,

$$1, 2, 3, 4$$

SOLUTION:

$$\sum_{i=1}^{4} x_i = 10, \quad \sum_{i=1}^{4} x_i^2 = 30, \quad s = \sqrt{\frac{1}{4-1}\left(30 - \frac{1}{4}(10)^2\right)} = 1.29$$

A PRACTICE PROBLEM The following represents the number of goals scored in six soccer games. Find the standard deviation of this data

$$1, 2, 1, 3, 5, 4$$

Note that if, for instance, we double all the x-values, then clearly $\bar{x}$ will be doubled too. It is also evident from formula (1.3) that s will also be doubled. This means that if we change the scale of measurements, both mean and the standard deviation will change accordingly but their ratio will remain unchanged. This is a motivation to define the so-called coefficient of variation, a scale-invariant measure (see below). It is interesting to note that, if we shift the data by adding or subtracting a constant from all data values, the mean will be increased or decreased accordingly, but the variance will remain unchanged. This is understandable since shifting or moving the data will not change the spread.

In summary, if the data are transferred using a linear relationship

$$y = a + bx$$

then

$$\bar{y} = a + b\bar{x}, \quad s_y = bs_x.$$

Here s_y and s_x refer to the standard deviation of x_i's and y_i's, respectively.

Chapter 1: Elements of Statistics

PRACTICE PROBLEMS

1. Find the mean and the standard deviation of the following:

$$1, 2, 3, 4, 5, 6, 7, 8, 9, 10$$

2. Without repeating the calculation find the means and the standard deviations of the following data sets:

   (a) 2, 3, 4, 5, 6, 7, 8, 9, 10, 11
   (b) 2, 4, 6, 8, 10, 12, 14, 16, 18, 20
   (c) 3, 5, 7, 9, 11, 13, 15, 17, 19, 21

Coefficient of Variation

The coefficient of variation, denoted by cv, is the ratio of s to $\bar{x}$, that is,

$$cv = s/\bar{x}$$

cv stands for the relative measure of the "average distance of the data values from the center." For example, if we compare two data sets, such as $\{999999, 1000000, 1000001\}$ and $\{1,2,3\}$, we find that standard deviation for both data sets is 1. However, just by looking at these values it is easy to see that such a dispersion is insignificant for data in the millions such as the first data set, but is quite significant for the second data set. Indeed in this case, the cv for the first data set is only 10^{-6}, whereas for the second data set is 1.

Note also that, as mentioned, cv is "scale invariant" in the sense that it is the same regardless of what scale is used to measure the observations. This is useful when data measured in one scale, for example pounds is changed to another scale, for example, kilos. For example, if we triple all the measurements, the mean and the standard deviation of the new data set are three times the original data but $cv = (3\bar{s})/3\bar{x}$ stays the same. We hope students could verify this for any constant.

A PRACTICE PROBLEM The following data sets present times (seconds) for men's 100- and 200-m races in the 2016 Rio Olympics. Find the coefficient of variation for each race and explain what conclusion may be drawn from them.

100 m:	9.81	9.89	9.91	9.93	9.94	9.96	10.04	10.06
200 m:	19.78	19.80	20.08	20.09	20.43	20.43	20.59	20.60

We end this part by noting that, although one can consider several other numerical summaries for data description, for most practical problems the mean and standard deviation used as numerical summaries of data are proved to be sufficient. Of course, there are situations where additional numerical summaries are necessary for better inference. There are also cases where some or all data are censored (their exact values are not known but it is known that they are either larger or smaller than a given value) so that numerical summaries such as sample mean and sample variance cannot be calculated. For such cases, other numerical measures quantifying the same characteristics are usually used. In the rest of this section, some of these measures are introduced and discussed.

Median

Another useful numerical measure for central tendency is the median, which represents the "middle value" in the data. It is a value with the property that 50% of data values are less than or equal to and 50% are larger than or equal to that value. When sample size is odd, after rearranging the data in ascending order, the sample median equals the value in the middle of the list. If there are even number of values in the data set, then the median is defined as the average of the two "middle values" in the rearranged list. More precisely,

$$\text{Median} = \begin{cases} x'_{\frac{N+1}{2}} & \text{if } N \text{ is odd} \\ \frac{1}{2}(x'_{\frac{N}{2}} + x'_{\frac{N}{2}+1}) & \text{if } N \text{ is even} \end{cases} \quad (1.6)$$

For example, the medians of the data sets 2, 3, 4, 1, 5 and 2, 3, 4, 1, 5, 6 (after rearranging) are, respectively, 3 and 3.5.

EXAMPLE Suppose I own a small business and have 10 employees. I pay each employee $30,000 and pay myself $140,000 per year. Then the mean salary is 440,000/11 = 40,000, whereas the median salary is 30,000. It is important to note that here 10 out of 11 (91%) of the employees make less than mean salary. More examples are presented below.

A PRACTICE PROBLEM Find the median of the following data sets

(a) 1, 3, 1, 2, 4
(b) 1, 2, 3, 1, 5, 4

Range

The range of a sample is defined as $R = x'_N - x'_1$. This is easy to calculate and provides a simple measure for the spread of data.

EXAMPLE 1.2 For the data given in Example 1.1, median equals to

$$(48.3 + 48.6)/2 = 48.45$$

and the range is

$$R = 50.4 - 47.2 = 3.2.$$

EXAMPLE 1.3 The following table lists the records of the American League Home Run Leaders from 2003 to 2007.

Year	Player	Home Runs
2003	Alex Rodriguez	47
2004	Manny Ramirez	43
2005	Alex Rodriguez	48
2006	David Ortiz	54
2007	Alex Rodriguez	54

Chapter 1: Elements of Statistics

For this data $N = 5$, $x_1 = 47$, $x_2 = 43$, $x_3 = 48$, $x_4 = 54$, and $x_5 = 54$. Thus,

$$\bar{x} = \frac{47 + 43 + 48 + 54 + 54}{5} = 49.2$$

$$s^2 = \frac{(47 - 49.2)^2 + (43 - 49.2)^2 + (48 - 49.2)^2 + (54 - 49.2)^2 + (54 - 49.2)^2}{5 - 1}$$

$$= \frac{90.8}{4} = 22.7, \quad s = 4.76.$$

We can use the alternative formula and compute s^2 as

$$s^2 = \frac{1}{5-1}[(47)^2 + (43)^2 + (48)^2 + (54)^2 + (54)^2 - \frac{(47 + 43 + 48 + 54 + 54)^2}{5}] = 22.7.$$

Also, rearranging the data in ascending order we get

$$43 \quad 47 \quad 48 \quad 54 \quad 54$$

Hence the median and the range of this data are, respectively, 48 and $54 - 43 = 11$.

The sample mean and the sample median can also be computed when the values are presented in a frequency table. We will see an example of this in the next section after the introduction of frequency tables.

Trimmed Mean

At this point, one may wonder which of the measures for central tendency is "better." To answer this, first note that the median is not sensitive to the extreme values (outliers). For example, if we add 50 points to the largest measurement in the sample, the median remains unchanged, whereas mean may change significantly (check this with the data given above). However, since mean uses more information (all data values) it is considered a better measure, at least theoretically. Another measure that has become popular offers a compromise. It is called <u>trimmed mean</u>. It is computed by first ordering the data values from smallest to largest, deleting a specific number of values from each end of the ordered list, and finally averaging the remaining values. The trimming percentage is the percentage of values deleted from each end of the ordered list. For example, to find a 20% trimmed mean for data in Example 1.3, first we rearrange the data from the smallest to the largest. Then, since 20% of five is one, we drop the smallest and the largest values and average the rest. By doing this, we obtain $\frac{47+48+54}{3} = 49.67$. The trimmed mean, or trimmed total points is used in sports to eliminate scores given by the judges from the countries competing for the top positions. In other words, it is an attempt to eliminate the scores in either end of the scale, given by friends and enemies.

As you may have guessed, each of the measures introduced has its own advantages and disadvantages. For example, median and trimmed mean are less sensitive to outliers, that is, values in a data set that are much larger or much smaller than other values. Also, there are cases (e.g., sensored data) where mean cannot even be calculated. For example, suppose that 100 light bulbs are installed in a big room. If we register their lifelength as they die (fail), we get a set of ordered times or numbers. After registering 51 failure times, we know the value of the median, whereas to find the value of the mean we need to wait until all bulbs fail. This may require a long waiting period.

One interesting place where the difference between these measures can be demonstrated is in soccer, where most games are very low scoring. Here, a few high-scoring games in a tournament can change the mean value significantly. For example, in the 1954 World Cup Austria defeated Switzerland 7 to 5 (12 goals). In 1938, 1954, and 1982, games between Brazil and Poland, Hungary and Germany, and Hungary and El Salvador ended respectively 6 to 5, 8 to 3, and 10 to 1. Also in 1958 France defeated Paraguay 7 to 3.

EXAMPLE A track coach wants to determine an appropriate pulse range for her athletes during their workouts. She chooses five of her best runners and asks them to wear heart monitors during a workout. In the middle of the workout, she reads the following pulse rates for the five athletes: 85, 135, 140, 145, 325. Should she consider all five values in calculating the mean?

Three of the five values are fairly close together and seem reasonable for mid-workout pulse rates. The high value of 320 is an outlier. This outlier is unreasonable because no one can have such a high pulse rate without being in cardiac arrest. She should therefore assume that the 325 is a mistake, perhaps caused by a faulty heart monitor. The same is true about the low value of 85. She should not include these values when calculating the mean. Better yet, she should repeat her study after replacing the faulty heart monitor or just use 20% trimmed mean.

A PRACTICE PROBLEM Find the 10% trimmed mean of the following data

$$1, 1, 2, 3, 4, 2, 9, 6, 5, 7$$

Midrange

Although less popular than the mean, median and trimmed mean, the midrange provides an easy-to-grasp measure of central tendency. However, it is severely affected (even more than $\bar{x}$) by the presence of an outlier in the data. The midrange is defined as

$$\text{Midrange} = \frac{\text{Largest measurement} + \text{Smallest measurement}}{2}.$$

This measure for central tendency is comparable to range defined as

$$\text{Range} = \frac{\text{Largest measurement} - \text{Smallest measurement}}{1},$$

which is a measure for spread.

Mode

The mode of a data set is the value that occurs most often(has the highest frequency). It is not always a measure of central tendency, since its value may not fall in the "center" of the data. One situation in which the mode is of interest is in the manufacturing of the sporting goods. The most common shoe size or hat size is what one would like to produce more, not the average size. Note that there may be more than one mode if several values have the same frequency. There may even be no mode if no repeats exist, in the data set.

Chapter 1: Elements of Statistics

EXAMPLE The ordered list of the life expectancies in seven Western European countries, excluding Leichtenstein and Monaco is

$$77, 77, 78, 78, 78, 78, 80$$

Since 78 has the highest frequency (4), the mode (being the same as median and mean) of this data is 78.

In practice most data sets indicate a unimodal distribution (only one mode or one pick). Data from a population that is not homogeneous with respect to the attribute measured may exhibit several modes (multimodal). Examples include a sample of student's heights including both male and female, and student's grades when two or more groups of students with varying backgrounds take the same course. Note that the interpretation of standard deviation discussed in Section 1.4 is not applicable to the data with a bimodal or multimodal distribution.

EXAMPLE 1.4 A sample of size 10 was taken to determine the typical completion time of a three pointer (time interval between three pointers) in a typical high school basketball game. The collected data is

$$4.1, 3.2, 2.8, 2.6, 3.7, 3.1, 9.4, 2.5, 3.5, 3.8$$

We find the average completion time as follows:

$$\bar{x} = \frac{4.1 + 3.2 + \ldots + 3.8}{10} = \frac{38.7}{10} = 3.87 \text{ min.}$$

Notice that 9.4 may be considered an outlier. To be sure, we should check to make sure that there were no mistakes in recording or transcribing this value. In the presence of outliers, the median generally provides a more reliable measure of central tendency, so we order the data

$$2.5, 2.6, 2.8, 3.1, 3.2, 3.5, 3.7, 3.8, 4.1, 9.4$$

From this we obtain

$$\text{Median} = m = \frac{3.2 + 3.5}{2} = 3.35 \text{ min.}$$

Also, the midrange is given by

$$\frac{2.5 + 9.4}{2} = 5.95 \text{ min.}$$

This value is severely affected by the presence of an outlier; the midrange of nearly 6 min is a poor measure of central tendency for this application. Finally, no mode exists because there are no repeats in the data values.

We end this section by noting that comparison of different measures for central tendency could provide useful inference regarding the shape of distribution for measurements. For example, for continuous numerical data, a positively skewed distribution (right-skewed) has a mean greater than median and mode. For symmetric distribution all these measures should be the same or very close to each other. The following figures demonstrate the relationship between these three measures. Note that the direction of skewness does not always indicate the ordering among measures of central tendency, at least with respect to discrete distributions.

Relationship Between the Mean, Median, and Mode

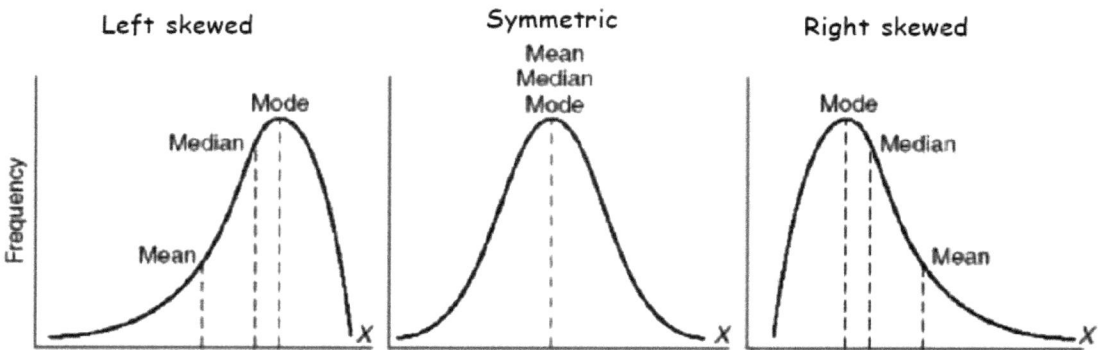

A PRACTICE PROBLEM In a recent competition the 12 members of a weight-lifting team lifted the following weights (in pounds):

$$173 \quad 200 \quad 200 \quad 180 \quad 190 \quad 190$$
$$190 \quad 190 \quad 175 \quad 179 \quad 177 \quad 188$$

Find all the measures of central tendency and compare them.

Alternative Description of Spread (Quartiles)

In previous sections, we discussed a few measures for the central tendency. As pointed out earlier, relying on a measure of the center alone can sometimes be misleading. An extreme (funny) example is the following. A sausage that is burned outside and is frozen inside cannot be considered "on average" cooked. The Census Bureau reports that in 2017 the median income of American households was $60,336. That is, half of all households had income below $60,336, and half had income higher than $60,336. But these figures do not tell us much about the distribution of income. In other words, they do not tell the whole story. Two nations may have the same median household income if one has extremes of wealth and poverty and the other has little income variation among households. A similar situation occurs in sport regarding the salaries of the professional players. For problems of this kind, one is interested in the spread, or variability of incomes as well as their centers values. For variability we already introduced the standard deviation as a measure. Standard deviation is probably the most "important" single measure in statistics. However, its interpretation for some situation, such as bimodal distribution (distribution with two picks), creates some difficulties. Summarizing this, the simplest useful numerical description of a distribution should include both a measure of center and a measure of spread.

One way to measure spread is to work with range, the difference between the largest and the smallest observations. For example, the percent of residents over age 65 in the United States ranges from 9.49% in Alaska to 19.06% in Florida. These observations show the full spread of the data provided that there are no outliers. Description of spread can be improved by looking at the spread of the middle half of the data known as quartile. The quartile mark out the middle half. To find quartile, count up the ordered list of observations, starting from the smallest. The first quartile lies one-quarter of the way up the list. The third quartile lies three-quarters of the way up the list. In other words, the first quartile is larger than 25% of the observations, and the third quartile is

Chapter 1: Elements of Statistics

larger than 75% of the observations. The second quartile is the median, which is larger than 50% of the observations. The rule for calculating the quartiles uses the rule for the median. To calculate the quartiles, follow the steps described below.

1. Arrange the observations in increasing order and locate the median in the ordered list of observations.
2. The first quartile Q_1 is the median of the observations whose positions in the ordered list is to the left of the location of the overall median.
3. The third quartile Q_3 is the median of the observations whose positions in the ordered list is to the right of the location of the overall median.

EXAMPLE 1.5 The number of points scored by a sample of 15 football players during the regular season were (arranged in order)

$$20\ 25\ 25\ 27\ 28\ 31\ 33\ 34\ 36\ 37\ 44\ 50\ 59\ 85\ 86$$

Here we have an odd number of observations, so the median is the middle data, the value 34 in the list. The first quartile is the median of the 7 observations to the left of the median. This is the fourth of these 7 observations, so $Q_1 = 27$. The third quartile is the median of the 7 observations to the right of the median, so $Q_3 = 50$. The overall median is left out of the calculation of the quartiles when there is an odd number of observations.

EXAMPLE 1.6 For the 10 players in an all-star basketball game, the number of the points scored is given below (again arranged in increasing order):

$$5\ 7\ 10\ 14\ 18\ \mid\ 19\ 25\ 29\ 31\ 33$$

There is an even number of observations, so the median lies midway between the middle pair. Its location is between the fifth and sixth values, marked by | in the list. The first quartile is the median of the first 5 observations, because these are the observations to the left of the location of the median. So, we find that $Q_1 = 10$ and $Q_3 = 29$. When the number of observations is even, all the observations enter into the calculation of the quartiles.

Some software packages use a slightly different rule to find the quartiles, so computer results may be a bit different from above calculations. However, the differences are always very small.

Five-Number Summary and Box Plots

The smallest and largest observations tell us a little about the distribution as a whole. However, they provide information about the extremes (e.g., best sport records) and the tail values that are not reflected in Q_1, m, and Q_3. To get a quick summary of both center and spread, we may combine all five numbers.

The five-number summary of a distribution consists of the smallest observation, the first quartile, the median, the third quartile, and the largest observation, written in ascending order.

In symbols, the five-number summary is given as

$$\text{Minimum}, Q_1, \text{Median}, Q_3, \text{Maximum}$$

These numbers offer a fair description of center and spread. The five-number summary for the football example is

$$20 \quad 27 \quad 34 \quad 50 \quad 86$$

For the basketball example, it is

$$5 \quad 10 \quad 18.5 \quad 29 \quad 33$$

A PRACTICE PROBLEM Find the five-number summary of the following data

$$1, 2, 1, 3, 4, 5, 2, 6, 7, 4, 8$$

The five-number summary of a distribution provides a different graphical description, the box plot (see example below). A box plot is a graph of the five-number summary. A central box spans the quartiles with a line marking the median. Lines extend out from the box to the smallest and largest observations.

Box Plots for Football and Basketball Examples

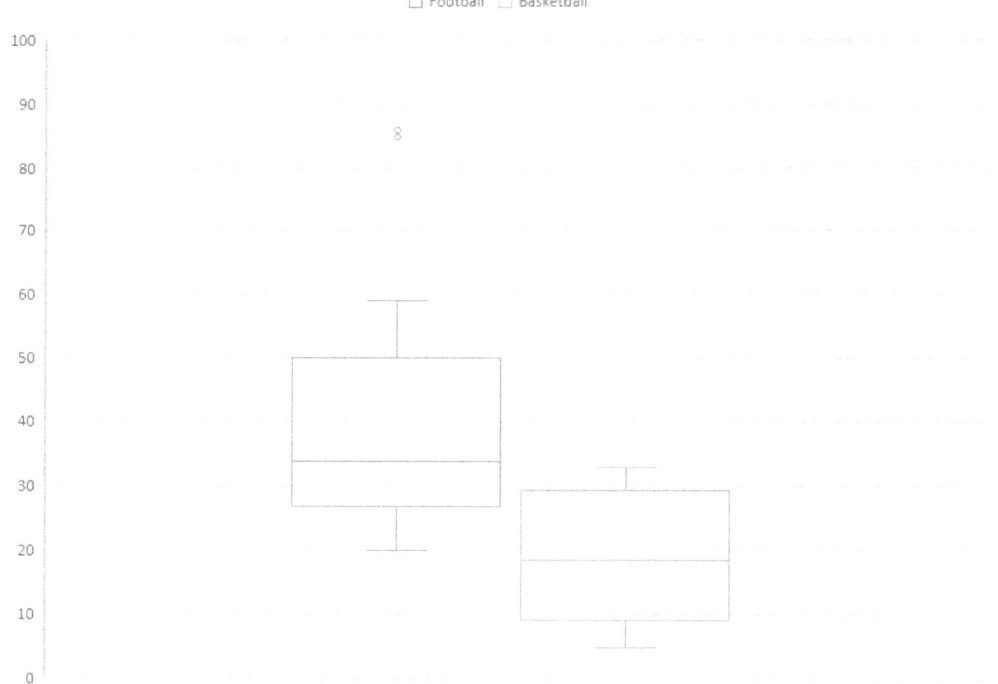

We can draw box plots either horizontally or vertically. In either case, we should include a numerical scale in the graph. Looking at a box plot, one should first locate the median, which marks the center of the distribution. Then look at the spread. The quartiles show the spread of the middle half of the data, and the extremes (the smallest and the largest observations) show the spread of the entire data set.

Because box plots show less detail than histograms or stem-and-leaf discussed below, they are best used for side-by-side comparison of more than one distribution (see the above figure).

1.2 FREQUENCY TABLE AND HISTOGRAM

The structure of the data can be further examined by constructing frequency tables and histograms. A frequency table displays the categories and their frequencies (the number of observations in each of the categories), and relative frequencies (the percentage of observations in each category). For example, suppose that 45 of a group of 100 tennis players use a Wilson racket. Then the frequency for the category "Wilson racket" is 45, and the relative frequency is $45/100 = 0.45$.

EXAMPLE 1.7 If we split the data of Example 1.1 into four categories 47–48, 48–49, 49–50, and 50–51, then we have a frequency table presented below. Conventionally, the value 49.0 is considered falling into the category 49–50, not 48–49.

Interval	Frequency	Relative Frequency
47–48	3	0.3
48–49	4	0.4
49–50	2	0.2
50–51	1	0.1
	10	1

Note that the sum of frequencies always equals the total number of observations in the data, and the sum of the relative frequencies always equals 1. The latter has an important implication that is useful when probability and its definition is discussed.

EXAMPLE 1.8 The blood pressure of 1,000 college students who exercise regularly is measured. To construct a frequency table, we split the values that measure the blood pressure into classes, say 90 to 95, 95 to 100, and so on, (mm Hg) and count how often the measured values fall into each of these classes. Table 1.1 shows the frequency distribution for this data.

The histogram provides a graphical summary of data. It is a plot consisting of several rectangles. The horizontal axis is divided into segments corresponding to the classes. The area of the rectangle centered on each segment is proportional to the frequency (or relative frequency) in the corresponding class. Figure 1.2 shows the histogram of the data in Table 1.1.

A further summary useful in practice is the cumulative relative frequency and the corresponding plots. The cumulative relative frequency is the sum of the relative frequencies for all the values less than (and sometimes equal to) the upper bound of the last class. For example, using Table 1.1 the cumulative relative frequency for 105 is $0.011 + 0.012 + 0.056 = 0.079$. The cumulative frequency curve is called the (empirical) distribution function. Table 1.2 and Figure 1.3 show the cumulative frequency table and cumulative histogram for the data in Example 1.8, respectively. Note that by forming differences from the cumulative table, one can obtain the original frequency table. For example, the frequency for class 95 to 100 is $23 - 11 = 12$. Although simple, these ideas are the basis of theoretical developments of the probability to be considered later in this book.

TABLE 1.1 Frequency Table for Blood Pressures Data

Classes	Frequency	Relative Frequency
90–95	11	0.011
95–100	12	0.012
100–105	56	0.056
105–110	101	0.101
110–115	163	0.163
115–120	180	0.180
120–125	141	0.141
125–130	129	0.129
130–135	106	0.106
135–140	56	0.056
140–145	24	0.024
145–150	10	0.010
150–155	7	0.007
155–160	4	0.004
Total	1000	1

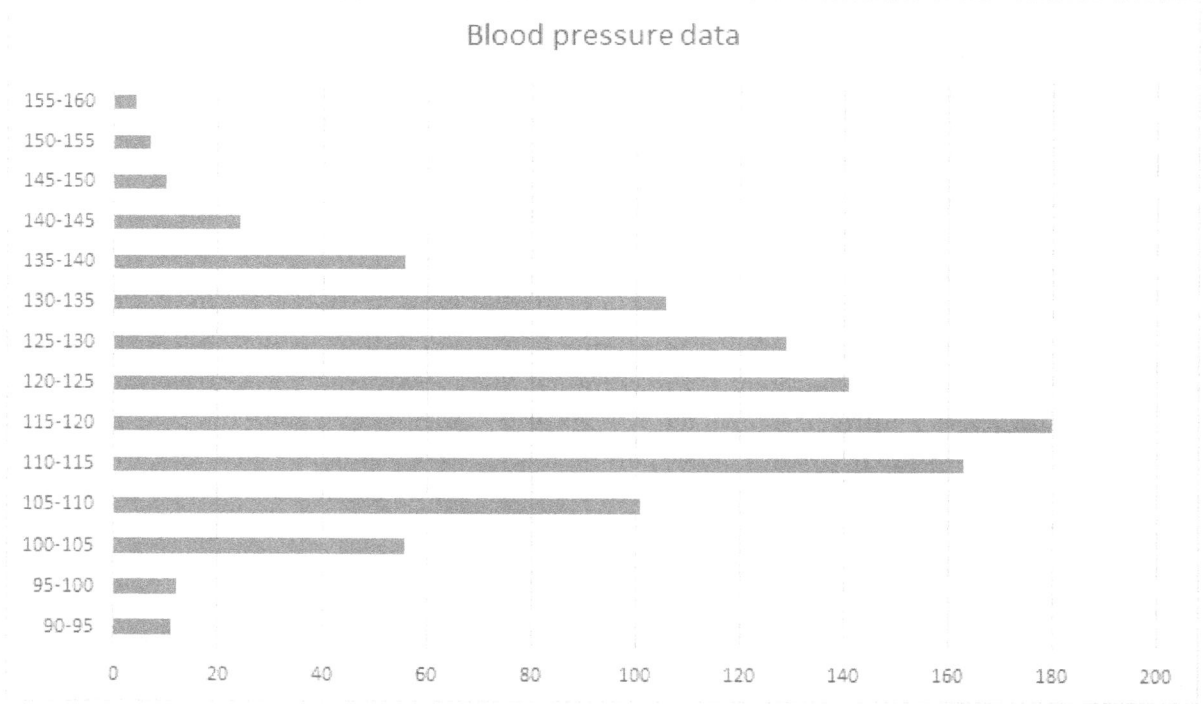

FIGURE 1.2 Histogram for Blood Pressures Data

TABLE 1.2 Cumulative Frequency Table for Blood Pressure Data

Categories (Classes)	Cumulative Frequency	Relative Cumulative Frequency
90–95	11	0.011
95–100	23	0.023
100–105	79	0.079
105–110	180	0.180
110–115	343	0.343
115–120	523	0.523
120–125	664	0.664
125–130	793	0.793
130–135	899	0.899
135–140	955	0.955
140–145	979	0.979
145–150	989	0.989
150–155	996	0.996
155–160	1000	1.000

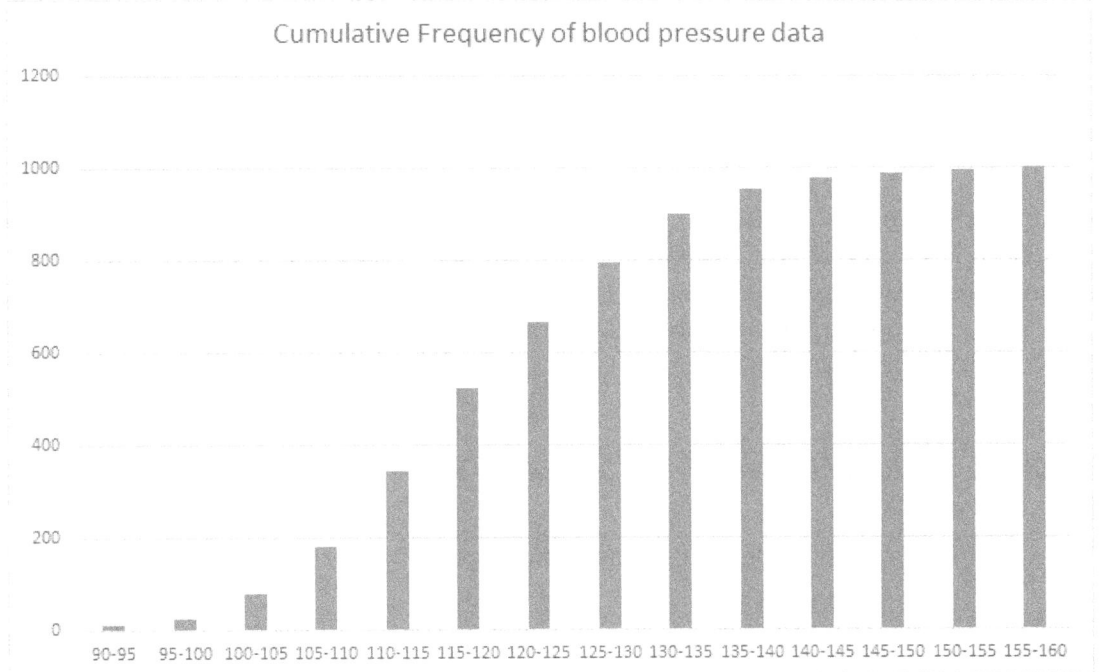

FIGURE 1.3 Cumulative Frequency Plot for Blood Pressure Data

The steps for constructing a histogram may be summarized as follows:

1. Identify the smallest and the largest values in the data.
2. Divide the interval between the smallest and the largest value (range) into several (usually between 5 and 20) equal subintervals called classes. Try to avoid having any value fall on a boundary of a subinterval. In other words, each value should fall into one and only one subinterval.
3. Compute the percentage, or proportion (relative frequency), of values in each subinterval.
4. Using a vertical axis (usually of about three-fourths the length of the horizontal axis), plot frequencies or relative frequencies. In practice, relative frequencies are usually preferred, since they add up to one and are consistent with theoretical development of the subject.

Note that to get a better description of data, one should choose a small number of classes for small data sets and a large number of classes for a large data sets. As the number of measurements in a data set are increased, a better description of the data may be obtained by decreasing the width of the class intervals. When the class intervals become small enough, a relative frequency histogram will appear as a smooth curve. Thus, looking at a histogram, we can imagine the appearance of the relative frequency histogram for a large data set (say a population).

Stem-and-Leaf

Aside from the histogram, there are several other ways of presenting data in graphical form. One interesting and popular method is the stem-and-leaf display, which, while preserving the original data, displays the possible shape of the distribution. The idea is based on presentation of numerical values in a format that is easy to understand and useful for inference. The following is an example. For another example, see the descriptive statistics for data from soccer presented in the next few pages.

EXAMPLE 1.9 Chatterjee et al, (1995) have analyzed data on 105 guards who played in 1992–1993 NBA season. Table 1.3 shows the stem-and-leaf display of points scored per minute (PPM) by these players. In this display, the unit for stem digit is 0.1 and for leaf digit is 0.01. Hence, for instance, when stem equals 1 and leaf equals 5, the corresponding PPM is 0.15, that is, this player scored 0.15 points for each minute he played (15 points for each 100 min). Here, the first column presents the frequency. By looking at this display one can infer that the distribution of PPM closely resembles a bell-shaped (normal) curve except perhaps for the last two values, 0.77 and 0.83. The latter presents PPM for Michael Jordan.

A PRACTICE PROBLEM The following data are the ages of a sample of male residents in Columbia county who play tennis regularly.

42 53 61 20 28 48 47 42 38 39 33 27 36

40 44 44 35 35 64 52 31 43 56 23 49

(a) Construct a stem-and-leaf display for the ages.
(b) Construct a relative frequency distribution and histogram for the data using five class intervals and the value 20 as the lower limit of the first class.

TABLE 1.3 Stem-and-Leaf Display of PPM of 105 National Basketball Association (NBA) Guards

Number of Cases	Stem	Leaves
1	1	5
5	2	1,3,3,4,4
6	2	5,6,7,8,8,9
18	3	0,0,0,0,0,0,1,1,1,2,2,2,3,3,3,4,4,4
15	3	5,5,6,6,6,7,7,8,8,8,8,9,9,9,9
23	4	0,0,0,0,0,1,1,1,2,2,3,3,3,3,3,3,3,3,4,4,4,4,4
14	4	5,5,6,6,7,7,7,7,8,8,8,9,9,9
7	5	0,1,1,2,2,4,4
10	5	5,5,5,6,7,7,7,7,8,8
3	6	0,2,3
1	6	6
0	7	
1	7	7
1	8	3

The following pages present some of the popular descriptive statistics obtained using the software Microsoft Excel. One example presents data from English soccer (Premiership) for the 2017–2018 season including 20 teams whose positions are determined on a point scale. The scale is as follows: 3 points per win, 1 point per draw (tie), 0 points per loss. For example, of the 38 games played, Manchester United (a world famous soccer club) won 24 and tied 8, giving them a point tally of 80. The teams and their points are listed below. The second and the third data sets present, respectively, the best annual records for men's 400 m dash 1860–2001 and the top 50 women times for the Boston marathon prior to the year 2001. In recent years, a lot has happened in the Boston marathon. We hope that students will include the most recent data and construct descriptive statistics.

Data From English Football(Soccer) Final Premiership 2000–2001 Season (Number of Games Played, N = 38, Win = 3 Points, Draw = 1 point, and Loss = 0 points)

	Team	Points
1	Manchester City	100
2	Manchester United	81
3	Tottenham	77
4	Liverpool	75
5	Chelsea	70

6	Arsenal	63
7	Burnley	54
8	Everton	49
9	Leicester	47
10	Newcastle	44
11	Crystal Palace	44
12	Bournemouth	44
13	West Ham	42
14	Watford	41
15	Brighton	40
16	Huddersfield	37
17	Southampton	36
18	Swansea	33
19	Stoke	33
20	West Brom	31

Stem – and – Leaf Plot

3	1,3,3,6,7
4	0,1,2,4,4,4,7,9
5	4
6	3
7	0,5,7
8	1
10	0

26		
34	Max	100
34	Min	31
42	Range	69
42	Mean	52.05
42	Mode	44
48	St. Dev.	19.17091
49	Sample Var.	367.5
51	Kurtosis	-0.257494
52	Skewness	0.9416976
52	Sum	1041
54	Count	20

Chapter 1: Elements of Statistics

57		
61		
66	Central Tendency	
68	Mean	52.05
69	Mode	44
70	Median	44
80		

Relative Frequency Distribution

Points	Frequency	Rel. Freq.	Cum. Freq.
$30 \leq 40$	5	0.25	0.25
$40 \leq 50$	8	0.4	0.65
$50 \leq 60$	1	0.05	0.7
$60 \leq 70$	1	0.05	0.75
$70 \leq 80$	3	0.15	0.9
$80 \leq 90$	1	0.05	0.95
$90 \leq 100$	0	0	0.95
$100 \leq 110$	1	0.05	1
	$N = 20$	1	

We finish this part by presenting an example that demonstrates a calculation of the sample mean from a frequency table. First we define the weighted mean.

Weighted Mean

When calculating a sample mean, we add all the measurements and divide the sum by the number of measurements, say N. This implies that all the measurements are assigned equal weights ($\frac{1}{N}$). If, for some reason, we assign unequal (varying) weights, the resulting mean is called a weighted mean.

EXAMPLE 1.10 The number of goals scored by a women's soccer team in the past 8 weeks is displayed in the frequency table below. For example, three goals were scored in each of 4 weeks. Find the sample mean

Number of the Goals Scored	Frequency
4	3
5	4
6	1

First note that we could present this data as $\{4,4,4,5,5,5,5,6\}$. Hence

$$\bar{x} = \frac{4+4+4+5+5+5+5+6}{8}$$
$$= \frac{4 \times 3 + 5 \times 4 + 6 \times 1}{8} = 4\left(\frac{3}{8}\right) + 5\left(\frac{4}{8}\right) + 6\left(\frac{1}{8}\right)$$
$$= \frac{38}{8} = 4.75.$$

It follows from this example that if the data are presented in a frequency table, then the sample mean can be calculated using the formula

$$\bar{x} = \frac{f_1 x_1 + f_2 x_2 + \ldots + f_k x_k}{N}$$

where $x_1, x_2, \ldots, x_k$ are distinct values in the sample, $f_1, f_2, \ldots, f_k$ are corresponding frequencies, and N is the sample size. Recall that

$$f_1 + f_2 + \ldots + f_k = N.$$

Hence if we use w_i to denote f_i/N for $i = 1, 2, \ldots k$, then w_i's are all nonnegative numbers and their sum is 1. In fact, if w_i's are all nonnegative numbers and their sum is 1, then the value

$$w_1 x_1 + w_2 x_2 + \ldots + w_k x_k$$

is a weighted average of $x_1, x_2, \ldots, x_k$ with w_i being the weight of x_i. Therefore, we see that the sample mean is actually a weighted average of all distinct values in the sample, with the weight for x_i being $w_i = f_i/N$. For ordinary average $w_i = \frac{1}{N}$, $i = 1, 2, \ldots, N$. One useful application of weighted average is when students calculate their GPA. Here x represents the grades and the frequencies are the number of credits.

A PRACTICE PROBLEM Suppose that in your statistics course the class grade is determined from four tests and the final according to the table below, that includes your grades. Find your grade for the course

Source	Grade	Weight (%)
Test 1	85	10
Test 2	80	15
Test 3	79	20
Test 4	91	25
Final	82	30

Finally, we note that it is also possible to construct frequency tables for categorical data. For example, we may register the blood types of the 25 players in a football team and construct a frequency table such as the one presented below.

Blood Type	Frequency	Relative Frequency	Cumulative Relative Frequency
A	5	0.20	0.20
AB	4	0.16	0.36
B	7	0.28	0.64
0	9	0.36	1.00

One interesting example involving frequency table is the ball sales through retail outlets in the United States. According to World Features Syndicate the number of balls of different types sold per day (year 2001) in United States were

Type of the Ball	Frequency
Golf	1.7 million
Tennis	344,600
Baseball(Softball)	27,400
Soccer	13,200
Basketball	1200

A PRACTICE PROBLEM Suppose that a swimming organization allocates competition points to the first five finishers assigning, respectively, 8, 5, 3, 2, and 1 points. Suppose that your friends have participated and placed third, first, third, fourth, seventh, and second in six successive races. Find your friend's average number of points.

1.3 POPULATION AND SAMPLING

So far, we have considered sample and its description. Samples are parts or subsets of a population. Population is like the universal set from which samples are taken. Population can also be described as the entire collection of individuals or objects that are of interest. In applications, we should always be clear about the population under investigation. For example, the population may be men between 18 and 25 years old in some country, in a special town, in a certain college–or it may consist just of students who play certain sports. A sample is a part (subset) selected from the population.

EXAMPLE 1.11 To estimate the amount of time students at a certain college spend to work out each week, a random sample of 55 students were surveyed. In this study, the entire student body of this college is the population, while the 55 surveyed students form the sample. Note that, rather than students, it is possible to think about all the work out times as population and the 55 recorded times as a sample.

Within the context of a problem, it is often possible to associate a number or a set of numbers to each member of the population. For example, to each student in a college we may associate two numbers: height and weight. In practice, once a sample is given, these numbers can be found for all the members in the sample by examining every member included in the sample. After doing this, the next step is to see how the relevant information contained in the sample can be summarized and be used to draw conclusions about population. It is important to remember that the main problem of statistics is to draw conclusions about population based on the information contained in the sample. This, as mentioned earlier, is called statistical inference or inferential statistics. Performing statistical inference and providing a measure for the reliability of inference requires basic knowledge of probability theory, discussed in Chapter 2 of this book. In practice, statistical inference is usually accompanied with a measure in terms of probability that presents its reliability.

Since sampling is an important part of inferential statistics and has a profound effect on their conclusions and their reliability, in the remainder of this section, we briefly discuss the idea of data gathering, sources of data, sampling issues, and errors involved in data gathering. Remember that bad samples or samples that are not representative of population could lead to incorrect conclusions about the population.

There are basically two main sources of statistical data, published and gathered. Published data are often very convenient and inexpensive to obtain. They are divided into two categories: primary and secondary. Primary data refer to the one that is collected and published by some organization. Examples of this are data published by, for example, the United States Bureau of the Census and Surveys. Surveys may be broken down into personal interviews, telephone interviews, and self-administered questionnaires. Personal interviews may be rather expensive and time-consuming, however, there is very little confusion for the participant because the experimenter is able to explain most questions. A big problem with personal interviews is that the experimenter must be extremely careful of biasing the response. Telephone interviews are a little more cost-efficient. However, they require a great deal of time and a bias sample is often recorded. Self-administered questionnaires mailed to participants are usually very inexpensive. However, there may be a very low response rate, which will usually lead to a biased sample. Secondary data are that which are collected by one organization and published by a different one. A popular source of secondary data is, for example, "The Statistical Abstract of the United States."

Gathered data may be from an observational or an experimental study. An observational study is one in which the experimenter strictly observes and records that he or she is studying. There is no manipulation of variables involved. Observational studies are often used in business and economics, along with most of the social sciences. An experimental study is one in which the experimenter manipulates one or more of the variables he wishes to study before observing and recording data. It is used when effect of a particular factor (e.g., treatment) is studied.

When the population the experimenter wishes to study is too large to observe or manipulate directly, a sample of that population is often taken in order to make inferences about the population as a whole. Two major advantages to sampling a portion of the population are cost and time. All in all it is much more efficient to only observe or manipulate a percentage of the whole. A disadvantage of sampling is that there may be some error involved in the sampling process and if samples are not large enough it may be difficult to draw conclusions about the population.

When sampling, one can use one of the following plans: Simple Random Sampling, Stratified Random Sampling, and Cluster Sampling.

Random Sampling is a sample selected in such a way that every possible sample with the same number of observations is equally likely to be chosen. In order to make this possible, each member of the population may be assigned a number and a sample of randomly selected numbers is picked. For example, to randomly choose two swimmers out of 10, we can number them from 1 to 10 and then randomly pick two numbers. The swimmers assigned with the two picked numbers will form the sample. In this sampling procedure, each swimmer has an equal chance of 2/10, or equivalently 20%, to be selected.

A Stratified Random Sample is obtained by separating the population into mutually exclusive sets, or strata, and then drawing random samples from each stratum. One advantage besides acquiring information about the entire population, is making inferences within each stratum or comparing strata. As an example, suppose among the 10 swimmers mentioned above, five are males and five are females. If two swimmers are to be selected and the coach wants to select one male swimmer and one female swimmer, then he can first divide the 10 swimmers into two strata: male swimmers and female swimmers. Then he can draw a simple random sample from each stratum. In this procedure, again each swimmer has an equal chance of 1/5 or 20% to be selected.

The Cluster Sample is a simple random sample of group or clusters of elements. This is useful when it is difficult or costly to develop a complete list of the population members or whenever the population elements are widely dispersed geographically. For example, in order to conduct a survey of customers' opinion on a certain kind of sports car, often the random sample is drawn from the customers in a particular geographical region or from those who purchased the car during a certain time period.

A sampling error refers to differences between the sample and the population that exists only because of the observations that happened to be selected for the sample. Sampling error is an error that we expect to occur when we make a statement about a population that is based only on the observations contained in a sample taken from the population.

A non sampling error is due to mistakes in the acquisition of data or due to the sample observations being selected improperly. There are three main types of non sampling errors: errors in data acquisition, non response error, and selection bias. Errors in data acquisition refer to the recording of incorrect responses. Non response errors occur when a response is not obtained from some members of the sample. A selection bias occurs when the sampling plan is such that some members of the target population cannot be selected for inclusion in the sample.

As a conclusion, we see that the nature of statistics allows the experimenter to make inferences about the whole population based solely on a portion of that population, the sample. When done correctly these conclusions may be very powerful. While studying statistics it is very important to learn the importance of a non biased sample, random sampling, and errors in sampling. Theoretical concepts related to each of these topics and more are discussed in Chapter 5.

1.4 INTERPRETING THE STANDARD DEVIATION

The sample mean and standard deviation are the most commonly used numerical summaries of data. As pointed out earlier, the standard deviation may be considered as the most important single measure in statistics since it quantifies the variation in data and hence directly affects the reliability of any inference. Remember that if there was no variability or variation there was no need for statistical inference. A measure of central tendency such as the sample mean can be combined with a measure of variability such as the standard deviation to provide information about how values of a data set are distributed. Often investigators are interested in the percentage of observations that fall within a specified number of standard deviations from the mean. In other words, they use the standard deviation to come up with a measure for relative standing. Interpretation of standard deviation may be summarized as follows.

1. **Comparison of the Two Samples.** When comparing two samples, the one with the larger standard deviation is the more variable of two. For example, when comparing two stocks the one with the larger standard deviation is more "risky." That is, it has a higher chance to deviate from its average price, or expected price. Also, when comparing two players based on some measure of performance the one with smaller standard deviation is more consistent.
2. **Summarization of a Single Sample** (applies to any set of data, whether population or sample). This can be explained by the following question: what percentage of observations are within 1 standard deviation of the mean? within two standard deviations of the mean? and, in general, within k standard deviations of the mean?

The following two rules can be used to at least partially answer these questions.

Chebyshev's Rule:

This rule is general and applies to any set of data regardless of the shape of the histogram. It can be summarized as

(a) It is possible that very few of the values will fall within $(\bar{x} - s, \bar{x} + s)$.
(b) At least three-fourths (75%) of the values will fall within $(\bar{x} - 2s, \bar{x} + 2s)$.
(c) At least eight-ninth (89%) of the values will fall within $(\bar{x} - 3s, \bar{x} + 3s)$.
(d) Generally, at least $1 - 1/k^2$ of the values will fall within $(\bar{x} - ks, \bar{x} + ks)$.

Note that, using (b), for example, we may make the following statement: at most 25% of the values will fall outside the interval $(\bar{x} - 2s, \bar{x} + 2s)$.

The Empirical Rule or Rule of Thumb:

This rule only applies to samples whose histogram is bell-shaped (mound-shaped). It can be summarized as

(a) Approximately 68% of the values will fall within $(\bar{x} - s, \bar{x} + s)$.
(b) Approximately 95% of the values will fall within $(\bar{x} - 2s, \bar{x} + 2s)$.
(c) Essentially, all the values will fall within $(\bar{x} - 3s, \bar{x} + 3s)$.

EXAMPLE 1.12 Suppose that the average number of years players play a particular sport is 13 years with a standard deviation of 4 years. If we make no assumptions about the distribution of the lengths (years) of time, what can be said about the fraction of players who play between 9 and 17 years?

Chapter 1: Elements of Statistics

SOLUTION: First note that

$$(\bar{x} - s, \bar{x} + s) = (13 - 4, 13 + 4) = (9, 17)$$
$$(\bar{x} - 2s, \bar{x} + 2s) = (13 - 8, 13 + 8) = (5, 21)$$
$$(\bar{x} - 3s, \bar{x} + 3s) = (13 - 12, 13 + 12) = (1, 25)$$

According to Chebyshev's rule, we may make the following statements:

1. It is possible that very few players will play between 9 and 17 years.
2. At least three-fourths of players will play between 5 and 21 years. Or at most one-fourth of players will play less than 5 years or more than 21 years.
3. At least eight-ninths of the players will play between 1 and 25 years.

EXAMPLE 1.13 Referring to Example 1.12, suppose that the histogram of years in a particular region of the country where this sport is popular is bell-shaped. Assuming that this claim is true, what proportion of the number of years played would be expected to fall within the intervals obtained above?

SOLUTION: According to the empirical rule, we may make the following statements:

1. Approximately 68% of the players play between 9 and 17 years (i.e., approximately 68% of the values fall within one standard deviation of the mean).
2. Approximately 95% of the players play between 5 and 21 years.
3. Essentially, all of the players play between 1 and 25 years.

So, if you find someone who has been playing (in that particular region) for, say, 24 years, one of the two inferences can be made. Either he is one of the approximately 2.5% of the players (half of 5% because of the symmetry) that play more than 21 years, or something about the information (the claim that the histogram is bell-shaped) is not true. Because, the chances are so small that someone will play for more than 21 years, you would have a good reason to have doubts about the claim regarding the shape of the histogram or perhaps other factors such as the appropriateness of the values of mean and the standard deviation.

We end this section by noting, once more, that Chebyshev's rule is always true, but provides a very conservative estimate of the fraction of measurements that fall into a particular interval. When applicable, the Empirical Rule provides a more accurate estimate of the fraction of measurements that fall into the interval. Since most of the measurements lie within two standard deviations, in many practical investigations the range $R = \bar{x} + 2s - (\bar{x} - 2s) = 4s$ is used to provide a quick estimate for s. That is, $s \approx R/4$.

A PRACTICE PROBLEM Sportszone has determined that daily demand for walking shoes has an approximate normal distribution, with a mean of 64 and a standard deviation of 8.

(a) For what percentage of days can we expect the number of shoes sold to be between 48 and 80?

(b) If the store begins each morning with a stock of 72 shoes, for what percentage of days will there be an insufficient number of shoes to meet the demand?

(c) Answer part (a) for the case where normality assumption is dropped.

1.5 MEASURES OF RELATIVE STANDING

1.5.1 Percentile Ranking

Percentile ranking is used frequently to present the relative standing. We hear, often, that some players or teams are in the upper 10%, or a student is in upper 5%(of her class or major) compared to all other students in the same class or major.

For any sample of size N (arranged in ascending or descending order), the p-th percentile is a number such that p% of the values are less than or equal to it and $(100 - p)$% are greater than or equal to it. In case, there are two values in the sample that both satisfy the condition, then the p-th percentile is defined as the average of these two values.

EXAMPLE 1.14 Suppose we are given four values: 1, 2, 3, 4. For this data, the 25th percentile, the first or the lower quartile, equals 1.5 (note that, in fact, any number greater than 1 and less than 2 could be considered a lower quartile). This is because in this case both values 1 and 2 satisfy the conditions for the 25th percentile (25% of values are less than or equal to it and 75% of values are greater than or equal to it). Hence their average, that is, 1.5, is defined as the 25th percentile. Similarly, we see that 2.5 is the 50th percentile (the same as median) and 3.5 is the 75th percentile (third or upper quartile).

EXAMPLE 1.15 Referring to the data given in Example 1.1, find the 25th percentile (or equivalently the lower quartile).

SOLUTION: Listing the data values in the ascending order we have

$$47.2, \ 47.5, \ 47.6, \ 48.2, \ 48.3, \ 48.6, \ 48.7, \ 49.0, \ 49.8, \ 50.4$$

Here the sample size is $N = 10$. Now 25% of 10 is 2.5 and 75% of 10 is 7.5. So there should be "two and a half" values that are less than or equal to the 25th percentile and "seven and a half" values that are greater than or equal to it. Naturally the third value from the lower end and the eighth value from the upper end will satisfy the requirements. It turns out that in this case the third value from the lower end is exactly the eighth value from the upper end. Hence this value, that is, 47.6, is the 25th percentile.

1.5.2 z-score

Is it true to say that we can only compare like with like? How can we compare two students from two different majors or two different universities? How can we compare two athletes from two different sports? z-score provides a reasonable answer to these questions.

Chapter 1: Elements of Statistics

Suppose that the mean value $\bar{x}$ and the standard deviation s of a sample $\{x_1, x_2, \ldots, x_N\}$ are both computed, then for each data value x_i, the sample z-score for x_i is defined as

$$z_i = \frac{x_i - \bar{x}}{s}.$$

Similarly, if we use μ and σ to denote, respectively, the population mean and the population standard deviation, then the population z-score for a value x in the population is defined as

$$z = \frac{x - \mu}{\sigma}.$$

By definition z-score of a value measures, how many standard deviations this value is away from (either above or below) the mean. If z-score is positive, the value is above the mean; if z-score is negative, the value is below the mean; and if z-score equals zero, then the value is exactly equal to the mean. So if, for example, you know your z-score in a test, you know whether you did better than average or else.

EXAMPLE 1.16 Suppose that the heights of three tennis players in a sample are, respectively, 5, 5.5, and 6 feet (or 150, 165, and 180 cm). Let x and y represent the heights in feet and centimeters, respectively. Then calculating the mean and the standard deviation the z-scores are as given the table below.

Height					Mean	Variance	Standard Deviation	z-Score for the Tallest One
(Feet)	x	5	5.5	6	5.5	0.25	0.5	$z = (6 - 5.5)/0.5 = 1$
(Cm)	y	150	165	180	165	22.5	15	$z = (180 - 165)/15 = 1$
	z-score	-1	0	1				

Suppose that people are shorter in a different country and the heights of three players in a sample are, respectively, 4.5, 5, and 5.5 feet (or 135, 150, and 165 cm), then we have

Height					Mean	Variance	Standard Deviation	z-Score for the Tallest One
(Feet)	x	4.5	5	5.5	5	0.25	0.5	$z = (5.5 - 5)/0.5 = 1$
(Cm)	y	135	150	165	150	22.5	15	$z = (165 - 150)/15 = 1$
	z-score	-1	0	1				

It follows from the this example that the z-score is invariant with respect to shift and scale and presents the relative standing of the values. If you are a student, it tells you, for example, where you stand in your class relative to the other students. Note that your standing will remain the same if your professor decides to add 10 points to everybody or perhaps increase the grades by 5% or both. This is because for transformed grades $y = ax + b$ we have $\bar{y} = a\bar{x} + b$ and $s_y = as_x$ so that $\frac{y - \bar{y}}{s_y} = \frac{x - \bar{x}}{s_x}$.

Two important facts about z-scores are that the mean of the z-scores is always 0 and the standard deviation of the z-scores is always 1.

Turning to a sport example, in 1910s, 1940s, and 1970s, the mean batting averages were 0.266, 0.267, and 0.261, respectively, with standard deviation of 0.037, 0.0326, and 0.0317. Suppose that we want to compare these three players and decide which one was ranked highest in relationship to his contemporaries. The players and their hit are as follow: Ty Cobb, 0.420 in 1911, Ted Williams, 0.4064 in 1941, and George Brett, 0.390 in 1980. To do this we calculate their z-scores. We find that z-scores are 4.151, 4.264, and 4.07, respectively. So Ted Williams was ranked as the best hitter, since he had a better relative standing. It is interesting to note that z-scores can also be used to compare two players from two different sports. This is particularly useful when we want to rank the best athletes in the world. The same is true when comparing, for example, two students from two different majors or two different universities (see example below). In short, since direct measurements do not always present the same thing (e.g., GPA in different majors), it is reasonable to judge players or students based on their relative standing. This eliminates factors such as bad teacher, bad books, hard tests, and so on.

As a different example involving teams, consider the English premiership data presented in Section 1.2. The mean and standard deviation of the points earned by all 20 teams are, respectively, 52 and 14. The z-score for Manchester United, with 80 points, is 2. The nearest team, Arsenal has a z-score of 1.29. This shows the strength of Manchester United and provides a reason for their popularity.

EXAMPLE 1.17 Suppose that we want to compare two students, one from Harvard and one from a local community college. Can we say that the one with the higher GPA is a better student? Obviously not. This is because we do not know what these numbers mean, what is their scale, and what they are representing. To clarify this, let us consider a comparison of two students taking the same course, in the same university, one seating in the morning and the other in the afternoon class. Suppose we have to decide, who is a "better" student? Clearly, we cannot base our decision on their respective grades. However we can base our decision on their relative standing. To clarify, consider the data in the following table:

Description	Grade	Class Average	Class Standard Deviation	z-Score
Student A in the morning class	80	85	5	$z = (80-85)/5 = -1$
Student B in the afternoon class	76	72	2	$z = (76-72)/2 = 2$

From this table, it is clear that although student A has a higher grade, she has a much lower z-score. This means that student B has a much better relative standing. In other words, the grade of student B is 2 standard deviations above the average grade in her class, whereas, the grade of student A is one standard deviation below the average in her class. This means that student B has done much better than most students in her class than student A in her class. In fact, assuming that the distribution of grade was normal in both classes, and using the empirical rule, we find that A has done better than 16% of students in her class, whereas B has done better than 97.5% of students in her class.

Chapter 1: Elements of Statistics　　　　　　　　　　　　　　　　　　　　　　　　　　　　　　　33

1.5.3 Interpretation of z-Score for Bell-Shaped Distribution

If the histogram of data is bell-shaped, then approximately 68% of the values in this data will have a z-score between -1 and 1 and approximately 95% of the values will have a z-score between -2 and 2. All, or almost all, of the values will have a z-score between -3 and 3. Note that this is equivalent to the Empirical Rule, since the z-scores for $\bar{x} - s$ and $\bar{x} + s$ are, respectively,

$$z = \frac{(\bar{x} - s) - \bar{x}}{s} = -1 \quad z = \frac{(\bar{x} + s) - \bar{x}}{s} = 1.$$

A large, positive z-score implies that the value is larger than almost all other values. A small, negative z-score implies that the value is smaller than almost all other values. A z-score that is 0 or near 0 implies the value is located near the mean. So, a student with a z-score of 1 is better than (assuming bell-shaped histogram and using the fact that bell-shaped histogram is symmetric) about $84\% (50\% + \frac{1}{2}(68\%))$ of the students in her class, and a student with a z-score of 2 is better than about 97.5% in her class. So, in relative sense the second may be considered a better student. As mentioned earlier, this kind of analysis could be used to rank players in a given sport or compare the players from different sports.

1.6 CASE STUDY: COMPARISON OF PROFESSIONAL PLAYERS AND TEAMS

1.6.1 Performance Measures

Professional sports such as baseball and basketball have grown in prominence in recent years with revenue, media attention, and player salaries rising dramatically. Several publications and websites publish the statistical totals for the year for each player and also each team.

What factors affect the (statistical) performance of a player? The first issue is the choice of measures of performance. Clearly performance is not unidimensional and one needs to consider many factors and define an index for performance. Some of the factors are related to each other and if the nature of relationship is discovered, it can be used to predict various measurements of the performance of the players and the teams. For instance, some experts have used the following index (passer rating) as a measure for rating quarterbacks (QB) in American Football.

$$\left(\frac{\frac{C}{A} - 0.3}{0.2} + \frac{\frac{Y}{A} - 3}{4} + \frac{\frac{T}{A}}{0.05} + \frac{0.095 - \frac{I}{A}}{0.04} \right) \cdot \frac{100}{6}$$

$A =$ Number of pass attempts　　$C =$ Number of completions
$Y =$ Number of passing yards　　$T =$ Number of touchdowns
$I =$ Number of interceptions

- "Perfect" QB Rating would be 158.3
- "Average" Rating set at 66.7 (1970s standards)
- 2018 Top QB Rating: Patrick Mahomes (Kansas City Chiefs): 137.4
- 2018 Worst QB Rating: Sam Bradford (Arizona Cardinals): 62.5

For soccer the Federation Internationale de Football Association (FIFA) World Ranking includes the national A-teams of all FIFA member associations who play international matches. Taken into consideration are all international matches played over a time span of the last eight years including

- World Cup Finals
- Continental championship final matches
- FIFA Confederations Cup matches

- World Cup preliminary final matches
- Continental championship preliminary matches
- Friendly matches

The factors taken into consideration are

- Winning, drawing, and losing
- Home or away match
- Regional strength (multiplication factor)

- Number of goals
- Importance of the match (multiplication factor)

For each team, only the seven best results per year are given full weighting. Results from the past are given progressively less weighting year by year until after 8 years they are dropped completely. In this way, current success is rated more highly than past results.

At the end of each year, two awards are given. The "Team of the Year" goes to the team that notches up the seven matches of the year reaping the overall highest average number of points during the year. "Best Mover of the Year" is the one that has made the most progress in the course of the previous 12 months.

In an article appeared in *Chance*, 1997, Vol. 10, No. 3, Bill and David Williams have discussed two indexes for on-ice performance of the players in the National Hockey League (NHL). There are many statistics available, including games played, goals scored and assisted, penalty minutes, short-on-goal, and so on. Clearly, selecting one or two of these variables will not be suitable for all players. The indexes presented are based on an advanced statistical method known as principal component that reduces the dimensionality of the data to OPI which measures offensive performance, and EPI, which measures defensive efficiency. Since understanding these indexes requires knowledge of advanced statistics, we do not present them here. However, we mention that such indexes are based on a linear combination of factors such as the ones mentioned above.

One notable aspect of the game of baseball is the wealth of numerical information that is recorded for the game. The effectiveness of batters and pitchers is usually assessed by particular numerical measures. A typical unidimensional measure of hitting effectiveness is the batting average calculated by dividing the number of hits by the number of at-bats, which gives the percentage of success for the batter. Batters are also evaluated on their ability to reach one, two, three, or four bases on a single hit. There are also similar basic measures for effectiveness of pitchers.

Jim Albert (http://www.math.bgsu.edu) has presented the so-called Sabermetrics that is a mathematical and statistical analysis of baseball records leading to a collection of improved measures for players performance. He refers, for example, to a measure suggested by Bill James (1982) for hitters, expressed as the following formula

$$\text{Runs} = \frac{(\text{Hits} + \text{Walks})(\text{Total Bases})}{\text{At-Bats} + \text{Walks}}.$$

Chapter 1: Elements of Statistics

In this section, we will consider a statistics regarding home runs that, not only contributes to the performance, but also makes players famous. So let us consider some famous baseball players and compare them based on one aspect of their game, that is, home runs.

George Herman "Babe" Ruth, who is ranked second in ESPN's list of top 100 North American athletes of the last century, is the legend among legends when it comes to baseball. However, in recent years many of his home run records have been surpassed. In fact, recently fans of baseball have been treated to an impressive home run display by the sluggers Mark McGwire, Sammy Sosa, and Barry Bonds. But how do these seasons compare with the great seasons of the past, like those of Ruth?

In what follows, we shall provide three tables containing numerical summaries in comparison of a few very popular top Baseball players including "Babe" Ruth, Mark Mcguire, Roger Maris, Sammy Sosa, and George Foster. We hope that students will be able to use the knowledge learned in this chapter to reach a reasonable conclusion on "who the best player is." We also excluded Barry Bonds so that interested students could research his statistics and carry on their own analysis.

Table 1.4 shows the total number of home runs hit by the player in the corresponding season, as well as the ratio of home runs to official at-bats (so when a player walks, no at-bat is charged). The lower the ratio, the better the season.

Table 1.5 provides the z-score for these players in the corresponding season. More precisely, the home run totals of all the players in MLB of that season with a sufficient number of at-bats (500 or more at-bats) are taken to form a sample. The mean and standard deviation of that sample are then computed, and further the z-score for the top player is calculated. In this table, the third column shows the home run total of the top player in that season, the fourth and fifth columns show, respectively, the mean value and the standard deviation of the corresponding sample, and the last column gives the z-score for the top player in that sample.

Table 1.6 shows the z-Score for the top players when they are compared with the top 10 home run totals in the corresponding season. In other words, now the corresponding samples are formed by the 10 top home run totals of that season.

TABLE 1.4 Total Home Runs (HR) and Ratio of HR to Official At-Bats

Player	Season	HR	HR: At-Bats
Mcguire	1998	70	1 : 7.27
Mcguire	1999	65	1 : 8.01
Ruth	1920	54	1 : 8.48
Ruth	1927	60	1 : 9
Maris	1961	61	1 : 9.67
Sousa	1998	66	1 : 9.74
Sousa	1999	63	1 : 9.92
Foster	1977	52	1 : 11.8

TABLE 1.5 Top Player's z-score in the Season

Player	Year	Home Run (HR) by Player	HR Average Major League Baseball/Year	Standard Deviation	Player's z-Score
Ruth	1920	54	5.54	6.95	6.97
Ruth	1927	60	8.26	11.04	4.69
Mcguire	1998	70	21.97	13.48	3.56
Sousa	1998	66	21.97	13.48	3.27
Foster	1977	52	17.22	10.8	3.22
Maris	1961	61	20.4	13.91	2.92

TABLE 1.6 z-Scores When Compared With Top 10 Home Run (HR) Totals in the Season

Player	Year	HR by Player	HR Average -Top 10 Players/Year	Standard Deviation	Player's z-Score
Ruth	1920	54	17.9	12.95	2.79
Foster	1977	52	38.1	5.72	2.43
Ruth	1927	60	28.5	14.38	2.19
Maris	1961	61	45.1	7.78	2.04
Mcguire	1999	65	48.5	8.40	1.96
Mcguire	1998	70	51.4	9.57	1.94
Sousa	1999	63	48.5	8.40	1.73
Sousa	1998	66	51.4	9.57	1.53

An interesting question, based on these summaries, is who should be considered the "best baseball player so far" up to the year 2001?

A PRACTICE PROBLEM Consider the calculation of z-scores presented above. Add data for most recent year (e.g., home runs for Barry Bonds) and recalculate the z-scores.

1.6.2 How Good Was Michael Jordan?

Data for basketball may be found, for example, in the Pro Basketball Bible, nba.com., and espn.com. Data published include number of games in which the player appeared, minutes per game (MPG), points per minute (PPM) played, field goal percentage (FGP), free throw percentage (FTP), assists per minute (APM), rebounds per minute (RPM), and so on. First we need to define a measure for performance. Here we are going to use points scored per minute played (PPM). Note that when comparing the basketball player, we should consider the position they play. This is because, for

Chapter 1: Elements of Statistics

some position, the primary responsibility is not to score, but to distribute the ball. So here, we will look at guards. Also, rather than comparing different players, here we just want to find out how good Michael Jordan was compared to the other guards.

As mentioned in Example 1.9, Chatterjee et al. (1995) have analyzed data on 105 guards who played in the 1992–1993 NBA season. This season was selected because Michael Jordan played in that season. The following table shows the stem-and-leaf display of points scored per minute (PPM) by these players. From inspection of the table, it is clear that the distribution of PPM is very close to a normal distribution so the empirical rule is applicable. Here the mean and standard deviation are, respectively, 0.4236 and 0.1159 points per minute. So we can calculate the z-score for Michael Jordan whose PPM was 0.8291. Note that in the stem-and-leaf display, the unit of the leaf digit is 0.01. Hence, for instance, when the stem value equals 1 and leaf value equals 5, the corresponding PPM is 0.15.

Let X denote the PPM, then approximately $X \sim N(0.4236, 0.1159^2)$. Thus for Michael Jordan we have

$$z = \frac{0.8291 - 0.4236}{0.1159} = 3.5.$$

It is interesting to note that according to the empirical rule, approximately 99.7% of guards have PPM to within three standard deviations of mean. So a z-score of 3.5 is extremely unusual and is very impressive.

Number of Cases	Stem	Leaves
1	1	5
5	2	13344
6	2	567889
18	3	000000111222333444
15	3	556667788889999
23	4	00000111223333333344444
14	4	55667777888999
7	5	0112244
10	5	5556777788
3	6	023
1	6	6
0	7	
1	7	7
1	8	3

A PRACTICE PROBLEM Consider data for a different season and a player of your choice. Check to see whether data follow a normal distribution. In any case, calculate the z-score for the selected player.

1.7 PAIRED DATA AND LINEAR CORRELATION COEFFICIENT

So far we have only considered descriptive statistics for a sample obtained by measuring a single attribute. The sample correlation coefficient is used when the data consist of paired values (x_1, y_1), $(x_2, y_2), \ldots, (x_N, y_N)$ measured on two attributes. For example, we may like to measure the degree of association between the number of hours of study and the grade in the test for which student studied, the height of a basketball player and his or her number of points per minute of play, or study the relation between amount of exercise and the cholesterol level of a person. One interesting example related to the sports postulates that athleticism or being coordinated have association with testosterone level, which in turn has association with the length of the fingers.

Suppose N individuals are interviewed and $x_1, x_2, \ldots x_N$ denote the number of hours they spend per week to workout and $y_1, y_2, \ldots, y_N$ denote their cholesterol levels. The resulting data contain N ordered pairs $(x_1, y_1), (x_2, y_2), \ldots, (x_N, y_N)$ where the first number x_i in the pair (x_i, y_i) stands for the number of hours the i-th individual spends to work out per week and the second number y_i in the pair is his or her cholesterol level. Here i ranges from 1 to N. Our goal is to quantify the association between these two attributes.

1.7.1 Correlation Coefficient

The most popular measure for this purpose is the (Pearson's) correlation coefficient. The correlation coefficient, usually denoted by r, is a measure of the linear "relation" or "association" between x's and y's. Let $\bar{x}, \bar{y}, s_x$, and s_y denote the mean of x's, mean of y's, standard deviation of x's, and standard deviation of y's, respectively. Then the sample correlation coefficient is obtained as

$$\begin{aligned} r &= \frac{\sum_{i=1}^{N}(x_i-\bar{x})(y_i-\bar{y})}{\sqrt{\sum_{i=1}^{N}(x_i-\bar{x})^2 \sum_{i=1}^{N}(y_i-\bar{y})^2}} = \frac{\sum x_i y_i - \frac{1}{N}(\sum x_i)(\sum y_i)}{\sqrt{\sum x_i^2 - \frac{1}{N}(\sum x_i)^2}\sqrt{\sum y_i^2 - \frac{1}{N}(\sum y_i)^2}} \\ &= \frac{\sum x_i y_i - \frac{1}{N}(\sum x_i)(\sum y_i)}{(N-1)s_x s_y} = \frac{1}{N-1}\left(\sum_{i=1}^{N}\left(\frac{x_i - \bar{x}}{s_x}\right)\left(\frac{y_i - \bar{y}}{s_y}\right)\right) \\ &= \frac{1}{N-1}\sum_{i=1}^{N} z_{x_i} z_{y_i} \end{aligned} \quad (1.7)$$

where z_{x_i} and z_{y_i}, respectively, represented z-scores of x_i and y_i.

Here, we assume that neither s_x nor s_y is zero (note that if one of them is zero, say $s_x = 0$, then x's do not change no matter how y's vary, and hence question of a relation between x's and y's does not need to be addressed). If r defined in (1.7) is positive, then we say that the pairs are positively correlated; if $r < 0$ then we say that the pairs are negatively correlated. Note that r does not depend on the units of measurement of x or y. It does not also depend on label (x or y) used for each attribute.

Before working with a more formal computational method, we first explore the data set to see what we can learn. We can often learn of a relationship between two variables by constructing a graph called a scatterplot or scatter diagram, which (x, y) are plotted with a horizontal x-axis and a vertical y-axis. Each pair (x, y) is plotted as a single point.

Chapter 1: Elements of Statistics

When examining a scatterplot, our goal is to first study the overall pattern of the plotted points. If there is a pattern, we note its direction. That is, as one variable increases, does the other seem to increase or decrease? We then try to specify whether there are any outliers (which are points that lie very far away from all of the other points).

Examples of scatterplots are shown in Figure 1.4. The graphs in Figure 1.4(b), 1.4(c), and 1.4(d) depict a pattern of increasing values of y that correspond to increasing values of x. As you proceed from (b) to (d), the dot pattern becomes closer to a straight line, suggesting that the relationship between x and y becomes stronger linearly. The scatterplots in (e), (f), and (g) depict patterns in which the y-values decrease as the x-values increase. Again, as you proceed from (e) to (g), the relationship becomes stronger. In contrast to the first six graphs, the scatterplot of (a) shows no pattern and suggests that there is no correlation (or relationship) between x and y. Finally, the scatterplot of (h) shows a pattern, but it is not linear.

Type of Correlation

Positive correlation: Both variables tend to increase (or decrease) together.

Negative correlation: The two variables tend to change in opposite directions, with one increasing while the other decreases.

No correlation. There is no apparent relationship between the two variables.

Nonlinear relationship: The two variables are related, but the relationship results in a scatter diagram that does not follow a straight-line pattern.

It is easy to verify the following facts about the correlation coefficient r, and we leave the verification to the students as exercises. We only mention that when x is the same as y, it follows from (1.7) that $r = 1$.

THEOREM 1.18 It is always true that $-1 \leq r \leq 1$, that is $|r| \leq 1$.

THEOREM 1.19 If y is a linear function of x (the strongest association), that is, if there are constants $a \neq 0$ and b such that
$$y_i = ax_i + b \quad i = 1, \ldots N$$
then
$$r = \begin{cases} 1 & \text{if } a > 0 \\ -1 & \text{if } a < 0. \end{cases}$$

THEOREM 1.20 The correlation coefficient for $\{(x_i, y_i); i = 1, \ldots N\}$ is the same as that for $\{(ax_i + b, cy_i + d); i = 1, \ldots N\}$ as long as $ac > 0$.

Note that, from these facts it follows that the correlation defined as (1.7) measures the degree of linear relationship between the two variables. Also, correlation remains unchanged if linear transformations such as shifting and scaling is applied to the data set.

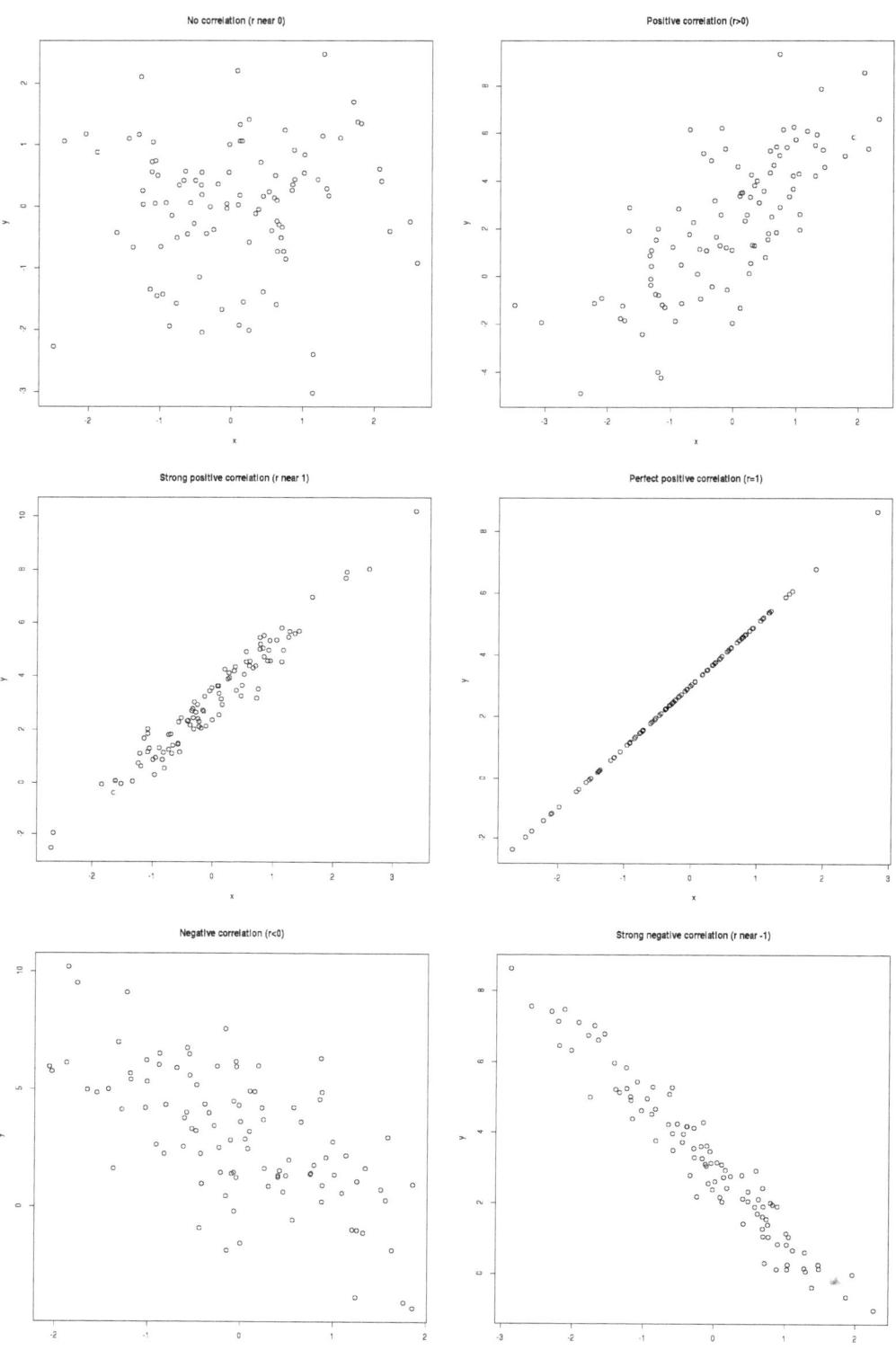

FIGURE 1.4 (*Continued*)

Chapter 1: Elements of Statistics

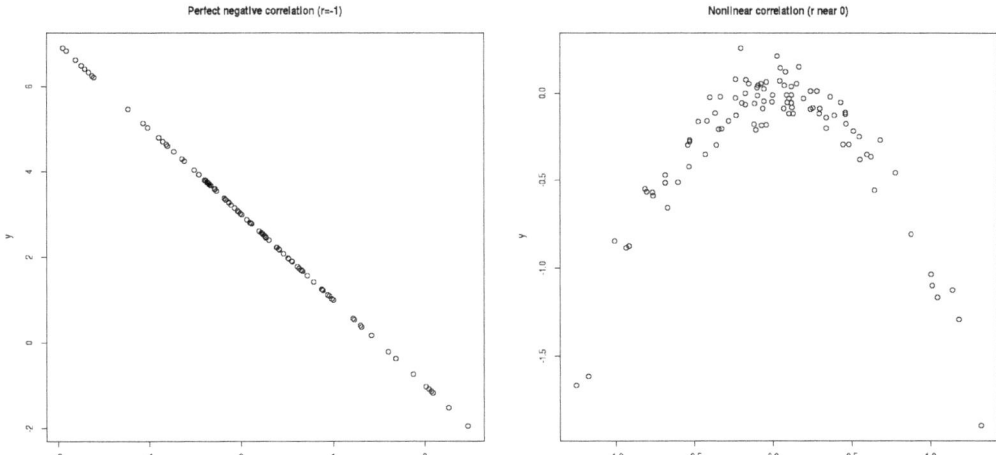

FIGURE 1.4 Scatterplots Representing Different Values of r and Their Implications

EXAMPLE 1.21 Find the sample correlation coefficient for the following four pairs $(1,0), (1,1), (-1,2), (0,0)$.

SOLUTION:

x_i	y_i	x_i^2	y_i^2	$x_i y_i$
1	0	1	0	0
1	1	1	1	1
-1	2	1	4	-2
0	0	0	0	0
$\sum x_i = 1$	$\sum y_i = 3$	$\sum x_i^2 = 3$	$\sum y_i^2 = 5$	$\sum x_i y_i = -1$
$\bar{x} = 0.25$	$\bar{y} = 0.75$			

$$r = \frac{-1 - \frac{1}{4}(1)(3)}{\sqrt{3 - \frac{1}{4}(1)^2}\sqrt{5 - \frac{1}{4}(3)^2}} = \frac{-\frac{7}{4}}{\sqrt{(11/4)(11/4)}} = -\frac{7}{11}.$$

Note that since we need to calculate s_x and s_y for prediction (discussed later) we may first find these as

$$s_x = \sqrt{\frac{1}{3}(3 - \frac{1}{4}(1)^2)} = \sqrt{11/12}, \quad s_y = \sqrt{\frac{1}{3}(5 - \frac{1}{4}(3)^2)} = \sqrt{11/12}$$

and then find r as

$$r = \frac{-1 - \frac{1}{4}(1)(3)}{3(\sqrt{11/12})(\sqrt{11/12})} = -7/11.$$

EXAMPLE 1.22 Referring to Example 1.1, if we use x's to denote the years and y's the time in seconds, then we have a set of pairs

$$(1980, 50.4), \ (1984, 49.8) \ (1988, 48.6) \ (1992, 49.0) \ (1996, 48.7)$$

$$(2000, 48.3), \ (2004, 48.2) \ (2008, 47.2) \ (2012, 47.5) \ (2016, 47.6)$$

These data are plotted in Figure 1.5. Here we have

$$N = 10, \ \bar{x} = 1998, \ s_x = 12.11, \ \bar{y} = 48.53, \ s_y = 1.01$$

and

$$r = \frac{-102.2}{9(12.11)(1.01)} = -0.928.$$

Here r is negative, representing a negative correlation and indicating the fact that the records are improving (time in seconds is decreasing) with time. The value of r is also very close to -1. This means that the correlation between the records and the years is very strong. That is the relation is almost linear. This is also demonstrated in Figure 1.5 where the plots are very close to a straight line.

EXAMPLE 1.23 Consider the data given in Example 1.3. We have a data set containing five pairs

$$(2003, 47), \ (2004, 43), \ (2005, 48), \ (2006, 54), \ (2007, 54)$$

for which

$$N = 5, \ \bar{x} = 2005 \ s_x = 1.58, \ \bar{y} = 49.2, \ s_y = 4.76, \ r = 0.83$$

Here r is positive, indicating that there is a positive correlation between the years and the home run records, that is, the records are improving with time. However, the value of r is rather distant from 1, indicating that the correlation is not very strong. In other words, the records may go up and down, showing a zigzag pattern as demonstrated in Figure 1.6.

EXAMPLE 1.24 In this example, the Atlanta Braves number of home runs, for the year 2012, is considered. The number of years played in the major league and the number of home runs may or

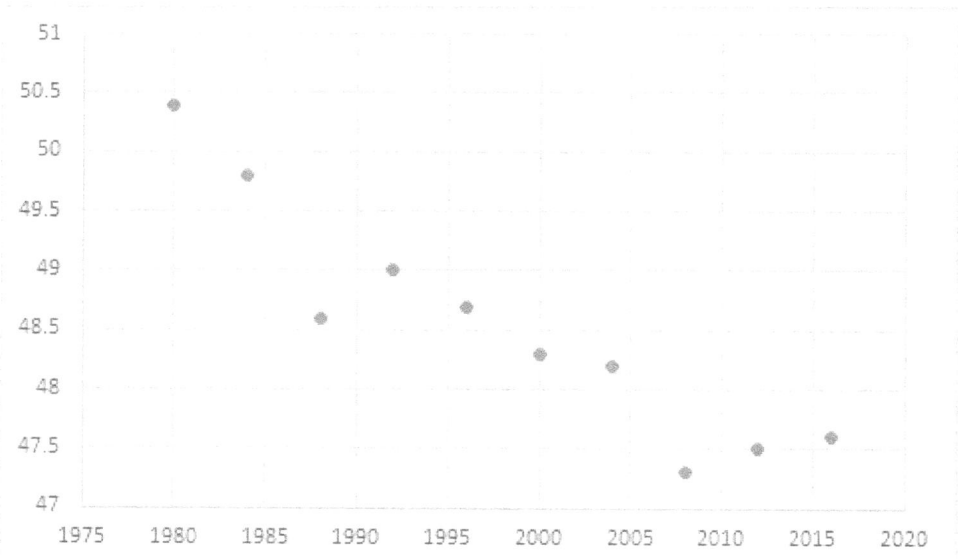

FIGURE 1.5 Paired Data in Example 1.22

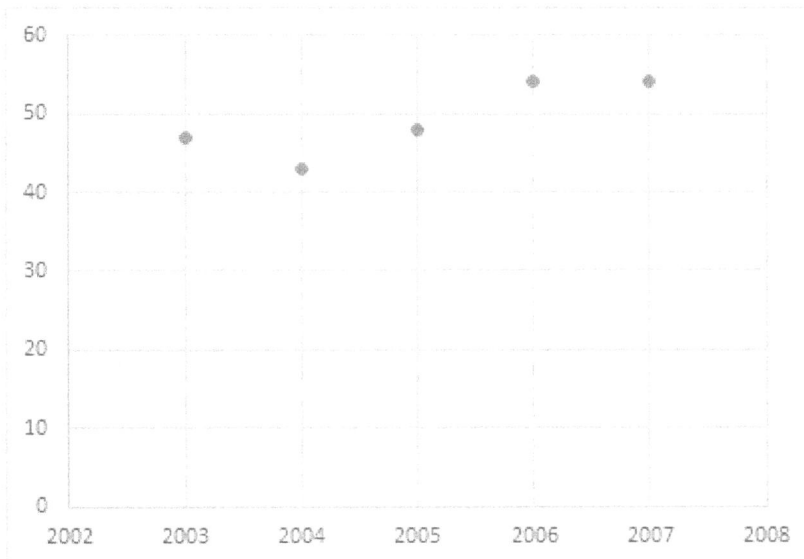

FIGURE 1.6 Paired Data in Example 1.23

may not have correlation. Given the following 10 players, their home runs and the number of years in the major leagues, compute the correlation coefficient to find out if there is a linear relationship.

Player	Years Played	Home runs
A. Jones	17	36
C. Jones	19	36
A. Galaaraga	19	28
J. Lopez	15	24
B. Jordan	15	17
B. Surhoff	19	14
R. Sanders	17	11
Q. Veras	7	5
B. Bonilla	16	5
W. Joyner	16	5

SOLUTION:

x	No. of Years	17	19	19	15	15	19	17	7	16	16
y	No. of Home Runs	36	36	28	24	17	14	11	5	5	5

$$\bar{x} = \frac{17+19+19+15+15+19+17+7+16+16}{10} = 16$$

$$\bar{y} = \frac{36+36+28+24+17+14+10+5+5+5}{10} = 18$$

$$r = \frac{\sum_{i=1}^{N}(x_i - \bar{x})(y_i - \bar{y})}{[\sum_{i=1}^{N}(x_i - \bar{x})^2 \sum_{i=1}^{N}(y_i - \bar{y})^2]^{\frac{1}{2}}} = \frac{195}{\sqrt{(112)(1357)}} = 0.50$$

The value of r does not indicate a strong linear relationship.

A PRACTICE PROBLEM Show that the sample correlation coefficient for the following data is 0.95

$$(0,0), (1,1), (2,3), (3,3)$$

We end this part by mentioning an interesting study that has appeared in *The College Mathematics Journal* (Jan. 2002, Vol. 33, No. 1). It is titled "Is Presidential Greatness Related to Height?" In this article, the author (Paul M. Sommers) asks the following question: what makes a president great or near great? Seemingly, nothing in the presidents' political backgrounds provides clues to their greatness. But he shows that, curiously, greatness does depend on height; that is, they are correlated.

Finally, it should be pointed out that correlation does not imply causality. In fact, correlations can appear to be significant for many reasons. Most of these reasons can be grouped into one of three general categories: (1) coincidence, (2) variables under consideration might be directly influenced by some common underlying cause, and (3) one of the variables may actually be a cause of the other.

1.7.2 Using Correlation for Prediction

Clearly one reason for calculating correlation is to use that for prediction. In fact, forecasting and predicting (estimation related to the future events) are important parts of statistics. For example, many colleges use applicant's performance in high school, SAT, or entrance examination to predict his or her success in college. In the stock market, investment advisors seek indicators that can be used to forecast stock market performance. One interesting prediction regarding stock market related to the sports is the following. According to the Super Bowl omen, a Super Bowl victory by a team with NFL origins (in NFL prior to 1970) will be followed by a year in which the New York Stock Exchange index rises. Otherwise, it falls. This indicator has been correct in more than 80% of the years.

Sometimes researchers try to predict a response, such as percent of body fat, by the use of quantities such as the skin fold measurements, or predict the maximal oxygen consumption, by a quantity such as distance run. In studies of this type, the predictor variables (e.g., skin fold measurements) are less time-consuming, less expensive, and more feasible for mass testing than the response variable; thus, a prediction equation, (regression equation) is developed. From this equation, the values of response variable are predicted using the values of the predictor variable. Note that using this approach, we do not need to make measurements on all the levels of predictor variable. We only need enough measurements to produce a reasonable equation (model).

Prediction is based on correlation. The stronger the relationship between two variables, the more accurately we can predict one using the other. If the correlation were perfect ($r = 1$ or $r = -1$), we could predict with complete accuracy. Of course, we do not encounter perfect relationships in the real world, but in introducing the concept of prediction (regression) equations, it is often advantageous to begin with a hypothetical example of a perfect relationship.

Chapter 1: Elements of Statistics

Suppose that Jim runs 5 miles a day and burns 900 calories. The total number of calories he burns running in 1 week is then 7(900) = 6300. If there are no other factors, we can predict with complete accuracy the number of calories burned. By plotting the number of days (the x, or predictor variable), the predicted number of calories (the y, or response variable) can be obtained. Thus, if we know that another runner (e.g., Peter) runs 4 miles a day and has the same age and weight as Jim, we can calculate y by replacing 4 for x in the equation

$$y = 180x$$

and obtain $y = 720$.

Now suppose that all the runners warm up on treadmill and stop after burning 100 calories. The formula becomes

$$y = 100 + 180x$$

To predict the number of calories burned by Jim, we use this formula and obtain

$$y = 100 + 180(5) = 1,000$$

This formula is the general formula for a straight line and is expressed as follows:

$$y = a + bx$$

where y = the predicted calories, or response; a = the intercept; b = the slope of the line, and x = the predictor.

In this example, the b factor was ascertained by common sense. The slope of the line (b) signifies the amount of change in y that accompanies a change of 1 unit of x. Therefore, any x unit (miles) is multiplied by 180 to obtain the y value. In actual problems, we will not intuitively know what b is, so we must calculate it using the following formula obtained by a method known as least squares:

$$b = r(s_y/s_x) = \frac{\text{Sum of } x_i y_i - \frac{(\text{Sum of } x_i)(\text{Sum of } y_i)}{n}}{\text{Sum of } x_i^2 - \frac{(\text{Sum of } x_i)^2}{n}}$$

$$= \frac{\Sigma x_i y_i - \frac{1}{n}(\Sigma x_i)(\Sigma y_i)}{\Sigma x_i^2 - \frac{1}{n}(\Sigma x_i)^2}$$

Here r = the correlation between x and y, s_y = the standard deviation of y, and s_x = the standard deviation of x. The a in the equation indicates the intercept of the line on the y-axis. In other words, a is the value of y when x is zero. On a graph, if you extend the line sufficiently, you can see where the line intercepts y. The a is a constant because it is added to each of the calculated bx-values. Once again, in our example we know that this constant is 100. In other words, this is the value of y even if there were no running (x) and only a warm up. But to calculate the intercept a, we must first calculate b; a is then obtained from the formula

$$a = \bar{y} - b\bar{x} = \frac{(\Sigma x_i^2)(\Sigma x_i) - (\Sigma x_i)(\Sigma x_i y_i)}{n\Sigma x_i^2 - (\Sigma x_i)^2}$$

where $\bar{y}$ = the mean of the y values, b = the slope of the line, $\bar{x}$ = the mean of the x-values. Deviation of these formulas is discussed in Chapter 8 titled "Regression." The equation discussed above is usually referred to as regression line.

EXAMPLE 1.25 Consider the following data:

$$(0,0), (1,1), (1,2), (2,3), (3,4)$$

Find r and the equation for regression line. Estimate y (second value) for $x = 2.5$.

SOLUTION:

x_i	y_i	x_i^2	y_i^2	$x_i y_i$
0	0	0	0	0
1	1	1	1	1
1	2	1	4	2
2	3	4	9	6
3	4	9	16	12
$\sum x_i = 7$	$\sum y_i = 10$	$\sum x_i^2 = 15$	$\sum y_i^2 = 30$	$\sum x_i y_i = 21$
$\bar{x} = 1.4$	$\bar{y} = 2$			

$\bar{x} = 1.4$, $\bar{y} = 2$

$$s_x = \sqrt{\frac{1}{4}(15 - \frac{1}{5}(7)^2)} = 1.14, \quad s_y = \sqrt{\frac{1}{4}(30 - \frac{1}{5}(10)^2)} = 1.58$$

$$r = \frac{21 - \frac{1}{5}(7)(10)}{(4)(1.14)(1.58)} = 0.972$$

$$b = (0.972)(1.58/1.14) = 1.347, \quad a = 2 - (1.347)(1.4) = 0.1142$$

$$y = 0.1142 + 1.347x$$

For $x = 2.5$, $y = 0.1142 + (1.347)(2.5) = 3.4817$

A PRACTICE PROBLEM Show that the regression equation for the data containing four pairs (0,0), (1,1), (2,3), (3,3) is

$$y = 0.1 + 1.1x$$

Next, let us use a practical example for demonstration. Suppose that the correlation between body weight and dynamometer strength is 0.67. Let x and y represent body weight and strength, respectively, and suppose that we are given the following information

$$\bar{x} = 98.00 \quad \bar{y} = 167.00, \text{ and } r = 0.67$$
$$s_x = 9.44 \quad s_y = 33.52$$

To find the regression line, first we calculate b as

$$b = r(\frac{s_y}{s_x}) = 0.67(\frac{33.52}{9.44}) = 2.38$$

Chapter 1: Elements of Statistics

Next, find a as

$$a = \bar{y} - b\bar{x} = 167 - 2.38(98) = -66.24$$

The regression equation is then

$$y = -66.24 + (2.38)x$$

For any body weight, we can, using this equation (model), calculate the predicted strength score. For example, a boy weighing of 100 lb (x) would have a predicted strength score (y) of $y = -66.24 + (2.38)100 = 171.8$. The main difference between this example and that of running and the number of calories discussed above is that, there was no error of prediction in the latter because the correlation was 1.00. When we predict strength from body weight, however, the correlation coefficient was less than 1.00, so there is an error of prediction. So when predicting one variable using another, the predictions are not the same as the observed values. If we predict the y values for all the available x-values and calculate the prediction errors for each y, we can use them to assess the model. To do this, we subtract the observed values from the predicted values and calculate the standard deviation of the errors. This is called the standard error of estimate. Usually a best model is developed by minimizing this value.

A simple way of obtaining the standard error of estimate is to use the formula

$$s_{y \cdot x} = s_y \sqrt{1 - r^2}.$$

For the body weight, strength data this is equal to

$$33.5\sqrt{1 - 0.67^2} = 24.9.$$

The standard error of estimate is interpreted the same way as the standard deviation. If distribution of errors is close to normal, then the predicted value (strength) in our example, plus or minus the standard error of estimate, will occur approximately 68 times out of 100. Thus, we predict that a 104-lb boy will score between 181.3 ± 24.9. To express it another way, the "prediction range" will be between 156.4 and 206.2 lb 68 times out of every 100 (or the chances are two in three). Note that the larger the correlation, the smaller the error of prediction. Also, the smaller the standard deviation of the response, the smaller the error. In the previous problem, if we had a correlation of 0.85, for example, the standard error of estimate would be

$$s_{y \cdot x} = 33.5\sqrt{1 - .85^2} = 17.6.$$

As mentioned the line of best fit is sometimes called the least squares line. This means that the calculated regression line is one about which the sum of squares of the vertical distances of every point from the line is minimal. We will not develop this point here. This subject will be discussed in detail in Chapter 8.

PRACTICE PROBLEMS

1. The following data represent the Olympic records prior to the year 2012. Find the regression line and predict the record for women's 100-m given that the record for men's

100 m run is 9.79. This was the world record set by Maurice Greene, 1999. The present record is 9.78 and was set in September 2002.

Olympic Records

Event:	Time (Men)	Time (Women)
	x	y
100 m Race	9.63 s	10.62 s
200 m Race	19.30 s	21.34 s
400 m Race	43.30 s	48.25 s
800 m Race	1.681 min	1.891 min
1500 m Race	3.535 min	3.899 min
10,000 m Race	27.020 min	29.291 min

Also, find the regression equations relating x and y (separately) to the distance. In both cases predict the best times for a 5,000 m race.

2. According to the National Weather Service, the number of deaths in the United States as a result of lightning in a time period 2012-2016 were

Year (x)	2012	2013	2014	2015	2016
Number of Deaths (y)	29	23	26	28	40

Find the regression line. As an example of statistical inference see if you can predict the number of deaths for the year 2017. Find the exact value from their website and compare that with your prediction.

EXAMPLE 1.26 Find the equation of regression line for predicting maximal oxygen consumption ($\dot{V}O_2$ max) from scores on the 12-min run. The information needed are given below

x(12-min run) y($\dot{V}O_2$ max)

$\bar{x} = 3,120$ yd $\bar{y} = 52.6$ ml $\bullet$ kg $\bullet$ min^2

$s_x = 334$ yd $s_y = 6.3$ ml $\bullet$ kg $\bullet$ min^{-1}, $r = 0.79$.

Using the result, what is the predicted $\dot{V}O_2$ max for a subject who ran 3,230 yd in 12 min? For a subject who ran 2,940 yd in 12 min? What is the standard error of estimate for the prediction equation? How would you interpret the predicted $\dot{V}O_2$ max for the subjects who ran 3,230 yd in 12 min?

SOLUTION: Before presenting the calculation, we note that the debate about precise mechanism by which athletes improve their performance continues in scientific literature. Numerous physiological variables affect human performance. However, the standard used to judge human capacity to perform prolonged or endurance exercise is maximal oxygen uptake (VO_2 max). The magnitude of increase in exercise capacity after training depends on the magnitude of increase in oxygen

Chapter 1: Elements of Statistics

consumption. In fact, scientists believe that VO_2 max represents the body's ability to transport oxygen from the air we breathe to active muscles.

Now returning to calculations we have

$$b = (0.79)(6.3/334) = 0.0149, \quad a = 52.6 - (0.0149)(3,120) = 6.108$$

$$y = 6.108 + (0.0149)x$$

$$\text{Subject 1}: \quad y = 6.108 + (0.0149)(3,230) = 54.235$$

$$\text{Subject 2}: \quad y = 6.108 + (0.0149)(2,940) = 49.914$$

$$s_{y \cdot x} = (6.3)\sqrt{1 - (.79)^2} = 3.8626$$

There is a 68% chance that VO_2 max for subject 1 is in the range 54.235 ± 3.8626. We leave finding answers to the other two questions for students.

A PRACTICE PROBLEM A researcher conducted a study examining the relationship between VO_2 max (y) and 1.5 mile run times (x). The resulting analysis led to $\bar{y} = 45.0$ mL/kg/min, $\bar{x} = 12.0$ min, $s_y = 4.0, s_x = 3.0, r = 0.85$.

(a) Find the linear regression equation.
(b) Calculate the predicted VO_2 max for the run time of 11 min.

We end this part by mentioning a point regarding the predictions using the best-fit lines. To demonstrate, we consider record times of 1 mile both for men and women. Examination of data reveals that for men these records have gradually decreased between the years 1942 and 2000, from around 246 s to around 224. For women there has been a faster improvement, from almost 278 s to around 253 s during the period 1967 to 2000. Both improvements are approximately linear. The slope of the line for women records times, is $(253 - 278)/(2000 - 1967) = -0.76$. For men, it is $(224 - 246)/(2000 - 1942) = -0.049$. Since, the former has a higher absolute value, one may conclude that in the future women would run the mile faster than men. But this is not a valid prediction because it is based on extending the best-fit lines beyond the range of the actual data. In fact, the men's first record set in year 1942 is still better than the women's best record set in year 2000.

Partial Correlation

The correlation between two variables is sometimes misleading and may be difficult to interpret when there is little or no correlation between the variables other than that caused by their common dependence on a third variable.

For example, for children many attributes increase regularly between the ages of 6 to 18 years. Examples include height, weight, strength, mental performance, athletic abilities, reading skills, and so on. Over a wide age range, the correlation between any two of these measures will almost certainly be positive and will probably be high because of the common maturity factor with which they are highly correlated. In fact, the correlation may drop to zero if the variability caused by age differences is eliminated. We can control this factor of age in one of two ways. We can select only children of the same age, or we can partial out the effects of age statistically by holding it constant.

The symbol for partial correlation is $r_{12 \cdot 3}$. It measures the correlation between variables 1 and 2 with variable 3 held constant (we can partial out any number of variables, e.g., $r_{12 \cdot 345}$). The calculation of partial correlation among three variables is quite simple. Let us refer to the correlation of shoe size and achievement in mathematics. This is a good example of spurious correlation, which means that the correlation between the two variables is due to the common influence of another variable (age or maturing). When the effect of the third variable (age) is removed, the correlation between shoe size and achievement in mathematics diminishes or vanishes completely. We label the three variables as follows: 1 = math achievement, 2 = shoe size, and 3 = age. Then, $r_{12 \cdot 3}$ is the partial correlation between variables 1 and 2 with 3 held constant. We can make up some correlation coefficients between the three variables. Let $r_{12} = 0.80; r_{13} = 0.90$; and $r_{23} = 0.88$. The formula for $r_{12 \cdot 3}$ is

$$r_{12 \cdot 3} = \frac{r_{12} - r_{13} r_{23}}{\sqrt{1 - r_{13}^2} \cdot \sqrt{1 - r_{23}^2}} = \frac{0.80 - (0.90 \times 0.88)}{\sqrt{1 - 0.90^2}\sqrt{1 - 0.88^2}} = \frac{0.80 - 0.792}{\sqrt{1 - 0.81}\sqrt{1 - 0.77}} = 0.038$$

Thus, we see that correlation between mathematics achievement and shoe size drops to about zero when age is partialed out.

The primary value of partial correlation is that it is used to develop a multiple regression equation with two or more predictor variables. In the selection process, when a new variable is "stepped in," its correlation with the response is determined with the effects of the preceding variable partial out. The size and the sign of a partial correlation may be different from the zero-order (two-variable) correlation between the same variables. Again this subject, will be discussed in more detail in Chapter 8.

1.7.3 Interpretation of Correlation Coefficient

Recall the example relating the stock market to the Super Bowl. The theory is that, the stock market goes up if the National Football Conference (NFC), or a former NFL team now in the American Football Conference (AFC), wins the Super Bowl. The market goes down otherwise. Is there really a strong correlation to help us with predictions, or this is just a coincidence? Clearly stock market has nothing to do with the outcome of a football game. People buy stock because it generally has an upward trend. Also the Super Bowl is usually won by NFC teams. In fact, during the last 30 years people have developed varieties of indicators for prediction of the stock market, which in reality have nothing to do with it. For example, the George Steinbrenner indicator monitors his firing of Yankee managers. On the five occasions that he fired Billy Martin, the Dow rose an average of 5 points the following day; the eight times Steinbrenner fired someone else, the Dow fell on average of 3 points the next day.

Recall also the example of Presidential greatness and its relation to the height, which claims that greatness does depend on height. Saul Feldman of Case Western University, whose research interest is height, has studied Presidential candidates and reported that in a period of 40 years starting 1932 except for one case, the taller of the two candidates has always won the electoral victory. The question that arises as a result of this study is the following: was Nixon's victory over Humphrey due to the fact that he was half an inch taller? Could height explain the four victories by Roosevelt? Of course, height of a candidate could influence some votes. But in a more recent elections Jimmy Carter beat the taller Gerald Ford and Bill Clinton beat the taller George Bush.

Finally, if you are a science major and had linear algebra or are familiar with vector analysis you may find the following connection interesting. Recall that if u and v are vectors in a two-dimensional space and θ is the angle between them, then the dot product or Euclidean inner product $u \cdot v$ is defined as

$$u \cdot v = \|u\| \|v\| \cos \theta \quad \text{or} \quad \cos \theta = \frac{u \cdot v}{\|u\| \|v\|}$$

where $\|u\|$ represents the length (norm) of vector u. If u and v are orthogonal $\theta = \pi/2$ and $u \cdot v = 0$. If $u = v$ or $u = -v$ then $\cos \theta = 1$ or $\cos \theta = -1$, respectively. Also

$$\left(\frac{\|u\|}{\|v\|} \cos \theta\right) v$$

represents vector component of u along v (projection of u on v). Thus for smaller values of θ (larger values of $\cos \theta$) the projection is bigger. Thinking of $\cos \theta$ as a correlation we see that a larger value of $\cos \theta$ leads to a better prediction of v using u. In fact, projection of u on v presents that part of v that can be predicted by knowing u.

1.7.4 Regression-to-the-Mean

In sports, the tendency to not to do very well after a successful period or to do well after a poor performing period is predicted by a statistical phenomenon known as regression-to-the-mean. If regression variables represent performances of a player or a team during the first and the second periods, then it is shown that

Performance in the Second Period = Average Performance + b (Performance in the First Period - Average Performance)

If performance in the first and the second periods have the same variance, then $b = r$, that is, the correlation between the performances in the two periods. This problem is similar to one studied by Galton who was interested in the relationship between the heights of tall fathers and their sons. He noticed that sons are usually shorter than their fathers although they are usually taller than average. If correlation between the heights (father and son) is around 50%, then the best prediction for the son's height is around halfway between his father's height and the average height.

1.7.5 Inference and Data Mining

As pointed out earlier inference based on data collected or to be collected is the major concern of statistics and its applications. In recent years, due to availability of high-speed computers, many large databases have become available for analysis. When using such databases, one problem faced by the researchers is to separate useful information or data from the junk. This has led to a relatively new area known as data mining.

In the statistical literature, the name data mining is used in two different contexts. One refers to the process of analyzing large databases and others to data grubbing, data dredging, or fishing. It is probably no exaggeration to say that most statisticians are concerned with primary data analysis. That is, data are collected with a particular question or set of questions at mind. Data mining, being different in this sense, is entirely concerned with secondary data analysis. In fact, data mining may be defined as the process of secondary analysis of large databases aimed at finding unsuspected relationships, which are of interest or have value to the database owners.

Also, statistics, as is presented in most textbooks, might be described as being characterized by data sets, which are small and clean. This does not apply to data mining, whose concern is the secondary analysis of large databases. Here many new problems arise, partly as a consequence of the sheer size of the data sets involved, and partly because of issues of pattern matching. Data mining is a new discipline lying at the interface of statistics, database technology, pattern recognition, machine learning, and so on.

Now the data mining as fishing expeditions refers to a process where a researcher takes a body of data, tests countless theories, and reports the one that is the most statistically significant since he or she believes that only statistically significant results are worth publishing. Most statisticians know very well that if only worthless theories are examined, the researcher will eventually stumble on a worthless theory that seems to be supported by the data. For example in 1989, *The New York Times* reported that if the Down Jones Industrial Average increased between the end of the November and the time of the Super Bowl, the football team whose city comes second alphabetically would probably win the Super Bowl. The Dow increased from 2115 to 2235 between November 30, 1989, and the day of the January 1990 Super Bowl, and San Francisco beat Cincinnati. Clearly, this is a coincidence uncovered by data mining. The behavior of the Dow and the names of the cities should have no effect on the outcome of a football game. The end of November was selected because this is the starting time that worked the best.

We end this chapter by mentioning a few sports-related websites helpful for learning and practicing the material covered.

Sports : Webquests - ESL Resources
www.henry4school.fr/Sports/sports/wq.htm

March Madness
In this web quest, learners use the Internet to locate the teams in the NCAA tournament. They then research the statistics of the teams participating in the tournament. Finally using statistics such as number of games played, number and percentage of games won, and number of NCAA championships, learners try to predict the winner of the tournament.

Through the web quest, learners can calculate percents, and probabilities. At the end of the web quest, learners will be able to relate mathematics and probability to basketball and understand the relationship between these topics and the game. https://www.ncaa.com/march-madness

Statistics in Sports
In this web quest, learners develop a profile of a professional player. They first choose a player who has been playing for at least 5 years. Form the Internet, they obtain statistics for that player. Using these statistics, they then find the mean, mode and range of either, points scored per game

Chapter 1: Elements of Statistics

(basketball), batting average (baseball), goals scored (hockey, soccer). Finally, they plot the data and try to predict the player's performance for the next season.

Olympic Web Quest
In this web quest, learners research swim times in the Olympic games. Once they have obtained the statistics, they graph the information. They then determine the regression lines and make a prediction about the times in the upcoming Olympics.

Monday Night Madness
Learners choose two NFL teams to research. They also gather statistics about the weather in the cities the teams play in. Using all the statistics that they have gathered, learners decide in which city they would like to see the teams play. Then use the statistics to determine the weather and predict the winner of the game.

Salary of NHL Players
Learners determine which characteristic/statistic of a hockey player are most important. Each characteristic will be given a different value based on importance to the learner. Learners then select their favorite hockey team and use the Internet to research the statistics of the individual players on the team. Using their own scale of importance, the learners rank the value of each individual player to the team. They then decide on a salary cap for the most valuable player and determine a salary for each player on the team.

IAAF
The International Association of Athletics Federations (IAAF) is the international governing body for the sport of athletics. It was founded on July 17, 1912 as the International Amateur Athletic Federation by representatives from 17 national athletics federations at the organization's first congress in Stockholm, Sweden. Since October 1993, it has been headquartered in Monaco. https://www.iaaf.org/home

1.8 USING R

R is a programming language and free software environment for statistical computing and graphics supported by the R Foundation for Statistical Computing. The R language is widely used among statisticians and data miners for developing statistical software and data analysis. In this section, one will learn to solve problems using R functions and scripts and interpret outputs. There are a lot built-in functions to find numerical characteristics of your sample (named dataset in R). However, you probably prefer to simply read the data into R instead of typing numbers manually. Therefore, let's start with reading our samples into R environment.

There are different popular data formats to store your samples, for example XLS or XLSX (Microsoft Excel worksheet), CSV (comma separated value), SAV (IBM SPSS data file), and so on. In this book, let's use CSV format as example, but readers will find packages in R to handle other various formats in the R package library repository named The Comprehensive R Archive Network (CRAN).

Recall Example 1.1, we observed winning times in the Men's 100-m Freestyle Olympic swimming competition from year 1980 to 2016 and stored in a file named *Example1.1.csv*. Now, we can simply import that CSV data file using following commend. Please note you need to always provide a dataset name to your imported data thus you can use it in your future code.

```
# Please replace Example1.1.csv with full path of your own
# data file in CSV format
dat1 = read.csv('Example1.1.csv')
```

Now you have your sample imported and you can refer its name *dat1* in your future program.

Next, we will calculate the numerical summaries of our sample *dat1*, for example, mean, median, range, variance, standard deviation, and correlation coefficient.

```
# Find the mean value of time
mtime = mean(dat1$time)
mtime
```

The function mean() calculates average value of a **vector** or **column** in your dataset in R. That's why I provide *dat1$time* as its argument, as dat1$time means the column time in my dataset dat1. To obtain all column names of your dataset, you can use *colname()* function, such as *colname(dat1)*. The second line **mtime** just display the calculated average time. Similarly, we can find median, variance, standard deviation, and correlation coefficient.

```
# Find the median value of time
medtime = median(dat1$time)
# Find the variance of time
vartime = var(dat1$time)
# Find standard deviation of time
sdtime = sd(dat1$time)
# Another way to find standard deviation of time
sdtime_alter = sqrt(var(dat1$time))
# Correlation between Year and Time
yt = cor(dat1$year, dat1$time)
```

There is a little different to calculate range in statistics and R program. Range was defined as the difference between the largest and the smallest observation. Yet R will simply tell the largest and the smallest values instead.

```
# Range function in R
range(dat1$time)
```

We will end this section by demonstrating how to build your simple linear model in R.

```
# A simple linear model
```

 # fitting Time using Year
model = **lm**(dat1$time ~ dat1$year)

Then you will see following output:

> **model**

Call:
lm(**formula** = dat1$time ~ dat1$year)

Coefficients:
(Intercept)   dat1$year
   203.22364   −0.07742

That suggested the model *Time* = 203.224 − 0.077 × *Year*.

CHAPTER 2

Elements of Probability

In chapter 1 we provided an overall view of statistics, what it is intended to accomplish, and why it plays an important role in the modern world. We also learned a few things about the first step of the inference, such as descriptive statistics, that may be used to deduce (infer) the nature of sampled population.

This chapter represents a shift in gears. Here we discuss a less familiar but more abstract concepts of probability that includes making the deductive leap from a sample to a population, or from population to a possible sample. The disciplines of probability and statistics are complementary in the same manner that theory and observation are. In a way this connection is related to Einstein's views on the antithesis between the deductive and empirical components of science. Several authors quoted Einstein as saying "...purely logical thinking cannot yield us any knowledge of the empirical world; all knowledge of reality starts from experience and ends with it."

Probability theory investigates (systematically) the laws concerning phenomena influenced by chance. It is the branch of science concerned with the study of scientific techniques for making quantitative inference about uncertainty. The theory starts with introducing a measure for degree of certainty or uncertainty about a particular outcome. In recent decades, probability theory has become exceedingly important. One reason for this stems from the fact that almost all real-life problems and decisions involve some degree of uncertainty. This is even true about science since, except for mathematics, most disciplines operate based on certain theories and hypotheses. Some known examples for this are evolution theory, big bang theory, tectonic theory, and so on.

In 1620, Francis Bacon argued that learning about the world can only take place by observation. With the rapid expansion of experimental sciences, scientists began to make use of the existent probability theory. In a monumental piece of experimental research between 1856 and 1863, Mendel laid bare the statistical laws of genetics. Without any precedent, Mendel perceived that the genetic mechanism operated like a random device.

It should be pointed out that although uncertainty is not comforting, it could make our lives interesting and exciting. Indeed, the world would be a dull place if everything was predictable. Among many real-life examples of this, people like to watch sports and get excited because the outcomes are not perfectly predictable. Of course, uncertainty can also cause grief and suffering.

Turning to a more precise description, the probability theory provides tools for making quantitative statements about uncertainty and allows one to draw conclusions from such statements using the rules of logic. It is (as opposed to the word probability) a mathematical theory (a model of reality) that enables us to calculate the likelihood of one outcome based on its relationship to other outcomes and their likelihood. Areas of applications include reliability calculation, risk analysis, quality control, business, social sciences, and so on.

Probability theory is also a useful tool for making inference. For example, suppose that as you come out of your apartment you notice a man looking inside your car through the window. You will probably infer that the man is trying to rob your car. How did you arrive at that conclusion? You considered the possible set of circumstances that might have produced the sample of event that you observed. Was the man trying to rob your car? Was the man interested in a car like yours and was looking to see how the interior of the car looks? Was the man trying to figure out whether this was his friend's car? Of these and many other possibilities for the event you observed, you picked the outcome that you thought was most probable. That is, you based your inference on your assessment of a set of probabilities.

This chapter is devoted to basic concepts and rules of the probability theory. Some interesting applications of this theory are also discussed in this chapter.

2.1 CHANCE EXPERIMENTS

An experiment is any action or process that generates observation. When the possible outcomes of an experiment are at least two and there is uncertainty about which outcome will result, the experiment is referred to as chance experiment. Flipping a coin and rolling a die are examples of chance experiment. In practice, often all the possible outcomes of a process are known, but it is not known exactly which outcome will be produced until the process is over. For example, when flipping a coin we cannot predict the outcome but we are sure that it will be either head or tail. As a sport example, if two soccer teams (A and B) are to play a game, then we know that there are only three possible outcomes: Team A wins, or Team B wins, or a tie is reached. However, we do not know for sure which outcome will occur until the game is over. So we refer to this as a chance experiment, or simply an experiment since these are the only situations considered here. Chance experiments are often repeatable. For example, we may flip a coin or roll a die as many times as we wish or the two soccer teams mentioned may have played many games in the past or may play many games in the future. In real life, experts and fans may predict the result of the game based on the past records. They may think that Team A has "a good chance" to win. Some may be even more "precise" and say that Team A has a "90% chance" of winning. When they say such a thing, they are talking about "probabilities." Fans also occasionally talk about the "odds" for or against a team winning. Odds and probabilities are related. Their relation is discussed in Section 2.3.

Before presenting the probability theory in a more systematic format, we should like to be clear about the frequent use of coins, dice, and cards. Since such items are often associated with gambling, some students might think that probability is all about gambling and related activities. Coins, dice, and cards are, in fact, used simply because they present (simulate) many real-life situations. For example, binary situations such as true–false, yes–no, on–off, male–female, defective–nondefective, and generally something and not that thing are all similar to head–tail in the experiment of flipping a coin. Thus, it is easier to work with coins since experiments with them are repeatable and understandable.

Whenever we have an experiment, we use the term "population," or "sample space," to mean the collection or universal set of all possible outcomes. For example, for the chance experiment of rolling a die it is $\{1,2,3,4,5,6\}$. The sample space or population is usually denoted by Ω. Ω could be the set of the integers from 1 to 10, a group of individuals with certain characteristics, or a set of sport records. Suppose that we perform a chance experiment E, which consists of picking an element ω of the population randomly and associating with it a number $x(\omega)$. For example, we may roll a die, or measure the blood pressure of a randomly selected person in Ω, and so on. Remember that Ω denotes the sample space, that is, the collection of all possible outcomes or realizations. The function X, which associates the value $x(\omega)$ to each ω, is called a random variable. A realization of X means an actual performance of the experiment E, by which we get a special ω and the associated value of X, namely $x(\omega)$. A simple example of this is the chance experiment of flipping a coin. Here the sample space contains two outcomes "Head" and "Tail." That is, $\Omega = \{\text{Head, Tail}\}$ or $\Omega = \{H, T\}$. If we associate 0 to tail and 1 to head, then we have defined a random variable, "number of heads." The sample space for X is then $\Omega = \{0, 1\}$. If we flip two coins, the same space is $\Omega = \{HH, HT, TH, TT\}$. If X presents the number of heads, then the sample space for X is $\Omega = \{0, 1, 2\}$. If we roll two dice and let X represents the sum of the two numbers observed, then the sample space is $\Omega = \{2, 3, 4, 5, 6, 7, 8, 9, 10, 11, 12\}$.

EXAMPLE 2.1.1 Referring to Example 1.2.1, Ω is the group of 1,000 students. Hence, each individual in this group is an element ω of the population Ω. The experiment E is to measure the blood pressure of these students. In this case, the number $x(\omega)$ assigned to each ω is the measured value of the blood pressure.

EXAMPLE 2.1.2 Two soccer teams play a game. Suppose that the objective of the experiment is to observe the result. Then the sample space or the population of all possible outcomes is

$$\Omega = \{\text{Team A wins, Team B wins, Tie}\}.$$

Here we may assign a number to each outcome. For example, we may let

$$X(\text{Team A wins}) = 1, \quad X(\text{Team B wins}) = -1, \quad X(\text{Tie}) = 0.$$

Note that, rather than telling you that team A won, we tell you that x took the value 1.

In most soccer tournaments, teams get 3 points for a win, 1 point for a tie, and 0 points for a loss. So, a different assignment may be

$$Y(\text{Team A wins}) = 3, \quad Y(\text{Team A ties}) = 1,$$

$$Y(\text{Team A loses}) = 0.$$

Note that here both X and Y are random variables. We can introduce many other random variables on the same sample space. See whether you can think of one. Random variables are discussed in detail later in Section 2.5.

EXAMPLE 2.1.3 Suppose that two people are playing tennis starting with the score of 0:0. After one exchange the sample space is

$$\Omega = \{(15{:}0),(0{:}15)\}.$$

After two exchanges it is

$$\Omega = \{(30{:}0),(15{:}15),(0{:}30)\}.$$

For this example, we may define a random variable and simplify the scoring system. For example, we may set $X(0) = 0, X(15) = 1, X(30) = 2$, and so on. Then the score of (30:0) would be simply (2:0).

If you are familiar with the scoring system of tennis, specify the sample space after three exchanges and values of the random variable defined above.

EXAMPLE 2.1.4 In tennis, players get two chances to make a proper service on each point. As a result, tennis players usually possess two types of serve. One labeled the "strong" serve is traditionally used on the first service. This serve generally gives a high probability of winning the point to the server, given that the serve is good. The second serve, labeled the "weak" serve is usually reserved for the service following an initial fault. If we denote these by s and w, respectively, then the sample space includes four possible outcomes $\Omega = \{sw, ss, ww, ws\}$. Here we can define a random variable presenting the number of "strong" serves. This leads to a sample space $\Omega = \{0, 1, 2\}$.

2.2 EVENTS AND THEIR ALGEBRA

Consider a chance experiment and the collection of its possible outcomes (sample space) Ω. In practice, one may be interested in the collection of outcomes with a certain property or attribute. Such collections are called events. For example, when flipping two coins the sample space is

$$\Omega = \{HH, HT, TH, TT\}.$$

Suppose that we are interested in outcomes that include at least one head. The collection of outcomes with this property is the following event:

$$\{HH, HT, TH\}.$$

In short, an event A is a subset of the population (or sample space) Ω (i.e., $A \subseteq \Omega$). Examples include tossing an even number in the chance experiment of rolling a die, or looking for all individuals whose blood pressure is greater than 140 mm Hg. A occurs if a realization leads to an ω in A. Based on this definition Ω itself and the empty set (ϕ) are both "events," Ω is the sure event that always occurs and ϕ is the empty event that never occurs. Each ω in Ω is called an elementary (simple) event, or outcome.

EXAMPLE 2.2.1 Consider the chance experiment of rolling a die. The sample space is

$$\Omega = \{1, 2, 3, 4, 5, 6\}.$$

Chapter 2: Elements of Probability

Consider the three events, "even," "greater than 4," and "divisible by 5." They include, respectively, the following outcomes:

$$A = \{2,4,6\}, \quad B = \{5,6\}, \quad C = \{5\}.$$

If we roll a 5, then both B and C occur, but not A.

EXAMPLE 2.2.2 Referring to the Examples 1.2.1 and 2.1.1, let A be the event that the value of blood pressure is less than or equal to 120. Then A contains 523 elements or values (i.e., the blood pressures of 523 people are in this subset).

EXAMPLE 2.2.3 Referring to the Example 2.1.2, let C be the event that Team A does not lose the game, then

$$C = \{\text{Team A wins, Tie}\}.$$

Sample space plays a crucial role for calculating the probability of occurrence of events of interest. Unfortunately, determination of all possible outcomes of a chance experiment is not always straightforward. One basic tool that has proved very helpful is the tree diagram. To see an example of application, consider the chance experiment of flipping two coins. The four possible outcomes can be displayed in the tree diagram presented below.

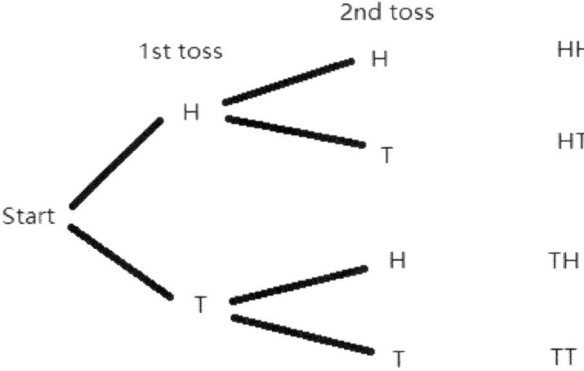

The two branches leading out from the start correspond to the two possibilities for the first coin (e.g., a dime). These are sometimes referred to as first-generation branches. There are two further second-generation branches corresponding to the possibilities for the second coin (e.g., a quarter). If three coins, rather than two, are tossed, there are eight possible outcomes since two new branches open from existing branches (four from adding an additional H to each outcome just considered and an additional four from adding a T). Perhaps you have already noticed a pattern in this problem. For one coin there are two outcomes, for two coins there are four (or 2^2), and for three coins there are 2^3 outcomes. Thus, it is easy to see that, for example, for four coins the number of outcomes is $2^4 = 16$. This is useful since in many situations we need to know how many outcomes are in the sample space.

EXAMPLE 2.2.4 A sports store stocks shorts in small, medium, large, and extra large, and all are available in blue or red. What are the possible choices and how many choices are there? Construct a tree diagram.

SOLUTION:

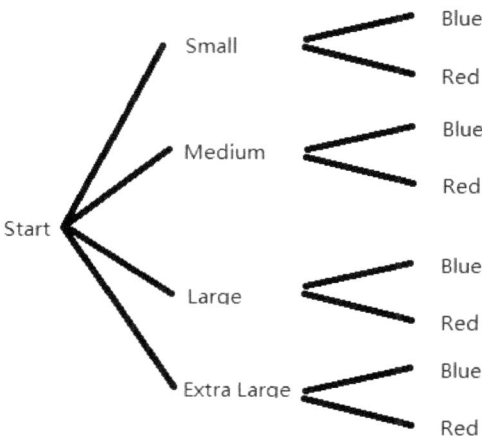

Thus, there are 8 possible choices (outcomes). There are 4 ways a size can be chosen and 2 ways a color can be chosen. The first element in the ordered pair represents a size choice, and the second element represents a color choice. Note that without listing all the possible outcomes we can figure out that there are $4 \times 2 = 8$ of them.

Algebra of Events

If A and B are two events, then we may combine them using certain operations to create new events. The operations used here are similar to those in set theory and logic. They include OR, AND, and NOT as defined below.

$A \cup B$ is the event "A <u>or</u> B." It occurs if, either A or B or both occur (at least one event occurs).

$A \cap B$ is the event "A <u>and</u> B." It occurs if, both A and B occur.

A' is the event "<u>not</u> A." It occurs if A does not occur.

$A \subseteq B$ means that A has B as a consequence. That is, if A occurs so does B.

More precisely, $A \cup B$ contains all the outcomes in A or in B or in both A and B (or all the outcomes that either have property A or property B or both), $A \cap B$ contains all the common outcomes of A and B, and A' contains all the outcomes that are not in A. Also $A \subseteq B$ means that every outcome of A is in B too. A and B are called incompatible (disjoint or mutually exclusive), if occurrence of "A <u>and</u> B" is impossible ($A \cap B = \phi$). In other words, A and B cannot occur together. For example, when rolling a die we cannot have both an even and an odd number observed, or a number less than three and greater than five cannot happen together. Note that although events are like sets and their algebra are very similar to each other, interpretation of the above operation is somewhat different. For example, here $A \subseteq B$ means that if A occurs, so does B, but the opposite is not necessarily true. The following pictures, known as Venn Diagrams, illustrate the operations discussed.

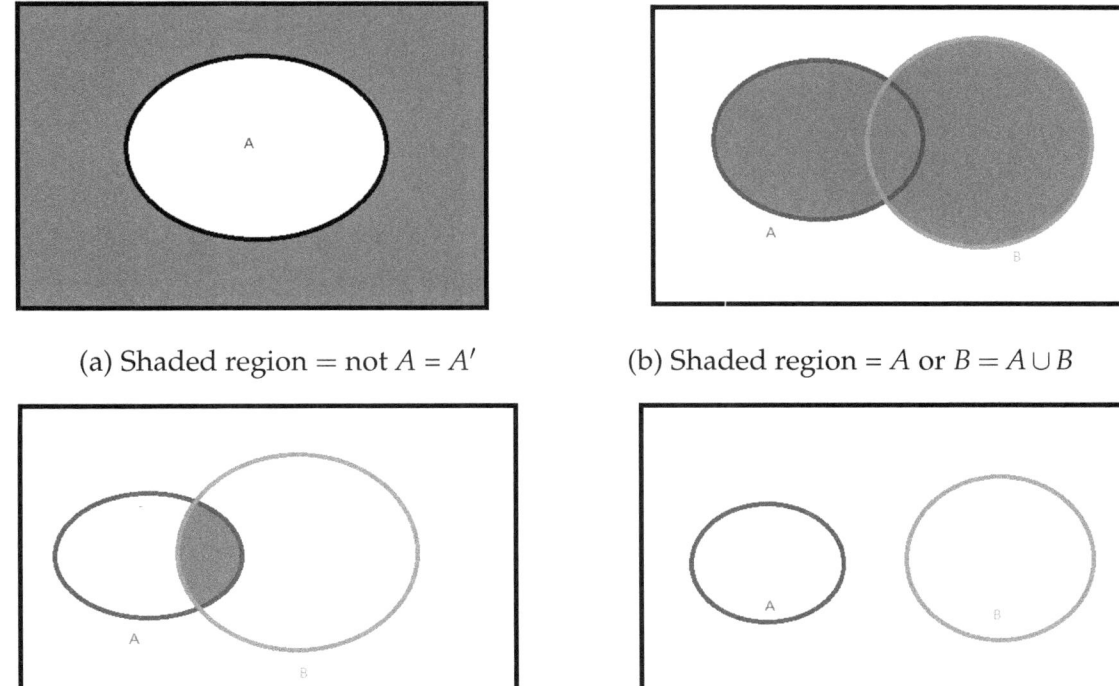

(a) Shaded region = not $A = A'$

(b) Shaded region = A or $B = A \cup B$

(c) Shaded region = A and $B = A \cap B$

(d) Two disjoint events

EXAMPLE Consider, once more, the chance experiment of rolling a die. The sample space is $\Omega = \{1,2,3,4,5,6\}$. Let $A = \{2,4,6\}, B = \{1,2\}$, and $C = \{3,5\}$. Then $A \cup B = \{1,2,4,6\}, A \cap B = \{2\}, A' = \{1,3,5\}$, and $A \cap C = \phi$.

EXAMPLE Consider the chance experiment of flipping two coins. The sample space is $\Omega = \{HH, HT, TH, TT\}$. Let $A = \{HH, HT, TH\}$ and $B = \{HH, TT\}$ and $C = \{TT\}$. Then $A \cap B = \{HH\}$, $A' = \{TT\} = C$, and $A \cap C = \phi$.

One important part of engineering work that draws on the areas of probability and statistics is reliability analysis and failure probability calculation of complex systems that are made up of many components. The simplest forms include components in series and in parallel. A set of components is considered to be in series if the system works only if each of the components work. In other words, the system fails whenever at least one component fails. A set of components is considered to be in parallel if the system works whenever at least one of the components works. In other words, the system fails only if all of the components fail.

Now consider a system with only two components, 1 and 2. Let A_1 and A_2 be events that components 1 and 2 work, respectively. Then the event that the system works is $A_1 \cap A_2$ for components in series and $A_1 \cup A_2$ for components in parallel. Note that the presentation can easily be extended to systems with more than two components. Note that here A_1' means that the component 1 has failed.

EXAMPLE 2.2.5 Referring to Example 2.1.2, let E_1 be the event that Team A wins, E_2 be the event that Team B wins, and E_3 be the event that Team B does not lose. Then

$$E_1 = \{\text{Team A wins}\},$$

$$E_2 = \{\text{Team B wins}\},$$

$$E_3 = \{\text{Team B wins, Tie}\}.$$

Here we have

$$E_1 \cup E_2 = \{\text{Team A wins, Team B wins}\}.$$

Hence, $E_1 \cup E_2$ is the event that one of the two teams win. Since

$$E_1 \cap E_2 = \phi,$$

that is, E_1 and E_2 are disjoint, these teams cannot both win. Also

$$E_2 \subseteq E_3,$$

demonstrating the fact that if Team B wins (E_2 occurs) then, E_3 will occur too.

One important and useful concept in studying the sample spaces and events is the partitioning. A set of events $A_1, \ldots, A_r$ is a decomposition (partition) of the population or sample space Ω, if $A_1 \cup A_2 \cup \ldots \cup A_r = \Omega$ and $A_i \cap A_j = \phi$ for each pair $i \neq j$.

EXAMPLE 2.2.6 Referring to Example 2.2.5, since

$$E_1 \cap E_3 = \phi \text{ and } E_1 \cup E_3 = \Omega.$$

E_1 and E_3 form a decomposition or partition of the sample space.

EXAMPLE 2.2.7 Referring to Example 2.2.2, let B be the event that the value of blood pressure is more than 120 but less than or equal to 150. Let also C be the event that the value of blood pressure is more than 150. Then $A \cup B \cup C = \Omega$, $A \cap B = \phi$, $A \cap C = \phi$, and $B \cap C = \phi$. Hence, A, B, and C form a decomposition (partition) of the sample space Ω.

The operations defined above can be extended to more than two events. Let $A_1, A_2, \ldots, A_k$ denote k events, then

1. The event A_1 or A_2 or $\ldots$ or A_k ($A_1 \cup A_2 \cup \ldots \cup A_k$) consists of all outcomes in at least one of the individual events $A_1, A_2, \ldots, A_k$.
2. The event A_1 and A_2 and $\ldots$ and A_k ($A_1 \cap A_2 \cap \ldots \cap A_k$) consists of all outcomes that are simultaneously in every one of the individual events $A_1, A_2, \ldots, A_k$.
3. These k events are disjoint if no two of them have any common outcomes. That is, $A_i \cap A_j = \phi$, for all $i \neq j$.

Cartesian Product

We end this part by discussing another operation of set theory that can be applied to events.

The Cartesian product of two sets, A and B, denoted by $A \times B$ is the set of all ordered pairs (a, b) whose first component is in A and second component is in B. If, for example, $A = \{a,b\}$ and $B = \{1,2,3\}$, then $A \times B = \{(a,1),(a,2),(a,3),(b,1),(b,2),(b,3)\}$. If $\Omega = \{H,T\}$, represents the sample space for flipping a coin, then $\Omega \times \Omega = \{(H,H),(H,T),(T,H),(T,T)\}$. In other words, if Ω is the set of outcomes of tossing a coin once, then $\Omega \times \Omega$ is the set of outcomes of tossing a coin twice. Similarly if $\Omega = \{1,2,3,4,5,6\}$, then

$$\Omega \times \Omega = \begin{Bmatrix} (1,1), & (1,2), & (1,3), & (1,4), & (1,5), & (1,6), \\ (2,1), & (2,2), & (2,3), & (2,4), & (2,5), & (2,6), \\ (3,1), & (3,2), & (3,3), & (3,4), & (3,5), & (3,6), \\ (4,1), & (4,2), & (4,3), & (4,4), & (4,5), & (4,6), \\ (5,1), & (5,2), & (5,3), & (5,4), & (5,5), & (5,6), \\ (6,1), & (6,2), & (6,3), & (6,4), & (6,5), & (6,6). \end{Bmatrix}$$

So if Ω is the set representing the outcomes of rolling a die once, then $\Omega \times \Omega$ is the set presenting the outcomes of rolling a die twice or equivalently rolling two dice. Clearly, if A has m elements or outcomes and B has n elements or outcomes, then $A \times B$ has mn elements or outcomes. This, as we shall see, is related to the so-called multiplication principle.

A PRACTICE PROBLEM Sports may be classified into three groups: combat sports, object sports, and independent sports (C, O, and I). Also there are three ways in which performance is evaluated in popular sports: judged, measured objectively, or scored objectively (J, M, and S). Consider the following events:
C = {boxing, wrestling, modern pentathlon}, O = {baseball, basketball, soccer, tennis}, I = {swimming, weightlifting, modern pentathlon}, J = {boxing, wrestling, gymnastics}, M = {weightlifting, speed skating}, S = {speed skating, baseball, soccer, tennis, volleyball}.

Find the union and intersection of all possible pairs.

2.3 PROBABILITY

2.3.1 Introduction and Definitions

As mentioned earlier, probability is a measure of "chances" or certainty about the occurrence of an event of interest. The measure could be objective or subjective or even a mixture of these. This means that there is more than one possible way for defining, choosing, or introducing such measures or for assigning probabilities. An important step in the study of probability was the introduction of rules or postulates that help with identification of inconsistent probability assignments. These rules provide guidelines for probability selections free of contradictions.

Suppose that to each event like A, a number "probability of occurrence of A," $P(A)$, is assigned that obeys the following three postulates or axioms:

I. $P(A) \geq 0$
  II. $P(\Omega) = 1$
 III. $P(A \cup B) = P(A) + P(B)$, if A and B are incompatible (disjoint)

Using these postulates, one can construct a deductive theory and include all the rules needed to study a chance experiment. Some straightforward consequences of these postulates are the following:

  IV. $P(A_1 \cup \ldots \cup A_k) = P(A_1) + \ldots + P(A_k)$, if $A_1, \ldots, A_k$ are pairwise incompatible
   V. $A \subseteq B \Rightarrow P(A) \leq P(B)$
  VI. $0 \leq P(A) \leq 1$
 VII. $P(A') = 1 - P(A), \quad P(\phi) = 0$

If Ω has infinitely many elements (has an infinite number of elements), then III has to be fulfilled both for finite and infinitely many disjoint events.

To demonstrate how IV to VII follow from the postulates let us, for example, consider VII. We know that A and A' are disjoint and that $A \cup A' = \Omega$. Thus using postulates II and III we have

$$1 = P(\Omega) = P(A \cup A') = P(A) + P(A') \Longrightarrow P(A') = 1 - P(A).$$

Also since $\Omega \cup \phi = \Omega$ and $\Omega \cap \phi = \phi$, using once more postulates II and III, we obtain

$$1 = P(\Omega) = P(\Omega \cup \phi) = P(\Omega) + P(\phi) \Longrightarrow P(\phi) = 0.$$

This completes the proof. Note that the above axioms do not completely determine the assignment of probabilities to outcomes. They only serve to rule out assignments inconsistent with our intuitive notions of probability. For example, when flipping a coin once we assign probability to occurrence of a head, the probability of occurrence of a tail is one minus that number; no more, no less. So the question that arises at this point is how do we assign probabilities to different outcomes of a chance experiment? In what follows, some frequently used definitions and methods are presented that could provide a partial answer to this question. Note that a complete answer to the question posed should come from the discipline related to the data or subject under study.

Classical (Equally Likely or Theoretical) Definition of Probability

If Ω has only finitely many elements or outcomes (has a finite number of elements), say n, and if by symmetry each of these elements (outcomes) has the same chance to occur, that is, if elements (outcomes) are equally likely, then an appropriate assignment for each possible outcome is $\frac{1}{n}$. If A has m elements (outcomes), then

$$P(A) = \frac{m}{n} = \frac{\text{Number of outcomes in } A}{\text{Number of possible outcomes}}. \tag{2.1}$$

In practice, it is not always easy to establish whether simple events (outcomes) are equally likely or not. To resolve this, if we have no reason to say that simple events (outcomes) are not equally likely, that is, if we have no reason to believe that any outcome is more or less likely than any other outcome, then we assume that they are equally likely (equally probable). This is sometimes

Chapter 2: Elements of Probability

referred to as **principle of insufficient reason**. For example, if we assume a given die is a fair die (or have no reason to say that it is not), then the probability of event $A = \{2,4,6\}$ is

$$P(A) = \frac{3}{6} = \frac{1}{2},$$

and probability of event $B = \{5,6\}$ is $P(B) = \frac{2}{6} = \frac{1}{3}$.

It should be pointed out that the classical definition of probability has a logical problem in that it uses the probability (not yet defined) in the definition of probability. Since equally likely means that outcomes have the same chance or probability of occurrence. Another problem relates to the fact that the only way to produce convincing evidence that, for example, a given coin is fair is by experimentation, that is by flipping it. For fair coin the relative frequency of the occurrence of event H is expected to tend to $P(H) = \frac{1}{2}$ if we perform the experiment (in this case flipping the coin) over and over (law of large numbers). Based on this observation, rather than 1/2 we may use the relative frequency as the probability of occurrence of an event since it is based on experiment. Calculation of probability based on this approach is discussed below and is demonstrated in Example 2.3.2.

A PRACTICE PROBLEM A fair coin is flipped six times, which of the following outcomes do you think is most likely to occur?

$$HTHTHT, \quad TTTHHH, \quad TTTTTT, \quad HHTTHH$$

Can you draw any conclusion from this?

Objective (Empirical) Definition of Probability

Definition of probability based on relative frequency is usually referred to as "empirical probability or statistical probability." Formally, for an event A it is defined as

$$P(A) = \frac{\text{Number of times event } A \text{ has occurred}}{\text{Total number of times the experiment has been performed}}$$

$$= \frac{\text{Frequency of } A}{\text{Total Frequency}} = \frac{f}{n} = \text{Relative Frequency of } A.$$

This definition of probability has wider applications compared to the classical definition. It is based on experiment or past data and therefore is more appealing. If we roll a die 1,000 times and observe one hundred 5's, it is reasonable to assume (initially) that the probability of observing a 5 is 0.10. Of course, we can always update the probability with further experimentation or observations. The following are some examples from sports. According to Newsweek (January 25, 1999) Michael Jordan played in 930 games in a regular season in the period 1984 to 1998. During this period, he tried 21,886 field goals and made 10,962. He also had 8,115 free throw attempts and made 6,798 of them. Using these statistics, the empirical probability for Jordan making a field goal and a free throw are, respectively $10,962/21,886 = 0.501$ and $6,798/8,115 = 0.838$. In soccer stopping a penalty kick is considered quite a feat. According to the *USA Today* (February 19, 2003), this is the ninth hardest thing to do in sports. The mouth of a soccer goal is 24 feet wide and 8 feet high. That is, 192 square feet compared to a ball about 9 inches in diameter. In the 2018 FIFA World

Cup, out of total of 39 penalty kicks attempts 26 were successful. Thus, the probability of success (or success rate) was 26/39 = 0.667.

This definition of probability makes sense to many practitioners since if, for example, we perform an experiment 10 times and seven times a particular event occurs, it makes sense to expect that it would occur seven times in the next 10 experiments.

To clarify the difference between the two definitions of probability, consider the experiment of flipping a coin and determining the probability of observing a head. If we have no reason to say that head is more or less likely than a tail, then we assume that they are equally likely and, hence, the required probability is 1/2. In the second approach, we do some flipping and then specify the probability using the result of the experiment. For example, if we flip a coin 100 times and observe 54 heads, then the required probability is 0.54. To get a better estimate of this probability, we can flip the coin 1,000 times, 10,000 times, and so on. According to the "Law of Large Numbers," the ratio will tend to the true value, for example, to 1/2 if the coin is really a fair coin.

Ten-Year Game

Records indicate that the number of goals scored in the final games in a soccer tournament was exceeded 5 just five times in the past 50 years. What is the probability that the number of goals scored in the next final game will exceed 5?

Based on the data, the relative frequency of more than five goals in a final game is

$$\frac{\text{Number of years with more than five goals}}{\text{Total number of years}} = \frac{5}{50} = \frac{1}{10}.$$

Based on this definition, the probability of more than five goals in the next final game is 0.10. Because the number of final games with more than five goals occurs on average once every 10 years, it is sometimes called a "10-year game." This is the same as saying that the probability of having a final game with more than five goals in any given year is $\frac{1}{10} = 0.10$. Note that we are assuming that the final game occurs once a year.

EXAMPLE 2.3.1 Referring to Example 2.2.1, if one element (one individual) is to be selected randomly from the population Ω and his or her blood pressure is measured, then each individual has the same chance of being selected. That is, all these elements are equally likely to be chosen or at least we have no reason to say that the contrary is true. In this case, since there are 523 elements (outcomes) in event A, the probability for A to occur is

$$P(A) = \frac{523}{1,000}.$$

EXAMPLE 2.3.2 A random sample of 80 adults were surveyed. They were asked whether they regularly watch NFL football games. They were also asked whether their favorite team had ever won the Super Bowl. The results are listed below.

Chapter 2: Elements of Probability

	Team won	Team did not win	Row total
Watch games	12	43	55
Do not watch games	12	13	25
Column total	24	56	80

Suppose a person is to be randomly selected from this sample. Let A be the event that the selected person's favorite team won the Super Bowl. Then $P(A) = 24/80 = 0.3$, or 30%. If B stands for the event that the selected person watches games regularly but his favorite team did not win, then $P(B) = 43/80 = 0.56$, or 56%. Note that here we cannot use equally likely definition and because there are only two possibilities argue that $P(A) = 1/2$. In other words, here we have reasons (empirical evidence) that A and A' are not equally likely, although indirectly a probability $\frac{1}{80}$ is assigned to each possible outcome.

EXAMPLE 2.3.3 Referring to Example 2.1.2, suppose these two teams have played 100 games in the past of which Team A has won 55 games, Team B has won 30 games, and tied in 15 games. The probabilities are then assigned as

$$P(\text{Team A wins}) = 0.55,$$

$$P(\text{Team B wins}) = 0.30,$$

and

$$P(\text{Tie}) = 0.15.$$

Based on these, we have

$$P(\text{Team A does not lose}) = P(\text{Team A wins}) + P(\text{Tie}) = 0.70.$$

As mentioned, the relative frequency interpretation of probability is referred to as objective interpretation because it rests on results of the experiments. In practice, this interpretation is not as objective as it might seem, since the limiting relative frequency (when an experiment is performed infinitely many times) of an event will not be known. Moreover, it refers to the repeatable experiments. Thus, we may have to assign probabilities based on our beliefs about the limiting relative frequencies. Additionally there are many real-life situations where we have no control on occurrence of the event of interest. For example, we cannot create natural disasters such as an earthquake. In fact, in many cases repetitions may not be possible. For instance, no frequency interpretation can be given to the event that New York Jets will win the Super Bowl next year. Other situations where this definition may not be applicable is when no or very few data regarding the event of interest is available. An example of the former is the probability calculation of a nuclear failure. For the latter case, if after six rolls of a die 4 is not yet observed, we cannot assign a probability $0/6 = 0$ to it since observing 4 has certainly a non zero probability. For cases such as occurrence of an earthquake or failure of a nuclear power plant, we could only use experts judgment and subjectively assing probabilities.

Summarizing these, probability is a number between 0 and 1 (inclusive) that provides an indication of how likely is an event to occur. Here 0 refers to impossible and 1 refers to definite and everything else is, of course, between. Mathematics works mostly with items with probability 1.

That is why you see theorems in mathematics books. Physics works with items with probability very close to 1. That is why you see laws in physics books. Then there are subjects such as Biology, Geology, and so on. where you see theories or hypotheses, where probabilities are not very close to 1.

We will now discuss an important and a familiar application of probability; the risk assessment of traveling. Our goal is to calculate the probabilities of interest using the definitions presented above.

Risk Assessment

One important application of objective probability is in risk analysis. Remember that risk is part of our lives and cannot be avoided. In fact, not taking risk itself is a risk, perhaps a serious one and maybe the greatest of all. Let us concentrate on a specific example. According to Insurance Institute for Highway Safety (IIHS), 37,133 Americans were killed in automobile accidents in 2017. Other years statistics are also close to this number. In fact, despite a significant increase in the number of automobiles, the number of fatalities has decreased from about 44,599 to 37,423 between 1990 and 2008. If we take the 37,133 as an average, we may ask what is the probability that a randomly selected individual will die in an automobile accident in any single year? Here, the empirical probability can be found by dividing the 37,133 annual deaths by the 325 million population of the United States in 2017, resulting in

$$37,133/325,000,000 = 114 \times 10^{-6}.$$

This is about 114 out of every 1,000,000 people.

For commercial airplanes, the number of accidents vary significantly from year to year. From the research, there were 4,641 deaths from 2004 to 2013. Let us assume that in a typical year there are $4,641/10 = 464$ deaths in commercial flights (excluding private, commuter, corporate, and military flights). Then we have

$$464/325,000,000 = 1/900,000 = 1.43 \times 10^{-6}.$$

This is about 143 out of every 100,000,000 people compared to 11,400 for every 100,000,000 people for automobile fatalities.

In 2017, passengers traveled approximately 1 trillion miles (in fact it was 964.3 billion miles according to Bureau of Transportation) taking 1 billion trips (stops are counted as separate trips and actual miles are multiplied by number of passengers) by commercial airplanes. Thus, the risk during a single year expressed in deaths per miles flown and per trip taken are, respectively

$$464/10^{12} = 4.64 \times 10^{-10} \text{(deaths/mile)},$$

and

$$464/10^9 = 4.64 \times 10^{-7} \text{(deaths/trip)}.$$

Now if we assume that an average person drives about 11,400 miles and takes about 815 car trips per year (2 to 3 a day), then the corresponding risks are, respectively

$$114 \times 10^{-6}/11400 = 1 \times 10^{-8} \text{(deaths/trip)},$$

and
$$114 \times 10^{-6}/815 = 1.4 \times 10^{-7} (\text{deaths/trip}).$$

Using these we can now compare the risks. Comparison based on deaths per mile leads to
$$1 \times 10^{-8}/4.64 \times 10^{-10} = 100/4.64 \approx 22.$$

That is, risk of death per mile is 22 times higher for car. However, comparison based on deaths per trip results in
$$4.64 \times 10^{-7}/1.4 \times 10^{-7} = 4.64/1.4 = 3.3.$$

That is, risk of death per trip is about 3.3 times higher for airplanes.

Note that in the above analysis we assumed that an average person drives 11,400 and takes 815 trips with each trip being 14 miles. Moreover, an average person flies 4,050 miles and takes 4 trips with each trip being 1,000 miles. So, for a person driving 1000 miles (71 trips) the risk can also be calculated as
$$1 - (1 - 1.4 \times 10^{-7})^{71} = 99.4 \times 10^{-7} \approx 21(4.64 \times 10^{-7}).$$

Summarizing these, if we assume that a typical person drives 11,400 miles and flies 4,000 miles per year, then we have the following risks for automobile and plane, respectively:
$$1 \times 10^{-8} \frac{\text{deaths}}{\text{mile}} \times 11,400 \frac{\text{miles}}{\text{person}} = 114 \times 10^{-6} \frac{\text{deaths}}{\text{person}},$$

and
$$4.64 \times 10^{-10} \frac{\text{deaths}}{\text{miles}} \times 4000 \frac{\text{miles}}{\text{person}} = 1.86 \times 10^{-6} \frac{\text{deaths}}{\text{person}}.$$

A PRACTICE PROBLEM Select the person of your choice and research how many trips or miles he or she travels. Calculate the risk of driving versus flying, for this person.

WHICH DEFINITION OF THE PROBABILITY SHOULD WE USE? So far we discussed different definitions of probability. But when, for example, should we use equally likely definition? If we have no reason to doubt that outcomes are not equally likely, we will assume that they are. This, for example, applies to problems such as flipping a coin, rolling a die, drawing cards from a deck of cards, or choosing a student from a group of students randomly. But it cannot be applied to, for example, weather prediction or being dead or alive tomorrow (we hope they are not equally likely), since otherwise we should expect half of the people dead by tomorrow. For these cases, relative frequency (statistics) is more appropriate and may be used. As a different example, think about buying insurance for your car. During a time period you may make a claim or no claim. But can an insurance company assume that these are equally likely? Of course not. So, insurance companies rely on statistics to find the probability that you will make a claim. The statistics they use include things such as your age, type of the car, and so on.

Now, if no past statistics are available or if an experiment is not repeatable (e.g., natural disasters), then one may consider a subjective assessment, such as an expert opinion, to assign probabilities to different possible outcomes. For example, what is the probability of complete failure of a nuclear power plant close to your home? For this, there is no past data (statistics), so experts may examine the components of the system and subjectively assign or estimate likelihoods. Then combine these likelihoods to arrive at a quantity presenting or estimating the risk (probability) of such events. In summary, subjective probabilities result from intuition, educated guesses, and estimates. Here, probabilities are considered to be measures of personal belief (or knowledge about the subject) and, hence, are usually different for two different people. As a sport example, given an athlete's extent of injuries, a doctor may think that he or she has 80% chance of full recovery. For the same athlete, a trainer with a different set of experiences may suggest a different probability estimate.

Finally, some words of caution about probability. First, probability is an undefined term, or it has many different meanings, to put it differently. The two most usual interpretations are (1) it is an intrinsic property of a physical system and (2) a measure of belief in the truth of some statement. Second, there are many examples demonstrating the fact that assigning likelihoods to events, especially rare events, may lead to serious problems. Following one's intuition often results in numbers very different from the true likelihoods. For example, if you have 30 students in your class, what is the probability that at least two of them would have the same birthday? You may think that this is unlikely. However, to anybody's surprise the probability is more than 70% (why?). As a sport-related example, consider baseball and football hall of famers who share birthdays. Is this highly likely or does it have a small chance? George Halas and Red Schoedienst were both born on February 2, Fran Tarkenton and Lang MacPhail on February 3, and Hank Aaron and Roger Stauback on February 5. Noting that these are just examples from early February, one can see that this is a very likely event.

Different Definitions of Probability

Type	Definition	Description
Classical (Theoretical)	$\frac{\text{Number of Outcomes in the Event}}{\text{Number of Possible Outcomes}}$	Finite Number of Equally Likely Outcomes
Objective (Empirical)	$\frac{\text{Frequency of the Event}}{\text{Total Frequency}}$ = Relative Frequency	Based on Available Statistics or Relative Frequency From an Experiment
Subjective (Judgmental)	No Definition	Based on Intuition, Available Knowledge or Educated Guesses
Relationships: As the number of experiments is increased empirical probability will approach the actual probability (The Law of Large Numbers). As our knowledge about the matter under investigation is increased, our guess gets closer to the actual probability		

PRACTICE PROBLEMS

1. Specify which interpretations of probability are being used in each of the following statements:

(a) The probability is 0.85 that the Duke basketball team will be ranked among the top 10 teams.
(b) The probability is 0.05 that there is life on other planets.
(c) The probability that a soccer player will score his next penalty kick is 0.70.
(d) The probability of buying a new car with a major defect is 0.005.

2. Your roommate is willing to bet $15 against $5 that in the World Cup, England will beat the Netherlands. What is his subjective probability that England will win?

Odds

We end this part by defining odds, a term used in sports frequently. Consider an event A with probability of occurrence $P(A)$. The odds for and against A are defined as

$$\text{Odds}(A) = \text{Odds in favor of A} = \text{Odd for } A = \frac{P(A)}{1-P(A)} = \frac{P(A)}{P(A')},$$

$$\text{Odds}(A') = \text{Odds against A} = \frac{1-P(A)}{P(A)} = \frac{P(A')}{P(A)}.$$

DEFINITION: The odds for success is the ratio of probability of success to the probability of failure.

Note that, whenever possible, odds are expressed as ratios of whole numbers. For example, if team A has 1/6 chance to win and 5/6 to lose a game against a much better team, then odds against A winning the game is $\frac{5}{6}/\frac{1}{6} = \frac{5}{1}$ or 5 to 1. The odds in favor of A winning is obviously 1 to 5. An odd of one-fifth can be interpreted to mean that the probability of A occurring is 1/5 the probability of A not occurring; equivalently, the odds are "5 to 1" that A will not occur.

Now, in terms of a fair game, if we bet $1 on team A winning a game we will lose $1 if they lose and be paid $5 (and in addition our bet of $1 is returned) if they win. More generally, if the odds for an event A are p to q, then a game is fair if our bet of $p is lost if event A does not happen, but $q is won (and in addition our bet of $p is returned) if A does happen. In this case, it is often said that house pays q to p on event A happening.

EXAMPLE Suppose that two fair dice are rolled once. What are the odds for rolling a sum of 6? If we bet $5 that a sum of 6 will turn up, what should the house pay (plus returning our $5 bet) if a sum of 6 does turn up for the game to be fair?

SOLUTION: Let A be the event that sum 6 will turn up. Then, $P(A) = 5/36$ and odds in favor of rolling a sum 6 are

$$\frac{P(A)}{P(A')} = \frac{5/36}{31/36} = 5/31 \text{ or 5 to 31}.$$

Thus, the house should pay us $31.

The odds ratio (OR), by definition is a ratio of two odds. That is,

$$OR_{AvsB} = \frac{\text{Odds}(A)}{\text{Odds}(B)}.$$

For example, suppose that the probabilities of suffering an injury for male and female soccer players are respectively 0.10 and 0.25 (statistics support the fact that female players are much more likely to suffer injury), then

$$OR_{MvsF} = \frac{0.10}{1-0.10} \Big/ \frac{0.25}{1-0.25} = 1/3.$$

In other words, the odds of injury for female are three times the male soccer players.

A more interesting example is the OR for winning a home versus an away game. Most fans believe that this ratio is always greater than one. Denoting win and loss by W and L, respectively, we have

$$OR_{WvsL} = \frac{\text{Prob(home win)}/\text{Prob(home loss)}}{\text{Prob(away win)}/\text{Prob(away loss)}}$$

which is expected to be larger than 1. In fact, some recent investigations revealed that this number varies between 1.2 (for major league baseball) to almost 2.5 (for college basketball).

Next, consider the above problem in the opposite direction: that is, we are now given the odds for an event, and we want to find the probability of that event.

From Odds to Probability

If the odds for event A are a to b, then the probability of A occurring is

$$P(A) = \frac{a}{a+b}.$$

EXAMPLE If in repeated rolls of two fair dice, the odds for rolling a 9 before rolling a 7 are 2 to 3, then the probability of rolling a 9 before rolling a 7 is

$$\frac{p}{p+q} = \frac{2}{2+3} = \frac{2}{5} = 0.40.$$

EXAMPLE If in repeated rolls of two fair dice, the odds against rolling an 8 before rolling a 7 are 6 to 5, then the probability of rolling an 8 before rolling a 7 is

$$\frac{5}{5+6} = 5/11 = 0.455.$$

EXAMPLE Suppose that you are playing the following game with nine of your friends. Ask each friend to choose a whole number between 1 and 100. The aim of the game is to pick a number different from others. This may look easy. After all, there are only 10 people. But even if everyone picks a number randomly, the chance of two people picking the same number is more than one in three, why?

Because the probability that all 10 numbers picked are different is

$$\frac{99}{100} \times \frac{98}{100} \times \cdots \times \frac{91}{100} = 0.628.$$

In fact, because people usually do not pick the numbers randomly (e.g., numbers above 50 are more popular), the probability is in fact almost 50% that two people will pick the same number.

The more people play this game, the more the chances of a coincidence escalate. With 20 people, the odds of a coincidence are about 7 to 1 (strong enough for you to do this as a mind-reading stunt).

Chapter 2: Elements of Probability

Odds and Gambling

Gambling has grown rapidly in the last few decades especially in relation to the sports competitions. People who gamble on outcomes of games and tournaments may be divided into three possibly overlapping categories:

1. The casual gamblers who gamble for possible enjoyment may not be aware of odds or strategic subtleties and may not even be very knowledgeable about players and the teams.
2. The compulsive gamblers who are happy primarily when in the act of gambling and do not care about the sport itself or even the outcome of the competitions.
3. The professional gamblers who look to gambling as another profession. Such individuals understand odds, the games involved, and also know players, teams, and their potential well, although they may not have a good background in mathematics and the calculus of the odds.

One interesting fact is the following: many individual gamblers, who understand the odds and their implications, may not fully understand the difference between the true and house odds. To clarify this, consider the following table comparing the true and house odds for the game of Roulette in a well-known place such as Las Vegas, Monte Carlo, and Atlantic City.

True and House Odds in Las Vegas Roulette

Type of bet	True odds	House odds
Color(red or black)	20 : 18	1 : 1
Parity(even or odd)	20 : 18	1 : 1
18#'s (1-18 or 19-36)	20 : 18	1 : 1
12#'s(columns or dozens)	26 : 12	2 : 1
6#'s(any 2 rows)	32 : 6	5 : 1
4#'s(any 4 number square)	34 : 4	8 : 1
3#'s(any row)	35 : 3	11 : 1
2#'s(adjacent)	36 : 2	17 : 1
Single #'s	37 : 1	35 : 1

If a $1 bet is made on red, then the chance of winning a dollar is 18/38, whereas the chance of losing is 20/38. So in the long run (or on average), one will lose $2 for every 38 games he or she plays. With so many players and so many tables to play, it is not hard to imagine what could happen in the long run.

Turning to a sport example, betting works as follows: The bookmaker sets odds on each of the possible outcomes. Suppose that an odds of 39 to 1 is set for a player or a team. If this player wins, the person who has bet on this player winning will receive amount given in the odds, plus his or her money. This means a gain of $39. If the chance of winning for this player is 1/40, then the bet is called a fair bet. As in gambling clubs, if we convert the odds to the probabilities, we find that total probability assigned to the players or teams is less than 1. The difference benefits the bookmakers and in long run makes them the winners. Remember that bookmakers have various ways to guarantee making money. People who know probability theory usually see no merit whatever in betting at racetracks, casinos, or in buying lottery tickets. After all, multimillion-dollar gambling places are built using gambling proceeds. Why would somebody bet with people who make a lot of money gambling?

2.3.2 Permutations and Combinations

For calculating the probability of an event A in a situation, where it is reasonable to assume that outcomes are equally likely, we need to find the total number of equally likely, mutually exclusive (disjoint) arrangements in the sample space and also in A. This sort of enumeration can often be facilitated by certain combinatorial formulas, which will be developed now. These formulas are based on the following two basic principles:

ADDITION PRINCIPLE : If an event A_1 can occur in a total of n_1 ways and if a different event A_2 can occur in a total of n_2 ways, then the event A_1 or $A_2 (A_1 \cup A_2)$ can occur in $n_1 + n_2$ ways provided that A_1 and A_2 are disjointed. That is, they cannot occur simultaneously.

MULTIPLICATION PRINCIPLE : If an event (operation) A_1 can occur in a total of n_1 ways and if after that a different event (operation) A_2 can occur in total of n_2 ways, then the event A_1 and A_2 ($A_1 \cap A_2$), the whole operation, can occur in $n_1 n_2$ ways.

These two principles may be illustrated by letting A_1 correspond to the drawing of a heart and A_2 correspond to the drawing of a spade from a deck of 52 cards. Each of these events can occur in 13 ways. The number of ways a heart or a spade can be drawn is therefore,

$$13 + 13 = 26.$$

To illustrate the second principle, suppose that two cards are drawn from the deck. In how many ways this can be done if we wish to draw one spade and one heart. There are $13 \times 13 = 169$ ways of doing this, since any of 13 spades can be paired with any of 13 hearts. This can easily be seen by using a tree diagram.

The two principles may clearly be generalized to take account of more than two events. Thus, if three mutually exclusive events A_1 or A_2 and A_3 can occur in n_1, n_2, and n_3 ways, respectively, the event A_1 or A_2 or A_3 can occur in $n_1 + n_2 + n_3$ ways, and the event A_1 and A_2 and A_3 can occur in $n_1 n_2 n_3$ ways.

Sir Arthur Eddington, an astronomer, once wrote in a satirical essay that if a monkey was left alone long enough with a computer and typed randomly, any great novel could be replicated. However, he came short of mentioning that such monkey needs to live billions of years and type billions of letters every day.

A PRACTICE PROBLEM What is the probability that this passage could have been written by a monkey? Ignore capitals and punctuation and consider only letters and spaces.

"I cannot think about life without sports."

We shall now use the second of these principles to enumerate the number of arrangements of a set of objects. Let us start with a simple example and consider the number of arrangements of the letters a, b, and c. Think about three positions where these letters can be placed. We can pick any one of the three letters and place that in the first position, either of the remaining two can be placed in the second position, and the third position must be filled by the unused letter. The filling of the first position is an event that can occur in three ways; the filling of the second position is an event that can occur in two ways; and filling the third position is an event that can occur in one way. The three events can occur together in a $3 \times 2 \times 1 = 6$ ways. The six arrangements or *permutations*, as they are called, are

$$abc, acb, bac, bca, cab, cba.$$

In this simple example, the elaborate method of counting was hardly worthwhile because it is easy enough to write down all six permutations. But if we had asked for the number of permutations of six letters, we should have had

$$6 \times 5 \times 4 \times 3 \times 2 \times 1 = 720$$

permutations to write down.

It is obvious now that, in general, the number of permutations of n different objects is

$$n(n-1)(n-2)(n-3)\ldots(2)(1).$$

This product of an integer by all the positive integers smaller than it is usually denoted more briefly by $n!$ (read n factorial). Thus $2! = 2$, $3! = 6$, $4! = 24$, $5! = 120$, and so on.

$$\text{Since } n! = n(n-1)!$$

it is common to define $0!$ as 1, so that the relation will be consistent when $n = 1$.

Let us now enumerate the number of permutations that may be made from n objects if only r of the objects are used in any given permutation. Reasoning as before, the first position may be filled in n ways, the second may be filled in $n-1$ ways, and so forth. When we come to the rth position, we shall have used $r-1$ of the objects so that $n-(r-1)$ will remain from which we can choose. The number of permutations of n objects taken r at a time is therefore $n(n-1)(n-2)\ldots(n-r+1)$. The symbol $P_{n,r}$ is used to denote this number. Thus,

$$P_{n,r} = n(n-1)(n-2)\ldots(n-r+1) = \frac{n!}{(n-r)!}. \tag{2.2}$$

Thus, the number of permutations of the four letters $a, b, c,$ and d taken two at a time is $P_{4,2} = 4 \cdot 3 = 12$. On putting $r = n$ in (2.2), we get the result stated earlier: the number of permutations of n objects taken n at a time is $n!$.

With the aid of (2.2), we can now solve the following problem involving combinations: in how many different ways r objects can be selected from n different objects?

We noted that $P_{n,r}$ counts all the possible selections as well as all the arrangements of each selection or combination. Two combinations are different if they are not made up of the same set of objects. Thus, abc and abd are different three-letter combinations, while abc and bac are different permutations of the same combination. Let the symbol $\binom{n}{r}$ denote the number of different combinations. Then it is clear that $P_{n,r}$ equals $\binom{n}{r}$ times $r!$, since each combination of r objects result in $r!$ arrangements. Therefore,

$$\binom{n}{r} = \frac{P_{n,r}}{r!} = \frac{n(n-1)(n-2)\ldots(n-r+1)}{r!} = \frac{n!}{r!(n-r)!}. \tag{2.3}$$

Another common symbol for this number is $C_{n,r}$. For example, the number of combinations of five objects taken three at a time is

$$\binom{5}{3} = \frac{5 \cdot 4 \cdot 3}{3!} = \frac{60}{6} = 10.$$

The number $\binom{n}{r}$ may be given a different interpretation. For example, since we are selecting r objects, for each selection corresponds a selection with $n-r$ objects. Thus, $\binom{n}{r}$ equals the number of ways in which n objects may be divided into two groups, one group containing r objects and the other group containing the other $n-r$ objects. We can use this to solve a more general problem. Suppose that we wish to divide n objects into three groups containing n_1, n_2, n_3 objects, respectively, with $n_1 + n_2 + n_3 = n$. In how many different ways can we do this?

First we divide the n objects into two groups containing n_1 and $n_2 + n_3$ objects. This can be done in $\binom{n}{n_1}$ ways. Then we divide the second group into two groups containing n_2 and n_3 objects. This can be done in $\binom{n_2+n_3}{n_2}$ ways. If the second principle of enumeration is used, the total number of ways of doing the two divisions together is

$$\binom{n}{n_1}\binom{n_2+n_3}{n_2} = \frac{n!}{n_1!(n_2+n_3)!} \cdot \frac{(n_2+n_3)!}{n_2!n_3!} = \frac{n!}{n_1!n_2!n_3!}.$$

This type of argument may be carried further to find the number of ways of dividing n objects into k groups containing $n_1, n_2, \ldots, n_k$ objects with $n_1 + n_2 + \ldots + n_k = n$. This number is readily found to be

$$\frac{n!}{n_1!n_2!\ldots n_k!}. \tag{2.4}$$

Thus, the number of ways of dividing four objects into three groups containing one, one, and two objects is

$$\frac{4!}{1!1!2!} = 12.$$

The Expression (2.4) also has a second interpretation. It is the number of different permutations of n objects when n_1 of the objects are alike and of one kind, n_2 are alike and of a second kind, and so forth. For example $4!/(1!1!2!) = 12$ represents the number of permutations of the letters

a,b,c,c. Here is a different way of arriving at Expression (2.4). Consider n different objects (e.g., the letters a, b, c,....,p) arranged in a definite order. Consider a division of this set of objects into k groups, the first group containing n_1 objects, the second n_2, and so forth. Now, in the original arrangement of objects, replace all the objects selected for the first group by 1's, all those selected for the second groups by 2's, and so forth. The result will be a permutation of n_1 1's, n_2 2's,..., n_k k's. A little reflection will convince one that every division of the letters into the k groups corresponds to a different permutation of the integers and that this is the total set of permutations, because if there is another, there would be another division of the letters into k groups.

Note that, in this section we derived three formulas. These are useful formulas and their derivation serves to illustrate the application of the two principles of enumeration stated at the beginning of the section. The formulas will aid in solving many problems, but we should also point out that they are useless in many others, and one must then fall back on the elementary principles.

EXAMPLE 2.3.4 Two cards are drawn from an ordinary deck, what is the probability of drawing one spade and one heart?

SOLUTION: Since nothing is said about the order in which the spade and the heart should occur, this is a combination problem. To compute the probability, we must find the total number of possible outcomes, n, of two-card draws. n is the number of points in the sample space and $1/n$ is the probability assigned to each point. We must also find the number of the points n_A that have the specified attribute. The total number of two-card combinations that can be made up from 52 cards is $\binom{52}{2} = 1,326 = n$. We have seen before that there are $13 \cdot 13 = 169 = n_A$ different combinations with the required attribute. Therefore,

$$P(A) = n_A/n = 169/1326 = 13/102.$$

EXAMPLE 2.3.5 A basketball team has five distinct positions. Out of 10 players, how many starting teams are possible if each of the following conditions is imposed:

(1) The distinct positions are taken into consideration?

(2) The distinct positions are not taken into consideration?

(3) The distinct positions are not taken into consideration, but only one of the two point guards, Hardaway or Carter (but not both) must start?

SOLUTION: (1): $P_{10,5} = 30,240$, (2): $\binom{10}{5} = 252$, (3): $2\binom{8}{4} = 140$.

EXAMPLE 2.3.6 There are 15 teams in a tournament. Each team will play with every other team once. How many games are there?

SOLUTION: $\binom{10}{2} = 105$. A different solution may be obtained by noting that team 1 will play with other 9 teams resulting in 9 games. Team 2 will play 8 more games (excluding the one with team 1), and so on. So we have $9+8+7+\cdots+2+1 = 105$ games.

A PRACTICE PROBLEM In table tennis to win a standard game, a player must either be the first to reach 21 points with a margin of 2 or win 2 consecutive points following a tie at 20 all before the opponent does. Considering the first case shows that for the game to finish 21-0, 21-1, $\cdots$, 21-19, the number of possibilities are those given in the table below. Note that for the game to end, for example, 21-4, the winner should take the last point plus 20 points from the first 24 points played. If you like challenge, try to calculate the number of possibilities for the second case, that is, going through 20-20.

21 − 0	1
21 − 1	21
21 − 2	231
21 − 3	1771
21 − 4	10626
21 − 5	53130
21 − 6	230230
21 − 7	888030
21 − 8	3108105
21 − 9	10015005
21 − 10	30045015
21 − 11	84672315
21 − 12	225792840
21 − 13	573166440
21 − 14	1391975640
21 − 15	3247943160
21 − 16	7307872110
21 − 17	15905368710
21 − 18	33578000610
21 − 19	68923264410

EXAMPLE 2.3.7 Soccer Tournament. We now consider a relatively difficult problem that has a simple and elegant solution. Suppose that n soccer teams are entered in a tournament. In the first round, the teams are paired one against another at random. The loser of each pair is eliminated

Chapter 2: Elements of Probability

from the tournament, and the winner of each pair continues into the second round. If the number of teams n is odd, then one team is chosen at random before the pairings are made for the first round, and that team automatically continues into the second round. All the teams in the second round are then paired at random. Again, the loser of each pair is eliminated, and the winner of each pair continues into the third round. If the number of teams in the second round is odd, then one of these teams is chosen at random before the others are paired, and that team automatically continues into the third round. The tournament continues in this way until only two teams remain in the final round. They then play against each other, and the winner of this match is the winner of the tournament. We assume here that all n teams have equal ability, and want to determine the probability p that two specific teams A and B will ever play against each other during the tournament.

SOLUTION: The number of possible pairs of teams is $\binom{n}{2}$. Each of the two teams in every match is equally likely to win that match and all initial pairings are made randomly. Therefore, before the tournament begins, every possible pair of teams is equally likely to appear in each particular one of the $n-1$ matches to be played during the tournament. Accordingly, the probability that teams A and B will meet in some particular match that is specified in advance is $1/\binom{n}{2}$. If A and B do meet in that particular match, one of them will lose and be eliminated. Therefore, these same two teams cannot meet in more than one match.

It follows from the preceding explanation that the probability p that teams A and B will meet at some time during the tournament is equal to the product of the probability $1/\binom{n}{2}$ that they will meet in any particular specified match and the total number $n-1$ of different matches in which they might possibly meet. Hence,

$$p = (n-1)/\binom{n}{2} = 2/n.$$

EXAMPLE 2.3.8 Three women and five men get together to play volleyball. Suppose that the eight people are divided randomly into two teams of four each. What is the probability that all three women end up on the same team?

SOLUTION:

$$\frac{\binom{3}{3} \times \binom{5}{1}}{\binom{8}{4}} = 5/70 = 0.0714.$$

A PRACTICE PROBLEM A die is rolled six times. What is the probability that the outcomes are all different?

Do you like a challenge? If so, try to confirm that there are 296 ways to make changes with a dollar.

SUMMARY : Suppose we have four distinct objects A, B, C, and D. Using all four we have only 1 combination, but 24 permutations as shown below.

Combinations

(Order Does Not Matter)

	A	
B		D
	C	

Permutations (Order Matters)

ADCB	ADBC	ACDB	ACBD	ABDC	ABCD
DACB	DABC	DCAB	DCBA	DBAC	DBCA
CADB	CABD	CDAB	BDCA	BCAD	CBDA
BADC	BACD	BDAC	BDCA	BCAD	BCDA

A box contains N distinct objects. If we draw n objects, we have

		Does Order Matter?	
		Yes	No
With		Permutation	Combination
Replacement?	Yes	N^n	*
	No	$\frac{N!}{(N-n)!}$	$\frac{N!}{(N-n)!n!}$

		Does Order Matter?		
		Yes	No	
With		Permutation	Combination	
Replacement?	Yes	$\frac{1}{N^n}$	$\frac{n!}{(n_1!)(n_2!)(n_3!)\cdots(n_k!)}$	$\frac{1}{(N^n)}$
	No	$\frac{1}{\left[\frac{N!}{(N-n)!}\right]}$	$\frac{1}{\left[\frac{N!}{(N-n)!n!}\right]}$	

A PRACTICE PROBLEM If you are familiar with the game of poker verify the following:

Hand	Probability	Odds
Royal Flush	1.54×10^{-6}	649,739:1
Straight Flush	1.39×10^{-5}	72,192:1
Four of a Kind	2.40×10^{-4}	4,164:1
Full House	1.44×10^{-3}	693:1
Flush	1.97×10^{-3}	508:1
Straight	3.92×10^{-3}	253:1
Three of Kind	0.0211	46:1
Two Pair	0.0475	20:1
One Pair	0.423	1.37:1

2.3.3 Misleading Use of Probability and Statistics

Mark Twain in his book published in 1924 mentions a famous line attributed to Benjamin Disraeli "There are three kinds of lies: lies, damned lies, and statistics." Also, we frequently hear people say "that is just statistics," "you can prove what you wish to prove with statistics." In fact, both statistics and statisticians have a poor image in the minds of many people. This is because most data can easily be manipulated in an unethical and unscientific fashion to draw desired conclusions. In other words, it is easy to distort the truth. This is because most people are unfamiliar with concepts and the language of statistics. A poor intuition about probability, especially small probabilities, is an additional problem.

An interesting comparison with medical science and doctors is as follows. If you do not find your doctor's instructions helpful, you will probably blame him or her and not the medical science. But when it comes to statistics most people blame the science or the methodology used not the user. There are good and bad engineers, and good and bad lawyers, despite the fact that both professions require a license to practice.

One problem that arises when using probability is that most of us, in our daily decisions, rely on our intuition about probability. The intuition that has been cultivated over many years, based on a wide range of experiences. Unfortunately our intuition can be completely unreliable. For example, why do people tend to think that parents with four baby girls in a row are due to a baby boy? Why do we often get more concerned about low risks like nuclear power plants than high risks like driving? Why do people tend to think a lottery ticket with the numbers 17, 15, 18, 19, 44, and 51 has a better chance of winning than one that has the numbers 9, 8, 7, 6, 5, and 4 when both actually have the same chance? Researchers have studied questions like these for many years, and their discoveries are illuminating.

Fortunately in sports and the related contexts people rely on statistics more than anything else. But as is discussed in several sections of this book the way statistics are presented may be misleading. A good example of this is Simpson's paradox discussed in Section 2.7 related to what is called hot hands in basketball. In what follows we will illustrate the point by presenting an example.

Looking at NBA 2018 to 2019 records on ESPN I noticed that almost all the teams in the Western division had a double digit number of wins and losses. So, it was hard to pick a team as the best. I thought this was a good example for the discussion of this section. Suppose that in the area you live people like to bet on outcomes of the games played between the teams in a particular tournament or conference (in this case Eastern division). You can take, for example, 64 of these people and send them a letter with a forecast for the next six games. Since there are a total of 64 possible outcomes such as WLLWWL and LLWWWL, you can send each of these outcomes to one individual. By doing so one person will get all the forecasts correct, and six people will get five correct forecasts. These people may then start believing that you must have some special information about the games. After all, the probability of correctly guessing outcomes of six successive tosses of a fair coin is only $(0.5)^6 = 0.031$. If each week you do this by sending forecasts to several different groups of 64 people and start new groups each week, you may be able to create reputation and generate clients to make significant profit.

One real story regarding the prediction is about to the price of an average family home in the year 2000. In mid-1980s, housing prices in a large part of United States rose by more than 20% per year. Based on this, some "experts" suggested that the price of an average family home would exceed $1 million in the year 2000. However, when year 2000 arrived prices were nowhere near their prediction. This again led some people to doubt and blame statistical forecasting, not the so-called "experts." A similar thing happened to stock market where experts were predicting 15,000 for Dow Jones Industrial Average and 6000 for Nasdaq for the year 2002 while in fact the opposite happened, creating a lot of blame for methods used for prediction.

Summarizing these, statistics do not lie, but people do. This science together with the methodology used, like a paring knife, can be a very valuable tool when it is properly used. When it is misused, it can lead to some very bad results. Some of the ways to intentionally or inadvertently distort the truth with statistics are described in the book, *How to Lie with Statistics* by Huff (1954).

2.4 CONDITIONAL PROBABILITY AND INDEPENDENCE

This section includes discussion of an extremely important concept, namely, the independence. This concept plays a role similar to that of orthogonality in geometry and calculus. Its presence makes the mathematical treatment of the statistical problem significantly easier. However, its absence limits our ability to learn about future using the knowledge of the past, and also learning about one attribute using the knowledge we have about the others.

2.4.1 Conditional Probability, Addition, and Multiplication Rules

So far we discussed some rules for calculating the probability of an event A with no reference to the occurrence of any other event. In this section, we will discuss probability calculation for an event A assuming that an event B has occurred. Consider the following question: what is the probability that a randomly selected individual living in a small community will develop a heart problem? Suppose that in the past 10% of the population had developed heart problem. So we may not be knowing anything about the individual selected except that he or she is a member of this community, say 10%. An important question that arises then is the following: would your answer be different if you know that the selected individual smokes or has parents with heart problems?

In other words, would this extra piece of information change the probability? As we all know, the answer for this case is yes. So, here we are dealing with two probabilities: (1) unconditional probability of the event that a randomly selected individual will develop a heart problem and (2) conditional probability of the same event given that (knowing that) the selected individual smokes or has parents with a heart problem.

In many instances, the knowledge that one event has occurred changes the probability for occurrence of another event. For instance, in Example 2.3.2, if we knew that the selected person watches the games, then the probability that his or her favorite team won the Super Bowl becomes 12/55 instead of 24/80. We call this the conditional probability of the event that the selected person's favorite team won the Super Bowl given that this person watches the games. As another example, we may be interested in knowing the conditional probability that a randomly selected person would develop diabetes given the person is overweight.

In general, for two events A and B, the conditional probability, denoted by $P(B|A)$, is the probability of the occurrence of B under the additional information that A has already occurred. It can be computed from the formula defined as

$$P(B|A) = \frac{P(A \cap B)}{P(A)}, \quad P(A) \neq 0. \tag{2.5}$$

To see why this formula works, consider the Figure 2.4.1. Since A has already occurred, the outcome is in the circle representing event A. Now for B to occur the outcome should fall in B too, that is, in the shaded area, representing $A \cap B$. Thus, the conditional probability is the ratio of the area common to both circles and the area of the circle A.

For example, when rolling a die the sample space is $\Omega = \{1,2,3,4,5,6,\}$. Let $B = \{2\}$. Then the probability of occurrence of B is $\frac{1}{6}$. This is an unconditional probability. Suppose now that someone inform us that the number on the die was even, that is, $A = \{2,4,6\}$ has occurred. Then we may

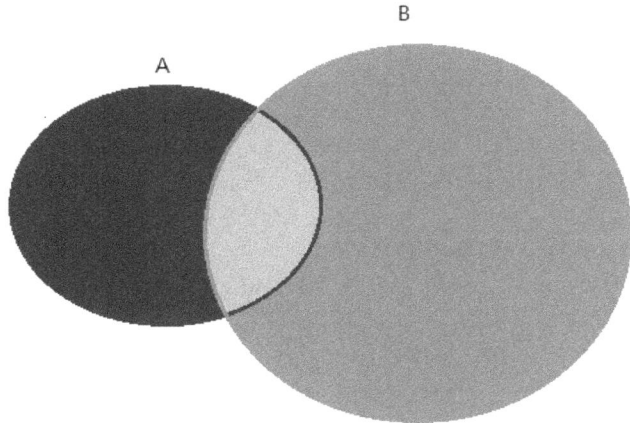

FIGURE 2.4.1 Conditional Event and Probability

want to find the probability that the observed number was 2. This is clearly $\frac{1}{3}$. We can find this by using the Formula (2.5) for conditional probability. Here we have $P(A) = \frac{1}{2}$ and $P(B|A) = \frac{P(A \cap B)}{P(A)} = \frac{P(B)}{P(A)} = \frac{1/6}{1/2} = 1/3$.

EXAMPLE 2.4.1 Suppose that in the mathematics department in a certain university 8% of male students are more than 6 feet tall, 2% are members of a basketball team, and 1% are both more than 6 feet tall and members of a basketball team. If we select a male student randomly, what are the probabilities that

(a) He is a member of a basketball team?
(b) He is a member of a basketball team given that he is more than 6 feet tall?
(c) He is more than 6 feet tall given that he is a member of a basketball team?

SOLUTION: Let A and B be respectively the events that a randomly selected student is more than 6 feet tall and a randomly selected student is a member of a basketball team. Then $P(A) = 0.08$, $P(B) = 0.02$, and $P(A \cap B) = 0.01$. Hence,

(a) $P(B) = 0.02$.
(b) $P(B|A) = \dfrac{P(A \cap B)}{P(A)} = \dfrac{0.01}{0.08} = 0.125$.
(c) $P(A|B) = \dfrac{P(A \cap B)}{P(B)} = \dfrac{0.01}{0.02} = 0.50$.

EXAMPLE (Bertrand's paradox) A boy has three fair coins, one is a regular coin, one has two heads, and one has two tails. A coin is chosen at random and flipped coming up heads. What is the probability that the chosen coin was the regular coin.

SOLUTION: Let H and R denote the events of observing H and coin being regular, respectively. Then,

$$P(R|H) = \frac{P(R \cap H)}{P(H)} = \frac{(1/3)(1/2)}{1/2} = 1/3.$$

One intuitive way to understand this problem is the following: the experiment is the same as picking one of the six sides of coins at random. If the side is a head, then each of the three heads are equally likely to occur. Two of the heads have heads on the other side of the coin and one has tails. Note that for this problem the naive answer is 1/2, since the coin could be either the fair coin or the two-headed coin, but this answer is not correct.

A PRACTICE PROBLEM Members of a wellness club have been classified by the change in general level of stamina they have experienced in the last 12 months and by the type of physical activity. The results are given in the accompanying table.

Chapter 2: Elements of Probability

Change in Stamina	Physical Activity		
	Tennis	Running	Walking
> 20%	1	2	3
10% to 20%	13	12	20
0% to 10%	20	20	20
−10% to 0%	2	2	16
< −10%	12	0	0

Assume that one person is selected at random from the 143 members. Find

(a) The conditional probability that the exercise was "walking," given that stamina increased more than 20%.
(b) The conditional probability that tennis was the exercise, given that stamina decreased by more than 10%.

EXAMPLE 2.4.2 Consider the random experiment of rolling a die. Let $A = \{1,3,5\}$ and $B = \{2,4\}$. Then A and B are disjoint, that is,

$$P(A \cap B) = P(\phi) = 0.$$

and hence $P(B|A) = 0$. As discussed before in this case, occurrence of A makes occurrence of B impossible. Note that in this case $P(A \cup B) = P(A) + P(B)$, which is a special case of a relationship known as the addition rule introduced below.

EXAMPLE Five teams are in a tournament. The probabilities are respectively 0.1, 0.2, 0.25, 0.3, and 0.15 for teams 1 to 5 to win the championship. If team 2 declared ineligible for the championship, what is the probability that team 5 will win the championship.

SOLUTION: Recall that the total probability is 1. Thus, the required probability is $\frac{0.15}{0.8} = 0.1875$. We hope students try to apply the formula for conditional probability by defining proper events.

Addition Rule

For any two events A and B in the same sample space we have

$$P(A \cup B) = P(A) + P(B) - P(A \cap B). \tag{2.6}$$

Imagine a Venn diagram (Figure 2.4.1) with two circles representing A and B and $A \cap B$ representing the overlap. It is easy to see that the area inside the two circles, that is, $P(A \cap B)$ is the sum of the areas of each circle minus the overlap, as it is counted twice. To see this more formally, since $A \cup B = A \cup (B \cap A')$ and $A \cap (B \cap A') = \phi$, it follows from the postulate III that

$$P(A \cup B) = P(A) + P(B \cap A').$$

Noting that $B = (B \cap A) \cup (B \cap A')$ and $(B \cap A) \cap (B \cap A') = \phi$ we have $P(B) = P(B \cap A) + P(B \cap A')$ using the same postulate and therefore $P(B \cap A') = P(B) - P(A \cap B)$. Upon replacing for $P(B \cap A')$ in above expression, we obtain (2.6).

A PRACTICE PROBLEM In a class with 40 students there are 25 female students, 15 students majoring in biology, and 10 female students majoring in biology. If we choose a student randomly, what is the probability of choosing a student that is either female or biology major.

EXAMPLE 2.4.3 In a residential community close to a well-known sport organization, 60% subscribe to Sport Illustrated, 80% subscribe to ESPN magazine, 50% subscribe to both. If we select one resident randomly, what is the probability that it subscribes to

(a) At least one of the two publications?
(b) Exactly one of the two publications?

SOLUTION: Consider the following events:

A: subscribes to the Sports Illustrated ,   B: subscribes to ESPN magazine

Then we have $P(A) = 0.6$, $P(B) = 0.8$, $P(A \cap B) = 0.5$.

(a) P(subscribes to at least one) = $P(A \cup B) = P(A) + P(B) - P(A \cap B) = 0.9$

(b) P(subscribes to exactly one) = P(subscribes to at least one) − P(subscribes to both)
 $= P(A \cup B) - P(A \cap B) = 0.9 - 0.5 = 0.4$.

The addition rule can be generalized to more than two events. Its general form is

$$P(\cup A_i) = \sum_i P(A_i) - \sum_{i<j} P(A_i \cap A_j) + \sum_{i<j<k} P(A_i \cap A_j \cap A_k) \cdots$$

EXAMPLE There are n players and each player has a ball. Suppose that they put their balls in a box and each player chooses a ball at random. We want to find the probability that at least one player gets her own ball.

Let A_k be the event that the kth player gets her own ball. Then by the inclusion–exclusion principle,

$$P \text{ (at least one player gets her own ball)} = P(\cup A_i) = \sum_i P(A_i) - \sum_{i<j} P(A_i \cap A_j) + \sum_{i<j<k} P(A_i \cap A_j \cap A_k) - \cdots$$

Now, there are n balls so that

$$P(A_i) = \frac{1}{n}, \; P(A_i \cap A_j) = \frac{1}{n(n-1)}, \; P(A_i \cap A_j \cap A_k) = \frac{1}{n(n-1)(n-2)}.$$

Chapter 2: Elements of Probability

Therefore,

$$\sum_i P(A_i) = 1, \quad \sum_{i<j} P(A_i \cap A_j) = \binom{n}{2} \frac{1}{n(n-1)} = \frac{1}{2!}.$$

$$\sum_{i<j<k} P(A_i \cap A_j \cap A_k) = \binom{n}{3} \frac{1}{n(n-1)(n-2)} = \frac{1}{3!}.$$

In fact,

$$P(\text{at least one player gets her own ball}) = 1 - \frac{1}{2!} + \frac{1}{3!} - \cdots (-1)^n \frac{1}{n!} \approx$$

$$1 - \left(1 + \frac{-1}{1!} + \frac{(-1)^2}{2!} + \frac{(-1)^3}{31} + \cdots\right) = 1 - e^{-1} = 0.63.$$

Multiplication Rule

The Formula (2.5) defining the conditional probability can be written as

$$P(A \cap B) = P(B|A)P(A). \tag{2.7}$$

This relationship is known as the *multiplication rule*. It plays an important role in calculating the probability of a joint occurrence of two events. The following are examples of application of this formula:

EXAMPLE 2.4.4 A box contains 5 black balls and 8 white balls. If we draw two balls one-by-one and without replacement, what is the probability that both balls drawn are black?

SOLUTION : Let B_1 be the event that the first ball is black and B_2 the event that the second ball is black. Here we have $P(B_1) = 5/13$. Now since drawing is done without replacement, after drawing the first ball 12 balls are left in the box of which 4 are black. Thus we have $P(B_2|B_1) = 4/12$ and using (2.7) we obtain

$$P(B_1 \cap B_2) = P(B_1)P(B_2|B_1) = (5/13) \cdot (4/12) = \frac{20}{156}.$$

We hope students try the problem using a tree diagram.

EXAMPLE 2.4.5 Suppose that a basketball player makes her first foul shot with probability 0.7. Also suppose that the probability that she makes her second shot depends on the outcome of the first shot. Specifically, if the first shot is successful then the probability for the second shot to be successful is 0.8, whereas if the first shot is not successful, the second shot is successful with probability 0.65. Now, consider two of her successive shots. What is the probability of making both shots? Further, what is the probability that she makes the second shot?

SOLUTION: Let A be the event that she makes the first shot and B be the event that she makes the second shot. Then $P(A) = 0.7$ and $P(B|A) = 0.8$. Hence the probability that she makes both shots is $P(A \cap B) = (0.8)(0.7) = 0.56$.

To answer the second question since $B = (A \cap B) \cup (A' \cap B)$ and $(A \cap B) \cap (A' \cap B) = \phi$, we have

$$P(B) = P(A \cap B) + P(A' \cap B)$$
$$= P(B|A)P(A) + P(B|A')P(A')$$
$$= 0.8 \times 0.7 + 0.65 \times 0.3 = 0.755.$$

In words, there are two possibilities for the second shot to be successful, after making the first and after missing the first. We hope students try to set up a tree diagram for this problem.

PRACTICE PROBLEMS

1. An Olympic team includes 1/3 female and 2/3 male athletes. The probability that a randomly selected athlete will test positive for steroids is 0.15 for male and 0.10 for female. Set up a tree diagram and include all the possible events and their probabilities. Using the multiplication rule, calculate probabilities for each of the possible outcomes.
2. Jim has a 30% chance to play in the next soccer match. Based on the past data, he has scored in 20% of the games he played. What is the probability that Jim will play in the next match and score a goal.

2.4.2 Independent Events

In practice there are many cases when the occurrence of event A has no influence on probability of the occurrence of B. When this is the case we say that A and B are independent and write

$$P(B|A) = P(B) \tag{2.8}$$

because the additional information "A has occurred" did not change the probability or odds for B. In other words the extra information is not relevant to occurrence of B. When A and B are independent we also have the formula (see (2.5))

$$P(A \cap B) = P(A) \cdot P(B), \tag{2.9}$$

known as multiplication rule for independent events. Note that (2.9) is a special case of general multiplication rule.

EXAMPLE The table below presents the heights of 40 randomly selected basketball players (20 from the NBA and 20 from the WNBA).

	Under 6'	6' – 7'	Over 7'
NBA	0	47.5%	2.5%
WNBA	12.5%	37.5%	0

Let A be the event that the basketball player is in the NBA. Let B be the event that he is over 7'. Are A and B independent events?

SOLUTION: $P(B|A) = \frac{P(B \cap A)}{P(A)} = \frac{0.025}{0.50} = 0.05$, $P(B) = 0.025$
Since $P(B|A) \neq P(B)$, A and B are not independent. Note also that $P(A \cap B) \neq P(A)P(B)$

Chapter 2: Elements of Probability

It is interesting to compare (2.9) with the way area of a rectangle is calculated. If $P(A)$ is the length and $P(B)$ is the width of the sides of a rectangle, then its area equals $P(A) \cdot P(B)$ or $P(A \cap B)$. Of course, this is only true for rectangle because its sides are orthogonal to each other.

EXAMPLE 2.4.6 Suppose that in Example 2.4.4 drawing the balls is done with replacement. Then whatever the color of the first ball, the chance of drawing a black ball in the second draw is 5/13 (events are independent). In this case we have

$$P(B_2|B_1) = 5/13 = P(B_2), \quad P(B_1 \cap B_2) = P(B_1)P(B_2) = (5/13)(5/13) = 25/169.$$

EXAMPLE 2.4.7 This example is modification of Example 2.4.5. We assume now that for a certain player all four possible outcomes $\{SS, SF, FS, FF\}$ are equally likely to occur. Here S and F stand for success (hit) and failure (miss), respectively. Thus, SS stands for making both shots, SF for making the first and missing the second, and so on. Since these outcomes are equally likely to occur, the probability for each outcome is 1/4. Now for $B = \{SS, FS\}$, we have $P(B) = 2/4 = 1/2$. Also $A = \{SS, SF\}$ and $A \cap B = \{SS\}$, so that $P(B|A) = P(A \cap B)/P(A) = \frac{1/4}{1/2} = 1/2$. Hence in this case, $P(B|A) = P(B)$ and therefore A and B are independent. It is easy to see that in this case (2.9) does hold too.

EXAMPLE Three racquetball players A, B, and C are playing a round robin match. Suppose that the probability that A beats B is 0.55, the probability that B beats C is 0.9, and the probability that A beats C is 0.6 and that the outcomes of matches are independent. Then the probabilities of winning the tournament are given by

$$P(A \text{ wins}) = (0.55)(0.6) = 0.33, \quad P(B \text{ wins}) = (0.45)(0.9) = 0.405$$

$$P(C \text{ wins}) = (0.4)(0.4) = 0.04.$$

Thus B has a higher chance to win the tournament even though A is favored to beat both B and C. Note that these numbers does not add to 1, because there is one more possibility. The probability that nobody wins is

$$1 - (0.33 + 0.40 + 0.04) = 0.225.$$

Therefore, the probability that somebody wins is 0.775.

EXAMPLE 2.4.8 Consider two volleyball teams, Team One and Team Two. The past records show that in p percent of times, Team One defeated Team Two. Let E be the experiment of letting these two teams play a game. The two possible outcomes are

$$A = \text{Team One wins the game}$$

$$A' = \text{Team One loses the game}$$

Then,

$$P(A) = p, \quad P(A') = q = 1 - p.$$

Now, let E_n be the n repetitions of the experiment E, and assume that these teams play n games under the same conditions. If we believe in independence of each repetition of E from all the others, we have $p^k \cdot q^{n-k}$ as the probability of the event $(\underbrace{A, A, \ldots, A}_{k-\text{times}}, \underbrace{A', \ldots, A'}_{(n-k)-\text{times}})$. More generally, the probability of the event having "A" in k places out of n and "A'" in the remaining $n-k$ places, can easily be obtained noting that such placement can be done in $\binom{n}{k}$ different ways, where $\binom{n}{k}$ is the number of combination of n objects k at the time. Recall that

$$\binom{n}{k} = \frac{n(n-1)\cdots(n-k+1)}{(1)(2)\cdots(k)} \tag{2.10}$$

is the number of possibilities, to put k "A" on n possible places. Thus, the probability of Team One winning k games and losing $n-k$ games is given by

$$p_k = \binom{n}{k} p^k \cdot q^{n-k}. \tag{2.11}$$

The formulation of this problem using the concept of random variable is as follows: Let X be the number of games that Team One wins. Consider the event $\{X = k\}$, that is, "The number of wins is equal to k." Then we get

$$P(X = k) = p_k.$$

The random variable and its properties will be discussed in Section 2.5.

We end this section by making a few remarks about the use of probability based on one's intuition. As discussed earlier, our intuition about probability is closely related to our understanding of randomness. We learned in this section that independent events have no memory. For example, in rolling a fair die, a run of three 5's does not change the likelihood of five or other numbers on the next roll. However, our intuition may direct us to think otherwise. Psychologists test this observation with survey questions such as the following:

Suppose that a fair coin was tossed four times and each time it came up heads. If you had to bet $10 on the next toss, would you bet on heads or tails? Based on what has happened, some people may think that a tail is due now or think that a streak of heads is under way and therefore heads and tails are not equally likely on the next toss. However, there is a converse to this misunderstanding. Consider the following two sequences (which could represent coin tosses, births of boys/girls, hits and misses for a good baseball player, or free throw results for a bad basketball player).

HMHHHMMMMHMHHMMMHHHMH

HMHMHMMMHHMHMHMMHHHMH

These sequences have the same number of H's and M's. The first sequence has longer strings of the same symbol, whereas the second sequence tends to alternate more frequently. In fact, the first sequence is generated with a 0.5 probability of the symbols alternating, which is what we would expect in the toss of a fair coin. The second sequence is generated with a 0.65 probability of the symbols alternating, which makes it a biased sequence. And yet surveys show that two-thirds of the respondents believe that the second sequence is more "random" than the first. In other words, a significant fraction of respondents sensed that randomness means frequent switching between two outcomes and that chance should limit consecutive strings of events.

Summarizing these, a common misperception about random event (known as gambler's fallacy) is that they should be self-correcting. For example, people tend to believe that a string of good luck will follow a string of bad luck in a casino. In other words, long-run frequency of an event should apply even to the short-run. Tversky and Kahneman in their 1989 article titled "The Cold Facts About the Hot Hand in Basketball" that appear in Chance Journal discuss a related misconception called the "Belief in the Law of Small Numbers" according to which people believe that even small samples are highly representative of the population from which they are drawn.

A PRACTICE PROBLEM Recall the practice problem in the end of Section 2.4.1. Check to see whether being a male athlete and testing positive for steroids are dependent or independent.

2.4.3 More on The Multiplication Rule

As discussed earlier, when events are not independent we can use the multiplication rule, namely

$$P(A \cap B) = P(B)P(A|B) = P(A)P(B|A).$$

In this section, we will discuss further applications of this rule.

EXAMPLE 2.4.9 A company sells three brands of fitness machines and offers a two-year warranty for each. Information regarding the sale and the likelihood that each brand will require repair while under warranty is summarized in the table below:

Brand	Sale(%)	Warranty Repair Work Required(%)
1	50	25
2	30	20
3	20	10

(a) What is the probability that a randomly selected purchaser has bought a brand that will need repair while under warranty?
(b) What is the probability that a randomly selected machine will need repair while under warranty?
(c) If a machine needs repair under the warranty, what is the probability that it is brand 1? Repeat this for brands 2 and 3.

SOLUTION: Consider the following events:

A_i : The event that brand i is purchased $\quad i = 1, 2, 3$.

B: The event that the randomly selected machine needs repair.

For this problem, it is useful to use a probability tree shown in Figure 2.4.2.

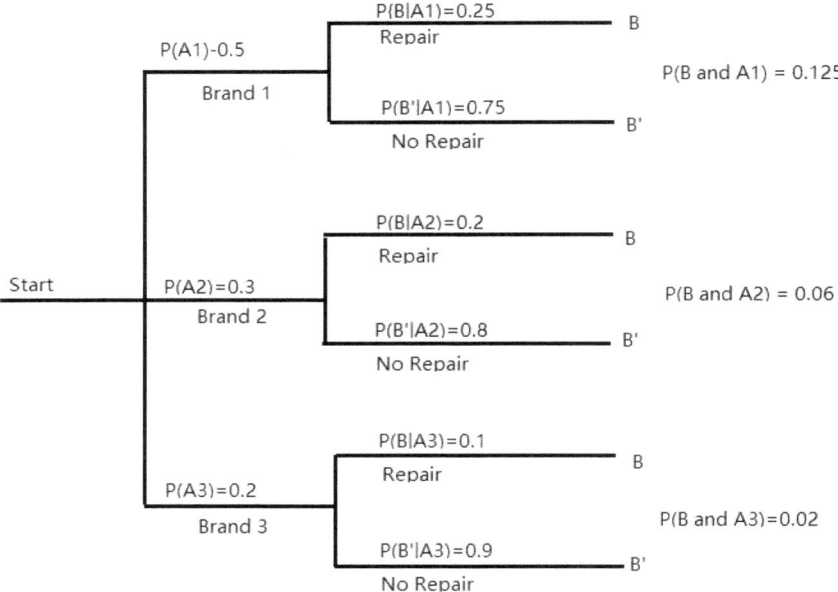

FIGURE 2.4.2 Tree Diagram for Example 2.4.9

(a) $P(A_1 \cap B) = P(A_1)P(B|A_1) = (0.50)(0.25) = 0.125$.

(b) Noting that $B = (A_1 \cap B) \cup (A_2 \cap B) \cup (A_3 \cap B)$ and the three events in the right-hand side are disjoint (only one brand is purchased), we have $P(B) = P(A_1 \cap B) + P(A_2 \cap B) + P(A_3 \cap B) = 0.125 + 0.060 + 0.020 = 0.205$.

(c) $P(A_1|B) = \frac{P(A_1 \cap B)}{P(B)} = \frac{0.125}{0.205} = 0.61, \quad P(A_2|B) = 0.29, P(A_3|B) = 0.10$.

2.4.4 The Law of Total Probability

It is interesting to compare the conditional probabilities $P(B|A_1)$ called prior and $P(A_1|B)$ called posterior. The former usually has the direction of time, whereas the latter moves in opposite direction to the time. In other words, in the first case we know the past and try to find the probability of occurrence of an event in the future, whereas in the second case we want to know the probability of occurrence of an event in the past that had effect on occurrence of what happened. The relationship between these two conditional probabilities has provided an important theorem known as Bayes' theorem discussed in Section 2.4.5.

Recall that the n disjoint or mutually exclusive events A_1 to A_n are called exhaustive if

$$A_1 \cup A_2 \ldots \cup A_n = \Omega.$$

Let $A_1 \ldots A_n$ be mutually exclusive and exhaustive events. That is, they form a partition of the sample space. Then for any event B we have

$$P(B) = \Sigma_{i=1}^{n} P(A_i \cap B) = \Sigma_{i=1}^{n} P(A_i)P(B|A_i).$$

Chapter 2: Elements of Probability

Part (b) of the Example 2.4.9 provides a demonstration for this. See also the following diagram:
This expression is known as the law of total probability.

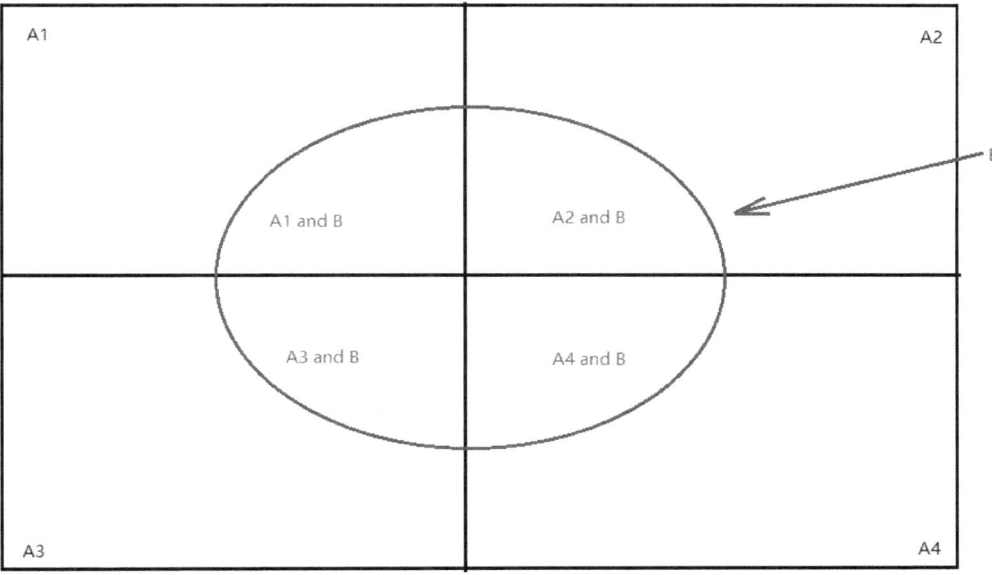

Application to Tennis

The rules of tennis give a player two chances to make a proper service on each point. As a result, tennis players usually practice two types of serve. One, labeled the "strong" serve, is traditionally used on the first service. This serve generally gives a high probability of winning the point to the server, given that the serve is good. The second serve, labeled the "weak" serve, is usually reserved for the service following an initial fault. Although this serve gives a reduced probability of winning the point given that it is good, it has the advantage of having a higher probability of actually being good. The relative efficacy of the two serves can vary markedly from player to player. It is an important area of study in tennis since players are expected to win their service games and failure to do so results in a "break" of their serve that can lead to loosing of the set.

A probability model for winning a service point may be developed for this situation using the following definitions:

E: The event of the server earning the point,
S: The event of a non-faulted strong serve,
W: The event of a non-faulted weak serve.

If a player follows the usual strategy of using an initial strong serve followed by a weak serve if the initial serve is faulted $(S \cap W)$, it follows that the probability of the server winning the point is

$$P(E) = P(S)P(E|S) + (1 - P(S))P(W)P(E|W).$$

If a player follows other strategies, that is, $S \cap S$, or $W \cap W$, or $W \cap S$ then the probabilities are respectively

$$P(E) = P(S)P(E|S)(2 - P(S)),$$

$$P(E) = P(W)P(E|W)(2 - P(W)),$$

$$P(E) = P(W)P(E|W) + P(S)P(E|S)(1 - P(W)).$$

A PRACTICE PROBLEM In a tennis game it is reasonable to assume that

$$P(E|S) \geq P(E|W)$$

at least for men. If so, show that strategy $W \cap S$ is always inferior to strategy $S \cap W$. Also let $U = 1/(1 + P(S) + P(W))$ and $V = P(E \cup W)/P(E \cap S)$ and show that $S \cap W$, $S \cap S$, and $W \cap W$ are, respectively, optimal if $1 \leq V \leq U$, $V \leq 1 \leq U$, and $1 \leq U \leq V$ are satisfied for each case.

2.4.5 Bayes' Theorem

As mentioned earlier, conditional probability may be used to predict future using what happened in the past (prior probability) or further past, based on what happened in the past which is just the opposite (posterior probability). For example, what happens in a court is an example of the latter. Is there any relationship between these two types of probabilities? The answer to this question, as is shown below, is yes.

Let $A_1, \ldots, A_n$ be a collection of n mutually exclusive and exhaustive events with $P(A_i) > 0$, $i = 1, \ldots, n$. According to the Bayes' theorem for any arbitrary event B for which $P(B) > 0$, we have

$$P(A_k|B) = \frac{P(A_k \cap B)}{P(B)} = \frac{P(A_k)P(B|A_k)}{\sum_{i=1}^{n} P(A_i)P(B|A_i)} \quad k = 1, \ldots n.$$

Bayes' theorem is very important. In practice we may know $P(A_k)'s$ (prior probabilities). They can be probabilities of some medical conditions or diseases. Conditional probabilities of event B given that event A_k has occurred may be known from experience. The event B may be fever, cough, bleeding, or other symptoms. From this theorem we can calculate posterior probabilities $P(A_k|B)$ for each k. That is, the probabilities for each of the possible diseases or medical conditions given that the patient has a particular symptom.

Note the difference between the prior and posterior probabilities. As pointed out one moves in the direction of time, the other in opposite direction of time. For example, in a court room questions are about what happened in the past, that is, people are concerned with posterior probabilities. In other words prior probabilities are about future, whereas posterior probabilities are about the past.

Thomas Bayes who developed the above theorem was a priest. According to the inscription on his tomb stone in Bunhill Field cemetery in London, Thomas Bayes died on April 7, 1761, at age of 59.

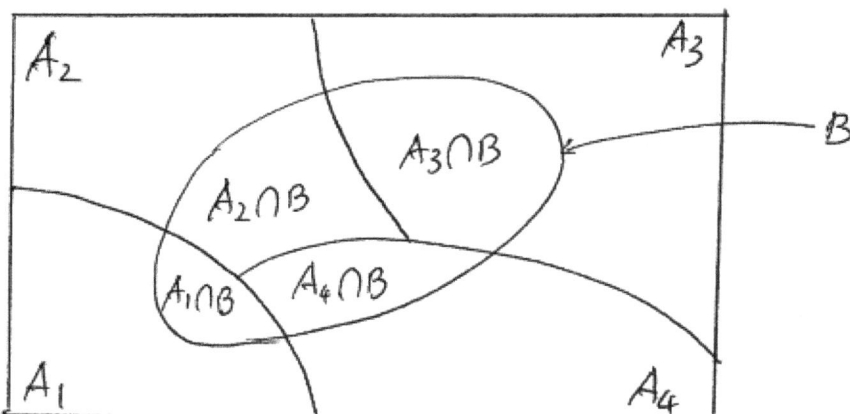

FIGURE 2.4.3 Bayes' theorem provides a relationship between posterior probability (the left-hand side) and conditional and prior probabilities (the right-hand side).

It is easy to see that this relationship is obtained using the formula for the conditional probability and the law of total probability.

Example From Basketball

A professional basketball player who is an expert in 3-point shots, attempts all his 3-pointers from 3 spots numbered 1, 2, and 3. Suppose that in each possession he will be in one of these spots and based on the past games the probabilities of being in these spots are, respectively 35%, 25%, and 40%. His statistics show that he has been more successful from spot 2. In fact, based on his past performance his rates of success are 40% from spot 1, 50% from spot 2, and 30% from spot 3.

(a) Find the probability that he will hit a 3-pointer in the next possession.

(b) If he had a 3-pointer in the last possession, find the probability that it was made from spot 3.

SOLUTION: Consider the following events:

H : The event that he will make the shot.

S_i : The event that he will be in spot $i, i = 1, 2, 3$.

Clearly the 3-pointer made was either from spot 1 or 2 or 3 and these three events are disjoint. Using this we get.

(a) $P(H) = P(S_1 \cap H) + P(S_2 \cap H) + P(S_3 \cap H) = P(S_1)P(H|S_1) + P(S_2)P(H|S_2) +$

$P(S_3)P(H|S_3) = (.35)(.40) + (.25)(.50) + (.40)(.30) = .385.$

(b) $P(S_3|H) = \frac{P(S_3 \cap H)}{P(H)} = \frac{P(S_3)P(H|S_3)}{.385} = \frac{.12}{.385} = .31.$

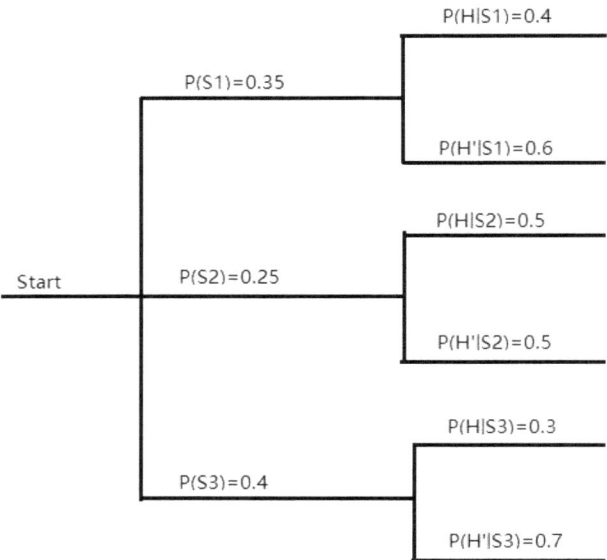

FIGURE 2.4.4 Solution chart

EXAMPLE 2.4.10 An insurance company has 20% high-risk and 80% low-risk policies for players of a certain sport. Suppose that the probability for a high-risk player to get injured is 50% and the probability for a low-risk player to get injured is 10%. If a player is injured, what is the probability that he is a high-risk policy holder?

SOLUTION: Let H and I represent the events of high risk and injury, respectively. Then,

$$P(H|I) = \frac{P(H \cap I)}{P(I)} = \frac{P(H)P(I|H)}{P(H)P(I|H) + P(H')P(I|H')} = \frac{(0.2)(0.5)}{(0.2)(0.5) + (0.8)(0.1)} = 0.555.$$

Example From Medical Diagnostic

One of the most interesting application of Bayes' theorem occurs in medical diagnostics. Let D be the event that a randomly selected person has a terrible disease. Assume that a test for this disease is 95% reliable in the following sense. If the person has the disease, then the test will show positive with probability 0.95. In other words, there is a 5% chance for a false negative. This is called the *sensitivity* of the test. If the person does not have the disease, the test will show negative also with probability 0.95. This is called *specificity* of the test. Note that for most test sensitivity and specificity are different. Let $+$ and $-$ denote the events of positive and negative test results, respectively. Then,

$$P(+|D) = 0.95, \quad P(-|D) = 0.05, \quad P(+|D') = 0.05, \quad P(-|D') = 0.95.$$

Suppose that 0.5% of a certain population suffer from this disease. This number represents the prevalence of the disease. If for a randomly selected person test result comes back positive, what is the probability that this person really has the disease?

Most people answer without hesitation that the person has the disease with probability 0.95, because this test is 95% reliable. But this is not true. We know that $P(D) = 0.005$. Applying the

Chapter 2: Elements of Probability

Bayes' theorem, we get

$$p_1 = P(D|+) = \frac{P(+|D)P(D)}{P(+|D)P(D) + P(+|D')P(D')}$$

$$= \frac{0.95 \times 0.005}{0.95 \times 0.005 + 0.05 \times 0.995} = 0.087.$$

This surprising result can be explained to nonmathematicians (e.g., physicians) in the following way. Suppose that the test were applied to 100,000 individuals of which about 99,500 are healthy and about 500 suffer from the disease. The test gives a false-positive result in 5% healthy individuals, so 4,975 of the healthy individuals are classified as having the disease. From 500 ill individuals, 95% will be detected to have the disease, which equals 475 individuals. Altogether we then have $4,978 + 475 = 5,450$ positive tests, but only 475 of them belong to ill individuals. That is, most positives are false positive. The ratio $475/5,450$ is equal to 0.087.

Bayes' formula allows us to cumulate experience. If the test were positive for a given person, the person would have the disease with probability 0.087. What happens if the test is applied again to this person? If the second test is independent of the first test, then the new probability p_2 can be calculated using the same formula as p_1, only instead of $P(D) = 0.005$ we have $P(D) = 0.087$ and $P(D') = 0.913$. This gives $p_2 = 0.645$. Continuing with independent tests with positive results, we obtain probabilities given in the table:

Number of Positive Test Results i	0	1	2	3	4	5
Probabilities, p_i	0.005	0.087	0.645	0.972	0.998	1.000

For five positive test results, the required probability is $p_5 = 0.999920$, which gives, after rounding, the value 1.000 included in the table. It is important to note that the assumption of independence of tests may not be realistic and therefore the true probabilities may be a little smaller.

A PRACTICE PROBLEM Suppose that a lie detector has the following properties. If the suspect is telling the truth, the test will say so with probability 0.90 and if he is lying, the detector will say so with probability 0.99. Suppose that 95% of the people are telling the truth. Show that the probability that a person is actually lying when the lie detector says he is, equals $99/289$.

We end this part by including a diagram that presents the addition and multiplication rules for different types of events and an example that demonstrates probability calculation for the independent events.

EXAMPLE As noted, independence makes calculation of probability easy. The following is a demonstrating example of how independence of two events is established. Suppose that two dice are thrown together. Define the following events:

$A = \{(3,1),(3,2),(3,3),(3,4),(3,5),(3.6)\}, \quad P(A) = 1/6.$

$B = \{(1,6),(2,5),(3,6),(4,3),(5,2),(6,1)\}, \quad P(B) = 1/6.$

$C = \{(2,6),(3,5),(4,4),(5,3),(6,3)\}, \quad P(C) = 5/36.$

Here we have $P(A \cap B) = \{(3,4)\} = 1/36 = P(A)P(B),$

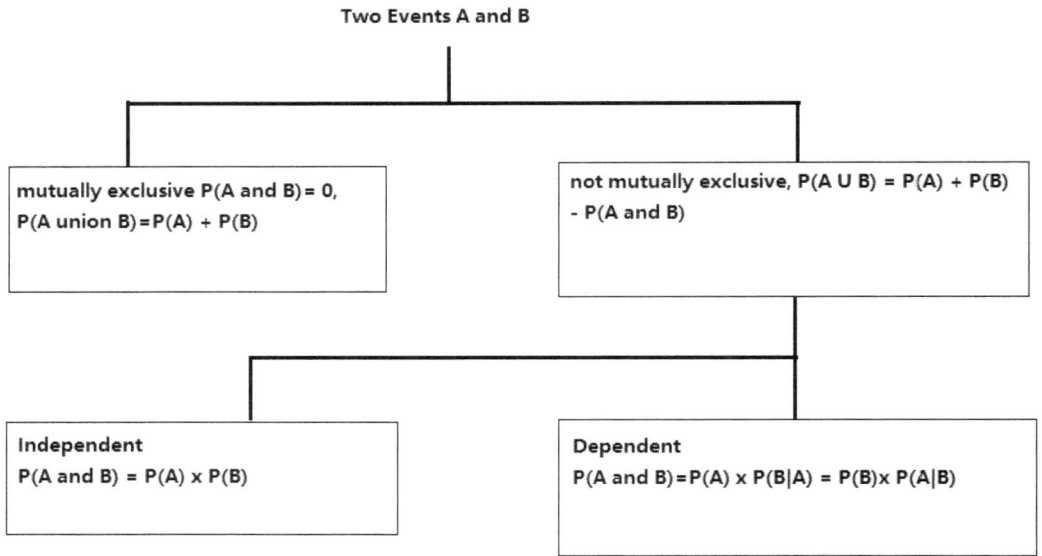

FIGURE 2.4.5 Relationships of events

$$P(A \cap C) = P\{(3,5)\} = 1/36 \neq (1/6)(5/36).$$

Thus A and B are independent, whereas A and C are not.

2.4.6 Independence of More Than Two Events

DEFINITION: Events $A_1,\ldots,A_n$ are mutually independent if for every $k, k = 1,2,3,\ldots,n$ and every subset of indices $i_1, i_2, \ldots, i_k$

$$P(A_{i_1} \cap A_{i_2} \cap \ldots \cap A_{i_k}) = P(A_{i_1})P(A_{i_2})\ldots P(A_{i_k}).$$

Two special cases are,

$$P(A_i \cap A_j) = P(A_i)P(A_j), \quad i \neq j,$$

$$P(A_i \cap A_j \cap A_k) = P(A_i)P(A_j)P(A_k), \quad i \neq j \neq k.$$

One mathematically interesting fact is the following: mutual independence implies pairwise independence, but the reverse implication is not always correct.

EXAMPLE 2.4.11 Suppose three friends A, B, and C alternately shoot free throws (A then B then C then A, etc.). They shoot till someone misses. Suppose that each person has the probability 0.5 of making each free throw. What is the probability that A is the first to miss? What is the probability that B is the first to miss? What is the probability that C is the first to miss?

Chapter 2: Elements of Probability

REGULAR SOLUTION: Let H_i and M_i, $i = A, B, C$ denote hit and miss by player i. Then for A to miss first we should have outcomes like

$$M_A, \; H_A H_B H_C M_A, \; H_A H_B H_C H_A H_B H_C M_A, \cdots$$

Due to the independence, the required probability is then

$$(1/2) + (1/2)^4 + (1/2)^7 + \cdots = \frac{1/2}{1 - (1/2)^3} = 4/7.$$

For B to miss first we should have outcomes like

$$H_A M_B, \; H_A H_B H_C H_A M_B, \; H_A B_B H_C H_A H_B H_C H_A M_B, \cdots$$

Due to the independence, the required probability is then

$$(1/2)^2 + (1/2)^5 + (1/2)^9 = \cdots = \frac{(1/2)^2}{10(1/2)^3} = 2/7.$$

Also for C to miss first we have

$$(1/2)^3 + (1/2)^6 + (1/2)^9 + \cdots = \frac{(1/2)^3}{1 - (1/2)^3} = 1/7.$$

SMART SOLUTIONS

1. Suppose that p_i represents the probability that i is the first to miss, where $i = A$, B, or C. Starting with A two things could happen. Either A misses the first free throw or he makes it and so do B and C. Then A tries again in which case his chance of being first to miss from that point on is again p_A. This leads to a simple equation:

$$p_A = 1/2 + (1/2)(1/2)(1/2) p_A \;\; \to \;\; p_A = 4/7.$$

Also after A making the first free throw, B will be in the same position as A from that point. Thus,

$$p_B = (1/2) p_A \;\; \Rightarrow \;\; p_B = 2/7.$$

Using the same logic $p_C = (1/2) p_B = (1/2)(1/2) p_A = 1/7.$

2. This solution is similar to (1). Using the same argument we have

$$p_A, \; p_B = (1/2) p_A, \text{ and } p_C = (1/2)^2 p_A.$$

Since eventually one person will miss, these probabilities should add to one. Thus,

$$p_A + 1/2 p_A + 1/4 p_A = 1 \;\; \to \;\; p_A = 4/7, \; p_B = 2/7, \text{ and } p_C = 1/7.$$

A PRACTICE PROBLEM Repeat Example 2.4.11 but now assume that each person tries two free throws until someone misses both shots.

EXAMPLE 2.4.12 Suppose that in a given year out of 270,000,000 million adults who played soccer, 40,000 suffered injury ending their soccer life. If we take this as an empirical probability of permanent injury, find the probability that a randomly selected adult player will suffer this type of injury in a 10-year period.

SOLUTION: The empirical probability of suffering such an injury in a given year is

$$40,000/270,000,000 = 1.5/10,000$$

That is, 15 players out of 100,000 players. To find the required probability, we can use the at least rule (or) and first find the probability of no injury in a single year. This is

$$1 - 1.5/10,000 = 0.99985$$

The probability of no injury in 10 consecutive years (assuming independence) is then

$$(0.99985)^{10} = 0.9986.$$

So the required probability is

$$1 - 0.9986 = 1.4/1,000 = 14/10,000.$$

This means that, about 14 out of 10,000 players can be expected to suffer this type of injury in a 10-year period.

A PRACTICE PROBLEM One of my friends and I take turns trying free throws until one of us makes one. If we are both 50% free-throw shooters, show that the probability that my friend makes the first shot (wins) if he tries first is 2/3.

2.5 ANALYSIS OF A TENNIS MATCH

Tennis is an object sport in which each competitor tries to control an object while the other competitor is in direct confrontation. The object of the game is to hit the ball into the opponent's court. If the ball is not returned to your side before it bounces twice, then you win the point. The only other way a point is won is if your opponent commits an error by returning the ball so that it strikes the net and fails to go over, or by hitting the ball out of bounds. The scoring of the game is rather unconventional, but nevertheless, it is accepted. The first point is 15, the second 30, and the third 40, and the fourth "game." The zero or no points is called "love." The first player to score four points wins the game. An exception to this rule is the so-called "Deuce" when the score is tied at 40 all (which is similar to 30 all). When the score is "Deuce," the player who wins the next point has the "Advantage." If this player wins the next point, she wins the game. If she loses the next point, the game goes back to "Deuce," and continues until one player wins the point on her advantage. The first player to win six games wins what is called a "set." The exception is when the set is tied at five games a piece. Then the first player to win the next two games wins the set, seven games to five. But, if the set is tied at six games all, a tie breaker is played. The first player to win seven points with a margin of two in the tie breaker wins the game and the set, seven games to six. Again, if in a tie breaker score becomes six points each the game continues until one player has a two-point advantage. A match is generally won by winning two sets out of three sets, or in case of some major men's tournaments, it is three out of five. We note that some of the rules may be different in different tournaments.

Tennis relates to probability mainly in the same way as most sports do. In this section, we will focus on some of the interesting probability questions involved in tennis. More precisely, suppose that two players, A and B, are playing against each other. Assume that player A has the probability x of winning any given point against player B. Our aim is to calculate the probability that player A will win the game, the set, and the match.

Realistically, we should consider two different probabilities for player A to win a point, one when she is serving, and one when receiving. The reason is that in tennis, players have a better chance of winning a point when serving, than when receiving. This is especially true for male players who play power games and could come up with lots of speedy services giving no chance to their opponent to even touch the ball (aces). Despite this and for simplicity we will assume that the probability for winning a point while serving is the same as when receiving. This assumption is not critical noting that players get approximately the same number of chances for serving and receiving during a match. In fact, in an article appeared in the *Australian Journal of Statistics*, 1983, Vol. 25, Pollard examined 5,503 points played in 35 championship matches. He showed that the assumptions of fixed probability and independence of points could not be rejected on statistical grounds. Considering this here we will let x represent the probability that player A wins the point, and $y = 1 - x$ represent the probability that player B wins the point.

Suppose that we are interested in the probability that the player A will win the game. The probability that player A will take the first four points and therefore the game is x^4, assuming that the events of winning these points are independent from each other. The probability that player A wins the game four points to one is $4x^4y$ noting that A must take the last point and from the remaining four points win three. Since the latter can happen in four different ways, we have $\binom{4}{3} x^3 y x = 4x^4 y$. The probability for player A to win the game four points to two is $10x^4 y^2$. Again using combination, Player A needs to win three of the first five points plus the last point. This probability is therefore $\binom{5}{3} x^3 y^2 x = 10 x^4 y^2$.

Putting these together, we have

$$P(\text{Player A wins the game with no deuce}) = x^4 + 4x^4 y + 10 x^4 y^2$$

since there are only three possibilities. Also

$$P(\text{Player A wins the game with deuce})$$
$$= P(\text{The deuce is reached}) \cdot P(\text{Player A wins the game after deuce}).$$

The first term equals to

$$P(\text{The deuce is reached}) = 20 x^3 y^3$$

since there are 20 ways to reach deuce (40 all) and each player should win three points out of six. To calculate P (Player A wins the game after deuce), we need to start with the state "deuce" and see what could happen after that.

When the deuce is reached, the "win by two" rule is the requirement. The following diagram shows the possible states of the game:

| A's game | Advantage A | Deuce | Advantage B | B's game |

Note that there is no limit to how long the game could go on, alternating between "Advantage A," "Deuce," and "Advantage B." The required probability denoted as p can be computed based on the following reasoning. Starting from deuce either A takes the next two points or A and B split the next two points and start all over again. Thus, we have

$$p = P(\text{Player A wins the next two points})$$
$$+ P(\text{Player A and Player B split the next two points}) \cdot p$$
$$= x^2 + (xy + yx)p = x^2 + (2xy)p.$$

This implies

$$p = \frac{x^2}{1 - 2xy} = \frac{x^2}{x^2 + y^2} \tag{2.12}$$

using the fact that $y = 1 - x$. It is interesting to note that p can also be written as (by simply dividing)

$$\frac{x^2}{1-2xy} = x^2(1 + 2xy + 4x^2y^2 + 8x^3y^3 + \ldots).$$

The expression in the bracket, which is a geometric progression with quotient $2xy \leq 1$ presents the probabilities corresponding to the possibilities after the deuce. For example, x^2 is the probability that A will take the next two points, $2x^3y = 2xyx^2$ is the probability that A will take a point, lose a point (or other way round), and then take the next two points.

Now if we put all the pieces together, we have the following conclusion. Let x denote the probability that player A wins a point, and $y = 1 - x$ denote the probability that player B wins that point. Then $g(x)$, the probability of winning a game for player A is

$$g(x) = P(\text{Player A wins the game}) = P(\text{Player A wins the game with no deuce})$$
$$+ P(\text{Player A wins the game with deuce}) = x^4 + 4x^4y + 10x^4y^2 + 20x^3y^3 \frac{x^2}{x^2 + y^2}$$
$$= x^4[1 + 4y + 10y^2 + 20xy^3/(x^2 + y^2)] = x^4[1 + 4y + 10y^2/(x^2 + y^2)]$$
$$= x^4[5 - 4x + 10(1-x)^2/(x^2 + (1-x)^2)]. \tag{2.13}$$

A PRACTICE PROBLEM Plot the function $g(x)$ versus x. What conclusion can you draw from this? Is $g(x)$ an increasing function? One-to-one? Also calculate $g'(0.5)$ and explain what the resulting value means.

Now that an expression for all the ways of winning a game has been created, we may want to find out for what values of x are each of the four situations (50:0, 50:15, 50:30, winning through deuce) most likely to occur? For this, we must compare the probabilities expressed by the different terms of expression (2.13). For example, for 50:0 versus 50:15, we need to slove the inequality $x^4 > 4(x^4)(1-x)$. Simplifying the expression we get $0.25 \geq 1 - x$ and finally $x \geq 0.75$.

This means that if $x \geq 75\%$, then the most likely score for player A to win the game is 50:0.

A PRACTICE PROBLEM For what values of x is 50:15 more likely than 50:30 and 50:30 is more likely than winning through the deuce?

EXAMPLE 2.5.1 Suppose $x = 0.60$ and $y = 0.40$, that is, player A has a 60% probability of winning any given point. What is the probability for player A to win the game?

SOLUTION: Using the above formula with $x = 0.60$ and $y = 0.40$, we get

$$P(\text{Player A wins the game}) = 0.7357.$$

Hence, the probability for player A to win the game is about 74%. Note that the probability for player A to win the game without and with the deuce are, respectively, 0.085 and 0.6506. Also starting from deuce, A's chance of winning the game is 0.6923. We hope students will try to confirm these calculations.

Note that in this example the probability for player A to win the game is almost 14% more than the probability of winning a point. That is, 20% (60% - 40%) edge in a point opens up to 74% to 26% = 48% edge in a game. So no two such players really need to compete. Table 2.5.1 shows the probabilities for player A (and B) to win the game for $x = 0.5, 0.6, 0.7, 0.8, 0.9$, and 1 with $y = 1 - x$. The results for $x < 0.5$ can be derived easily due to the symmetry.

When $x = 50.5\%$ and $y = 49.5\%$, that is, one player has 1% edge over the other, we can apply the idea of derivative from in calculus and find the edge in a game, in a set and the match (see the first practice problem above). In fact, edge in a game is precisely derivative of Expression (2.13) calculated at point $x = 0.50$. We leave these as an exercise but recommend trying this to appreciate the power of mathematics and understand the meaning of the derivative better. Similar analysis can be applied to find expressions for probabilities of winning a set and the match. Such analysis will show that with a probability of more than 60% of winning each point, the probability of winning a five-set match is almost 100%.

According to above analysis, the better player will usually dominate the opponent. However, in reality, it is not uncommon for a player who loses a match to win a rematch. This is because the

TABLE 2.5.1 Probability of Winning a Game

x	y	P(A Wins the Game)	P(B Wins the Game)
0.5	0.5	0.5	0.5
0.6	0.4	0.736	0.264
0.7	0.3	0.901	0.099
0.8	0.2	0.978	0.022
0.9	0.1	0.999	0.001
1	0	1	0

probability of winning a point does not stay fixed during the game and also between competitions. In fact, it fluctuates as the players' level of practice or concentration varies. For instance, suppose that players A and B play every day of the week. Suppose that player A's probability of winning a point is 55% on Monday (after a weekend of rest), but is 37.5% the other days of the week. Thus, on average, his or her probability of winning a point is $(55\% + 6 \times 37.5\%)/7 = 40\%$. Despite this, player A is more likely to win the Monday match. In other words, player A would win 1/7th of the matches instead of almost none. Still, even if the probability of winning a point fluctuates from match to match, it is not unreasonable to assume that the probability of winning a point stays constant throughout a given match. In general, if player A wins the match, he or she is most likely the better player, at least for that day. However, player B can take consolation in the fact that the loser of a tennis match may only be slightly weaker than her opponent. Player B may only need to win one more point in 10 to be even with or better than player A. With a little practice, player B may be able to win the next match.

PRACTICE PROBLEMS Let g denote the probability of winning a game.

1. Show that the probability of winning a set s without tie-break (traditional rule, e.g., Wimbledon) is

$$s = s(g)$$
$$= g^6 \left[1 + \binom{6}{1}(1-g) + \binom{7}{2}(1-g)^2 + \binom{8}{3}(1-g)^3 \right.$$
$$\left. + \binom{9}{4}(1-g)^4/(1-2g(1-g)) \right]$$
$$= g^6 \left[1 + 6(1-g) + 21(1-g)^2 + 56(1-g)^3 + \frac{126(1-g)^4}{1-2g(1-g)} \right].$$

Plot $s(g)$ and make tables similar to Table 2.5.1

2. Show that the probability for player A to win a set after a tie-break is

$$g_T(x) = x^7 \left[1 + \binom{7}{1}y + \binom{8}{2}y^2 + \binom{9}{3}y^3 + \binom{10}{4}y^4 \right.$$
$$\left. + \binom{11}{5}y^5 + \binom{12}{6}xy^6/(1-2xy) \right]$$
$$= x^7 \left[1 + 7y + 28y^2 + 84y^3 + 210y^4 + \frac{462y^5}{1-2xy} \right]$$
$$= x^7 \left[330 - 1155x + 1540x^2 - 924x^4 + \frac{462(1-x)^5}{x^2 + (1-x)^2} \right].$$

Plot $g_T(x)$.

Chapter 2: Elements of Probability

3. Again let g denote the probability of winning a game. Show that the probability of winning a set with tie-break $s_T(g)$ is

$$s_T(g) = g^6 \left[1 + \binom{6}{1}(1-g) + \binom{7}{2}(1-g)^2 + \binom{8}{3}(1-g)^3 + \binom{9}{4}(1-9)^4 \right.$$

$$\left. + \binom{10}{5} g(1-g)^5 + 2\binom{10}{5}(1-g)^6 g_T(x) \right]$$

$$= g^6 \left[1 + 6(1-g) + 21(1-g)^2 + 56(1-g)^3 + 126(1-g)^4 + 252g(1-g)^5 \right.$$

$$\left. + 504(1-g)^6 g_T(x) \right]$$

where $g_T(x)$, as in previous problems, is the probability of winning a set after a tie-break. Plot $s_T(g)$.

4. Let s_T denote the probability of winning a set with tie-break. Show that the probability of winning a match m (best of five) is

$$m = s_T^3 + \binom{3}{1} s_T^3 (1-s_T) + \binom{4}{x} s_T^2(1-s)s(g)$$

$$= s_t^3 + 3s_T^3(1-s_T) + 6s_T^2(1-s_T)^2 s(g).$$

Plot m.

A CHALLENGING PROBLEM Assume that the probability of winning a point is higher when serving than when receiving. Show that, for the server, the probability of winning the game starting from 0 to 30 is always greater than that from 15 to 40. Show, also, that the same is true for 15 to 30 compared to 30 to 40.

2.5.1 Table Tennis

Let g denote the probability of winning a standard table tennis game.
To win a standard game, a player must either

(1) Be the first to win 21 points with a margin of 2.

(2) Win 2 consecutive points following a tie at 20-20 before the opponent does.

Let x denote the probability of winning a point. Then g as a function of x is given by

$$g(x) = P(21-0) + P(21-1) + \cdots + P(21-19) + P(20-20)\frac{P(A \text{ wins 2 consecutive points.})}{P(A \text{ or } B \text{ wins 2 consecutive points.})}$$

$$= \sum_{j=0}^{19} \left\{ \binom{20+j}{j} x^{21}(1-x)^j \right\} + \binom{40}{20} x(1-x)^{20}(1-x)^{20} \frac{x^2}{x^2+(1-x)^2}.$$

Table below presents probabilities related to all possible cases for several different values of x. The last three rows are the probabilities of winning under case (1), case (2), and the total, $g(x)$.

x:	0.500	0.505	0.510	0.525	0.550	0.575	0.600	0.650	0.700	0.750	0.800	0.900	1.000
21-0	0.000	0.000	0.000	0.000	0.000	0.000	0.000	0.000	0.001	0.002	0.009	0.109	1.000
21-1	0.000	0.000	0.000	0.000	0.000	0.000	0.000	0.001	0.004	0.012	0.039	0.230	0.000
21-2	0.000	0.000	0.000	0.000	0.000	0.000	0.001	0.003	0.012	0.034	0.085	0.253	0.000
21-3	0.000	0.000	0.000	0.000	0.001	0.001	0.002	0.009	0.027	0.066	0.131	0.194	0.000
21-4	0.000	0.000	0.000	0.001	0.002	0.003	0.006	0.019	0.048	0.099	0.157	0.116	0.000
21-5	0.001	0.001	0.001	0.002	0.003	0.007	0.012	0.033	0.072	0.123	0.157	0.058	0.000
21-6	0.002	0.002	0.002	0.004	0.007	0.012	0.021	0.050	0.094	0.134	0.136	0.025	0.000
21-7	0.003	0.004	0.004	0.006	0.012	0.020	0.032	0.067	0.108	0.129	0.105	0.010	0.000
21-8	0.006	0.007	0.007	0.011	0.018	0.030	0.045	0.082	0.114	0.113	0.073	0.003	0.000
21-9	0.009	0.011	0.012	0.016	0.027	0.041	0.058	0.093	0.110	0.091	0.047	0.001	0.000
21-10	0.014	0.016	0.017	0.023	0.036	0.052	0.069	0.098	0.099	0.068	0.028	0.000	0.000
21-11	0.020	0.022	0.024	0.031	0.046	0.062	0.078	0.096	0.084	0.048	0.016	0.000	0.000
21-12	0.026	0.029	0.031	0.040	0.055	0.070	0.083	0.090	0.067	0.032	0.009	0.000	0.000
21-13	0.033	0.036	0.039	0.048	0.063	0.076	0.084	0.080	0.051	0.020	0.004	0.000	0.000
21-14	0.041	0.043	0.046	0.055	0.069	0.078	0.082	0.068	0.037	0.012	0.002	0.000	0.000
21-15	0.047	0.050	0.053	0.061	0.072	0.078	0.077	0.055	0.026	0.007	0.001	0.000	0.000
21-16	0.053	0.056	0.058	0.065	0.073	0.074	0.069	0.044	0.018	0.004	0.000	0.000	0.000
21-17	0.058	0.060	0.062	0.067	0.071	0.069	0.060	0.033	0.011	0.002	0.000	0.000	0.000
21-18	0.061	0.063	0.064	0.068	0.068	0.062	0.051	0.025	0.007	0.001	0.000	0.000	0.000
21-19	0.063	0.064	0.065	0.066	0.063	0.054	0.042	0.018	0.004	0.001	0.000	0.000	0.000
Case (1):	0.437	0.462	0.488	0.564	0.684	0.789	0.870	0.964	0.0994	0.999	1.000	1.000	1.000
Case (2):	0.063	0.064	0.065	0.066	0.061	0.051	0.038	0.015	0.003	0.000	0.000	0.000	0.000
Total	0.500	0.526	0.553	0.629	0.746	0.940	0.909	0.978	0.997	1.000	1.000	1.000	1.000

A PRACTICE PROBLEM In October 2002, the United States Association of Table Tennis changed the format of the table tennis games. 38 mm balls were replaced with 40 mm. The three games to 21 were replaced by five games to 11 with each players serving twice rather than five times. Repeat the above calculations for a game played to 11.

2.6 RANDOM VARIABLES AND THEIR DISTRIBUTION

As we know, a major part of science is developed to deal with attributes that are quantifiable. Due to this fact methods of statistical analysis require that we concentrate on certain numerical aspect of the data. The concept of a random variable allows us to pass from the experimental outcomes themselves to the numerical function of the outcomes.

Random variables, as is clear from the name, are variables that take values according to a chance mechanism. As such, some of their values are more likely than the others. If the likelihoods corresponding to possible values of a random variable are known, then the distribution of the random variable is known. Let us start the discussion of this section by presenting some examples. Consider the chance experiment of flipping two coins. The sample space is

$$\Omega = \{TT, TH, HT, HH\}.$$

Suppose that the goal of the experiment is to register the number of heads observed. Clearly this number is either 0, 1, or 2. It is a variable whose value depends on the outcome of a chance experiment. We call this a random variable. Random variables are usually named X, Y, and so on.

Chapter 2: Elements of Probability

Random variables are used to pass from the experimental outcomes to numerical values using the idea of function. They are also used to simplify the sample space and to change qualitative data to quantitative data so that mathematical ideas and results can be utilized.

In most chance experiments an investigator will concentrate on one or more variable quantities. For instance, in Example 2.4.8, the variable X (the number of games that Team One wins in n games) can be the quantity of interest. Such a quantity is a random variable. The term "random" here indicates that the value this variable takes cannot be predicted in advance, that is varies by "chance." As a second example, when rolling two dice, the random variable could be the sum of the two numbers. Formally, a random variable is any rule that associates or assigns a number to each outcome in the sample space. In general, there are two types of random variables, discrete and continuous.

Now, to understand a situation, other than a random variable and it's possible values, we need to know something about the probabilities of the values taken by a random variable. To see why, suppose that we wish to purchase several thousand identical devices and are informed by the manufacturer that the device has an average life span of 7 years. It is obviously hard to figure out what is going to happen to the one we purchased. In fact, we would be much better informed if, in addition to knowing the average life span knew that, for example, 95% or more of such devices survive 2 years, 90% or more survive 2 and 1.5 years, 75% or more survive 3 years, 50% or more survive 6 years, 25% or more survive 10 years and 10% or more survive 12 years. Such information can be summarized in a table including a random variable (life span of a randomly selected device), it's possible values, and probabilities related to different length of life or survivals. We call this a distribution.

DISCRETE RANDOM VARIABLE: A random variable X is called discrete if it takes on only a finite or countably infinite number of values. For simplicity, let these values be $0, 1, 2, \ldots$. For example, if we flip a fair coin twice and let X represent the number of heads, then possible values of X are 0, 1, and 2. This is an example where X takes a finite number of values. As a second example, let X be the variable given in Example 2.4.5, then possible values for X are $0, 1, \ldots, n$. The function $p_k = P(X = k)$ is called the probability function (probability mass function) of X and

$$F(x) = P(X \leq x) = \sum_{k \leq x} p_k \tag{2.14}$$

is called the (cumulative) distribution function of X. Obviously, $F(x) = 0$ for $x < 0$, F is piecewise constant and has jumps at certain values of X. Also, if m is the greatest value of X, then $F(x) = 1$ for $x \geq m$.

EXAMPLE 2.6.1 Referring to the example of flipping a fair coin twice. Let X denote the number of heads. Then we have

$$p_0 = P(X = 0) = P(TT) = 0.25,$$
$$p_1 = P(X = 1) = P(HT) + P(TH) = 0.5,$$
$$p_2 = P(X = 2) = P(HH) = 0.25.$$

Thus, the probability function or probability mass function of X is

x	0	1	2
$P(X=x)$	0.25	0.5	0.25

This is a special case of a distribution known as Binomial distribution. The cumulative distribution function can easily be obtained as

$$F(0) = P(X \leq 0) = P(X=0) = 0.25,$$
$$F(1) = P(X \leq 1) = P(X=0) + P(X=1) = 0.75,$$
$$F(2) = P(X \leq 2) = P(X=0) + P(X=1) + P(X=2) = 1.$$

Incorporating all the information we have

$$P(X \leq x) = F(x) = \begin{cases} 0 & x < 0, \\ 0.25 & 0 \leq x < 1, \\ 0.75 & 1 \leq x < 2, \\ 1 & 2 \leq x. \end{cases}$$

Note that $F(x)$ is a step function and its value remains constant over the intervals between the possible values of X. For example, $F(1.75) = F(1.5) = F(1)$, since probabilities for all the values other than 0, 1, and 2 are zero.

A PRACTICE PROBLEM Suppose that you and your friend flip a fair coin. If heads comes up you win a dollar. If tail comes up you lose a dollar. If X represents the amount of money you win then

x	1	−1
$P(X=x)$	0.5	0.5

Find $F(x)$ and plot that.

EXAMPLE 2.6.2 Refer to Example 2.4.8 and consider the case $n = 2$. For this case the possible values of X are 0, 1, and 2 and their probability are, respectively

$$P(X=0) = (1-p)^2, \quad P(X=1) = 2p(1-p), \quad P(X=2) = p^2.$$

Thus, the probability mass function is

x	0	1	2
$p(X=x)$	$(1-p)^2$	$2p(1-p)$	p^2

The plot of probability mass function for $p = 0.6$ is shown in Figure 2.6.1. For this example the cumulative distribution function $F(x)$ is

$$F(x) = \begin{cases} 0 & x < 0, \\ (1-p)^2 & 0 \leq x < 1, \\ (1-p)^2 + 2p(1-p) & 1 \leq x < 2, \\ 1 & 2 \leq x. \end{cases}$$

Chapter 2: Elements of Probability

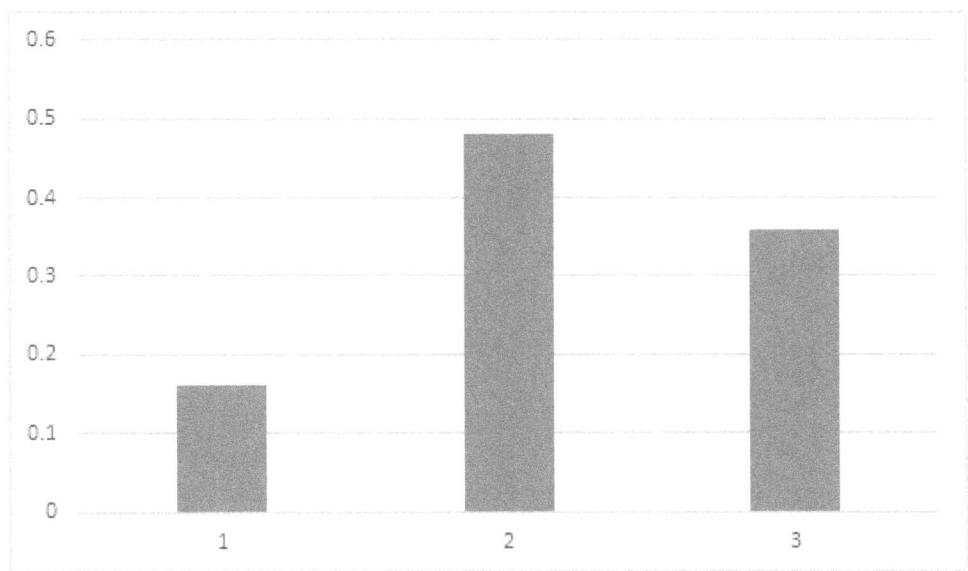

FIGURE 2.6.1 An Example of Probability Function for a Discrete Random Variable (Example 2.6.2 with $p = 0.6$)

Figure 2.6.2 presents the graph of $F(x)$ for $p = 0.6$.

EXAMPLE 2.6.3 In the experiment of rolling two fair dice, let X represent the sum of two numbers, then we have

Outcomes and Sums for Rolling Two Dice

		Second Die					
		1	2	3	4	5	6
First Die	1	1 + 1 = 2	1 + 2 = 3	1 + 3 = 4	1 + 4 = 5	1 + 5 = 6	1 + 6 = 7
	2	2 + 1 = 3	2 + 2 = 4	2 + 3 = 5	2 + 4 = 6	2 + 5 = 7	2 + 6 = 8
	3	3 + 1 = 4	3 + 2 = 5	3 + 3 = 6	3 + 4 = 7	3 + 5 = 8	3 + 6 = 9
	4	4 + 1 = 5	4 + 2 = 6	4 + 3 = 7	4 + 4 = 8	4 + 5 = 9	4 + 6 = 10
	5	5 + 1 = 6	5 + 2 = 7	5 + 3 = 8	5 + 4 = 9	5 + 5 = 10	5 + 6 = 11
	6	6 + 1 = 7	6 + 2 = 8	6 + 3 = 9	6 + 4 = 10	6 + 5 = 11	6 + 6 = 12

x	2	3	4	5	6	7	8	9	10	11	12
P(X = x)	1/36	2/36	3/36	4/36	5/36	6/36	5/36	4/36	3/36	2/36	1/36

For this example, find $F(x)$ and plot it.

A PRACTICE PROBLEM In the chance experiment of rolling two fair dice, let Y and Z represent respectively the product and the maximum of two numbers. Find the probability mass functions for Y and Z and plot them.

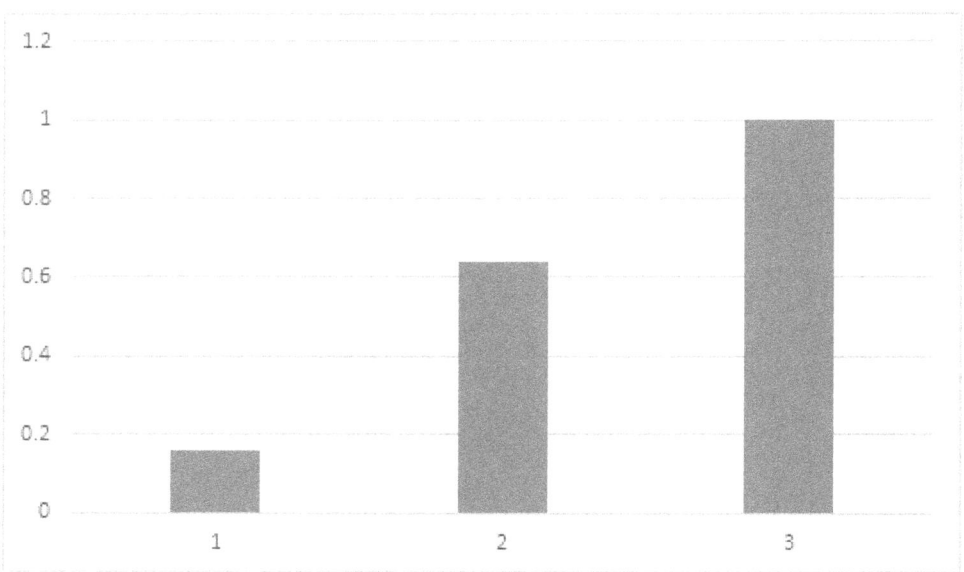

FIGURE 2.6.2 An Example of Cumulative Distribution Function (Example 2.6.2 with $p = 0.6$) for a Discrete Random Variable

The Dice Experiment
(http://www.math.uah.edu/stat/applets/DiceExperiment.html)

This website provides a tool for better understanding of discrete random variables. There are many experiments to choose from of which one is described here. The experiment consists of rolling n dice, each governed by the same probability distribution. You can specify the die distribution by clicking on the die probability button; this button brings up the die probability dialog box. You can define your own distribution by typing probabilities into the text fields of the dialog box, or you can click on one of the buttons in the dialog box to specify one of the following special distributions:

- **Fair:** each face has probability 1/6.
- **1-6 flat:** faces 1 and 6 have probability 1/4 each; faces 2, 3, 4, and 5 have probability 1/8 each.
- **2-5 flat:** faces 2 and 5 have probability 1/4 each; faces 1, 3, 4, and 6 have probability 1/8 each.
- **3-4 flat:** faces 3 and 4 have probability 1/4 each; faces 1, 2, 5, and 6 have probability 1/8 each.
- **Skewed left:** face i has probability $i/21$ for $i = 1, 2, 3, 4, 5, 6$.
- **Skewed right:** face i has probability $(7 - i)/21$ for $i = 1, 2, 3, 4, 5, 6$.

The following random variables are recorded on each update:

- The sum of the scores Y, the average score M, the minimum score U.
- The maximum score V, the number of aces (1's)Z.

Chapter 2: Elements of Probability

Any one of these variables can be selected with a list box. The information including probability distribution of the selected variable are shown in blue in the distribution graph and are recorded in the distribution table. When the simulation runs, the empirical distribution and other information are shown in red on the distribution graph and are recorded in the distribution table. The parameter n can be varied with a scroll bar.

EXAMPLE 2.6.4 To provide an example for a case where a discrete random variable takes infinite but countable values, consider an experiment of flipping a fair coin until a head is observed. Then we have outcomes such as H, TH, TTH, and $TTTH$. Let X represent the number of times the coin is flipped. Then the probability mass function takes the following form:

x	1	2	3	$\cdots$	n	$\cdots$
$P(X=x)$	$\frac{1}{2}$	$(\frac{1}{2})^2$	$(\frac{1}{2})^3$	$\cdots$	$(\frac{1}{2})^n$	$\cdots$

For example,
$P(X=3) = P(TTH) = P(T)P(T)P(H) = (1/2)(1/2)(1/2)$.

$F(2) = P(X \leq 2) = P(X=1) + P(X=2) = 1/2 + (1/2)^2 = 3/4$.

Note that, rather than using a table, we may like to express $P(X=x)$ as a function of x. In fact, for the example just discussed, we have

$$P(X=x) = (1/2)^{x-1}(1/2) = (1/2)^x, \quad x = 1, 2, \ldots$$

If $P(X=x)$ can be expressed as a function of x, then we may call that a probability model. This form or presentation is much more desirable and practical. In fact, this is one of the major goals of mathematics. Classical probability models are discussed in Chapter 3.

Continuous Random Variables: A random variable X is called continuous if possible values of X form an interval. Unlike discrete random variables, here the possible values are neither finite or countable. For example, if the capacity of the gasoline tank of a sport car is 12 gallons, then the amount of gasoline remaining in the tank at a random time, denoted by X, can be anywhere between 0 and 12 gallons. In this case, the possible values of X form an interval $[0, 12]$, and hence X is a continuous random variable. In general, a continuous random variable has a density function $f(x) \geq 0$, such that

$$P(a \leq X \leq b) = \int_a^b f(x)dx. \quad (2.15)$$

This means that, the probability that X will fall between a and b is the area under the density function $f(x)$, between a and b.

Note that for discrete random variables we have probability mass function and use summation to find probabilities, whereas for a continuous random variable we have probability density function and use integration to find probabilities.

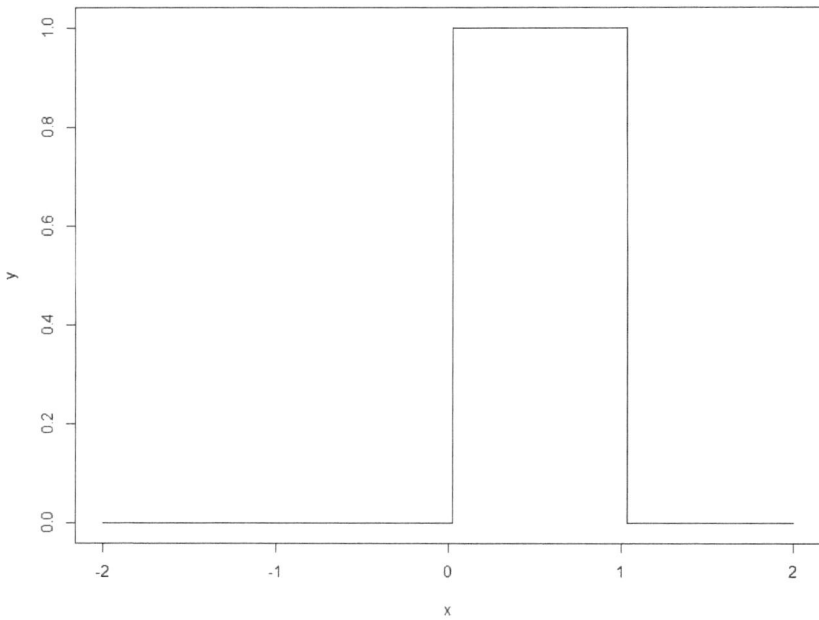

FIGURE 2.6.3 An Example of Probability Density Function (Example 2.6.5.)

EXAMPLE 2.6.5 Consider a probability density function defined as

$$f(x) = \begin{cases} 1 & 0 \leq x \leq 1, \\ 0 & \text{elsewhere}. \end{cases}$$

This is called rectangular distribution. Since the chance that X falls in intervals of equal length are equal (area under the density function is the same for intervals of equal length), this distribution is also named uniform distribution. The general form of the uniform distribution is discussed in Chapter 3. The probability density function of this distribution is shown in the Figure 2.6.3. To find the probability that X falls, for example, between $1/3$ and $1/2$ we proceed as follows,

$$P(1/3 \leq X \leq 1/2) = \int_{1/3}^{1/2} f(x)dx = \int_{1/3}^{1/2} dx = x\vert_{1/3}^{1/2} = 1/2 - 1/3 = 1/6.$$

The cumulative distribution function for uniform distribution is given by

$$P(X \leq x) = F(x) = \begin{cases} 0 & x \leq 0, \\ x & 0 < x < 1, \\ 1 & 1 \leq x. \end{cases}$$

Figure 2.6.4 presents the graph of $F(x)$.
This function was obtained using

$$F(x) = P(X \leq x) = \int_{-\infty}^{x} f(t)dt. \tag{2.16}$$

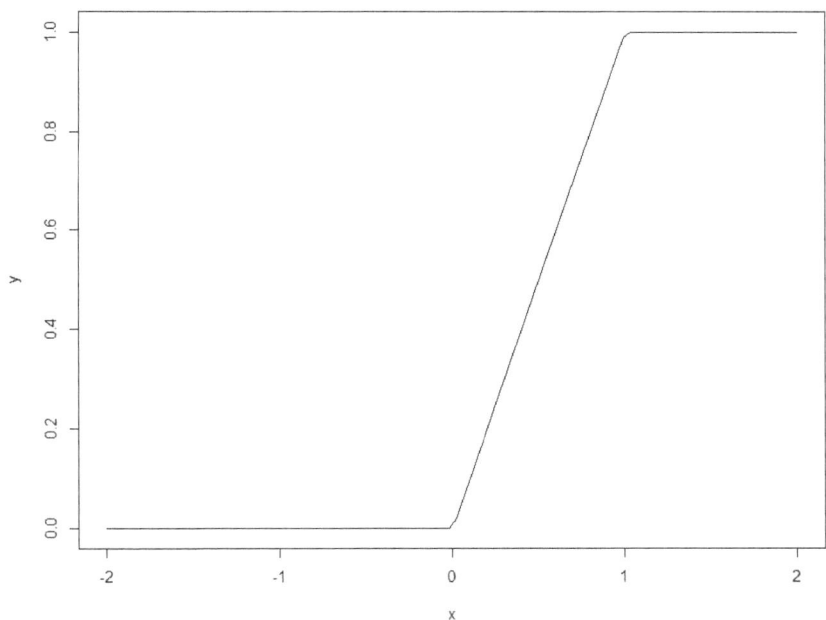

FIGURE 2.6.4 An Example of Cumulative Distribution Function (Example 2.6.5) for a Continuous Random Variable

Note that in terms of $F(x)$ we have

$$P(a \leq X \leq b) = F(b) - F(a),$$

$$F(-\infty) = 0, \quad F(+\infty) = \int_{-\infty}^{+\infty} f(x)dx = 1,$$

and $P(a \leq X \leq b)$ is the area under the curve $y = f(x)$ between $x = a$ and $x = b$. As an example for the uniform distribution in Example 2.6.5, we have

$$P(1/2 \leq X \leq 2/3) = F(2/3) - F(1/2) = 2/3 - 1/2 = 1/6.$$

It should be pointed out that for a continuous random variable the probability assigned to any specific value is zero since there is no area under the density function for an interval of length zero. Formally we have $P(X = a) = \int_a^a f(x)dx = 0$. This is quite different from a discrete random variable where probability mass is on certain points (finite or countably infinite).

EXAMPLE 2.6.6 The probability density function of a random variable X is given by

$$f(x) = \begin{cases} \lambda e^{-3x} & x > 0, \\ 0 & \text{elsewhere.} \end{cases}$$

Find λ and $F(x)$. Calculate $P(0.5 \leq X \leq 1)$.

SOLUTION:

$$\int_{\infty}^{-\infty} -\lambda e^{-3x}\,dx = \int_0^{\infty} \lambda e^{-3x}\,dx = 1 \Rightarrow -\frac{\lambda}{3}e^{-3x}\Big|_0^{\infty} = \frac{\lambda}{3} = 1 \Rightarrow \lambda = 3.$$

For $x > 0$,

$$F(x) = \int_{-\infty}^{x} f(t)\,dt = \int_0^{\infty} 3e^{-3t}\,dt = -e^{-3t}\Big|_0^{x} = 1 - e^{-3x}.$$

So, we have

$$F(x) = \begin{cases} 0 & x \leq 0, \\ 1 - e^{-3x} & x > 0. \end{cases}$$

Finally,

$$P(0.5 \leq X \leq 1) = F(1) - F(0.5) = (1 - e^{-3}) - (1 - e^{-1.5}) = 0.173.$$

A PRACTICE PROBLEM Consider a continuous random variable with probability density function

$$f(x) = \lambda x^2 \quad 0 \leq x \leq 1.$$

Find λ and $F(x)$. Calculate $P(1/3 \leq X \leq 2/3)$.

2.7 EXPECTATION AND VARIANCE

Suppose that we flip a fair coin 40 times. How many heads do we expect to observe? It is easy to see that the answer is 20. This value is called expected value or mean value of X (the number of heads). Of course, in practice, we will not always observe exactly 20 heads. However, if we repeat this experiment a large number of times, register number of heads and average them we should get a number very close to 20. How do we know this? From a well-known theorem called "The Law of Large Numbers."

Formally, let Ω be a finite population with n elements ω_i and X be a discrete random variable taking the distinct values $x_1, \ldots, x_k$. The population mean

$$\mu = E(X) = \frac{1}{n}\sum_{i=1}^{n} X(\omega_i) \qquad (2.17)$$

is called the expected value of X. If m_j is the number of elements of Ω having the same X-value x_j, then we can rewrite μ as

$$\mu = E(X) = \frac{1}{n}\sum_{j=1}^{k} m_j x_j = \sum_{j=1}^{k} x_j p(x_j) \qquad (2.18)$$

where $P(X = x_j) = p(x_j) = m_j/n$ and k is the number different values X takes ($k \leq n$). The same definition applies to the case where k is not finite. In general, if X is a discrete random variable with

Chapter 2: Elements of Probability

possible values $x_1, x_2, ...x_k$ and respective probabilities $p(x_1), p(x_2), ...p(x_k)$, then the expectation or expected value of the function $g(X)$ of the random variable X is given by

$$E(g(X)) = \sum_{j=1}^{k} g(x_j)p(x_j). \tag{2.19}$$

EXAMPLE Consider the experiment of rolling a fair die. Let X be the number observed. Then since the possible values 1, 2, 3, 4, 5, and 6 are equally likely, we have

$$E(X) = 1(1/6) + 2(1/6) + 3(1/6) + 4(1/6) + 5(1/6) + 6(1/6) = 3.5.$$

Also for $g(X) = X^2$, we have

$$E(X^2) = (1)^2(1/6) + (2)^2(1/6) + (3)^2(1/6) + (4)^2(1/6) + (5)^2(1/6) + (6)^2(1/6) = 89/6.$$

If $g(X) = (X - E(X))^2$ the resulting expectation is called (population) variance and is denoted by σ^2. It can be calculated as

$$\sigma^2 = \text{Var}(X) = E(X - E(X))^2 = E(X^2) - (E(X))^2$$
$$= \sum_{j=1}^{k}(x_j - E(X))^2 p(x_j) = \sum_{j=1}^{k} x_j^2 p(x_j) - \left(\sum_{j=1}^{k} x_j p(x_j)\right)^2.$$

The population standard deviation, ($\sigma \geq 0$) is the square root of $\text{Var}(X)$, that is $\sigma = SD(X) = \sqrt{\text{Var}(X)}$.

Consider the example of flipping a fair coin twice and defining X as the number of heads. We have

x	0	1	2
P(X=x)	0.25	0.50	0.25

Applying the above formulas we get

$$E(X) = (0)(0.25) + (1)(0.50) + (2)(0.25) = 1$$
$$E(X^2) = (0)^2(0.25) + (1)^2(0.50) + (2)^2(0.25) = 1.5$$
$$\sigma^2 = \text{Var}(X) = E(X^2) - (E(X))^2 = 1.5 - (1)^2 = 0.5$$
$$\sigma = SD(X) = \sqrt{0.5} = 0.707$$

Note the similarity of expected value and the center of gravity. Here probabilities are like masses. In fact, calculation of expected value is the same as the calculation of center of gravity. The only difference is that, here total mass is always equals 1.

EXAMPLE 2.7.1 Two teams A and B have played 100 games of soccer. For each game the three possibilities are assigned the following values:

$$x_1 = 1 \text{ (Team A wins)},$$

$$x_2 = -1 \text{ (Team B wins)},$$
$$x_3 = 0 \text{ (Tie)}.$$

Based on the results, the following probabilities are assigned to these three values

$$P(X = x_1) = p(x_1) = 0.55, \quad p(x_2) = 0.30, \quad p(x_3) = 0.15.$$

Find $E(X)$ and $\text{Var}(X)$.

SOLUTION:
$$E(X) = 1(0.55) + (-1)(0.30) + 0(0.15) = 0.25$$

and
$$\sigma^2 = (1 - 0.25)^2(0.55) + (-1 - 0.25)^2(0.30) + (0 - 0.25)^2(0.15) = 0.7875.$$

We may also find the standard deviation by taking the positive square root of σ^2;

$$\sigma = \sqrt{0.7875} = 0.8874.$$

AN EXAMPLE OF APPLICATION A basketball player has kept his statistics and found that he makes 40% of his 2-pointer attempts and 30% of his 3-pointer attempts. Find expected number of points in each case.

SOLUTION:
$$\text{2–Pointer:} \quad 2(0.40) + 0(0.60) = 0.8$$
$$\text{3–Pointer:} \quad 3(0.30) + 0(0.70) = 0.9.$$

So in a sequence of 10 attempts he expects to get 8 points if he tries 2-pointer and 9 points if he tries 3-pointer.

EXAMPLE 2.7.2 Referring to Example 2.5.2, we have

$$\mu = E(X) = 0 \cdot (1-p)^2 + 1 \cdot 2p(1-p) + 2 \cdot p^2 = 2p$$

and
$$\sigma^2 = \text{Var}(X) = 2p(1-p).$$

A PRACTICE PROBLEM Thousand raffle tickets are sold at $2 each for a local baseball tournament. Prizes are awarded as follows: two prizes of $100, four prizes of $50, and 10 prizes of $10. Show that the distribution of return for a person who buys one ticket is

x	98	48	8	-2
$P(X=x)$	2/1,000	4/1,000	10/1,000	984/1,000

Find the expected return and also the variance.

Chapter 2: Elements of Probability

In the continuous case, the expectation of a random variable X with density function $f(x)$ is given by

$$\mu = E(X) = \int_{-\infty}^{+\infty} xf(x)dx. \tag{2.20}$$

Analogously,

$$E(g(X)) = \int_{-\infty}^{+\infty} g(x)f(x)dx, \tag{2.21}$$

and especially

$$\sigma^2 = \text{Var}(X) = E(X - E(X))^2$$
$$= \int_{-\infty}^{+\infty} (x - E(X))^2 f(x)dx = E(X^2) - (E(X))^2. \tag{2.22}$$

EXAMPLE 2.7.3 Applying formulas (2.20) and (2.22) to the uniform distribution used in Example 2.6.5, we find

$$\mu = E(X) = \int_0^1 x\,dx = \frac{1}{2}$$

and

$$\text{Var}(X) = \int_0^1 (x - 1/2)^2 dx = \frac{1}{12}.$$

We can alternatively find Var(X) as

$$\text{Var}(X) = E(X^2) - (E(X))^2 = \int_0^1 x^2 dx - (1/2)^2$$
$$= 1/3x^3|_0^1 - (1/2)^2 = 1/3 - 1/4 = 1/12.$$

EXAMPLE 2.7.4 Money made from a high-school football game (in units of thousands) can be modeled as a continuous random variable having the probability density function

$$f(x) = \begin{cases} \frac{1}{18}(x+1) & -1 \leq x \leq 5 \\ 0 & \text{elsewhere} \end{cases}$$

Find the expected profit per game. Find also the variance.

SOLUTION:

$$E(X) = \int_{-\infty}^{\infty} xf(x)dx = \int_{-1}^{5} \frac{x}{18}(x+1)dx = \frac{x^3}{54} + \frac{x^2}{36}\Big|_{-1}^{5} = 3,$$

$$E(X^2) = \int_{-1}^{5} \frac{x^2}{18}(x+1)dx = \frac{x^4}{72} + \frac{x^3}{54}\Big|_{-1}^{5} = 11,$$

$$\text{Var}(X) = E(X^2) - (E(X))^2 = 11 - (3)^2 = 2.$$

It is easy to see that, if a and b are constant numbers, then

$$E(a \pm bX) = a \pm bE(X),$$

$$\text{Var}(a \pm bX) = b^2 \text{Var}(X). \tag{2.23}$$

This means changing the scale of the data will effect the variability and it's measure, whereas shifting will not have any effect on that. However, both transformations will affect the expected value or the mean value accordingly.

Interpretations of the Expected Value and Standard Deviation

Suppose that you and your friend play a simple game that involves flipping a fair coin. You win a dollar if head comes up and lose a dollar if tail comes up. Let X denote the amount of money you win. We have

x	1	-1
P(X=x)	$\frac{1}{2}$	$\frac{1}{2}$

with $E(X) = 0$ and $SD(X) = 1$. A game with expected value or expected return equal to zero is called a fair game. This means that if you play this game for a long period of time, you will break even. Further if several other pairs play the same game, then, according to the Chebyshev's rule, at least 75% of them will lose or win at most 2 dollars and 89% of them will win or loose at most $3. In fact, standard deviation provides an indication of how far you may move away from the expected value (expected return). Another interesting interpretation of standard deviation is related to its use in quantification and calculation of risk. Suppose now rather than 1 dollar you win or lose 100 dollars after each flip of a fair coin. Then,

y	100	-100
P(Y=y)	$\frac{1}{2}$	$\frac{1}{2}$

with $E(Y) = 0$, and $SD(Y) = 100$. For this situation too the expected return is zero. However, the standard deviation is 100 times the previous case. Also, now at least 75% of the players win or lose at most 200 dollars and at least 89% win or lose at most 300 dollars. In a way, standard deviation provides a measure for the risk. That is, although in both cases we expect to break even. In this case our win or loss will deviate from the expectation by a larger amount. Deviation from expectation is one definition often used for risk.

AN EXAMPLE OF APPLICATION Two equally good tennis players are about to play two games. Before the games, we are given a choice of the following payoffs for our favorite player:

Payoff 1: Win $1 for each win, Lose $3 for two losses.
Payoff 2: Win $1 if the results of two games are different.
Win $2 for two loses, Lose $3 for two wins.
Payoff 3: Win $3 for two wins, Win $1 for one win, and one loss.
Lose $4 for two loses.

Which payoff should we choose?

SOLUTION: Here the sample space is

$$\Omega = (WW, WL, LW, LL)$$

where W and L stand for win and loss, respectively. It is easy to see that the distributions corresponding to these payoffs are respectively:

Playoff 1:
x_1	1	2	−3
$P(X_1 = x_1)$	2/4	1/4	1/4

Playoff 2:
x_2	1	2	−3
$P(X_2 = x_2)$	2/4	1/4	1/4

Playoff 3:
x_3	1	3	−4
$P(X_3 = x_3)$	2/4	1/4	1/4

Note that X_1 and X_2 have an identical distribution. Now

$$E(X_1) = E(X_2) = E(X_3) = 1/4,$$

$$\text{Var}(X_1) = \text{Var}(X_2) = 15/4 \quad \text{Var}(X_3) = 26/4,$$

$$SD(X_1) = SD(X_2) = 1.94, \quad SD(X_3) = 2.55,$$

that is, all three payoffs have the same expected value or expected return. Hence, with each payoff, we should expect to gain the same amount on average, if we play the game a large number of times (using the chosen payoff each time) and winnings are accumulated arithmetically. Does this mean that we should be indifferent when choosing between payoff 3 and the other two? Not necessarily.

Note that the range of possible values for X_3 is wider than the range of possible values for X_1 and X_2. In particular, with payoff 3, it is possible both to win or lose more than with either of the other payoffs. Hence, if a person is risk averse, in the sense that they feel the pain of a monetary loss more than the joy of a monetary gain, they would probably choose payoff 1 or payoff 2 over payoff 3. On the other hand, if a person is risk tolerant, in the sense that the joy of a substantial gain outweighs the pain of a substantial loss, they would probably choose payoff 3 over payoff 1 or payoff 2.

Most people are risk averse and thus would probably choose payoff 1 or payoff 2. However, people are not risk averse all of the time. Hence, when payoff distributions are different, the choice between payoffs is determined to a large extent by personal preference. Nevertheless, if we are given partial information about an individual's preferences, we can often make assertions about what choices are consistent with these preferences. For example, if we are told that an individual prefers payoff 3 to payoff 2, then we can conclude that the individual must also prefer payoff 3 to payoff 1 because payoff 1 and payoff 2 have the same distribution.

Social scientists have long been intrigued with how people make decisions, and much research has been conducted on the topic. The most popular theory among early researchers, in the 1930s and 1940s, was that people made decisions to maximize their expected utility. This may or may not correspond to maximizing their expected dollar amount. The idea was that people would assign a worth or utility to each outcome and choose whatever alternative yielded the highest expected value.

EXAMPLE 2.7.5 Suppose that you are thinking of going to a game of your favorite team, but you are not sure whether you will be free on that night. The ticket for the game is $30 if you buy it in advance and $40 if you buy it 2 days before the game from a friend or an agency. Knowing yourself, you estimate your probability of being able to go to the game is p. What should you do?

SOLUTION: Here you can utilize the mathematical expectation. If you buy the ticket in advance you expect to lose $30(1-p)$ dollars. On the other hand if you wait your loss would be $(40-30)p = 10p$. These two losses are equal for $p = 3/4$. Thus, if $p > 3/4$ it is better to buy the ticket in advance. For $p \leq 3/4$ you should probably wait.

EXAMPLE 2.7.6 Men are known to forget "important" days such as their anniversary. Suppose John is not sure that today is his anniversary, so he is wondering whether he should buy a bunch of flowers or not. He decides to look at the problem like an expected value problem. How could he formulate the problem and make a decision?

SOLUTION: Let A be the event that it is his anniversary and B be the event of buying a bunch of flowers. Suppose also that $P(A) = p > 0$. Here he should consider four disjoint events $A \cap B$, $A \cap B'$, $A' \cap B$, and $A' \cap B'$ and think about some payoff for each. Suppose he could somehow come up with monetary values for attention, love, disappointment, and so on, to construct the following payoff table:

	Anniversary(A)	No Anniversary(A')
Flowers(B)	10	-1
No Flowers(B')	-10^4	10

If he buys flowers, then the expected return or payoff is $10p - (1-p)$. This is because the possible values are 10 and -1 with respective probabilities p and $1-p$. Now, to see for which values of p buying flowers result in a greater payoff, we should solve the following inequality:

$$10p - (1-p) > -10^4 p + 10(1-p) \Rightarrow 10021p > 11 \Rightarrow p > 1/911$$

So the expected payoff for buying flowers is larger if $p > 1/911$.

Two noticeable outcomes are the following:

1. Since there are 365 days in a year, the probability that any given day is his anniversary is $1/365 > 1/911$. So a man who has been married for several years should know this without any calculation.
2. 911 is a well-known date (and number) and has association with an event that cannot and should not be forgotten.

EXAMPLE 2.7.7 This example is based on Blaise Pascal's philosophical argument (1660) regarding God. Let E denote the event that God exists and B denote the event that a randomly selected person believes that God exists. Suppose that $P(E) = p > 0$ and consider the following payoffs:

Events	Payoff($)
$E \cap B$	10^5
$E \cap B'$	-10^{10} (eternal damnation)
$E' \cap B$	-10^4 (for not enjoying certain pleasures)
$B' \cap E'$	10^5

	God Exists	God Does Not Exists
Believer	10^5	-10^4
Non believer	-10^{10}	10^5

So the expected payoffs are $10^5 p - 10^4(1-p)$ for a believer and $-10^{10}p + 10^5(1-p)$ for a nonbeliever. These two expected payoffs are equal if $p = 1/90,911 = 0.000109$. Thus the expected payoff is higher for believers if $p > 0.000109$. In fact, Pascal's original argument uses $-\infty$ rather than -10^{10} and concludes that believing in God results in better payoff for any value of p, no matter how small.

PRACTICE PROBLEMS

1. You are a good basketball player, but your performance in a competitive game depends very much on your mood. Suppose that you tend to be in a good mood 4 days out of 6. When you are in a good mood, there is a 0.9 probability that you will make 100% of your free-throws and only a 0.1 probability that you will make 80% of them. When you are in a bad mood, there is a 0.2 probability that you will make 100%, a 0.3 probability of making 80%, and a 0.5 probability of making only 50% of your free-throws.

    Your coach offers you a choice. Either you play 30 min in two of the future games or else you play 6 min in each of the next 10 games (one each week). Your goal is to establish a good statistics regarding your game. Analyze this proposition and explain your choice.

2. This problem is related to the draft lottery. Many major professional sports leagues have a system by which individual teams are awarded exclusive negotiating rights to players entering the league. This is typically accomplished through a "draft." The teams alternately choose prospective players, and when a player has been chosen, or drafted by a certain team, no other team can offer that player a contract unless the drafting team waives its rights to the player. This system gives a great advantage to the teams who get to choose early, so most leagues order the draft in the reverse order of the teams' placement in the previous year's competition. Thus, the weakest teams have an opportunity to add the best available players to their rosters. For the National Basketball Association (NBA), the system adopted in 1990 ordered the 11 nonplayoff teams by using 66 ping-pong balls placed in a hopper. The team with the worst record in the league had 11 balls with its logo painted on them, the second worst team had 10 balls, and so on, with the nonplayoff team having the best record getting only one ball. A drawing was held to determine the

first three draft picks. The first draft pick was awarded to the team whose logo was on the first ball drawn. More balls were drawn until a new logo appeared, and the second pick was awarded to that team. Still more balls were drawn to determine which of the remaining nine teams would receive the third pick. At that point the lottery ended, and the remaining nonplayoff teams were assigned draft positions in the reverse order of their records. Analyze this system using the ideas of probability theory.

2.7.1 Lifetimes and Life Expectancies

There are many insurance and investment products in which the obligations of the insurer are contingent on the death (unable to play) or continued survival of a given person (player). For example, life insurance is a financial security product in which an insurance company agrees to pay a predetermined amount of money on the death of a particular individual in exchange for a regular stream of payments while the insured person is alive. Annuities, on the other hand, are financial security products in which an insurance company agrees to make regular payments to a particular individual while that individual is alive in exchange for a fixed amount of money at the commencement of the annuity contract. In sports we can replace these with the length of time a professional athlete can play. Here, being able to play is like being alive and not able to play is like being dead.

The future lifetime of any given individual is uncertain. However, the survival pattern for large groups of individuals is generally quite predictable. Consequently, the insurance mechanism is usually feasible for large groups of similarly situated individuals. Nevertheless, to determine the appropriate rates to charge for insurance protection, the insurer needs to have a pretty good idea what the survival pattern for any given group will be. How would the insurer do this? Let's summarize some of the important probability concepts needed for this.

The probability distribution for a random variable can be described by a cumulative distribution function or a survival function. For any random variable X, the cumulative distribution function or simply the distribution function of X is the function F that assigns to each value x the probability accumulated up to x,

$$F(x) = P(X \leq x) \quad \text{for all } x.$$

For a random variable X, the survival function of X is the function S that assigns to each value x the probability that $X > x$,

$$S(x) = P(X > x) \quad \text{for all } x.$$

Note that $F(x) + S(x) = 1$. Also a continuous random variable is a random variable X with the property that $P(X = x) = 0$ for all x. Equivalently, it is a random variable for which the distribution function F is continuous.

Life Expectancy (LE)

For discrete random variables, the concept of expected value is useful when analyzing payoffs. The following represents the application of the same concept for continuous random variables T. When T represents a future lifetime, the expected value of T is usually referred to as the life expectancy. Intuitively, the life expectancy (at birth) is the average of life for subjects in the given group. Alternately, it is the number of years a person of a given age is expect to live on average.

Chapter 2: Elements of Probability

For example, in the year 2015, the estimated life expectancy for Americans was about 79.3 years, 76.9 for males, 81.6 for females. In 2016, the highest life expectancies in the world were in Japan (83.7 years) and Switzerland (83.4 years). Among the lowest were in Sierra Leone (50.1 years) and Angola (52.4 years). According to a recent study "Health, United States 2017" (https://www.cdc.gov/nchs/data/hus/hus17.pdf), life expectancy in United States is at an all-time high. Racial and ethnic differences in life expectancy persisted during 2006 to 2016, but continued to narrow. For both sexes in 2016, the difference between the group with the highest (Hispanic) and lowest (non-Hispanic black) life expectancy was 7.0 years, compared to a 7.2-year advantage in 2006. Between 2006 and 2016, life expectancy at birth in the United States increased from 77.8 to 78.6 years for both sexes, but was higher for females than for males. We note that the age of expected death actually increase with age. For example, based on 1995 data in United States, the life expectancy at age 20 was about 57 years. Whereas at 60, it was about 21 years (77 compared to 81). The death rate (deaths per 1,000 people) at age 20 was 2 and at age 60 was 14.

The average length of life for members in a group under study is the total number of years lived by all members of the group, divided by the number of members initially in the group. For example, if a group consists of four members who die at exact ages 4, 6, 7, and 8, then the average length of life for members in this group is

$$(4+6+7+8)/4 = 6.25.$$

This suggests that the life expectancy when the distribution of T is given by a continuous survival function S is

$$E(T) = \int_0^\infty S(t)dt.$$

Using integration by parts, we can write this in a form that more closely resembles the definition of expectation given earlier. To see this, let

$$du = dt, \quad v = S(t),$$
$$u = t, \quad dv = S'(t)dt.$$

Then,

$$E(T) = \int_0^\infty 1 \cdot S(t)dt = tS(t)|_0^\infty - \int_0^\infty tS'(t)dt$$
$$= 0 + \int_0^\infty tf(t)dt.$$

The final equality follows from the fact that $S'(t) = \frac{d}{dt}(1 - F(t)) = -F'(t) = -f(t)$ and the assumption that $S(t) = 0$ for all t greater than some specific t^* (i.e., no one lives longer than t^*).

EXAMPLE Suppose that T has an exponential distribution with parameter $1/50$. Then,

$$S(t) = e^{-t/50}, \quad t \geq 0,$$

and

$$E(T) = \int_0^\infty e^{-t/50}dt = -50e^{-t/50}|_0^\infty = 50$$

as expected.

In an interesting research published in The Physician and Sports Medicine 28, No. 5, May 2000, Richard Cohn and his coworkers studied the life expectancy of major league baseball umpires to find answer to questions regarding the mortality risks of this profession. The objective was to determine whether the life expectancy of MLB umpires differs from that of general public. The conclusion was MLB umpiring is not associated with a shortened life expectancy. While this is most likely attributed to the profession having no inherent risk, it could also be explained by inherent risks being overcome by yet unidentified, unique factors.

2.8 SIMPSON'S PARADOX AND HOT HAND IN SPORT

2.8.1 Introduction and Examples

Many misleading uses of statistical methods can be found in newspapers, magazines, or even journal articles summarizing experimental research. One source of this problem is poor statistical reasoning. However, it is also possible for a perfectly correct statistical reasoning to give misleading or puzzling results. It is often difficult to decide which of several ways of analyzing data are most appropriate. The fact is, there is more to the wise use of statistics than a knowledge of classical statistical techniques. The aim of this section is to discuss some examples in which different analysis lead to apparently contradictory results and to resolve those contradictions, at least partially. Specifically, we discuss examples of Simpson's paradox, which can arise from treating nonrandom samples as random or from misinterpreting probabilities even though they have been computed correctly. Then use that to explain a phenomenon known as "Hot Hand," in sports such as basketball.

Simpson's paradox is named after Edward Hugh Simpson, a statistician who gave a careful discussion of it. The best way to gain an appreciation of the surprising results associated with Simpson's paradox is to see concrete examples. One example described by Simpson involves a rookie who is trying to break into a Major League Baseball lineup by replacing a veteran. For our first example, we also present such a scenario.

EXAMPLE 2.8.1 A rookie is told by the manager that starters are chosen on the basis of hitting ability. Since the rookie is batting 0.380 and the veteran is batting 0.320, the rookie is excited in anticipation of the first game. However, the rookie is dismayed to learn that the veteran is designated to start the season opener. When he questions the manager about this, he is told that the opponents are using a right-handed pitcher for the first game. Since the veteran is batting 0.300 against right-handers and the rookie is only batting 0.200 against right-handers, the rookie accepts his spot on the bench for game one. He knows that the opponents have scheduled a left-handed pitcher for the second game, and he bats 0.400 against left- handers. When to his horror the rookie learns that the veteran is also slated to start the second game, he immediately confronts the manager. Again, the manager defends his choice and explains that the veteran is batting 0.500 against left-handers and is hence a better choice than the rookie. The rookie is left to sit on the bench in disbelief.

Has the manager been deceiving the rookie? Can the numbers be correct? Is it really possible for the veteran to have a better batting average against right-handers and left-handers and yet have a worse batting average overall? After all, right-handed pitchers and left-handed pitchers are

Chapter 2: Elements of Probability

TABLE 2.8.1 Baseball Batting Averages

	Batting Average		
Player	Vs Left-Handed	Vs Right-Handed	Overall
Veteran	$50/100 = 0.50$	$270/900 = 0.30$	$320/1{,}000 = 0.32$
Rookie	$36/90 = 0.40$	$2/10 = 0.20$	$38/100 = 0.38$

the only kinds of pitchers that there are. The fact that this apparent contradiction is possible is the reason that we have a paradox on our hands. The data in Table 2.8.1 certainly confirms that the batting averages are correct.

Careful analysis of Table 2.8.1 reveals the source of the paradox. The basic idea is that most of the veteran's at-bats came against right-handers while most of the rookie's at-bats are against left-handers. Since the rookie is better against left-handers than the veteran is against right-handers, it is possible for the rookie to have a better overall average. A more careful explanation is possible once we discuss the concept of a weighted average. However, we first present a few more examples of Simpson's paradox.

EXAMPLE 2.8.2 The batting averages of the veteran and the rookie in Table 2.8.1 are not that far apart. Will that always be the case in examples of Simpson's paradox? Is it possible to have examples with extreme differences in the overall batting average? The data in Table 2.8.2 shows that even the greatest extremes are possible.

The numbers in Table 2.8.1 might have suggested that the manager's means of deciding his starters is a good one. However, the example in Table 2.8.2 certainly warrants a decision based on overall batting average and not batting averages against different kinds of pitchers. In general, neither method is foolproof. This is one reason why they do not simply hire statisticians as baseball managers.

EXAMPLE 2.8.3 Two treatments A and B are being considered for a sport injury. A doctor tells his patient that the success rate of treatment A is 36% and the success rate of treatment B is 45%. Which should the patient choose? What if the patient was given the results in Table 2.8.3? There the study is split according to gender, and treatment A has greater success for both men and women.

Which treatment is better?

TABLE 2.8.2 Extreme Baseball Batting Averages

	Batting Average		
Player	Vs Left-Handed	Vs Right-Handed	Overall
Veteran	$1/1 = 1.000$	$1/1{,}999 \cong 0.001$	$2/2{,}000 = 0.001$
Rookie	$1{,}998/1{,}999 \cong 0.999$	$0/1 = 0$	$1{,}998/2{,}000 = 0.999$

TABLE 2.8.3 A Comparison of Two Treatments

	Proportion Recovering		
Treatment	Among Females	Among Males	Overall
A	$40/50 = 0.80$	$50/200 = 0.25$	$90/250 = 0.36$
B	$70/100 = 0.70$	$20/100 = 0.20$	$90/200 = 0.45$

The round numbers in Examples 2.8.1, 2.8.2, and 2.8.3 make it pretty obvious that they are fictional. To eliminate any thoughts that Simpson's paradox never actually happens our next two examples are real.

EXAMPLE 2.8.4 Is smoking good for women? In 1972 to 1974 a one-in-six survey was carried out in Whickham, United Kingdom (Tunbridge et al. 1977). For the sake of this example, we are only concerned with the fact that 1,314 women were asked their age and whether or not they smoked. Twenty years later, a follow-up study was conducted (Vanderpump et al. 1995). Our interest here is in which of the women from the original study were still living. The results of the study are presented in Table 2.8.4. Notice that the women are categorized according to their age at the time of the first survey.

Although the overall survival rate was greater among smokers, the opposite is true in the individual age groups. Which part of the results do you suppose the tobacco industry would report if they did the study? Does the study tell us that smoking is not bad (or even good) for women?

EXAMPLE 2.8.5 The death rates in 1910 from tuberculosis in New York and Richmond have been separated according to race (Cohen and Nagel, 1934). Table 2.8.5 shows that the death rates for both Whites and non-Whites were greater in New York. However, the overall death rate was greater in Richmond.

Public health officials might have reached very different conclusions depending on which part of the table received their attention.

TABLE 2.8.4 Smoking and 20-year Survival Rates

	20-year Survival Rate (From Age at Start of Study)						
Habit	<35	35–44	45–54	55–64	65–74	>75	overall
Nonsmoker	$\frac{207}{213} = 0.972$	$\frac{114}{121} = 0.942$	$\frac{66}{78} = 0.846$	$\frac{81}{121} = 0.669$	$\frac{28}{129} = 0.217$	$\frac{0}{64} = 0$	$\frac{502}{732} = 0.686$
Smoker	$\frac{169}{174} = 0.971$	$\frac{95}{109} = 0.872$	$\frac{103}{130} = 0.792$	$\frac{64}{115} = 0.557$	$\frac{7}{36} = 0.194$	$\frac{0}{13} = 0$	$\frac{443}{582} = 0.761$

TABLE 2.8.5 Death Rates from Tuberculosis in 1910

	Death Rates From Tuberculosis		
City	Among Non-Whites	Among Whites	Overall
New York	$513/91,709 = 0.00560$	$8,365/4,675,174 = 0.00179$	$8,878/4,766,883 = 0.00186$
Richmond	$155/46,733 = 0.00332$	$131/8,0895 = 0.00162$	$286/127,623 = 0.00224$

2.8.2 Weighted Averages

In each of Tables 2.8.1 through 2.8.5, the last column seems to contradict the previous columns. The key to understanding how this can occur is to realize that, in each row, the last entry is a weighted average of the previous entries. Everyone knows how to take a simple average of two values x and y. It is given by $\frac{x+y}{2} = \frac{1}{2}x + \frac{1}{2}y$. The right-hand side of this equation emphasizes the fact that the average gets an equal contribution from each of x and y. More generally, given two nonnegative numbers α and β such that $\alpha + \beta = 1$, there is an associated weighted average of x and y given by $\alpha x + \beta y$. If $\alpha > \beta$, then x carries more weight in the average. The simple average is just the special case in which the weights are equal since $\alpha = \beta = 1/2$. Of course, weighted averages can also be computed for any number of values (not just two) provided the appropriate number of coefficients (α, β, γ, etc.), where sum to 1 is given.

The formal concept of a weighted average may seem strange at first, but it is definitely not unfamiliar. Consider what happens if someone invests in the stock market. Suppose, for example, that investments are made in three companies and Table 2.8.6 shows their rate of growth after 1 year.

The profit made on the investment is a weighted average based on the relative amounts invested in each company. If $1,000 is invested by putting $500 into company A, $200 into company B, and $300 into company C, then the percent increase obtained from the investment is

$$\frac{500}{1000}(0.20) + \frac{200}{1000}(0) + \frac{300}{1000}(-0.10) = 0.07.$$

That is, a 7% profit or $70 is made on the investment of $1000. Another familiar example of a weighted average is the computation of student's grades in a class. For example, if a syllabus states that tests comprise 50% of the grade, homework 20%, and the final exam 30%, then the overall grade is a weighted average.

With the concept of a weighted average in hand, the numbers in Tables 2.8.1 through 2.8.5 can be better understood. For example, the batting averages in the last column of Table 2.8.1

$$\frac{100}{1000}(0.500) + \frac{900}{1000}(0.300) = 0.320$$

$$\frac{90}{100}(0.400) + \frac{10}{100}(0.200) = 0.380$$

are weighted averages. The key to understanding how Simpson's paradox occurs is realizing that the weights are different for the veteran V and the rookie R. For V, the weights are $\alpha_V = 100/1000 = 0.1$ and $\beta_V = 900/1000 = 0.9$. For R, the weights are $\alpha_R = 90/100 = 0.9$ and

TABLE 2.8.6 Stock Growth

Company	% Increase
A	20
B	0
C	−10

$\beta_R = 10/100 = 0.1$. The Veteran's overall average gets most of its weight from at-bats against right-handers, while the Rookie's average gets most of its weight from at-bats against left-handers. Since the weights are completely different, comparing their overall averages is like comparing apples and oranges. It does seem reasonable to compare the Veteran and the Rookie versus left-handed pitchers since they have faced similar numbers. However, the Rookie's ability against right-handers is certainly not yet clear. The manager would probably want to see several more at-bats before making a long-term decision. The point is that statistical analysis requires more than merely looking at the data. The underlying application is relevant to deciding how to interpret the results.

In Table 2.8.2, it seems that the last column is the most important. Whereas, in Table 2.8.4, the columns representing each age group give the important results. One reason the overall survival rate in the last column of Table 2.8.4 may seem counterintuitive can be seen by thinking about the weighted average computation. Few of the older women in the study were smokers, but many of them had died by the time of the follow-up. Perhaps, there were not too many older smokers around to participate in the original study because smokers die at an earlier age. However, this possibility is not accounted for in the last column containing overall results.

Careful inspection of Tables 2.8.1 through 2.8.5 shows the source of the paradox. In each case, the weights used for the weighted average in the first row are very different from those used in the second row. Hence, the summary statistics in the final column containing overall averages seems counterintuitive. Statistics don't lie, but they can mislead. It pays to have some understanding of statistical analysis and to be able to scrutinize reported results. Another very readable explanation of Simpson's paradox and a real-world example can be found in Westbrook 1998.

2.8.3 Application to "Hot Hand" in Sport

"Hot Hand" in basketball is a phenomenon that most basketball fans accept and consider well understood. It is a topic that has generated a great deal of interest among researchers in areas such as psychology, statistics, and physical education. For a player who is hot, basket appears as if it is so wide that he or she will make shot after shot with ease. The opposite situation may be called "Cold Hand." It is interesting to note that such concepts may not be very meaningful in other sports. For example, in soccer (football) any player who scores only one goal may be considered hot or having a hot foot!

Although hot hand is meaningful in basketball it is demonstrated in some publications that statistically such phenomenon may not even exist. Amos Tversky, a psychologist who studied every basket made by the Philadelphia 76ers for more than a season has concluded the following:

(1) The probability of making a second basket (hit) did not rise following a successful shot (success did not breed success).
(2) The number of "runs" or baskets in succession, was no greater than what a standard random (Bernoulli trials) or coin-tossing model would predict.

To clarify item (2) consider a good player who makes half of his shots. For such a player, four hits in a row is expected to occur once in 16 sequences of four shots, noting that this is one of the 16 equally likely outcomes. The chance that this player hits at least three shots in a row is 3/16 assuming that trials are independent. Does this mean that the player has a hot hand? To add to this suppose that I flip a coin four times and by luck observe four heads in a row. Could I claim that I have a hot hand, say, for heads?

As is explained in Gould "The Streak of Streaks" appeared in Chance 1989, a great player like Larry Bird, will have more sequences of five hits than an average player—but not because he has greater will or gets in that magic rhythm more often. He has longer runs because his average success rate is so much higher and has a much better chance of having more frequent and longer sequences. Suppose that using one of the many available computer softwares we simulate Larry's game. If Larry shoots field goals at 0.6 probability of success, he will get five in a row about once in every 13 sequences since the chance of this event is $(0.6)^5 \cong 1/13$. If another player, by contrast, hits only 0.3 of his shots, he will get his five straight hits only about once in 412 times. In other words, we need no special explanation for the apparent pattern of long runs, except perhaps when the performance is far beyond what is expected from a player based on his past performance.

Similar studies in baseball indicate that nothing ever happened in baseball above and beyond the frequency predicted by coin-tossing models. The longest run of wins or losses are as long as they should be, and occur about as often as they ought to. Again, this may also be demonstrated using computer simulation. However, this rule has one exception that should never have occurred; DiMaggio's 56-game hitting streak in 1941. Purcell calculated that to make it likely (probability greater than 50 percent) that a run of even 50 games will occur once in the history of baseball up to now (and 56 is a lot more than 50 in this kind of league), baseball's rosters would have to include either four lifetime 0.400 batters or 52 lifetime 0.350 batters over careers of 1,000 games. In actuality, only three men have lifetime batting averages in excess of 0.350 and no one is anywhere near 0.400. He then concluded that DiMaggio's streak is the most extraordinary thing that ever happened in American sports.

Having stated all of the above, the question that still remains is that, how people define "Hot Hand," and whether this definition is based on a pattern or reality or is just a perception. Study of Larkey et al. (1989) reveals that different measures for Hot Hand leads to consider different players hot or winner. It is interesting to note that, although the precise meaning of a term like "Hot Hand" is unclear, its common use implies a shooting record that departs from coin tossing or Bernoulli trials with probability of success greater than that expected. An equally interesting point relates to the fact that fans who talk about hot hand usually refer to pattern of streak shooting, or something that is noticeable, memorable, and unlikely. Examples are observations such as HHHHHM or MHHHHH. But then, why an outcome like HMHMHM, which presents a sequence of hit followed by miss and vice versa is not considered a pattern and why is not given a name? As is pointed out in the Skeptic's Disctionary, the clustering illusion is the intuition that random events that occur in clusters are not really random events. The illusion is due to selective thinking based on a false assumption.

Here is another example that supports the fact people only recognize certain patterns and ignore many others. Let us replace hit by 1 and miss by 0. Then we have a sequence of 1s and 0s (a number in binary system). Examples are

$$0\ 0\ 0\ 0\ 0\ 0\ 1\ 1\ 1\ 1\ 1,$$
$$1\ 1\ 1\ 1\ 0\ 0\ 1\ 1\ 0\ 1\ 0.$$

The first one may be recognized as having a pattern. But what about the second one? In fact, this is a famous pattern. It is the binary presentation of the number 1946, the author's birth year. So to me this is a recognizable pattern, but to the others may not be (in fact it is not). A similar example happens in poker. Getting a royal flush is surprising even though the chance of any hand in poker, with no particular name, is extremely low.

It is also possible to argue in favor of hot hand. In an interesting paper, Wardrop (1998) presents many discussions concerning an inherent weakness in the methods used by Gilovich et al. (1985) and Tversky and Gilovich (1989a). Hooke (1989) discusses the inherent difficulty of using statistical methods to study complete phenomena such as a game of basketball. Hale (1999) has discussed this issue and has raised several questions. He has argued that hot hand is an internal phenomenon and that the sense of being "hot" does not predict hits or misses. When a player realizes he is hot, he tends to push the envelope and attempt more difficult shots, which then leads to predictably a failure. Or he raises the question that, precisely how unlikely does a streak of success need to be before we are prepared to count it as a legitimate instance of hot hand?

According to Hale (1999), there are three prominent arguments that conclude there are no hot hands in sports. The first argument of the hot hands critics creates a tradition in the very act of destroying it. By making, "success breeds success" a necessary condition of having hot hands, the critics have established a previously undefended and barely articulated account of hot hands, only to demolish it. Instead, he has argued that there are good reasons to reject "success breeds success" as a requirement for having hot hands. While it is true that many players believe that future success is more likely when they are already hot, either this is only a belief that their current state has causal efficacy into the future, or is it inductive reasoning that their current high rate of success is evidence of future success? Yet neither disjunction makes "success breeds success" part of the concept of having hot hands.

The next two arguments offered by the critics of the hot hand theory are of a well-known skeptical pattern: Set the standards for knowledge of something extremely high, then show that no one meets those standards. The canonical reply to this strategy, of which he availed himself, is to reject those standards in favor of more modest ones that charitably preserve our claims of knowledge. The skeptical insistence upon exceedingly rare streaks or statistically remote numbers of streaks as being the only legitimate instances of hot hands is arbitrary and severe. He has then argued that "being hot" denotes a continuum, one that is nothing other than deviation from the mean itself. This obviously comes in degrees.

Let us now present an argument based on annual performance of professional basketball players. Table 2.8.7 presents Chris Mullin's free throw percentages for years 1988 to 1998. Let us assume that Mullin's free throw shooting is simply a series of coin tossing or Bernoulli trials. We estimate the fixed probability for the trials by his overall percentage $\hat{p} = 0.851$ over the 10-year period. Then the expected value for his shooting percentage after any number of trials is 0.851. Noting that

$$\sqrt{\hat{p}(1-\hat{p})} = \sqrt{(0.851)(0.149)} = 0.361,$$

the standard deviation for such n trials is $0.361/\sqrt{n}$. The last column of the Table 2.8.7 presents the z-score, that is, the number of standard deviations from the mean for each year. The large numbers, 3.13 in 97–98 and -4.07 in 93–94 suggest that the variation in Mullin's performance is not simply due to randomness. In other words, these values would be highly unlikely to result from Bernoulli

TABLE 2.8.7 Chris Mullin's Free Throw Percentage 1988–1998

Year	Made/Attempted x/n	Free Throw % $\tilde{p}$	Standard Deviation $\sqrt{0.361/n}$	z-score $(\tilde{p} - 0.851)/\sqrt{0.361/n}$
97–98	154/164	0.939	0.0281	3.13
96–97	184/213	0.864	0.0247	0.526
95–96	137/160	0.856	0.0285	0.17
94–95	184/213	0.864	0.0247	0.526
93–94	165/219	0.753	0.0244	-4.017
92–93	183/226	0.809	0.0240	-1.75
91–92	350/420	0.833	0.0176	-1.02
90–91	513/580	0.884	0.0150	2.20
89–90	505/568	0.889	0.0151	2.50
88–89	493/553	0.892	0.0154	2.67
Total	2,823/3,316	0.851		

trials. The probability that Mullin's percentage would be as low as 0.753 as it was in 93–94 is less than 0.00003 if the Bernoulli trial model were correct. The results suggest that Mullin had a "hot" year in 97–98 and a "cold" year in 93–94 for reasons other than random variation. Note that a similar argument could be made for a performance, over a shorter time span. In fact, as noted earlier, Wardrop (1998) has presented such arguments.

To summarize the arguments against the hot hand, Tversky and Gilovich (1989) used several data sets to conclude that existing data does not support "Hot Hand." They have devised a clever experiment to obtain convincing evidence that knowledgeable basketball fans are much too ready to detect occurrences of streak shooting and the "Hot Hand" in sequences that are, in fact, the outcome of Bernoulli trials. To clarify this argument further they have considered the data concerning the free throws for nine regular players of Boston Celtics from 1980 to 1982. Then they asked the following question: when shooting free throws, does a player have a better chance of making the second shot after making the first shot than after missing the first shot? To answer this they have chosen a sample of 100 Cornell and Stanford students randomly. The responses were 68% yes implying hot hand and 32% no implying independence or negative association.

After analyzing the data, Tversky and Gilovich (1989) concluded that the data provided no evidence that the outcome of the second shot depends on the outcome of the first one. Adams (1992) using data on 83 players showed that the mean interval from making a field goal ($n = 372$) to making a field goal in 19 NBA games did not differ from the mean interval from making to missing ($n = 394$), which further challenges assumptions regarding hot hand.

In a more recent study Koehler and Conley, "The Hot Hand Myth in Professional Basketball" appeared in the *Journal of Sport and Exercise Psychology*, have offered further evidence against hot hand in a unique setting, the NBA Long Distance Shootout contest. They have concluded that declarations of hotness in basketball are best viewed as historical commentary rather than as a prophecy about future performance.

Having discussed opposing views, the final and perhaps more important question is, why arguments for and against hot hand seem convincing. In an attempt to answer this question, Wardrop (1995) has performed a very interesting analysis of data for Boston Celtics players. His analysis is based on, as he stated, the fact that the data available to lay persons may be very different from the data available to professional researchers. In addition, laypersons unfamiliar with a counterintuitive result, such as Simpson's Paradox, may give the wrong interpretation to the pattern in their data and their analysis. There are many problems of this type in probability theory where the right answer is counter intuitive.

Here we will borrow the data and its analysis presented in Wardrop for demonstration. To understand Wardrop's analysis consider the following data presented in Wardrop (1995).

Let us use the notations

$$\hat{p}_{hit} = \text{The proportion of first shot hits that are followed by a hit,}$$
$$\hat{p}_{miss} = \text{The proportion of first shot misses that are followed by a hit.}$$

Then we have

$$\hat{p}_{hit} = 251/285 = 0.881 \text{ and } \hat{p}_{miss} = 48/53 = 0.996 \text{ for Larry Bird,}$$

$$\hat{p}_{hit} = 54/91 = 0.593 \text{ and } \hat{p}_{miss} = 49/80 = 0.612 \text{ for Rick Robey,}$$

but for collapsed data we have

$$\hat{p}_{hit} = 305/376 = 0.811 \ (= (285/376)(251/285) + (91/376)(54/91)),$$

and

$$\hat{p}_{miss} = 97/133 = 0.729.$$

Note that, contrary to the hot hand theory in the sense of success breeds success, each player's shot is slightly better after a miss than after a hit, although as is shown by Wardrop, the differences are not statistically significant.

It is possible, of course, to ignore the identity of the player attempting the shots and examine the collapsed data in Table 2.8.8. For example, on 509 occasions either Bird or Robey attempted two free throws, on 305 of those occasions both shots were hit, and so on. For the collapsed data, $\hat{p}_{hit} = 0.811$ and $\hat{p}_{miss} = 0.729$. These values support the hot hand theory, that is, a hit was much more likely than a miss to be followed by a hit.

The data from Bird and Robey illustrate Simpson's paradox namely, $\hat{p}_{hit} < \hat{p}_{miss}$ in each component table, but $\hat{p}_{hit} > \hat{p}_{miss}$ in the collapsed table. It is easy to verify algebraically that the proportion of successes for a collapsed table proportions equals the weighted average of individual player's proportions, with weights equal to the proportion of data in the collapsed table that comes from the player. For the after-a-hit condition, for example, the weight for Bird is $285/376 = 0.758$, the weight for Robey is $91/376 = 0.242$, and the proportion of successes for the collapsed table, $305/376 = 0.811$, is

$$\frac{285}{376} \cdot \frac{251}{285} + \frac{91}{376} \cdot \frac{54}{91}.$$

As a result, even though both Bird and Robey shot better after a miss than after a hit, the collapsed values show the reverse pattern due to the huge variation in weights associated with

TABLE 2.8.8 Observed Frequencies for Pairs of Free Throws by Larry Bird and Rick Robey, and the Collapsed Table (Wardrop 1995, Table 1)

Larry Bird			
	Second		
First	Hit	Miss	Total
Hit	251	34	285
Miss	48	5	53
Total	299	39	338
Rick Robey			
	Second		
First	Hit	Miss	Total
Hit	54	37	91
Miss	49	31	80
Total	103	68	171
Collapsed Table			
	Second		
First	Hit	Miss	Total
Hit	305	71	376
Miss	97	36	133
Total	402	107	509

each player. In short, Wardrop has concluded that Simpson's paradox has occurred because the after-a-miss condition, when compared to the after-a-hit condition, has a disproportionately large share of its data originating from the far inferior shooter Robey.

2.9 COINCIDENCES

In this section, some interesting issues involving probability are discussed. One surprising fact about the probability is that, it is possible to construct many counter-intuitive examples of which some are the results of using inappropriate logic or reasoning, and others stem from misinterpretation or confusion regarding different concepts. Other reasons include looking for pattern that seem stunning and yet do not even exist. Examples of this type includes "Hot Hand" discussed in Section 2.8 and the well-known Abraham Lincoln/John F. Kennedy coincidences. According to Diaconis and Mosteller, 1989, *Journal of the American Statistical Association*, Vol. 84, a coincidence is a surprising concurrence of events, perceived as meaningfully related, with no apparent causal connection. Think about September 11 (9/11) for instance and consider the following patterns. 911 is the number used for emergency cases and $9 + 1 + 1 = 11$. The twin towers looked like the number 11. So perhaps everything that happened had something to do with the number 11. The first flight to hit the twin towers was flight 11, and there were 92 people on board $9 + 2 = 11$. Next, September 11 is the 254th day of the year $2 + 5 + 4 = 11$ (also $365 - 254 = 111$). There are 11 letters each in "New York City," "Afghanistan," "the Pentagon," and "George W. Bush." New York was the 11th state admitted to the union, 119 is the area code to both Iraq and Iran and flight 77 that crashed in Pennsylvania had 65 people on board $6 + 5 = 11$. Do these patterns mean anything? Here is another surprising fact. Recall the March 11, 2004, attack in Spain. There are exactly 911 days between this date and the September 11, 2001.

To provide some insight, let us start with a major question. Do you think coincidences are as surprising as some people would think? Have you ever experienced something that you consider extraordinary or even remarkable, something very unlikely (1 in 100,000 or 1 in 1,000,000)? Could you name one? Remember that people who win the lottery often attribute their winning to good luck or to some special system they used in choosing a lottery number. Yet someone was bound to win, and it is merely luck that a particular person held the wining entry. In fact, the laws of probability dictate that many coincidences are bound to happen, even though the particular form of coincidence is unpredictable. The following is considered a coincidence by many people. What do you think?

EXAMPLE 2.9.1 Presidential Coincidence One strange historical coincidence concerns presidents of the United States. Three of the first five presidents died on the same day of the year. And the date? It was none other than the July 4. Of all the dates to die on, that must surely be the most significant to any American. This might of course be part of the explanation as to why the coincidence happened.

You can imagine the early presidents being really keen to hang on until the anniversary of Independence, a date which meant so much to them, and giving up the ghost as soon as they knew they had reached it. This is apparently what happened to Thomas Jefferson, the third president. John Adams, the second president, actually died a few hours after Jefferson.

EXAMPLE 2.9.2 Consider professional American football hall of famers. How likely is to have two of them being from the same high school? Well here are some, Al Davis and Sid Luckman, George Halas and Leo Nomellin, Mel Renfiro and Arnie Weinmeister, Elroy Hirsch and Jim Otto, High McElhenny and Bill Walsh, and Bobby Layne and Doak Walker.

EXAMPLE 2.9.3 Birthdays We have already talked about this problem in the text and also in problem sets. Suppose that you have the same birthday with someone in your class or on your team. Would you consider such an event a coincidence? Is probability of such an occurrence very small?

There are 365 days in the year, so you need 366 people in your class to guarantee a birthday match. But is it true to say that you need about 180 people in your class to have a 50-50 chance of a birthday match? Surprisingly, you only need 23 classmates or teammates to have more than a 50% chance of having at least one birthday match. Suppose that there are 7,300 students in the university you attend. Pick 365 students. Suppose that none of them have the same birthday (the

Chapter 2: Elements of Probability

chance of this is extremely small). This is possible only if one was born on January 1, one on January 2.....and the last one on December 30. Then the next student you pick (number 366) must have the same birthday as one of the first 365. So if all days are equally likely we should have on average $\frac{7300}{365} = 20$ students with the same birthday everyday. That is 20 matches. Now, a semester is more than 100 days. Thus, there is a good chance that two students with the same birthday will fall in the same class.

Let us now discuss why we need only 23 students to have more than a 50% chance of at least one match. Suppose that there are 23 students in your class. The probability of no match (different birthdays) is then

$$\frac{364}{365} \times \frac{363}{365} \times \frac{361}{365} \times \cdots \times \frac{343}{365} = 0.49 = 49\%.$$

To see why, take one student. Let's say his or her birthday is on March 10. The probability that the next student was not born on March 10 is 364/365. Let say the second student's birthday is on July 2. The probability that the third student was not born on March 10 or July 7 is 363/365, and so on. Thus, we have

Probability of a Match $= 1 - 0.49 = .51 = 51\%$ (at least one match).

Following the same lines the probability for n individuals to have different birthdays is

$$(364/365)\,(363/365) \cdots ((366-n)/365)$$

and the probability of at least one match is one minus this probability. The following table gives probability of at least one match for different values of n.

n	10	15	20	23	30	35	40	50
Probability	0.117	0.253	0.411	0.507	0.706	0.814	0.891	0.970

From this table, the number of people needed to have a 50% chance of at least one birthday match is 23. Note that this is because the problem states that any two people out of 23 can share birthdays. In fact, it can be shown that one needs 253 people to have an even chance of finding someone to have his or her exact birthday.

Using the same approach it can be shown that, the number of people needed to have a 50% chance of three people having the same birthday is 88. For four, five, six people having the same birthday the number of people necessary is respectively 187, 313, and 460.

The number of people needed so that there is a 50% chance that two people will have a birthday within one day of each other is 14, and if you are looking for birthdays a week apart, the magic number is seven.

To see this, suppose there are n people in the group, and that A, represents the event "at least two people have birthdays within at most one day of each other." Again we first find $P(A')$, that is, the probability of the complementary event. If a person's birthday is on August 2, for example, then the second person's birthday must not fall on August 1, 2, or 3, giving 362 choices for the second person's birthday. For the third person, however, there are either 359 or 360 choices, depending on whether the second person's birthday is on August 4 or July 31 or some other day that has not been previously excluded from the possibilities. Using these approximate answers are

$$P(A') = \frac{365 \cdot 362 \cdot 359 \cdots (368-3n)}{365 \cdot 365 \cdots 365},$$

n	5	10	12	13	14	15	16
Probability	0.080	0.316	0.429	0.485	0.540	0.593	0.643

Looking at these numbers, one can see that unlike general perception these events are not really unlikely.

Let us now discuss the following question: Why do some people think that such events are unlikely and sometimes refer to them as coincidence? Why do people see patterns that do not even exist?

To examine this, the first thing to do is to distinguish between the chance of a particular unlikely event occurring and the chance of any unlikely event occurring. These two are usually confused. For example, what is your chance of winning a lottery? Very small. But what is the probability that someone wins the lottery? Reasonably large. In other words, although a particular event may be hardly likely, some similar events may be highly likely or even certain to happen.

To clarify the birthday problem, if you pick two specific students, say Lisa and John, out of 23 students in the class, the chance of them having the same birthday is 1 in 365. But the chance of having at least two students with the same birthday in the same class is 1 in 2. The two "coincidences," however, feel about the same.

In every other walk of life, there is a similarly huge difference between the chance of a specific coincidence happening such as "Next week you will score a goal playing soccer for your school" and "Next week somebody on your team will score a goal." Both will grab your attention, but the latter is far more likely. Putting this differently there are the laws of big numbers and the laws of small numbers. The world is simultaneously so large that anything could happen and so small that weird things seem to happen all the time.

There are also unlikely events that happen all the time and yet we do not classify them as coincidences. Just because something is unlikely it does not necessarily make it interesting or surprising. For example, what is the probability that your birthday is on May 13 and your sister's birthday is on November 26? Well, it is $(1/365)(1/365) = 1/130{,}000$. Who could have thought that something so unlikely as this would happen? You are probably asking yourself "so what."

Boring incidents are quickly forgotten but coincidences grab the attention and stick in the mind. Recall the presidential coincidence discussed in Example 2.9.1. What is the probability of such an event? It is easy to show that it is

$$\binom{5}{3}(1/365)^3(364/365)^2 < 1/4{,}889{,}467$$

This is so unlikely and hence may be called a coincidence. Note that this also provides an evidence in the favor of a theory that, if we look forward to something, we could postpone our deathday. In this example, presidents were looking forward to July 4th. In fact, statistics shows that more famous people die within few months after their birthdays than few months before that. The same is true about people who are close to their 100th birthday. Most of them die the year after this magic age. A good example of this is Queen Elizabeth's mother who died at the age of 101.

Finally turning to the September 11 question, the numerical message involving the number 11 is not actually a pattern that exists but merely a pattern we have found. In fact, the second flight was number 175 and like the rest of the information does not follow the 11th pattern. Also, names

Chapter 2: Elements of Probability

with the same number of letters in them will have different number of letters in language other than English.

We end this part noting that the section we are in is Section 2.9 for which $2+9=11$ and that using nine 11's we get a Palindrome number as $(111,111,111)(111,111,111) = 12,345,678,987,654,321$. This means that by looking closely we can find many patterns in any significant event in history.

2.9.1 Streaks

In many sports fans show interest in streaks and consider that a sign for "hot hand." Think about Bernoulli trials and their possible outcomes. Since outcomes like a long sequence of successes are among possible outcomes, one may wonder whether a streak is a real one. William Feller who has written few well-known books in probability theory and its application has developed a formula for streaks assuming Bernoulli trials. An approximation to Feller's approach, described by Warrack (Chance 1995, Vol. 8, No. 3) leads to formula

$$P(N,r) = [1 + (N-r)(1-p)]p^r$$

where P(N, r) presents the probability of obtaining at least one run of r successes in N trials, and p is the probability of success in each trial. Note that the expression without p^r gives the expected number of opportunities that a player will have to start a streak, once at the beginning of the season and then once after the $(N-r)(1-p)$ expected failures that leave enough games in the season for an r game streak to occur.

EXAMPLE 2.9.4 Suppose that Jeff, who plays basketball for his college, is a 50% free throw shooter. If he tries 10 free throws, what is his chance of having at least one run of 5 hits?

SOLUTION: Here $N=10, p=0.50$, and $r=5$. Thus,

$$P(10,5) = [1 + (10-5)(1-0.5)](0.5)^5 = 0.109 = 11\%.$$

Now consider an average NBA player who is an 80% free throw shooter. Using the formula, we get

$$P(10,8) = [1 + (10-8)(1-0.80)](0.8)^8 = 0.235$$

Thus, this player is expected to have at least one run of 8 hits in every four games he plays.

Warrack (1995), in an article titled "The Great Streak" published in Chance, Vol. 8, No. 3, using DiMaggio's career 0.325 batting average over $N=1,736$ games estimated p to be 0.777. Using these he has obtained

$$P(1,736,56) = 0.000274 = 1/3,700$$

as the probability of having a 56-game hitting streak. This is too small to be considered a random occurrence. That is, it looks like a real streak and confirms the greatness of the DiMaggio.

As is explained by Stern, the column editor of Chance (1997–98), DiMaggio's 56-game hitting streak during the 1941 season is considered most unusual by many observers. For that season he was successful in getting a hit in 35.7% of his attempts and averaged 3.89 attempts or at-bats per game. Following Stern, let us assume (for simplicity) that he had four at-bats per game and that each results in success with probability 0.35 independent from others. Then the probability of at least one hit during a randomly chosen game is $1 - (0.65)^4 \approx .82$ and the probability of 56 consecutive games with at least one hit is then about 0.00015 or 1/6,700. Note that this is the probability of getting a hit in each of a specified sequence of 56 games. In fact with $N = 154$, $r = 56$, and $p = 1 - (0.65)^4$, we have

$$P(154,56) = [1 + (154 - 56)(0.65)^4][1 - (0.65)^4]^1 = 0.00031 = 1/3,200$$

which is slightly different from 1/3,700 obtained earlier due to the simplifying assumption.

EXAMPLE 2.9.5 In January 7, 2003, the Los Angeles Lakers' Kobe Bryant hits a NBA single-game record nine consecutive three-points against Seattle Super Sonics. Bryant's total of 12 three-points was also a single-game record. Lakers' coach Phil Jackson was quoted as saying, "That was perhaps the greatest streak shooting I think I have every seen in my life." Find P(18,9) and discuss the result.

SOLUTION: Here is Bryant's sequence (H for hit and M for miss)

MHHHHHHHHHHMHHMMHMM.

To apply the formula for streaks we need to estimate p. There are two different estimates for p. One for the Seattle game which is 2/3(12-of-18) and one for Bryant's seasonal three-point percentage prior to Seattle game, which was 0.283. Using these estimates we obtain respectively

$$P(18,9) = [1 + (18 - 9)(1 - 0.67)](0.67)^9 = 0.108,$$

and

$$P(18,9) = [1 + (18 - 9)(1 - 0.283)](0.283)^9 = 0.0000867.$$

Note that if we looked at Bryant's three-point shooting in the Seattle game against the backdrop of his seasonal three-point percentage coming in, rather than just within the context of the Seattle game, his performance would almost certainly look impressive. But, within the Seattle game, the nine consecutive hits from three-point range were not statistically noteworthy. It is interesting to note that in a February 4, 2003, game against the Los Angeles Clippers, the New York Knicks' Latrell Sprewell went 9-for-9 on three-point shots, setting a new NBA record for most three-pointers in a game without missing any. Sprewell's season three-point hit rate prior to this game was 0.384 leading

$$P(9,9) = (0.384)^9 = 0.00018.$$

A PRACTICE PROBLEM Find statistics for Peter Rose who had a 44-game hitting streak in 1978 and calculate the corresponding probability following the above example. Note that this is the second longest streak in baseball.

An Alternative Formulation

An alternative formulation of streak problem is as follows: Consider $P(N+1,r)$ and its relation to $P(N,r)$. We have

$$P(N+1,r) = P(N,r) + (1-p)p^r(1 - P(N-r,r)).$$

Chapter 2: Elements of Probability

This is because either a run of at least r successes occurs in the first N trials or none occurs in the first N trials but one occurs involving the $(N+1)$st trial. Since we can not have r successes in less than r trials, it follows that:

$$P(1,r) = P(2,r) = \cdots P(r-1,r) = 0, \text{ and } P(r,r) = p^r.$$

To answer the questions in Example 2.9.4, we use the above relation to obtain:

$$P(10,5) = P(9,5) + (1-p)p^5(1-P(4.5)),$$
$$P(9,5) = P(8,5) + (1-p)p^5(1-P(3,5)),$$
$$P(8,5) = P(7,5) + (1-p)p^5(1-P(2,5)),$$
$$P(7,5) = P(6,5) + (1-p)p^5(1-P(1,5)),$$
$$P(6,5) = P(5,5) + (1-p)p^5(1-P(0,5)).$$

Adding both sides of these equalities we get

$$P(10,5) = P(5,5) + 5(1-p)p^5 - p^5 + 5(1-p)p^5 = p^5(6-5p).$$

For $p = 0.50$ this is equal to 0.109 (as before). Also

$$P(10,8) = P(9,8) + (1-p)p^8(1-P(1,8)),$$
$$P(9,8) = P(8,8) + (1-p)p^8(1-P(0,8)).$$

Adding both sides we get

$$P(10,8) = P(8,8) + 2(1-p)p^8 = p^8 + 2(1-p)p^8 = p^8(3-2p).$$

For $p = 0.8$ this is equal to 0.235 for the NBA player in the above example.

Now to see how the value of p affects the results we calculated $P(N,r)$ for several values of N and r. These are presented in the tables below for $p = 0.04$ and 0.6.

$p = 0.4$

$N \setminus r$	5	6	7	8	9
10	0.041	0.014	0.005	0.001	0.004
20	0.100	0.038	0.014	0.005	0.002
30	0.156	0.062	0.024	0.009	0.004
50	0.257	0.108	0.043	0.017	0.007
70	0.345	0.151	0.062	0.025	0.010
100	0.459	0.213	0.090	0.036	0.015

$$p = 0.6$$

N\r	5	6	7	8	9
10	0.233	0.121	0.062	0.030	0.014
20	0.478	0.291	0.170	0.097	0.054
30	0.644	0.428	0.266	0.159	0.093
50	0.835	0.628	0.426	0.271	0.166
70	0.924	0.758	0.551	0.368	0.233
100	0.976	0.873	0.689	0.490	0.324

We end this section by citing an excellent article by Lisa Belkin about coincidences and conspiracy theories published in *The New York Times*, August 11 (not September 11) 2002.

2.10 USING R

2.10.1 Probability of random variables

In this section, we will demonstrate how to use R to calculate the probability of a random variable, especially with known distributions. Most commonly used distribution functions have been pre-built in R already, and therefore one can use that directly when needed. For a full list of supported distributions, please refer to Chapter 8 in *An introduction to R* (https://cran.r-project.org/doc/manuals/r-release/R-intro.pdf).

Discrete random variable

In many cases, it will be very simple to find the probability of a given discrete random variable when its distribution is known. However, it may need some calculation, and R will be quite helpful to handle that.

EXAMPLE 2.10.1 A binomial random variable X with $n = 10$ and $p = 0.3$. We want to find what is the probability of $P(2 < X < 6)$.

SOLUTION In R, we have *pbinom* and *dbinom* functions defined for Binomial distribution. *dbinom* is probability mass function (PMF), while *pbinom* is cumulative distribution function (CDF). To solve the question, we can simply call *pbinom* to obtain the answer in the following:

```
# pbinom(t, n, p) will return P(X<t) for X ~ B(n,p)
ans = pbinom(6, 10, 0.3) - pbinom(2, 10, 0.3)
```

You will find $ans = 0.6066251$ as answer.

It will be very simple to find expectation and variance of any given discrete random variable. For popular distributions, we have existing formulas to use for computing the values, such as $E(X) = np$ and $Var(X) = np(1-p)$ for Binomial distribution. In fact, one can define arbitrary discrete random variable with provided probability distribution in R and thus calculate the expectation and variance.

Chapter 2: Elements of Probability

EXAMPLE 2.10.2 Let's use Example 2.6.2, for $p = 0.6$ for demonstration here. We will define the random variable X and its probability distribution P in the following R code, and then find its expectation and variance.

SOLUTION

```
p = 0.6 # Value of parameter
# The random variable using c() defining a vector
X = c(0, 1, 2)
# Provided probability distribution
P = c((1-p)^2, 2*p*(1-p), p^2)
# Expectation
mu = sum(X*P)
# Variance
var = sum((X-mu)^2*P)
```

You will get $\mu = 1.2$ and $\sigma^2 = 0.48$ as answers.

Continuous random variable

Similar to discrete cases, R supports a lot popular continuous distributions, such uniform distribution, exponential distribution, normal distribution, and so on.

EXAMPLE 2.10.3 Recall Example 2.6.5, we have a uniform random variable $X \sim U[0,1]$. We want to know $P(0.05 < X < 0.89)$.

SOLUTION

```
# punif(t, a, b) will return P(X<t) for X ~ U[a,b]
ans = punif(0.89, 0, 1) - punif(0.05, 0, 1)
```

You will see the answer is 0.84.

It will be tricky to calculate the expectation and variance of an arbitrary continuous random variable, because one need to define an R function for the integral. For example, we want to find the expectation of above uniform random variable, yet we don't know existing formula for that. Therefore, the only way is calculating the integral according to the general definition $E(X) = \int x f(x) dx$. Here we know $f(x) = 1, 0 \leq x \leq 1$ for above uniform random variable, thus the function we should define is $xf(x) = x$.

```
# Defining g(x) = xf(x) = x here for the integral
g = function(x){x}
# Find the integral
mu = integrate(g, 0, 1)
print(mu)
```

You will see the expectation is 0.5.

REFERENCE

Darrell Huff. 1954. *How to Lie with Statistics*. New York: W.W.Norton & Company.

CHAPTER 3

Some Special Distributions

In practice, certain distributions arise more often than the others. Such distributions have a given name and are known to the users in different fields of application. Some of the more well-known distributions of this kind are presented in this chapter. Discussion illustrating the relationship between some of these distributions together with their applications in sports and a few other disciplines such as reliability theory is also included. An important and widely used probability model known as Poisson Process, is also introduced in this chapter.

In Chapter 2, we presented some examples of both discrete and continuous distributions. For discrete random variable, a table including possible values of the random variable together with their likelihoods, constitutes the probability mass function. In some cases, such tables can be converted to formulas (models) through expressing the likelihoods as a function of possible values of the random variable. To illustrate this, consider, for example, the experiment of rolling two balanced dice and the probability mass functions of M, the maximum of the two numbers, X, the sum of the two numbers, and Y, the difference of two number. The mass functions are, respectively,

m	1	2	3	4	5	6
$P(M=m)$	1/36	3/36	5/36	7/36	9/36	11/36,

x	2	3	4	5	6	7	8	9	10	11	12
$P(X=x)$	1/36	2/36	3/36	4/36	5/36	6/36	5/36	4/36	3/36	2/36	1/36,

and

y	-5	-4	-3	-2	-1	0	1	2	3	4	5
$P(Y=y)$	1/36	2/36	3/36	4/36	5/36	6/36	5/36	4/36	3/36	2/36	1/36.

It is easy to see that each of these tables present a pattern that can be converted to the following formulas, respectively,

$$P(M=m) = \frac{2m-1}{36} \quad 1 \leq m \leq 6$$

$$P(X=x) = \begin{cases} (x-1)/36 & 2 \leq x \leq 7 \\ (13-x)/36 & 8 \leq x \leq 12 \end{cases}$$

or

$$P(X=x) = \frac{6-|x-7|}{36} \quad 2 \leq x \leq 12$$

and
$$P(Y=y) = \frac{6-|y|}{36} \quad -5 \leq y \leq 5.$$
Clearly, it is much easier to work with these forms of mass functions than tables including a large number of possible values and their likelihoods. They are mathematically more tractable and practically much shorter and more precise. A more illustrative example is the distribution of the number of heads when flipping a fair coin. If we flip the coin a few times, then a table format may be reasonable. But for a very large number of flips, table formats are no more practical. Analysis of this situation leads to a well-known distribution known as binomial distribution discussed in Section 3.1.

Bernoulli Random Variable

The simplest random variable is the one that takes only two values 0 and 1. It is usually referred to as a binary, or Bernoulli random variable and can be constructed by letting $X = 1$ if an event of interest (A) occurs and $X = 0$ if it does not occur (A'). Several probability models are developed based on Bernoulli random variables and in this sense Bernoulli random variable plays an important role in probability modeling. For this random, variable we have

x	0	1
P(X=x)	$1-p$	p

or
$$P(X=x) = \begin{cases} 1-p, & x=0 \\ p, & x=1 \end{cases}$$
and
$$E(X) = 0(1-p) + (p) = p, \quad \text{Var}(X) = 0^2(1-p) + 1^2(p) - p^2 = p(1-p).$$
This model is a special case of a model known as binomial distribution to be discussed in the next section.

A sequence of Bernoulli random variables is called Bernoulli process, named after the mathematician James Bernoulli. It is one of the simplest yet important random processes in probability theory. Essentially, the process is the mathematical abstraction of coin tossing. However, because of the wide applicability it is usually stated in terms of a sequence of generic trials that satisfy the following assumptions:

1. Each trial has two possible outcomes, generically called success and failure.
2. The trials are independent. Intuitively, the outcome of one trial has no influence over the outcome of another trial (in probability sense).
3. On each trial, the probability of success p and the probability of failure $1-p$ are fixed, and remain fixed throughout the experiment.

Example 70% of applicants for driver's license past the test in the first attempt. If $X = 1$ and $X = 0$ represent, respectively, pass and fail then we have the following Bernoulli model:

x	0	1
P(X=x)	0.30	0.70

Chapter 3: Some Special Distributions

3.1 BINOMIAL DISTRIBUTION (DISCRETE)

This section includes discussion of a simple but important distribution. It is based on Bernoulli random variables introduced above.

3.1.1 Binomial Distribution

Binomial distribution has proven to be very useful in practice. It applies to any situation where an experiment consists of a set of trials with each trial resulting in an event A or its compliment A'. Moreover, the trials are independent from each other and interest is centered around the number of times A occurs. Formally, it is assumed that the three conditions stated above are satisfied. Examples include yes–no, true–false, on–off, male–female, defective–nondefective, and so on. Suppose, for example, you are taking a multiple-choice exam with five possible answers where you get one point for a right response and zero points for a wrong response. You do not know the answer to eight of the questions so you just guess. Here, the experiment consists of eight identical trials, each with two possible outcomes. The trials are independent because whatever you pick for one question is not going to help you with other questions and your chance of picking the right answer is fixed (20%) for each question. Please see Figure 3.1.1 for its distribution.

Formally, let each of the n independent random variables $X_1, \ldots, X_n$ take the value "1" (success) with probability p and the value "0" (failure) with probability $q = 1 - p$. Then the random variable

$$X = X_1 + \ldots + X_n$$

which counts the number of "successes," has a distribution known as Binomial distribution $B(n,p)$ with probability mass function

$$P(X = x) = \binom{n}{x} p^x q^{n-x}.$$

Note that each X_i is either 1 or 0. So one possible outcome is

$$X = 1 + 0 + 0 + 1 + 1 + 1 + 0 + \cdots + 0.$$

Since $X_i's$ are independent, the probability of having a specific sequence of x ones and $n - x$ zeros is $p^x q^{n-x}$. Noting that the number of sequences that include x ones and $n - x$ zeros is $\binom{n}{x}$, the above probability mass function follows. See Section 2.5 for further details concerning the derivation of probability mass function.

EXAMPLE 3.1.1 A multiple-choice exam with five questions has four choices of which only one is correct. A student is attempting to guess the answers. The variable X is the number of questions answered correctly. X is binomial random variables with $n = 5$ and $p = \frac{1}{4}$. Suppose we want to

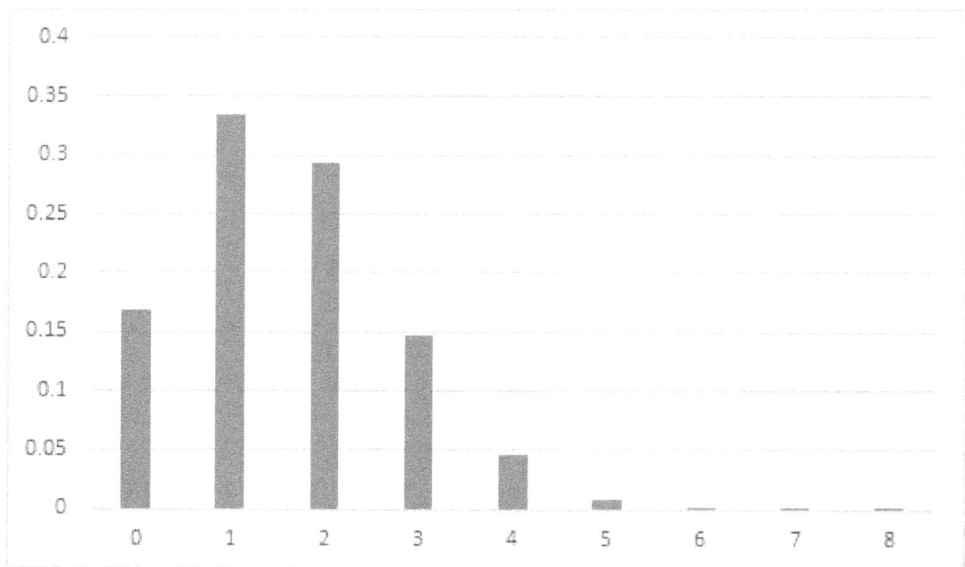

FIGURE 3.1.1 Illustration of Binomial Distribution, n = 8 and p = 0.2

answer the following questions: What is the probability that a student gets exactly three answers correct? At most three answers correct? At least four answers correct?

SOLUTION: All the questions posed can be answered by means of the probability mass function for X, which is

$$P(X=x) = \binom{5}{x}\left(\frac{1}{4}\right)^x\left(\frac{3}{4}\right)^{5-x}, \quad x = 0,1,2,3,4,5.$$

Thus,

$$P(X=3) = \binom{5}{3}\left(\frac{1}{4}\right)^3\left(\frac{3}{4}\right)^2 = 10\left(\frac{1}{64}\right)\left(\frac{9}{16}\right) = \frac{90}{1,024} = 0.0879$$

and

$$P(X \leq 3) = \binom{5}{0}\left(\frac{1}{4}\right)^0\left(\frac{3}{4}\right)^5 + \binom{5}{1}\left(\frac{1}{4}\right)^1\left(\frac{3}{4}\right)^4 + \binom{5}{2}\left(\frac{1}{4}\right)^2\left(\frac{3}{4}\right)^3 +$$

$$\binom{5}{3}\left(\frac{1}{4}\right)^3\left(\frac{3}{4}\right)^2 = \left(\frac{243}{1,024}\right) + 5\left(\frac{1}{4}\right)\left(\frac{81}{256}\right) + 10\left(\frac{1}{16}\right)\left(\frac{27}{64}\right) + 10\left(\frac{1}{64}\right)\left(\frac{9}{16}\right)$$

$$= \frac{1,008}{1,024} = 0.9844.$$

Also

$$P(X \geq 4) = 1 - P(X < 4) = 1 - P(X \leq 3) = 1 - 0.9844 = 0.0156.$$

EXAMPLE 3.1.2 Consider a box that contains 6 black balls and 4 white balls. Suppose that we are drawing 5 balls one-by-one and with replacement. Let X denote the number of black balls. Since,

Chapter 3: Some Special Distributions

with replacing the ball, the probability of drawing a black ball remains the same (equal to 6/10) and that drawing a black or white ball is independent of the previous drawings, X has a binomial distribution with $n = 5$, $p = 0.6$ and $q = 0.4$. Thus the distribution of X is

$$P(X = x) = \binom{5}{x} (0.6)^x (0.4)^{5-x} \quad x = 0, 1, 2, 3, 4, 5.$$

As an example,

$$P(X = 2) = \binom{5}{2} (0.6)^2 (0.4)^3 = 0.2304$$

is the probability of drawing two black and three white balls.

EXAMPLE 3.1.3 Consider the problem involving two volleyball teams. Suppose we let the two teams play n games. For $i = 1, 2, \ldots, n$ let

$$X_i = \begin{cases} 1 & \text{if team one wins the } i\text{-th game,} \\ 0 & \text{if team one loses the } i\text{-th game.} \end{cases}$$

The variable $X = X_1 + \ldots + X_n$ stands for the number of games that team one wins in the n repeated games. Assume that all these repeated games are independent from one another. Then X has the binomial distribution $B(n, p)$, and probability questions can be answered easily.

For binomial distribution we have

$$\mu = E(X) = E(X_1 = X_2 + \ldots + X_n) = E(X_1) + E(X_2) + \ldots + E(X_n) = np,$$
$$\sigma^2 = \text{Var}(X) = \text{Var}(X_1 + X_2 + \ldots + X_n)$$
$$= \text{Var}(X_1) + \text{Var}(X_2) + \ldots + \text{Var}(X_n) = npq. \tag{3.1}$$

Thus, in the long run, we expect team one to win about np games out of n games played. Also, when we flip a fair coin ($p = 1/2$) n times we expect to observe $np = n(1/2) = n/2$ heads. For this example, the standard deviation $\sqrt{n/4}$ together with Chebyshev's rule may be used to assess the possible deviations from the expectation.

EXAMPLE 3.1.4 In the 2018–2019 National Basketball Association (NBA) season, Malcolm Brogdon had an amazing 95.3% free throw percentage. Suppose that in a playoff game he gets 10 free throws. What is the probability that he would make 9 out of 10? At least 9? Find the expected value and the variance of number of hits. Assume that the result of the free throws are independent from each other.

SOLUTION: Here $n = 10$, $p = 0.953$ and X is the number of hits. Therefore

$$P(X = 9) = \binom{10}{9} (0.953)^9 (0.047) = 0.305,$$

$$P(X \geq 9) = P(X=9) + P(X=10) = \binom{10}{9}(0.953)^9(0.047)$$
$$+ \binom{10}{10}(0.953)^{10}(0.047)^0$$
$$= 0.305 + 0.618 = 0.923,$$
$$\mu = E(X) = np = 10(0.953) = 9.53,$$
$$\sigma^2 = \text{Var}(X) = npq = 10(0.953)(0.047) = 0.448.$$

Table B1 (in the end of the book) lists the probabilities, that is the values of $P(X=x)$, of the binomial distribution for some frequently occurring values of n and p. To find other values calculators or softwares such as R, MINITAB or SAS may be used.

EXAMPLE Suppose that when shooting at a target the probability of a hit is 0.80. Then the probability of at least one hit is, respectively, 0.80, 0.96, and 0.992 for 1, 2, and 3 tries. Thus to be 99% sure of hitting the target one should shoot at the target at least three times.

A Special Case: Consider a binomial distribution with $n=1$. The random variable X takes only the values 0 and 1 with respective probabilities $1-p$ and p. As mentioned earlier, in this special case, X is called a Bernoulli random variable. For this case, the probability mass function is

$$P(X=x) = p^x(1-p)^{1-x}, \quad x = 0, 1.$$

Also, the expected value and variance are, respectively,

$$E(X) = p, \quad \text{Var}(X) = p(1-p).$$

We like to point out once more that it is possible to define a Bernoulli random variable on any sample space. Just choose an event. If the event occurs, let $X=1$; otherwise $X=0$.

We end this section by noting that, in practice, it is important to make sure that conditions leading to binomial distribution are satisfied. For example, suppose that a soccer player is undergoing a well-structured training to learn how to shoot a penalty kick to the upper left-hand corner of the goal. He tries this 200 times everyday. If we want to figure out his chance of 75 successes in a given day, our answer would depend on whether we think enough learning can take place to change p as the player progresses from trial 1 to trial 200.

Practice Problems

1. Fifty percent of the TV-viewing public in Europe watched the Euro 2018 final. In a random sample of 200 TV viewers, what is the probability that more than 150 watched the coverage?
2. In the World Cup 2002 when South Korea played against Turkey, 4 out of 11 South Korean players on the field had the same last name, Lee. Suppose that from the 9 players on the bench three will be selected as substitutes. What is the probability that at least one will have the last name Lee?

3. In 2018 Federation Internationale de Football Association (FIFA) World Cup Russia, Harry Kane scored 1 headed goal of his total 6 goals. If we select 5 of his goals randomly, what is the probability that there is no headed goal.

3.1.2 Bernoulli Distribution and System Reliability

Recall that Bernoulli random variable X assumes only two values 0 and 1. In spite of their deceptively simple structure, Bernoulli random variables are exceedingly useful in a wide variety of problems, some of which were presented earlier and some will be presented now.

Consider a complex system (e.g., a space telescope, a nuclear reactor or a complicated computing machine) consisting of n components, each of which has only two states: a functioning state (1) and a failing state (0). Consider a random variable X_i, where $X_i = 1$ means the i-th component is functioning and $X_i = 0$ means the i-th component is not functioning (has failed). The reliability r_i of the i-th component is defined as

$$r_i = P(X_i = 1).$$

Since the state of this component is represented by a Bernoulli random variable, we can write this as

$$r_i = 0 \cdot P(X_i = 0) + 1 \cdot P(X_i = 1) = E(X_i).$$

Similarly, the state of the system as a whole is a Bernoulli random variable denoted by X, where $X = 1$ means that the system is functioning and $X = 0$ means that the system has failed. The reliability r of the system is defined as

$$r = P(X = 1) = E(X).$$

The state of the system X is called the system function. It is a function, possibly a complicated one, of the states of its components. Reliability Engineers classify an n-component system into one of three classes: a series system, a parallel system, or a k-out-of-n system. The following provides a description of each class:

1. **Series System.** A system of components connected in series functions if and only if each of its components functions. Consequently, the system function X is given by

   $$X = X_1 X_2 \cdots X_n.$$

   This equation simply states that $X = 1$ if and only if each factor on the right-hand side equals 1.

2. **Parallel System.** A parallel system functions if and only if at least one of its components functions. Consequently, the system function is given by

   $$X = 1 - (1 - X_1)(1 - X_2) \cdots (1 - X_n).$$

   Suppose $X_2 = 1$, so $1 - X_2 = 0$. This implies that the product $(1 - X_1)(1 - X_2) \cdots (1 - X_n) = 0$ and therefore $X = 1$. Thus if the second component is functioning, then so is the system. The same argument shows that $X = 1$ if $X_i = 1$ for at least one i.

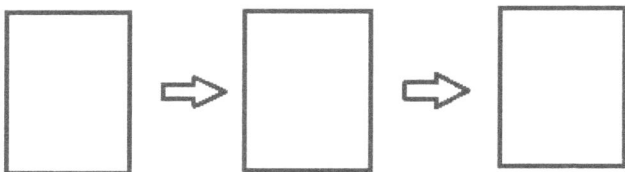

FIGURE 3.1.2 Series System

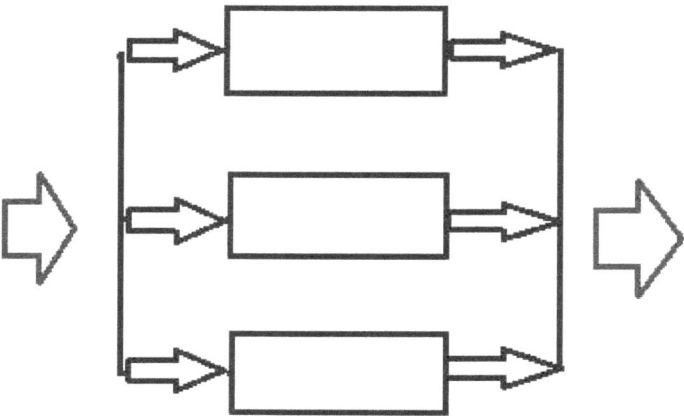

FIGURE 3.1.3 Parallel System

3. **k-Out-of-n System.** A k-out-of-n system functions if and only if at least k of the n components function. For example, an airplane functions if and only if at least two of its four engines (perhaps one in each side) function. This is an example of a 2-out-of-4 system. The function for a 2-out-of-3 system, is given by

$$X = X_1 X_2 X_3 + X_1 X_2 (1 - X_3) + X_1 (1 - X_2) X_3 + (1 - X_1) X_2 X_3.$$

Here $X = 1$ if at least two of the X_i's (out of three) are 1. As a different example, in a test with 100 multiple-choice questions students pass if at least 60 of their responses are correct.

Reliability

To find the reliability of the systems mentioned above we need to calculate $P(X = 1)$ or $E(X)$. Let us consider each case separately.

1. Series System

$$P(X=1) = P(X_1 X_2 \cdots X_n = 1) = P(X_1 = 1 \text{ and } X_2 = 1 \text{ and } \cdots \text{ and } X_n = 1)$$
$$= P(X_1 = 1, X_2 = 1, \cdots, X_n = 1)$$
$$= P(X_1 = 1)P(X_2 = 1) \cdots P(X_n = 1) = r_1 r_2 \cdots r_n.$$

2. Parallel System

$$P(X=1) = 1 - P(X=0) = 1 - P((1-X_1)(1-X_2)\cdots(1-X_n) = 1)$$
$$= 1 - P(X_1 = 0, X_2 = 0, \cdots X_n = 0)$$
$$= 1 - P(X_1 = 0)P(X_2 = 0)\cdots P(X_n = 0)$$
$$= 1 - (1-r_1)(1-r_2)\cdots(1-r_n).$$

3. k-Out-of-n System

$$P(X=1) = r_1 r_2 \cdots r_n + r_1 r_2 \cdots r_k (1-r_{k+1})(1-r_{k+2}) \cdots (1-r_n) + \cdots$$

For example, for 2-out-of-3 system we have

$$P(X=1) = r_1 r_2 r_3 + r_1 r_2 (1-r_3) + r_1 r_3 (1-r_2) + r_2 r_3 (1-r_1).$$

An Application

Suppose n components with reliabilities $r_1, r_2, \cdots, r_n$ are placed in series. A new component (number $n+1$) can be placed in parallel with any one of these n components. Where should this new component be placed in order to maximize the overall system reliability? Does the answer change with r_{n+1}?

Suppose that the new component is placed in parallel with component i. We then have the following system reliability

$$r = r_1 r_2 \cdots r_{i-1}(1 - (1-r_i)(1-r_{n+1}))r_{r+1} \cdots r_n$$
$$= r_1 r_2 \cdots r_{i-1} r_i r_{i+1} \cdots r_n + r_1 r_2 \cdots r_{i-1} r_{i+1} - r_1 r_2 \cdots r_i r_{i+1} \cdots r_n r_{n+1}.$$

Note that for varying i the first and the third terms remain unchanged. The decision depends only on the second term, and this term can be written as

$$r_1 r_2 \cdots r_{i-1} r_{i+1} \cdots r_n r_{n+1} = \frac{r_1 r_2 \cdots r_{i-1} r_i r_{i+1} \cdots r_n r_{n+1}}{r_i}.$$

Again, the numerator remains unchanged for different values of i. Therefore, the answer is to place the new component in parallel with the one with minimum reliability (this will make the denominator small and therefore the ratio large). Note that this answer is independent of the value of r_{n+1}.

3.2 GEOMETRIC AND NEGATIVE BINOMIAL DISTRIBUTIONS (DISCRETE)

3.2.1 Geometric Distribution

Suppose that we perform an experiment and are concerned only about the occurrence (success) or nonoccurrence (failure) of some event A. Assume, as in the discussion of the binomial distribution, that we perform the experiment repeatedly, the repetitions are independent, and on each repetition $P(A) = p$ and $P(A') = 1 - p = q$ remain unchanged. Suppose that we repeat the experiment until A occurs for the first time. For example, you may flip a coin until you observe, for the first time, a head; or dial your Internet connection as many times as necessary until you successfully log on. Here, we depart from the assumptions leading to the binomial distribution. There, the number of repetitions was predetermined, whereas here it is a random variable. In fact now the outcomes are of the form 1, 01, 001, 0001, and so on, where, as before, 1 and 0 represent success and failure, respectively.

Define the random variable X as the number of repetitions required up to and including the first occurrence of A. Thus the possible values of X are 1, 2, …. Noting that $X = x$ if and only if the first $(x-1)$ repetitions of the experiment result in A', while the x-th repetition results in A, we have

$$P(X = x) = q^{x-1} p, \quad x = 1, 2, \ldots \quad (3.2)$$

A random variable with the probability distribution (3.2) is said to have a geometric distribution $(G(x;p))$. The expression (3.2) defines a legitimate probability distributions since $P(X = x) \geq 0$, and

$$\sum_{x=1}^{\infty} P(X = x) = p(1 + q + q^2 + \ldots) = p/(1-q) = 1.$$

This is because $q < 1$ and the sum in the bracket is a geometric series. It is also easy to show that for geometric distribution (3.2) we have

$$\mu = E(X) = \frac{1}{p}, \quad \sigma^2 = \text{Var}(X) = \frac{q}{p^2}.$$

For example, for $p = 0.50$, $E(X) = 1/0.50 = 2$ as expected.

When applying the geometric distribution, the trials must meet the following requirements.

a. The total number of trials is potentially infinite
b. There are just two outcomes of each trial: success and failure
c. The outcomes of all the trials are independent
d. All the trials have the same probability of success

The geometric distribution is sometimes referred to as the waiting time (to the first success) distribution.

EXAMPLE 3.2.1 Suppose that a professional basketball player makes 60% of his shots from a certain spot. If he attempts shots from that spot, what is the probability that he will make his first shot in the third attempt? Assume that the outcome of successive shots are independent from each other.

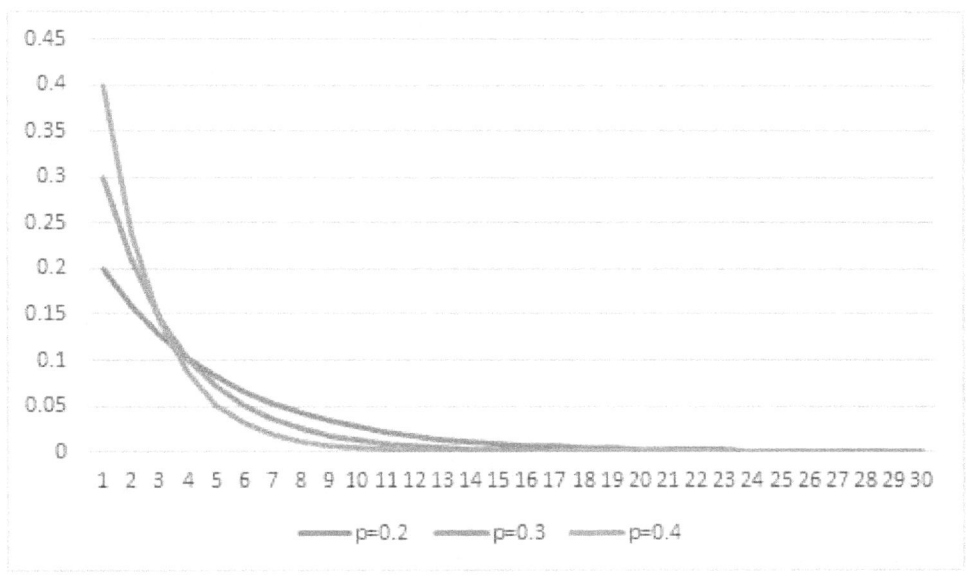

FIGURE 3.2.1 Geometric 0.2, 0.3 and 0.4.

SOLUTION: Here $p = 0.60$ and $q = 0.40$. Therefore,

$$P(X = 3) = (0.4)^2(0.6) = 0.096.$$

From this, we see that it is highly unlikely that his first successful shot will be the third, or equivalently, it is very likely that he will make a shot in his first or second try. In fact, here $E(X) = \frac{1}{0.60} = 1.67$. After all, his chance of making the first shot is 0.60 in the first try and $(0.4)(0.6) = 0.24$ in the second.

EXAMPLE 3.2.2. If the probability is 0.75 that an applicant for a driver's license will pass the road test on any given try, what is the probability that this applicant will pass the test on the fourth try? Find also $E(X)$ and Var(X)

SOLUTION: Here $p = 0.75$ and $q = 0.25$. Therefore,

$$P(X = 4) = (0.25)^3(0.75) = 0.0117.$$

$$E(X) = \frac{1}{0.75} = \frac{4}{3} = 1.33 \; , \;\; \text{Var}(X) = \frac{0.25}{(0.75)^2} = \frac{4}{9} = 0.44.$$

EXAMPLE 3.2.3 A director of a recreational center is interested in the reliability of a certain exercise machine. These machines typically fail at the instant they are switched on and not while they are in operation (much like light bulbs). Suppose that the probability of failure each time the machine is switched on is p and failure on a given trial has no effect on future failures unless the machine has already failed, in which case all future trials result in failure with certainty. The director is interested in the probability that the machine survives at least 100 ons and offs. For what values of p is this probability at least 90%?

SOLUTION: Let X be the number of times that a given machine survives switch ons and switch offs. Then, from the given information, X has a geometric distribution with parameter p, and the probability of interest is $P(X \geq 100)$, that is;

$$P(X \geq 100) = \sum_{x=100}^{\infty} p(1-p)^x = p(1-p)^{100} \sum_{n=0}^{\infty} (1-p)^n$$

$$= p(1-p)^{100} \frac{1}{1-(1-p)} = (1-p)^{100}.$$

Using this we get

$$(1-p)^{100} \geq 0.90, \text{ or } p \leq 0.0010531.$$

Thus, to ensure the desired reliability, the probability of failure on a given trial should be at most 0.1%.

Practice Problems

1. If the probability is 0.75 that a fan will believe a rumor about the star of his favorite team, find the probabilities that

    (a) the eighth person to hear the rumor will be the first to believe it;
    (b) the 15th person to hear the rumor will be the first to believe it.

2. A shooter has 75% chance that he can hit a standing target at 500 yards using his rifle. What is the probability he will make his first hit within 2 rounds?

The St. Petersburg Paradox

Suppose someone offers you the following game knowing that you are a 50-50 free throw shooter. You will repeatedly try free throws. You will receive an award of 2^x pennies, where X is the number of shots made before the first miss. How much would you be willing to pay for this game?

SOLUTION: The probability that the award will be 2^x pennies is equal to the probability that you will make x shots and then miss one, which is equal to $1/2^{x+1}$. Hence, the expected value of the award (in pennies) is equal to

$$\sum_{x=0}^{\infty} (2^x)(1/2^{x+1}) = \sum_{x=0}^{\infty} 1/2 = \infty.$$

That is, the average amount of the award is infinite! Hence, according to the interpretation of expected value, it seems that you should be willing to pay an infinite amount.

Now suppose that it is agreed that the award will be truncated at 2^{30} cents (which is just over 10 million dollars!), that is, the award will be frozen once it exceeds 2^{30} cents. For this case, the award is equal to $2^{min(30,X)}$ pennies, where X is as before. How much would you be willing to pay for this new award?

The expected value of the new award (in cents) is equal to

$$\sum_{x=0}^{\infty} (2^{min(30,x)})(1/2^{x+1}) = \sum_{x=0}^{29} (2^x)(1/2^{x+1}) + \sum_{x=30}^{\infty} (2^{30})(1/2^{x+1})$$

Chapter 3: Some Special Distributions

$$= \sum_{x=0}^{29}(1/2) + (2^{30})(1/2^{30}) = 30/2 + 1 = 16.$$

That is, truncating the award at just over 10 million dollars changes its expected value enormously, from infinity to 16 cents!

3.2.2 Negative Binomial (Pascal) Distribution

In connection with repeated Bernoulli trials, we are sometimes interested in trial (number) on which the kth success occurs. For instance, we may be interested in the probability that a basketball player will make his third three-pointer on the seventh attempt, a soccer team will score their second goal on the tenth attack, the 10th child exposed to a contagious disease will be the third to catch it, the fifth person to hear a rumor will be the second to believe it, or that a burglar will be caught for the second time on his eighth attempt. If we are interested in the second success, the outcomes will look like 11, 011, 0011, 1001, 0101, and so on.

Let X denote the number of trials up to and including the kth success. If the k-th success is to occur on the x-th trial, there must be $k-1$ successes in the first $x-1$ trials, and the probability of this event is

$$\binom{x-1}{k-1}p^{k-1}(1-p)^{x-k}.$$

Since the probability of a success on the kth trial is p, the probability that the kth success occurs on the xth trial is p times the above probability. That is,

$$\binom{x-1}{k-1}p^{k}(1-p)^{x-k}.$$

DEFINITION: A random variable X has a negative binomial distribution, if and only if its probability mass function take the form

$$P(X=x) = \binom{x-1}{k-1}p^{k}(1-p)^{x-k}, \quad x=k, k+1, k+2, \cdots$$

Thus, the number of the trial on which the kth success occurs is a random variable having a negative binomial distribution with the parameters k and p. The name "negative binomial" derives from the fact that the values of $P(X=x)$ for $x=k, k+1, k+2, \cdots$, are the successive terms of the binomial expansion of $\left(\frac{1}{p} - \frac{1-p}{p}\right)^{-k}$. In statistics books, negative binomial distribution is also referred to as Pascal distribution. Note that geometric distribution is a special case of a negative binomial when $k=1$.

The negative binomial distribution does basically the same thing as the binomial, except that now we are asking about the probability of a particular sample size, given that we have observed x successes, whereas we had expected to observe u, say, successes.

EXAMPLE 3.2.4 If the probability is 0.40 that a child exposed to a certain sport will like it, what is the probability that the 10th child exposed to the sport will be the third to like it?

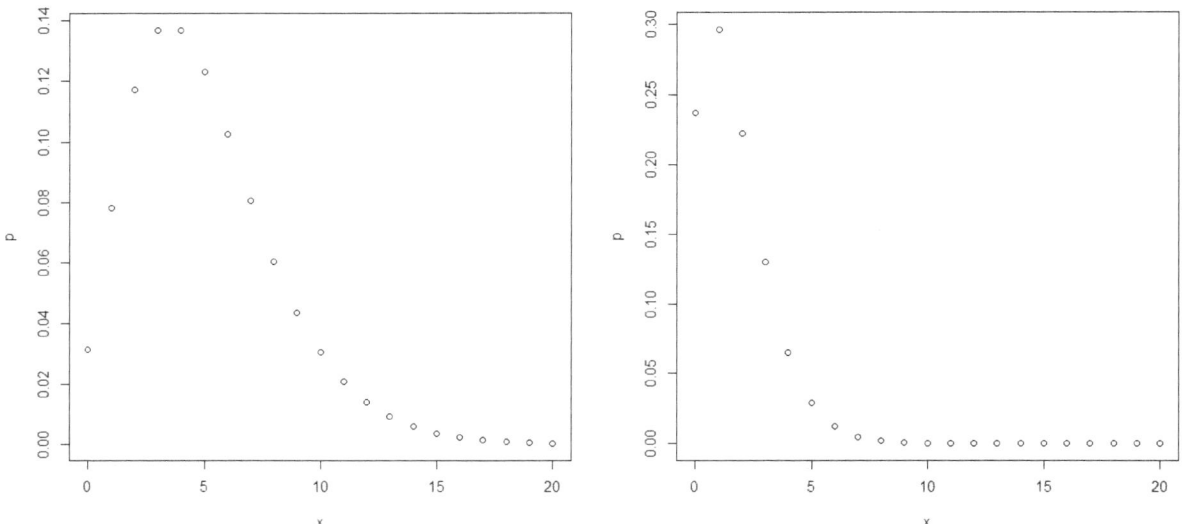

FIGURE 3.2.2 Negative Binomial (5, 0.5) and (5, 0.75)

SOLUTION: Substituting $x = 10, k = 3$, and $p = 0.40$ into the formula for the negative binomial distribution, we get

$$P(X = 10) = \binom{9}{2}(0.40)^3(0.60)^7 = 0.0645.$$

EXAMPLE 3.2.5 A basketball player is trying 3-pointer shots from a certain spot. If his rate of success (based on past attempts) is 0.30, what is probability that he will make his second 3-pointer in the fifth attempt? Assume that the outcomes of the successive shots are independent from each other.

SOLUTION: Substituting $x = 5, k = 2$, and $p = 0.30$, we obtain

$$P(X = 5) = \binom{4}{1}(0.30)^2(0.070)^3 = 0.1235.$$

The mean and the variance of the negative binomial distribution are, respectively,

$$E(X) = \frac{k}{p} \text{ and } \text{Var}(X) = \frac{k}{p}\left(\frac{1}{p} - 1\right).$$

A PRACTICE PROBLEM A soccer player attempts penalty shots. He has scored 70% of his penalty shots during the season. What is the probability that he will score his fifth penalty shot in the 10th attempt?

EXAMPLE 3.2.6 If the probability is 0.75 that an applicant for a coaching license will pass the license test on any given try, what is the probability that this applicant will finally pass the test on the fourth try?

Chapter 3: Some Special Distributions

SOLUTION: Substituting $x = 4$, $k = 1$, and $p = 0.25$ into the formula for the geometric distribution, we get

$$P(X = 4) = 0.25(1 - 0.25)^{4-1} = 1.055.$$

Of course, this result is based on the assumption that the trials are all independent, an assumption whose validity may be questionable. For this case we also have

$$E(X) = \frac{1}{p} = \frac{1}{0.25} = 4, \quad Var(X) = \frac{1}{0.25}(10.25 - 1) = 12.$$

A PRACTICE PROBLEM If the probabilities of having a boy or girl are each 0.50, find the probabilities that

  (a) a family's fourth child is their first son
  (b) a family's seventh child is their second daughter
  (c) a family's tenth child is their fourth or fifth son

We end this section by noting that in the literature regarding soccer and its modeling, the negative binomial is used as an appropriate model both for the number of goals scored and for passing movements (number of passes). We can justify this by looking at the way this game is played. In soccer, the ball is passed from player to player on the same team until possession is lost, an infringement of the rules occurs, the ball goes out of play, or a goal is scored. Teams fail so many times before they succeed for the first time, second time, and so on, in the certain number of tries.

3.3 HYPERGEOMETRIC DISTRIBUTION (DISCRETE)

This section includes discussion of another important discrete distribution. This distribution may arise in many different situations. One case is when sampling is done without replacement. In this sense, it is related to the binomial distribution where sampling is done with replacement.

3.3.1 Hypergeometric Distribution

Consider a set of N elements of which k are looked upon as successes (e.g., wins) and the other $N - k$ as failures (e.g., losses). For example, think of a box with k black balls and $N - k$ white balls where drawing a black ball is a "success." Suppose that, like the binomial distribution, we are interested in the probability of getting x successes in n trials. But now, unlike binomial experiment, we are choosing the balls without replacement so that the probability of success changes from trial to trial. We will use the notation $C_{l,m}$ or $\binom{l}{m}$ to denote the total number of combinations when choosing m elements out of l without replacement. Here both m and l are nonnegative integers such that $m \leq l$. Recall that

$$C_{l,m} = \frac{l!}{m!(l-m)!} = \binom{l}{m}$$

where $l!$ stands for the factorial of l.

Using this formula, there are $C_{k,x}$ ways for choosing x of the k successes, and $C_{(N-k),(n-x)}$ ways for choosing $n-x$ of the $N-k$ failures. Hence using the multiplication principle, there are $C_{k,x}C_{(N-k),(n-x)}$ ways for choosing x successes and $n-x$ failures. Also, altogether there are $C_{N,n}$ ways for choosing n of the N elements. Thus, it follows that the probability of "x successes in n trials" is the ratio

$$\frac{C_{k,x}C_{(N-k),(n-x)}}{C_{N,n}}.$$

assuming that outcomes (selections) are equally likely. In general, a random variable X has a hypergeometric distribution if and only if its probability distribution take the form

$$P(X=x) = \frac{C_{k,x}C_{(N-k),(n-x)}}{C_{N,n}} = \frac{\binom{k}{x}\binom{N-k}{n-x}}{\binom{N}{n}} \qquad (3.3)$$

for $x = 0, 1, 2, \ldots, n$, $x \leq k$ and $n - x \leq N - k$. In this case, X is referred to as a hypergeometric random variable. A popular notation for this distribution is $H(x; n, N, k)$.

Thus, for sampling without replacement, the number of successes in n trials is a random variable having a hypergeometric distribution with the parameters n, N, and k. The mean and variance of the hypergeometric distribution are, respectively,

$$\mu = E(X) = \frac{nk}{N}$$

and

$$\sigma^2 = Var(X) = \frac{nk(N-k)(N-n)}{N^2(N-1)}.$$

EXAMPLE Consider once more the example of the box with six black balls and four white balls. Suppose that we are drawing three balls one-by-one but now without replacement. Let X be the number of black balls. Then we have

$$P(X=x) = \frac{C_{6,x}C_{4,(3-x)}}{C_{10,3}} = \frac{\binom{6}{x}\binom{4}{3-x}}{\binom{10}{3}}, \quad x = 0, 1, 2, 3.$$

For example,

$$P(X=2) = \frac{\binom{6}{2}\binom{4}{1}}{\binom{10}{3}} = 0.5$$

Chapter 3: Some Special Distributions

is the probability of choosing 2 black balls from 6 black balls and 1 white ball from 4 white balls. Also,

$$E(X) = \frac{(3)(6)}{10} = 1.8, \text{ and } Var(X) = \frac{(3)(6)(4)(7)}{(10^2)(9)} = 0.56.$$

If sampling with replacement is used rather than sampling without replacement, then instead of hypergeometric we will have a binomial distribution with the same expected value. Note that drawing 3 balls one-by-one is equivalent to drawing 3 balls together.

EXAMPLE 3.3.1 Among the 100 applicants for the job of swimming coach, only 30 are actually qualified. If 8 of these applicants are randomly selected for an "in-depth" interview, find the probability that only 3 are qualified for the job.

SOLUTION: In this case, $x = 3$, $n = 8$, $N = 100$, and $k = 30$. Using (3.3) we get

$$P(X = 3) = \frac{C_{30,3} C_{70,5}}{C_{100,8}} = 0.264.$$

That is, the probability that only 3 of 8 are qualified for the job is about 26.4%.

A PRACTICE PROBLEM In a statistics class there are 30 students, 18 male, 12 female, 10 mathematics majors, and 20 statistics majors. If a professor selects 5 students randomly, what is the probability that

(a) Three are male and two female students?
(b) Four are mathematics majors and one is a statistics major?

EXAMPLE 3.3.2 A basketball team has 5 distinct positions. Out of 12 players available, 5 play as guards. How many starting teams are possible if the coach wants to play two guards? If the coach simply picks 5 players out of the 12 at random, what is the probability that the starting team includes exactly two guards?

SOLUTION: Here $x = 2$, $n = 5$, $N = 12$, and $k = 5$. The answer to the first question is simply the numerator of the fraction given in (3.3). That is, $C_{5,2} C_{7,3} = 350$. Hence there are 350 possible ways to form a starting team that includes two guards. To answer the second question, we use (3.3) and obtain

$$P(X = 2) = \frac{C_{5,2} C_{7,3}}{C_{12,5}} = 0.442.$$

Practice Problems

1. From a shipment containing 20 basketballs, 8 are selected at random. If the shipment contained 4 flat basketballs, what is the probability that

    (a) All 8 selected balls are good?
    (b) At least 4 are good?

2. It was reported in a box of 1,000 ammunitions there were 50 contaminants that will be failed to shoot. A soldier got 100 rounds in that box, what is the probability that

   (a) All 100 rounds were fired without any issue?
   (b) At most 3 were contaminants?

3.4 POISSON DISTRIBUTION (DISCRETE)

Suppose that a special event is occurring with the rate (mean) of λ times per unit time or per unit area (e.g., number of car accidents in a day, number of incoming telephone calls per minute, number of goals scored per game, number of trees per acre, and number of earthquakes in a given region per year). Let X denote the number of the events per unit time or unit area, and so on. It is shown that (see Section 3.9) under certain assumptions and conditions X has Poisson distribution, that is, it has the probability mass function of the form

$$P(X=x) = P_0(x;\lambda) = \frac{\lambda^x}{x!}e^{-\lambda}, \quad x = 0, 1, 2, \ldots \tag{3.4}$$

where $e \approx 2.71828$ is an irrational number. For the Poisson distribution we have

$$E(X) = \lambda \quad \text{and} \quad \text{Var}(X) = \lambda \tag{3.5}$$

and therefore a coefficient of variation equal to $1/\sqrt{\lambda}$. Table B2 lists probabilities for different values of λ.

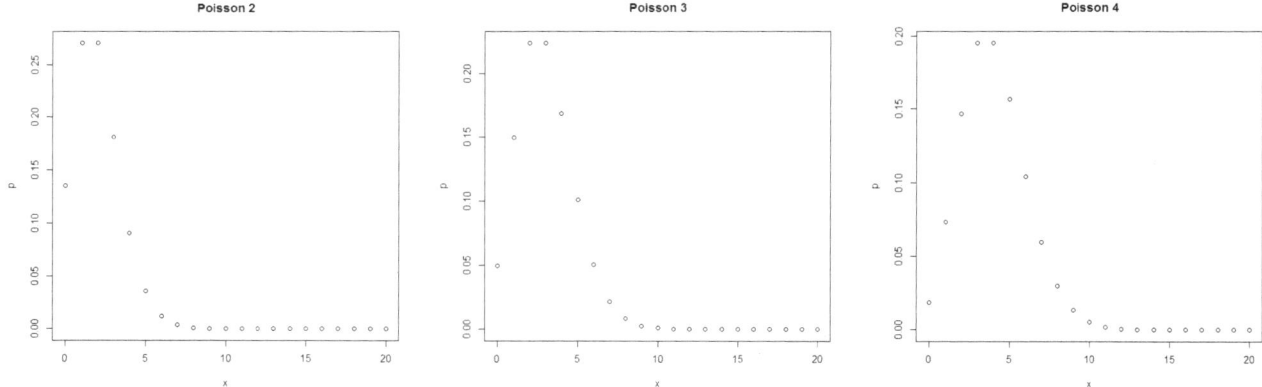

FIGURE 3.4.1 Poisson 2, 3, and 4

It can also be shown that the Poisson distribution arises as the limit of the binomial distribution as $n \to \infty$ and $p \to 0$. This connection is discussed in Section 4.2. This means that the Poisson distribution applies to rare events; when probability of success is small and number of

Chapter 3: Some Special Distributions

trials is large. It gives the probability of a certain number of occurrences of an event when the average number of occurrences is known.

EXAMPLE 3.4.1 In a recreation center injuries happen, on average, 5 times per month. Find the probability that during the next month injuries happen 4 times. Find also the probability that it will happen more than 4 times.

SOLUTION: Here $\lambda = 5$ and therefore $P(X = x) = \frac{e^{-5}5^x}{x!}, x = 0, 1, 2 \cdots$ The required probabilities are then

$$P(X = 4) = e^{-5}5^4/4! = 0.1755$$

$$P(X > 4) = 1 - P(X \leq 4) = 1 - \sum_{x=0}^{4} e^{-5}5^x/x!.$$

A PRACTICE PROBLEM In a busy intersection, two accidents happens per week. Find the probability of three accidents in a given week. Find also, the probability of five accidents in a 2 week period ($\lambda = 4$).

EXAMPLE 3.4.2 Suppose a new record for the 400-m run has just been set at this time. Let X denote the number of times this record gets broken in the next 10 years. It has been shown that X has a Poisson distribution. If λ could be estimated from the past data, then it can be used to predict the number of times records get broken in a time period of interest.

EXAMPLE 3.4.3 It has been shown that the number of goals scored in a soccer game during a fixed time interval can be modeled by Poisson distribution. During the course of a game, each team makes, on average, about 25 shots at the goal with about 10% success rate. Suppose that for the home team $\lambda = 1.77$. Find the probability that this team will score (a) two goals, (b) at most two goals.

SOLUTION:

(a) The probability that the home team scores two goals is

$$P(X = 2) = \frac{1.77^2}{2!}e^{-1.77} = 0.267.$$

(b) The probability that the home team scores at most two goals is

$$P(X \leq 2) = P(X = 0) + P(X = 1) + P(X = 2)$$
$$= \frac{1.77^0}{0!}e^{-1.77} + \frac{1.77^1}{1!}e^{-1.77} + \frac{1.77^2}{2!}e^{-1.77} = 0.739.$$

EXAMPLE 3.4.4 The chance that a fan watching a soccer game on a very hot day will suffer from heat exhaustion is 0.005. What is the probability that out of 5,000 fans, 200 will suffer from heat exhaustion on a very hot day?

SOLUTION: Here $\lambda = (50,000)(0.005) = 250$. Therefore,

$$P(X = 200) = e^{-250}(250)^{200}/200!$$

Note that we can also obtain the required probability from the binomial distribution, using $n = 50,000$ and $p = 0.005$ as; $\binom{50,000}{200}(0.005)^{200}(0.995)^{49,800}$.

A PRACTICE PROBLEM A baseball player averages 1.1 hits per game. Assuming that the number of hits per game has a Poisson distribution, calculate probabilities for 0, 1, 2, 3, 4, 5 hits.

EXAMPLE 3.4.5 Suppose that two teams A and B play a game in which A scores X points (e.g., goals in soccer or runs in baseball) and independently B scores Y points. Assume that X and Y have Poisson distributions with mean λ and μ, respectively. That is,

$$P(X = x) = e^{-\lambda}\lambda^x/x!, \ x = 0,1,2,\cdots \text{ and } P(Y = y) = e^{-\mu}\mu^y/y!, \ y = 0,1,2,3\cdots$$

Now, using these information the probability that A and B tie zero-zero is

$$P(X = 0, Y = 0) = P(X = 0)P(Y = 0) = e^{-\lambda}e^{-\mu} = e^{-(\lambda+\mu)}.$$

The probabilities that they tie game $1-1, 2-2, \cdots, i-i$ are, respectively,

$$\lambda\mu e^{-(\lambda+\mu)}, \ \lambda^2\mu^2 e^{-(\lambda+\mu)}/2!^2, \cdots, \lambda^i\mu^i e^{-(\lambda+\mu)}/i!^2.$$

Thus the probability of a tie game is

$$P(D = 0) = P(0) = e^{-(\lambda+\mu)}\sum_{i=0}^{\infty}(\lambda\mu)^i/(i!)^2 = \sum_{i=0}^{\infty}e^{-\lambda}\lambda^i/i!\, e^{-\mu}\mu^i/i! = e^{-(\lambda+\mu)}P(i,i)$$

where D denotes the goal difference and $P(i,i)$ denotes $\sum \lambda^i/i!\, \mu^i/i!$. Note that in practice (e.g., real soccer games) i is finite and hence the above probability can be calculated easily once λ and μ are estimated.

The probability that A beats B with one goal, for Example 3.2 is

$$P(1) = e^{-\lambda}\lambda/1!\, e^{-\mu}\mu^0/0! + e^{-\lambda}\lambda^2/2!\, e^{-\mu}\mu/1! + e^{-\lambda}\lambda^3/3!\, e^{-\mu}\mu^2/2! + \cdots$$

$$= \sum_{i=0}^{\infty}e^{-\lambda}\lambda^{i+1}/(i+1)!\, e^{-\mu}\mu^i/i!\, e^{-\mu}\mu^i/i! = e^{-(\lambda+\mu)}\sum_{i=1}^{\infty}\lambda^{i+1}\mu^i/(i+1)!i!$$

$$= e^{-(\lambda+\mu)}\, P(i+1,i).$$

It is interesting to note that

$$P(i,i) = \frac{d}{d\lambda}P(i+1,i).$$

A PRACTICE PROBLEM Suppose that for two soccer teams $\lambda = 1.6$ and $\mu = 1.4$. Calculate $P(0), P(1)$, and $P(2)$.

Chapter 3: Some Special Distributions

We end this part by noting that Poisson distribution has also been applied to sports such as hockey with success. Research indicates that the goals scored by and against teams in the National Hockey League (NHL) are surprisingly well described by Poisson distributions, and even more surprisingly, the goals for and against a team seem to be independent.

3.4.1 Binomial-Poisson Hierarchy

This section presents an example involving combination of two models. It describes the most classic hierarchical model. Let us consider an example. A large number of students in high schools play soccer, each continuing to play at college level with probability p. On the average, how many will play at the college level?

Clearly, the number of high school students playing soccer is a random variable. It is often taken to have a Poisson distribution with parameter (rate) λ. If we assume that students survival as players are independent, then we have Bernoulli trials. Therefore, if we let X be the number of survivors and Y be the number of students who play soccer in high school, then X given Y has a binomial distribution. Also, as mentioned Y has a Poisson distribution.

The advantage of the hierarchy is that complicated processes may be modeled by a sequence of relatively simple models placed in a hierarchy. Also, dealing with the hierarchy is no more difficult than dealing with conditional distribution that refers to the distribution of one variable given the other and marginal distribution that refers to distribution of one variable regardless of the value of the other.

Returning to the above example, the random variable of interest, X, the number of survivors, has the distribution given by (using the multiplication rule)

$$P(X=x) = \sum_{y=0}^{\infty} P(X=x, Y=y) = \sum_{y=0}^{\infty} P(X=x \mid Y=y) P(Y=y)$$

$$= \sum_{y=x}^{\infty} \left[\binom{y}{x} p^x (1-p)^{y-x} \right] \left[\frac{e^{-\lambda} \lambda^y}{y!} \right]$$

since X given Y is binomial(y,p) and Y is Poisson (λ). If we now simplify this last expression, canceling what we can and multiplying by λ^x/λ^x, we get (letting $t = y - x$),

$$P(X=x) = \frac{(\lambda p)^x e^{-\lambda}}{x!} \sum_{y=x}^{\infty} \frac{((1-p)\lambda)^{y-x}}{(y-x)!}$$

$$= \frac{(\lambda p)^x e^{-\lambda}}{x!} \sum_{t=0}^{\infty} \frac{((1-p)\lambda)^t}{t!}$$

$$= \frac{(\lambda p)^x e^{-\lambda}}{x!} e^{(1-p)\lambda} = \frac{(\lambda p)^x}{x!} e^{-\lambda p}.$$

So, X has a Poisson distribution with parameter λp. Thus, any marginal inference on X is with respect to a Poisson (λp) distribution, with Y playing no part at all. Introducing Y in the hierarchy was mainly to aid our understanding of the model. There is an added bonus in that the parameter of the distribution of X is the product of two parameters, each relatively simple to understand. Putting all these together the answer to the original question is

$$E(X) = \lambda p.$$

So, on the average, λp players will survive.

So far we presented some discrete random variables and their respective probability mass function. In the next sections, we will present some well-known continuous distributions and their probability density functions.

3.5 UNIFORM DISTRIBUTION (CONTINUOUS)

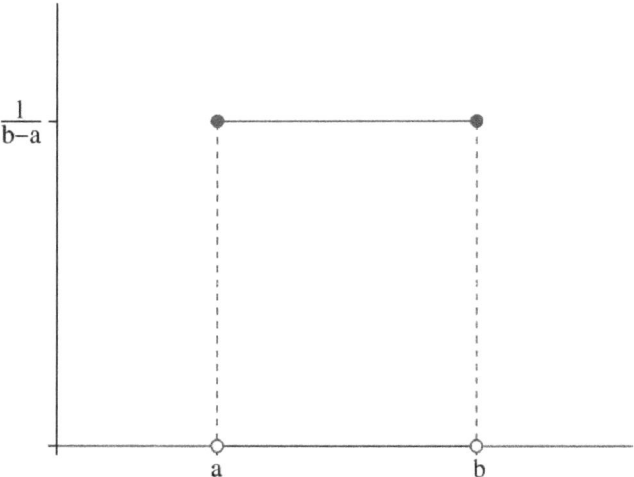

The uniform distribution is one of the most fundamental and frequently used distributions employed in probability calculations. It may occur in either discrete variable or continuous variable form, both of which will be briefly discussed in this section. In discrete case, it presents a situation where possible values of the random variable are equally likely. In continuous case, it presents a situation where the probability that the random variable assumes values in two intervals of equal length are equal. It corresponds to a homogenous beam with a uniform mass.

EXAMPLE *Discrete Case* Let X be the number observed when a fair die is rolled. The possible values are $\{1,2,3,4,5,6\}$ and the probability mass function is $P(X=x) = 1/6$ for $x = 1,2,3,4,5$, and 6. That is all possible values are equally likely. This is an example of discrete uniform distribution.

EXAMPLE *Continuous Case* If shuttle buses arrive at one of the stops every 14 min and waits there for 1 min and I arrive at that stop at some random time. The uniform distribution of shuttle bus arrival probability states that I have a 1/15 chance of catching the bus upon arrival. The longer I wait the probability will summate.

Chapter 3: Some Special Distributions

For the continuous case, the random variable X is uniformly distributed over the interval $a \leq x \leq b$ if it has the density function of the form (see the figure)

$$f(x) = \begin{cases} \frac{1}{b-a} & \text{if } a \leq x \leq b, \\ 0 & \text{elsewhere}. \end{cases} \qquad (3.6)$$

Also

$$F(x) = \begin{cases} 0 & x < a, \\ \frac{x-a}{b-a} & a \leq x \leq b, \\ 1 & x > b. \end{cases}$$

For uniform distribution we have

$$E(X) = \frac{a+b}{2} \quad \text{and} \quad \text{Var}(X) = \frac{(b-a)^2}{12}. \qquad (3.7)$$

EXAMPLE Suppose that the time you have to wait for the bus is uniform random variable varying between 0 and 10 min. What is the probability that you have to wait for more than 4 min? Find the expected value and the variance of the waiting time.

SOLUTION:

$$f(x) = \begin{cases} 1/10, & 0 \leq x \leq 10, \\ 0, & \text{elsewhere}. \end{cases}$$

$$P(X > 4) = \int_4^{10} 1/10 \, dx = 6/10,$$

$$E(X) = 5, \text{Var}(X) = 100/12.$$

EXAMPLE 3.5.1 Let us consider the best record of a given year in a sport such as men's 100-m run. Suppose that this sport is practiced continuously and records can be set at any time. The questions we then like to answer is: when should we expect a new record to occur? If we let X denote the time that record occurs, then it is shown that X has a uniform distribution over the time interval of 1 year. Suppose that the last record of men's 100 meter was set during the August, 2009, by Usain Bolt. We may like to know the probability that it was set during the third week of that month. The random variable representing the occurrence time of this record has a uniform distribution over August.

Note that the above example considers the "best record in a given period of time" and the variable X is the "moment that the record occurs in that period." Hence, we know that X must take a value in that time interval and its distribution is uniform over that time interval, representing the fact that "no moment is more likely than any other moment for the best record to occur." The example does not consider the record that beats the current or past records, which will lead to a much different story.

EXAMPLE 3.5.2 Suppose that you turn on your television to watch a soccer match. You are late and the game has already started. You notice that it is the 30th min and team A is winning one goal to none. You are wondering when the goal was scored. More precisely what is the probability that it was scored, for example, in the first 10 min. It is natural to assume that the distribution of the time the goal was scored is uniform over the interval (0,30). That is,

$$f(x) = \begin{cases} 1/30, & 0 \leq x \leq 30, \\ 0, & \text{elsewhere.} \end{cases}$$

Thus the required probability is $P(0 \leq X \leq 10) = \int_0^{10} 1/30 \, dx = 10/30 = 1/3$.

Finally, for the discrete case with n possible outcomes we have

$$P(X = x) = 1/n, \quad x = 0, 1, 2, \cdots, n,$$

$$E(X) = (n+1)/2, \quad \text{Var}(X) = (n-1)(n+1)/12.$$

For example, when rolling a single (fair) die

$$P(X = x) = 1/6, \quad x = 0, 1, 2, 3, 4, 5, 6,$$

$$E(X) = (6+1)/2 = 3.5, \quad \text{Var}(X) = (6-1)(6+1)/12 = 2.916.$$

One important application of discrete uniform distribution is random number generator used for simulation. Here, numbers are generated randomly with each digit having the probability of 1/10.

3.6 NORMAL DISTRIBUTION (CONTINUOUS)

The normal (Gaussian) distribution is probably the most important and most widely used distribution in statistics due to its vast applications and the celebrated "Central Limit Theorem" (Chapter 4). It has many applications in real life, including those in the study of sports.

A random variable X is normally distributed if it has the density function of the form

$$f(x) = \frac{1}{\sigma\sqrt{2\pi}} \exp(-\frac{1}{2}(\frac{x-\mu}{\sigma})^2), \quad (\sigma > 0). \tag{3.8}$$

This distribution involves two parameters μ and σ. In fact,

$$E(X) = \mu \quad \text{and} \quad \text{Var}(X) = \sigma^2. \tag{3.9}$$

In future, we will write $X \sim N(\mu, \sigma^2)$ to represent a random variable having a normal distribution with mean μ and variance σ^2.

Each pair of the parameters (μ, σ^2) represents a particular normal distribution. Of special importance is the normalized version of this distribution, $N(0,1)$. Its density function takes the form

$$\Phi(z) = \frac{1}{\sqrt{2\pi}} \exp(-\frac{1}{2}z^2). \tag{3.10}$$

It is called the standard normal distribution. Conventionally, a random variable that has the standard normal distribution is denoted by z. A standard normal variable z has a mean equal to

Chapter 3: Some Special Distributions

zero and a standard deviation equal to one. Since the function e^{-z^2} does not have an antiderivative, the calculation of probabilities are carried out using numerical techniques and results are tabulated. Table B3 presents one such tabulation and can be used for the probability calculation involving the standard normal distribution. Other than the end of the book, you can find the Table B3 in the very first page of your book. It provides probabilities for values less than a fixed quantity.

The importance of the standard normal distribution lies in the following fact. If X is distributed as $N(\mu, \sigma^2)$, then the random variable $z = \frac{X-\mu}{\sigma}$ is distributed as $N(0,1)$. That is, every normal distribution can be transformed to standard normal distribution by applying a shift and a scale. Therefore, only the case $N(0,1)$ needs to be tabulated. The distribution function of the standard normal distribution is denoted by Φ. Usually $\Phi(z)$ is only tabulated for $z \geq 0$. For negative arguments the values are calculated using the relation

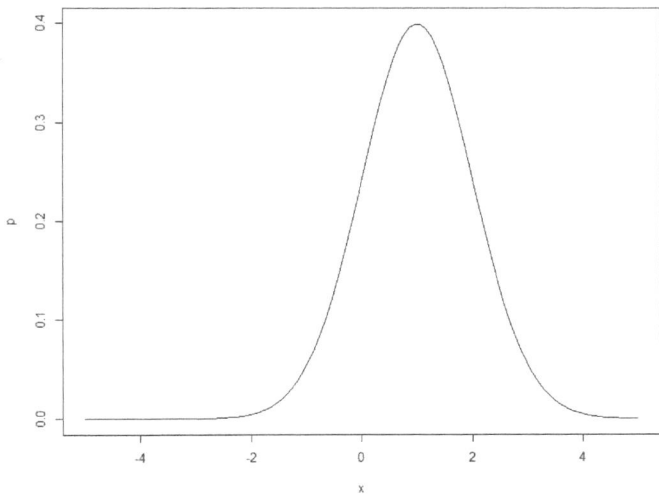

FIGURE 3.6.1 Normal Distribution

$$\Phi(-z) = 1 - \Phi(z) \tag{3.11}$$

which is the consequence of the fact that

$$\Phi(z) - \Phi(-z) = 2\Phi(z) - 1. \tag{3.12}$$

EXAMPLE The time spent on a treadmill by an individual in a recreation center has a normal distribution with mean of 20 min and standard deviation of 10 min. What is the probability that a randomly selected individual will spend

(a) Less than 10 min on the treadmill?
(b) More than 25 min on the treadmills?
(c) Between 15 and 30 min on the treadmills?

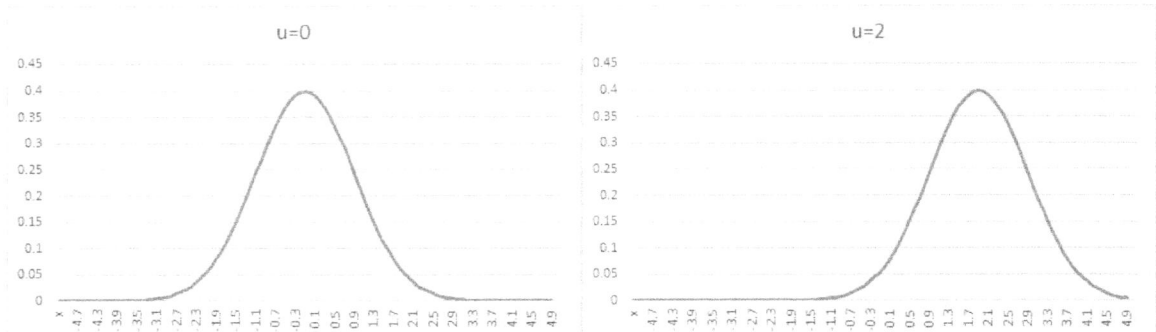

FIGURE 3.6.2 Normal Distribution With $\mu = 0$, and $\mu = 2$

SOLUTION: Let X denote the time a randomly selected individual spends on a tread mill. Then

(a) $P(X \leq 10) = P(\frac{X-20}{10} \leq \frac{10-20}{20}) = P(z \leq -1) = \Phi(-1) = 0.1587$.
(b) $P(X > 25) = P(\frac{X-20}{10} > \frac{25-20}{10}) = P(z \geq 0.5) = 1 - P(z \leq 0.5) = 1 - \Phi(0.5) = 1 - 0.5915 = 0.3085$.
(c) $P(15 \leq X \leq 30) = P(\frac{15-20}{10} \leq \frac{X-20}{10} \leq \frac{30-20}{10}) = P(-0.5 \leq z \leq 1) = P(z \leq 1) - P(z \leq -0.5) = 0.8413 - 0.3085 = 0.5328$.

Note that using relative frequency interpretation of probability, we can change the format of the questions. For example, we can ask what percentage of individuals will spend less than 10 min on the treadmill?

A PRACTICE PROBLEM Suppose that the grades in a statistics class follows a normal distribution with a mean of 75 and a standard deviation of 8 (a normal class). Find the probability that a randomly selected student

(a) Has a grade less than 71.
(b) Has a grade greater than 87.
(c) A grade between 71 and 87.

EXAMPLE 3.6.1 An analysis of the scores of professional football games has led to the conclusion that a team that is favored by k points will outscore its opponent by a random number of points, say X, which is approximately normally distributed with $\mu = k$ and $\sigma = 14$. For example, if a team is favored by 3 points then X, the difference in the points scored by this team and its opponent, has approximately a normal distribution $N(3, 14^2)$. Note that being precise, X is a discrete random variable since it only takes integer values. However, the normal distribution can be applied to approximate its true distribution.

EXAMPLE 3.6.2 Given the theory described in Example 3.6.1, what is the probability that a team favored by 2 points wins the game?

SOLUTION: As in Example 3.6.1, let X denote the difference in the points scored by this team and its opponent. X has approximately a normal distribution $N(2, 14^2)$. The team wins if and only if

Chapter 3: Some Special Distributions   171

$X > 0$. Let $z = (X - 2)/14$, then z has the standard normal distribution and $X > 0$ if and only if $z > -2/14$. Hence, the probability is approximately

$$P(z > -2/14) = 1 - P(z \leq -2/14) = 1 - \Phi(-2/14) = \Phi(2/14) = 0.9236.$$

A more simpler presentation is as follows:

$$P(X > 0) = P(\frac{X-2}{14} > \frac{0-2}{14}) = P(z > -2/14) = 1 - P(z \leq -2/14)$$

which, using Table B3, equals 0.9236.

EXAMPLE 3.6.3 Chatterjee et al. (1995) have analyzed data on 105 guards who played in the 1992–1993 NBA season. The stem-and-leaf display of points scored per minute (PPM) by these players is given below. From this display it is clear that the distribution of PPM is very close to a normal distribution. Here the mean and standard deviation are, respectively, 0.4236 and 0.1159 points per minute. We want to find the probability that a randomly selected guard will have a PPM higher than that of Michael Jordan, which was 0.8291. Note that in the stem-and-leaf display, the unit of the leaf digit is 0.01. Hence, for instance, when the stem value equals 1 and the leaf value equals 5, the corresponding PPM is 0.15.

SOLUTION: Let X denote the PPM, then $X \sim N(0.4236, 0.1159^2)$. Now

$$P(X > 0.8291) = P(z > \frac{0.8291 - 0.4236}{0.1159}) = P(z > 3.5) = 1 - P(z \leq 3.5)$$
$$= 1 - \Phi(3.5) = 1 - 0.09998 \approx 0.0002 = \frac{1}{5,000}.$$

Number of Cases	Stem	Leaves
1	1	5
5	2	13344
6	2	567889
18	3	000000111222333444
15	3	556667788889999
23	4	00000111223333333344444
14	4	55667777888999
7	5	0112244
10	5	5556777788
3	6	023
1	6	6
0	7	
1	7	7
1	8	3

One interpretation of this result is the following: if 50 guards are added to the NBA each year, on average it takes approximately 100 years to produce a player who can perform (like) equal to or better than Jordan. Of course, we are assuming that no new advances will be introduced to the game and everything including the players perform in the level of 1992–1993 season.

An important quantity used frequently in inferential statistics is the p-th percentile of the standard normal distribution, denoted by z_p. Let z have a standard normal distribution and p be a number between 0 and 1. Then z_p is the number such that

$$P(z \leq z_p) = \Phi(z_p) = p.$$

Another useful value is the two-sided percentiles y_p, defined by

$$P(|z| \leq y_p) = \Phi(y_p) - \Phi(-y_p) = p. \tag{3.13}$$

It is easy to see that $y_p = z_{(1+p)/2}$, or equivalently $z_p = y_{2p-1}$. In certain cases, α is used to denote $1-p$. Thus $p = 1 - \alpha$ and $y_{1-\alpha} = z_{1-\alpha/2}$, or equivalently, $z_{1-\alpha} = y_{1-2\alpha}$. For reference, we list some frequently used values of the z_p and y_{2p-1} in Table 3.6.1.

TABLE 3.6.1 Some Percentiles of the Standard Normal Distribution

p	z_p	$= y_{2p-1}$
0.90	1.28	$= y_{0.80}$
0.95	1.645	$= y_{0.90}$
0.975	1.96	$= y_{0.95}$
0.990	2.34	$= y_{0.98}$
0.995	2.58	$= y_{0.99}$

Turning to a different application of the normal distribution, we see from Table 3.6.1 that 1.645 is the 95th percentile of the distribution. This means, for example, that if in a normal class a student's z-score in a test is greater than 1.645, he or she falls in the upper 5% of the class. If the mean and the standard deviation of grades are given, then this information can be used to find the range of the grades that qualifies a student to fall in the upper 5% of the class. The following is an example:

EXAMPLE In a statistics and sports class the average grade is 83 and standard deviation is 5. The instructor decided that only students falling in the upper 20% of the class will be granted an A. Also students falling in the lower 10% will fail the class. Find the range for A and F in this class, assuming that the grades follow a normal distribution.

SOLUTION: Note that here $P(X > A) = 0.20$ or $P(\frac{X-83}{5} > \frac{A-83}{5}) = P(z > \frac{A-83}{5}) = 0.20$ or $P(z \leq \frac{A-83}{5}) = 80$. From Table B3 we have $\frac{A-83}{5} = 0.84$ and hence the minimum grade to earn an A is 87.2 (87.2–100). Similarly for F we have $P(X < F) = P(\frac{X-93}{5} < \frac{F-83}{5}) = P(z < \frac{F-83}{5}) = \Phi(\frac{F-83}{5}) = 0.10$. Again,

Chapter 3: Some Special Distributions 173

using Table B3 we get $\frac{F-83}{5} = z_{.10} = -1.28$ and hence any grade below 76.60 (0 – 76.60) is a failing grade.

EXAMPLE 3.6.4 The time it takes for middle school boys to run 1 mile is normally distributed with $\mu = 450$ s and $\sigma = 40$ s. All those falling in the slowest 10% are considered not to be in good shape and need additional training. What is the critical time above which one is deemed to need additional training?

SOLUTION: Let X denote the time it takes for a randomly selected high school boy to run 1 mile. Then as assumed, X has a normal distribution $N(450, 40^2)$. Let x denote the critical time we want to find. Then we should have $P(X > x) = 0.10$, or equivalently, $P(X < x) = 0.90$. This is further equivalent to

$$P(z < (x - 450)/40) = 0.90$$

where $z = (X - 450)/40$ has the standard normal distribution. Hence, we see that

$$\frac{x - 450}{40} = z_{0.90} = 1.28.$$

and therefore $x = 501.2$. That is, if a middle school boy takes 501.2 s or more to finish the 1 mile run, then he is among the slowest 10% and needs additional training.

A PRACTICE PROBLEM Among diabetics, the fasting blood glucose level X may be assumed to be approximately normally distributed with mean 125 mg/100 mL and standard deviation 10 mg/100 mL. According to the American Diabetics Association 126 is a critical value.

a. Find $P(X \leq 110)$.
b. What percentage of diabetics have blood glucose levels between 90 and 120?
c. Find $P(X > 126)$.
d. Find the point x that has the property that 25% of all diabetics have a fasting glucose level lower than x. That is, find the 25th percentile of the distribution.

Note that, for a standard normal distribution, the areas to within one, two, and three standard deviations are, respectively,

$$P(-1 \leq z \leq 1) = P(|z| \leq 1) = 0.683,$$
$$P(-2 \leq z \leq 2) = P(|z| \leq 2) = 0.955,$$
$$P(-3 \leq z \leq 3) = P(|z| \leq 3) = 0.997.$$

The same is true for any normal distribution. That is

$$P(\mu - \sigma \leq X \leq \mu + \sigma) = P(|\frac{X - \mu}{\sigma}| \leq 1) = 0.683,$$
$$P(\mu - 2\sigma \leq X \leq \mu + 2\sigma) = P(|\frac{X - \mu}{\sigma}| \leq 2) = 0.955,$$
$$P(\mu - 3\sigma \leq X \leq \mu + 3\sigma) = P(|\frac{X - \mu}{\sigma}| \leq 3) = 0.997.$$

In fact, we have already used these results in Chapter 1 under the name of the empirical rule.

A PRACTICE PROBLEM Shows that the first and the third quartiles of a normal distribution are, respectively, $\mu - 0.6756$ and $\mu + 0.6756$.

We now study further properties of the normally distributed random variables. One very important property of the normal distribution is described in the following theorem:

THEOREM 3.6.1 If X_1 is distributed as $N(\mu_1, \sigma_1^2)$ and X_2 is distributed as $N(\mu_2, \sigma_2^2)$ and X_1, X_2 are independent, then $X_1 \pm X_2$ is distributed as $N(\mu_1 \pm \mu_2, \sigma_1^2 + \sigma_2^2)$. This is true for any number of independent normal variables and any linear combination.

EXAMPLE 3.6.5 Continue with Example 3.6.4. Let us now further assume that the time it takes for high school girls to run 1 mile is also normally distributed, but with $\mu = 480$ s and $\sigma = 35$ s. Randomly select a high school boy and a high school girl. Let the boy run the first mile and then let the girl run. If we assume these two runs are independent, then the total time it takes for them to finish as a team is a normal variable with $\mu = 450 + 480 = 930$ s and $\sigma = \sqrt{40^2 + 35^2} = 53.15$.

EXAMPLE 3.6.6 A basketball coach estimates that the number of points scored by two of his players can be considered to be observations from a normal distribution. His choice of playing each player is based upon the considerations that any score greater than 5 is "satisfactory" and a score greater than 10 is "excellent."

Suppose that the number of points scored by player A, X_A, is distributed as

$$X_A \sim N(8.0, 1.5^2) = N(8.0, 2.25).$$

In addition, suppose that the number of points scored by player B, X_B, is distributed $X_B \sim N(9.5, 2^2)$ independently of player A. What are the probabilities of an "unsatisfactory" performance for each player? An "excellent" performance? What is the probability that B will perform better than A? What is the probability that B scores at least two more points than A?

SOLUTION: The probability that the performance of the player A proves to be "unsatisfactory" is

$$P(X \leq 5.0) = P(z \leq \frac{5.0 - 8.0}{1.5}) = P(z \leq -2) = 0.0228,$$

and the probability that his or her performance proves to be "excellent" is

$$P(X \geq 10.0) = P(z > \frac{10.0 - 8.0}{1.5}) = P(z > 1.33) = 1 - 0.9082 = 0.0918.$$

The probability that player B's performance proves to be "unsatisfactory" is

$$P(X \leq 5.0) = P(z \leq \frac{5.0 - 9.5}{2.0}) = P(z \leq -2.25) = 0.0122,$$

and the probability that player B's performance proves to be "excellent" is

$$P(X \geq 10.0) = P(z > \frac{10.0 - 9.5}{2.0}) = P(z > 0.25) = 1 - 0.5987 = 0.4013.$$

Chapter 3: Some Special Distributions

If $Y = X_B - X_A$, then $Y \sim N(9.5 - 8.0, 4.00 + 2.25) = N(1.5, 6.25) = N(1.5, 2.25^2)$. The required probability is therefore

$$P(X \geq 0) = P(z > \frac{0 - 1.5}{2.25}) = P(z > -0.6) = 1 - 0.2743 = 0.7257.$$

The probability that B scores at least two points more than A is

$$P(Y \geq 2.0) = P(z > \frac{2.0 - 1.5}{2.25}) = P(z > 0.2) = 1 - 0.5793 = 0.4207.$$

Practice Problems

1. Suppose that the outside diameter of a particular kind of bolt (in inches) has a normal $N(1, 0.01)$ distribution and the inside diameter of its associated nut has a normal $N(1.03, 0.01)$ distribution. The nut and bolt will fit as long as the nut is bigger than the bolt but not by more than 0.06 inches. Show that the probability of fit is 0.9652.
2. Two roommates play on a college basketball team. Suppose that in a given season roommate A's score is normally distributed with mean 80 and standard deviation 15 and roommate B's score is normally distributed with mean 75 and standard deviation 20. Show that the probability for roommate A to score more points than roommate B on a given season is 0.5793.

Sampling Distribution of the Sample Mean

Theorem 3.6.1 and its extension to more than two variables has an extremely important consequence and plays a central role in most investigations concerning the averages. If $X_1, \ldots, X_n$ are independent and each is distributed as $N(\mu, \sigma^2)$, then the sample mean

$$\bar{X} = \frac{1}{n}(X_1 + \ldots + X_n) \tag{3.14}$$

and is distributed as $N(\mu, \sigma^2/n)$. This is because, according to the theorem 3.6.1 the distribution of $X_1 + X_2 + \ldots + X_n$ is $N(n\mu, n\sigma^2)$ and also $E(aX) = aE(X)$ and $Var(aX) = a^2 Var(X)$. This is called the sampling distribution of the sample mean. It follows from this important result that averaging will significantly reduce the variation. So it is of no surprise if, for example, your doctor asks you to measure your blood pressure a few times and find their average. In fact, averaging is a tool to smooth the data. It is an operation that acts like a low-pass filter in the sense that it removes the high-frequency noise and retains the low frequency signal (trend). A celebrated theorem known as The Central Theorem proves that for large n and under certain conditions, the distribution of $\bar{X}$ is still normal even if we drop the normality assumption for the X_i's.

EXAMPLE 3.6.7 Continue with Example 3.6.4. Assume that we are interested in the average performance of high school boys. Suppose we randomly select 64 high school boys and let $\bar{X}$ denote the average of the time they will take to run the 1 mile. Assuming that their times are all independent, we see that $\bar{X}$ has the normal distribution with $\mu = 450$ s and $\sigma = 40/\sqrt{64} = 5$ s. Note that the mean of $\bar{X}$ takes the same value as the mean of the original population, but the standard deviation of the sample mean $\bar{X}$ is much smaller than that of the original population.

In some sense, this says that the sample mean $\bar{X}$ is "less random" or less variable than a single measurement or the original variable. The larger the sample size is, the "less random" the $\bar{X}$ gets. This idea is applied to many problems in inferential statistics.

3.7 RELATIVES OF THE NORMAL DISTRIBUTION: CHI-SQUARE (χ^2), t AND F-DISTRIBUTIONS (ALL CONTINUOUS)

Chi-square, t-, and F-distributions are all related to the standard normal distribution $N(0,1)$. This is one reason for their frequent applications. In fact, as will be discussed in chapter 4, almost all distributions are in a way related to the normal distribution through the celebrated Central Limit Theorem.

3.7.1 Chi-Square Distribution

Let $z_1, z_2 \ldots, z_k$ be k independent random variables each distributed as $N(0,1)$. Then

$$X = z_1^2 + z_2^2 + \ldots + z_k^2 \qquad (3.15)$$

has a χ^2 (chi-square) distribution with k "degrees of freedom," and is denoted by $\chi^2(k)$ or $\chi^2_{(k)}$. Since this distribution involves the square of random variables, it is useful for inference regarding the variation or variability. A random variable having a chi-square distribution assumes non-negative values only. Table B5 lists the percentage points of the chi-square distribution for many different degrees of freedom.

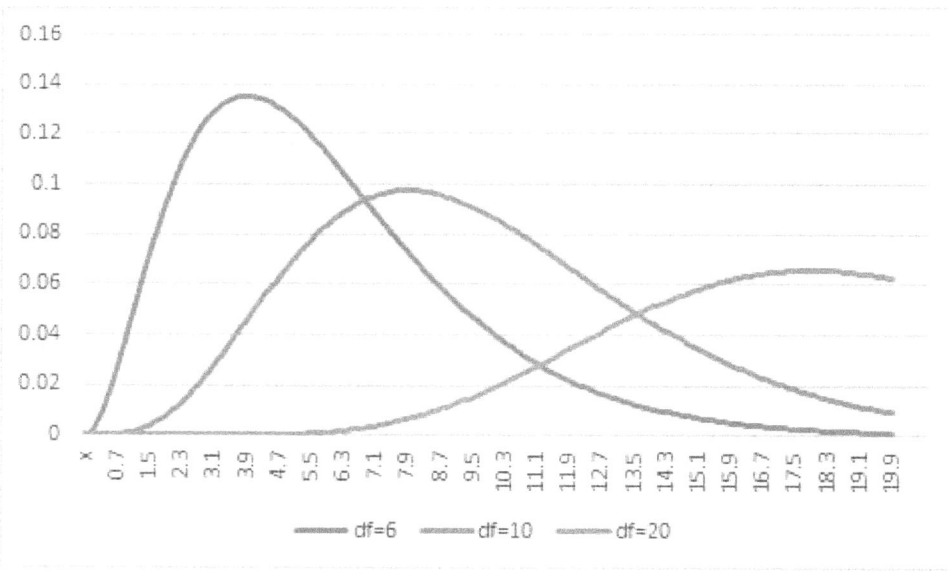

Chi-Square Distribution (6)(10)(20)

Chapter 3: Some Special Distributions

EXAMPLE 3.7.1 Continue with Example 3.6.6. Let us use X_i, $i = 1, 2, \ldots, 64$ to denote the time taken by each of 64 boys to finish the 1-mile run. Let z_i be the standardized difference between X_i and the population mean 450. That is,

$$z_i = \frac{X_i - 450}{40}, \quad i = 1, 2, \ldots, 64.$$

Then z_i's are independent and all have the standard normal distribution $N(0,1)$. Note that some z_i's may take positive values and others negative values. Some may even assume zero if X_i turns out to be exactly 450, although theoretically the probability of such occurrence is zero. Therefore, in order to measure the "sum of the squares of standardized differences" of this sample from the mean of population, we may consider

$$X = z_1^2 + z_2^2 + \ldots + z_{64}^2.$$

In this case X is a random variable with a chi-square distribution $\chi^2(64)$, or $\chi^2_{(64)}$.

As an example, think of a right triangle with sides adjacent to the right angle having random lengths of X and Y. Let T denote the length of the side opposite to the right angle. Assuming that X and Y have a standard normal distribution, we have $T^2 = X^2 + Y^2$ and hence T^2 is distributed as $\chi^2(2)$.

3.7.2 t-Distribution

The *t*-distribution, also known as student's *t*-distribution was introduced by William Goset (1976–1937) who is known mostly under the pseudonym student. He was hired by an ancestor of Mr. Guinness as a brewer in Guinness' brewery in Dublin as a statistician and later as head brewer in Guinness' brewery in London.

Formally, if X_1 is distributed as $N(0,1)$ and X_2 is distributed as $\chi^2(k)$ and X_1, X_2 are independent, then

$$T = \frac{X_1}{\sqrt{X_2/k}} \qquad (3.16)$$

has a *t*-distribution with k degrees of freedom ($t(k)$ or $t_{(k)}$). Like normal distribution, the density function of *t*-distribution is bell shaped and tends to the density function of normal distribution as k gets larger and tends to infinity. Practically, for $k \geq 30$ the $t(k)$ distribution can be replaced by normal distribution. The *t*-distribution is useful when inference regarding the mean of a normal distribution is of interest in a situation where the population standard deviation is unknown. Table B4 lists the percentage points of the *t*-distribution for many different degrees of freedom.

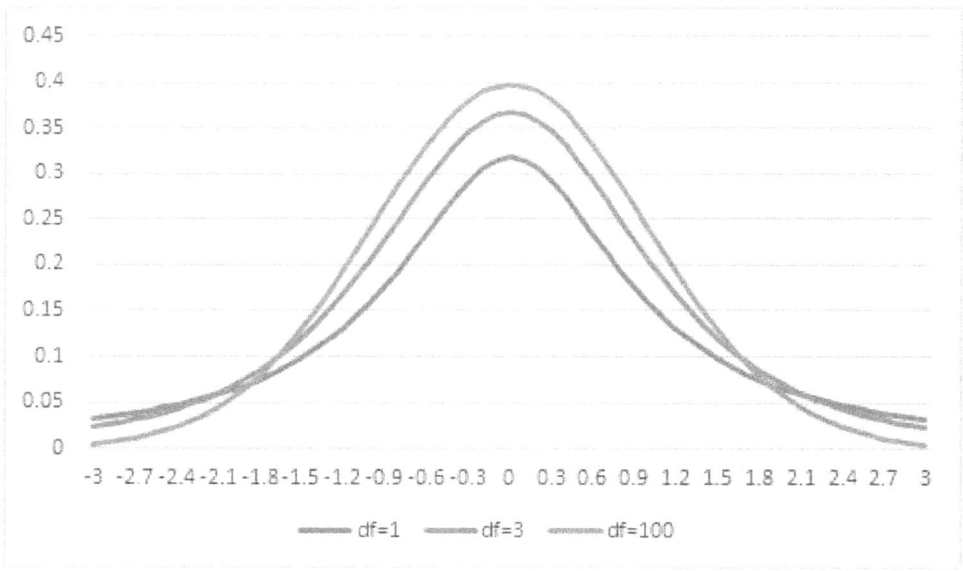

Student's Distribution (1)(3)(100)

EXAMPLE 3.7.2 Continue with Example 3.7.1. Suppose we wish to look at the ratio of the first boy's standardized difference z_1 to the "average standardized difference" $\sqrt{X/64}$. That is,

$$T = \frac{z_1}{\sqrt{X/64}}.$$

Then T has a t-distribution with 64 degrees of freedom. In this case, since 64 is greater than 30, the distribution of T can be approximated by a normal distribution.

3.7.3 *F*-Distribution

F-distribution (Fisher distribution) arises in the following situation: if X_1 is distributed as $\chi^2(k_1)$ and X_2 is distributed as $\chi^2(k_2)$ and if X_1, X_2 are independent, then the quotient

$$\frac{X_1/k_1}{X_2/k_2} \tag{3.17}$$

has F(Fisher) distribution with k_1, k_2 degrees of freedom and is denoted as $F_{(k_1,k_2)}$. This distribution is useful when inference regarding the variances of two populations is of interest. Table B6 lists the percentage points of the *F*-distribution for many different degrees of freedom.

Chapter 3: Some Special Distributions

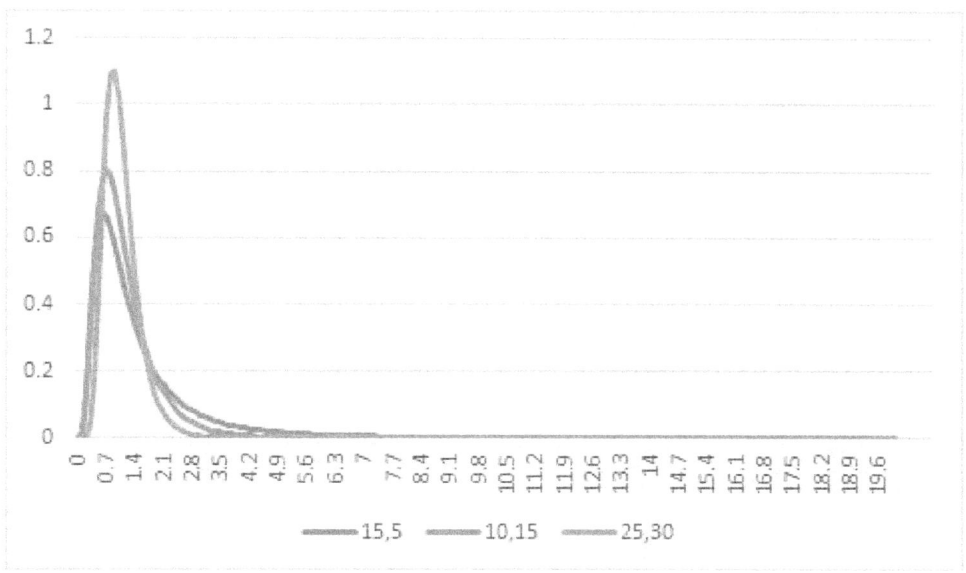

Fisher Distribution (15,5)(10,15)(25,30)

EXAMPLE 3.7.3 Continue with Example 3.7.1. Suppose we further select a random sample of 50 high school girls for 1-mile run. Also, like in Example 3.6.5, assume that the time it takes for high school girls to run 1-mile is also normally distributed with a mean of 480 seconds and a standard deviation of 35 s. Like in Example 3.7.1, we may consider the sum of the squares of the standardized differences for this group of girls. If we use Y to denote this variable, then Y has a $\chi^2(50)$ distribution. Suppose now that we are interested in the "averaged" ratio of Y to X (the sum of squares for the group of 64 boys in Example 3.7.1). That is

$$\frac{Y/50}{X/64}.$$

This ratio has an F-distribution with degrees of freedom 50 and 64.

3.8 EXPONENTIAL, GAMMA, BETA, LOGNORMAL, GUMBEL, WEIBULL AND FRÉCHET DISTRIBUTIONS (ALL CONTINUOUS)

In this section, we introduce several continuous distributions that have proved extremely useful both in modeling and in analysis of important applied problems.

3.8.1 Exponential Distribution

A random variable X has exponential distribution with parameter $\lambda > 0$ if its density function takes the form

$$f(x) = \begin{cases} \lambda e^{-\lambda x} &, \ x \geq 0, \\ 0 &, \ x < 0. \end{cases} \qquad (3.18)$$

Its distribution function takes the form

$$P(X \leq x) = F(x) = \begin{cases} 1 - e^{-\lambda x} &, \quad x \geq 0, \\ 0 &, \quad x < 0. \end{cases} \quad (3.19)$$

The mean and variance of exponential distribution are, respectively,

$$E(X) = 1/\lambda \quad \mathrm{Var}(X) = 1/\lambda^2.$$

This distribution is widely used in reliability theory, life testing, and also in queueing theory to represent waiting times or service times.

EXAMPLE The waiting time to purchase tickets has exponential distribution with $\lambda = 1/5$ min. Find the probability that one needs to wait more than 7 min to buy a ticket.

SOLUTION: Let X denote the waiting time. Then

$$P(X > 7) = 1 - P(X \leq 7) = 1 - F(7) = 1 - (1 - e^{-7/5}) = e^{-7/5} = 0.247$$

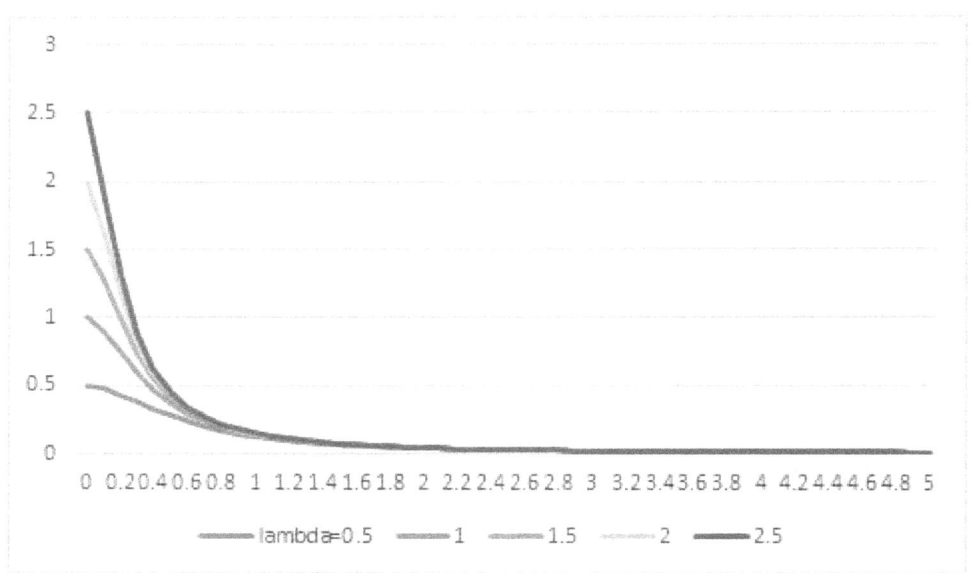

Exponential Distribution (0.5)(1)(1.5)(2)(2.5)

EXAMPLE 3.8.1 To show how an exponential distribution might arise in a sport, consider a soccer game and suppose that we are interested in calculating the probability of X goals being scored during a time interval of length t, and in time interval between the goals. Suppose that based on our past experience we can make the following assumptions:

(i) The probability of scoring a goal during a very small time interval t to $t + \Delta t$ is $\lambda \Delta t$ (i.e., it is proportional to the length of interval)
(ii) The probability of scoring more than one goal during a very small time interval negligible

Chapter 3: Some Special Distributions

(iii) The probability of scoring during a very small interval does not depend on what happened prior to the beginning of this interval.

Let $P_0(t)$ denote the probability of no goals being scored in the time interval 0 to t. Then we have

$$P_0(t + \Delta t) = P_0(t)(1 - \lambda \Delta t)$$

since the probability of no goals being scored in the interval 0 to $t + \Delta t$, equals the probability of no goals being scored during the time interval 0 to t ($P_0(t)$) times the probability of no goals being scored in the time interval t to $t + \Delta t$, which based on (iii) equals one minus the probability of one or more goals being scored in this interval, (that is, $1 - \lambda \Delta t$ based on (i) and (ii)).

Rearranging the terms, we get

$$\frac{P_0(t + \Delta t) - P_0(t)}{\Delta dt} = -\lambda P_0(t).$$

Letting $\Delta t \to 0$ and using the definition of derivative, and the fact that $P_0(0) = 1$ we obtain

$$\frac{dP_0(t)}{dt} = -\lambda P_0(t) \Rightarrow \frac{dP_0(t)}{P_0(t)} = -\lambda dt \Rightarrow P_0(t) = \bar{e}^{\lambda t}.$$

Using this we can derive the probability density function of waiting time T (a continuous variable) to the first goal.

It is easy to see that $F(t) = P(T \leq t) = 1 - P(T > t) = 1 - P$ (no goal in a time interval of length t) $= 1 - P_0(t) = 1 - \bar{e}^{\lambda t}$.

This is an exponential distribution. The number of goals scored in soccer and their distributions will be discussed in detail in Section 3.9.

This exponential distribution is useful for making inference regarding the lifetime of items such as electronic components. It is also useful in studying the size of an earthquake as is shown by Gutenberg and Richter.

EXAMPLE 3.8.2 In the NHL, games that are tied at the end of all three periods are sent to "sudden death" overtime. In overtime, the team to score the first goal wins. A study of all NHL overtime games played between 1970 and 1993 has shown that the length of elapsed time (in minutes) before the winning goal is scored has approximately an exponential distribution with $\lambda = 1/9.15 = 0.11$ (*Chance*, Winter 1995). Hence, for a randomly selected overtime NHL game, the probability of scoring the winning goal in, for example, 3 min or less is

$$F(3) = 1 - e^{-0.11 \times 3} = 0.281.$$

EXAMPLE 3.8.3 Suppose that the amount of time a player spends with the coach is exponentially distributed with a mean of 10 min. That is, $\lambda = 1/10$. What is the probability that a player will spend more than 15 min with the coach? What is the probability that a player will spend more than 15 min with the coach given that he was still with him after 10 min?

SOLUTION: Let X be the amount of time a randomly selected player spends with the coach. Then $E(X) = 1/\lambda = 10$ and we have

$$P(X > 15) = 1 - P(X \leq 15) = 1 - F(15) = 1 - (1 - e^{-15/10}) = e^{-15/10} = 0.223.$$

Also

$$P(X > 15 \mid X > 10) = \frac{P(X > 15 \text{ and } X > 10)}{P(X > 10)} = \frac{P(X > 15)}{P(X > 10)} = e^{-15/10}/e^{-10/10} = e^{-1/2} = 0.607.$$

It is interesting to note that

$$P(X > 15 \mid X > 10) = P(X > 5) = 0.607.$$

This means spending an extra 5 min with the coach after the initial 10 min has the same probability as spending 5 minutes with the coach. In fact, exponential distribution possesses an interesting property known as the lack of memory or the "memoryless" property. It can be shown that if $X \geq 0$ is a random variable that satisfies the following equation:

$$P(X > s) = P(X > s + t \mid X > t) \tag{3.20}$$

for any value of $t \geq 0$, that is, if X has no memory on the "starting point t," then X is an exponential random variable. Conversely, it is easy to see that every exponential random variable satisfies (3.20). In fact, if X is an exponential random variable. Then by (3.19) we have

$$P(X > s) = 1 - F(s) = e^{-\lambda s}.$$

On the other hand, if X has the density function (3.19), then

$$P(X > s + t \mid X > t) = \frac{P(X > s + t)}{P(X > t)} = \frac{P(X > t \cap X > t + s)}{P(X > t)} = \frac{e^{-\lambda(s+t)}}{e^{-\lambda t}} = e^{-\lambda s}$$

and equation (3.20) is satisfied. It is interesting to note that among discrete distributions the geometric distribution has the memoryless property.

A PRACTICE PROBLEM Suppose that an electronic component has an average life of 100 hours. Find the probability that one such component will last more than 100 h. If a component has already worked for 100 h, what is the probability that it will last a further 50 h?

3.8.2 Gamma Distribution

The exponential distribution is a special case of a more general model known as Gamma distribution. Gamma distribution provides a model for waiting time until the r-th occurrence of an event of interest. A random variable X has a Gamma distribution with parameters $\alpha > 0$ and $\lambda > 0$ if its density function takes the form

$$f(x) = \begin{cases} \frac{\lambda^\alpha}{\Gamma(\alpha)} x^{\alpha-1} e^{-\lambda x} & x \geq 0 \\ 0 & x < 0. \end{cases} \tag{3.21}$$

Here the Gamma function $\Gamma(\alpha)$ is defined by

$$\Gamma(\alpha) = \int_0^\infty y^{\alpha-1} e^{-y} dy \quad \text{for } \alpha > 0. \tag{3.22}$$

It is easy to verify that

$$\Gamma(1) = \int_0^\infty e^{-y} dy = 1$$

and

$$\Gamma(\alpha) = (\alpha - 1)\Gamma(\alpha - 1) \quad \text{for any } \alpha$$

Therefore, when α is a positive integer,

$$\Gamma(\alpha) = (\alpha - 1)!$$

which is the factorial of $(\alpha - 1)$.

Clearly, when $\alpha = 1$, the function $f(x)$ given by (3.21) becomes the same as the one given in (3.18). In other words, the exponential distribution is a special case of the Gamma distribution when $\alpha = 1$. In fact, it can be shown that if $X_1, X_2, \cdots, X_k$ are independent and all have exponential distribution, then $X_1 + X_2 + \cdots + X_k$ has a gamma distribution. For the soccer example discussed earlier, the waiting time to the first goal was found to have an exponential distribution. Thus the waiting time to the second, third, and so on, all have Gamma distribution. This relationship between exponential and Gamma distributions is similar to that of geometric and negative binomial. There are a number of insurance-related applications of the Gamma distribution. This distribution is, for example, used to model the average rate of claims filed by different policyholders of an insurance company. A further application of the Gamma distribution will be discussed in the next section.

The mean and variance of the Gamma distribution are, respectively,

$$E(X) = \frac{\alpha}{\lambda} \quad \text{and} \quad \text{Var}(X) = \frac{\alpha}{\lambda^2}.$$

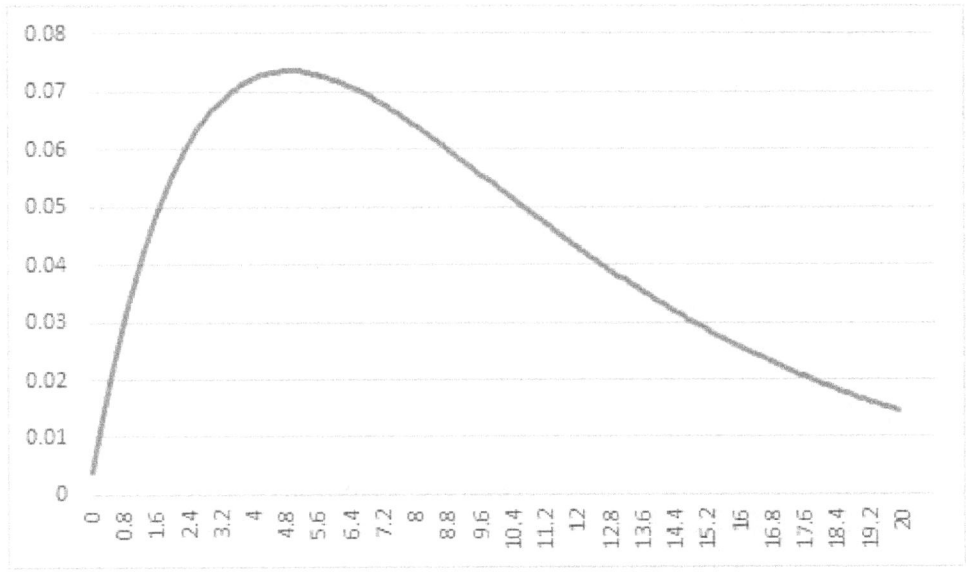

Gamma Distribution (Some Examples)

EXAMPLE Consider the time in minutes between goals scored in a soccer game. Suppose that this can be modeled using exponential distribution with the parameter $\lambda = 1/20$. Let X_1 and X_2 be the waiting times for the first goal and the waiting time between the first and the second goals. The total waiting time to the second goal is then $X = X_1 + X_2$ and has a Gamma distribution with $\alpha = 2$ and $\lambda = 1/20 = 0.05$. Its density function is

$$f(x) = \frac{(0.05)^2}{\Gamma(2)} x^{2-1} e^{-x/20}, \quad x \geq 0.$$

For example, the probability that the total waiting time to the second goal is between 30 and 40 min is

$$P(30 \leq X \leq 40) = \int_{30}^{40} \frac{1}{400} x e^{-x/20} dx = -(1+x/20)e^{-x/20} \Big|_{30}^{40} = -3e^{-2} + 2.5e^{-1.5} = 0.1.$$

Also

$$E(X) = 40 \;,\; \text{Var}(X) = 800.$$

It is interesting to note that the chi-square distribution is also a special case of the Gamma distribution. When $\alpha = \frac{v}{2}$ and $\lambda = \frac{1}{2}$, the resulting Gamma distribution is $\chi^2(v)$, where v is the number of degrees of freedom.

3.8.3 Beta Distribution

A random variable X has a beta distribution, and is referred to as a beta random variable, if and only if its probability density function takes the form

$$f(x) = \begin{cases} \frac{\Gamma(\alpha+\beta)}{\Gamma(\alpha)\Gamma(\beta)} x^{\alpha-1}(1-x)^{\beta-1}, & \text{for } 0 < x < 1, \; \alpha > 0 \;,\; \beta > 0 \\ 0 & \text{elsewhere.} \end{cases}$$

The uniform distribution $f(x) = 1$ for $0 < x < 1$ is a special case of the beta distribution when $a = \beta = 1$. In recent years, the beta distribution has found an important application in nonparametric or distribution-free inference statistical inference where result obtained are general and are applicable to any arbitrary distribution (Chapter 10). It can be applied to study the percent of defective units in a manufacturing process, the percent of errors made in data entry, and the percent of fans satisfied with the performance of the team they support.

Chapter 3: Some Special Distributions

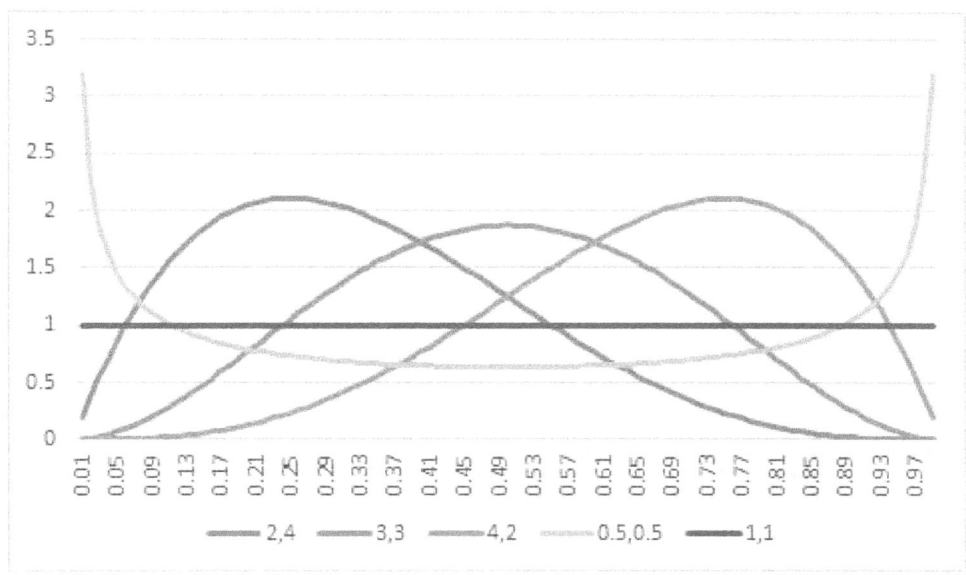

Beta Distribution ((2,4)(3,3)(4,2)(.5, 0.5)(1,1))

EXAMPLE The percent of fans who telephone a sport organization for information or services in a given week is a beta random variable with $\alpha = 4$ an $\beta = 3$. For $0 < x < 1$ the density and distribution functions are, respectively,

$$f(x) = \frac{6!}{2!3!}x^{4-1}(1-x)^{3-1} = 60(x^3 - 2x^4 + x^5),$$

and

$$F(x) = \int_0^x f(u)du = \int_0^x 60(x^3 - 2x^4 + x^5)dx = 60(x^4/4 - 2x^5/5 + x^6/6).$$

For example

$$P(X \le 0.40) = F(0.40) = 0.1792.$$

The mean and variance of the beta distribution are, respectively,

$$\mu = \frac{\alpha}{\alpha + \beta} \, , \, \sigma^2 = \frac{\alpha\beta}{(\alpha+\beta)^2(\alpha+\beta+1)}.$$

For the above example

$$E(X) = \frac{4}{4+3} = 0.5714,$$

$$Var(X) = \frac{(4)(3)}{(4+3)^2(4+3+1)} = 0.0306.$$

3.8.4 Lognormal Distribution (Another Relative of the Normal Distribution)

A nonnegative random variable X is said to have a lognormal distribution if random variable $Y = ln(X)$ has a normal distribution. The probability density function of a lognormal random variable when $\ln(X)$ is normally distributed with parameters μ and σ is

$$f(x) = \begin{cases} \frac{1}{\sqrt{2\pi}\sigma x} e^{-[\ln(x)-\mu]^2/(2\sigma^2)}, & x \geq 0 \\ 0, & x < 0. \end{cases}$$

Note that here μ and σ are not the mean and standard deviation of X but of $\ln(X)$. The mean and variance of X are, respectively,

$$E(X) = e^{\mu+\sigma^2/2}, \quad \text{Var}(X) = e^{2\mu+\sigma^2} \cdot (e^{\sigma^2} - 1) = (E(X))^2(e^{\sigma^2} - 1)$$

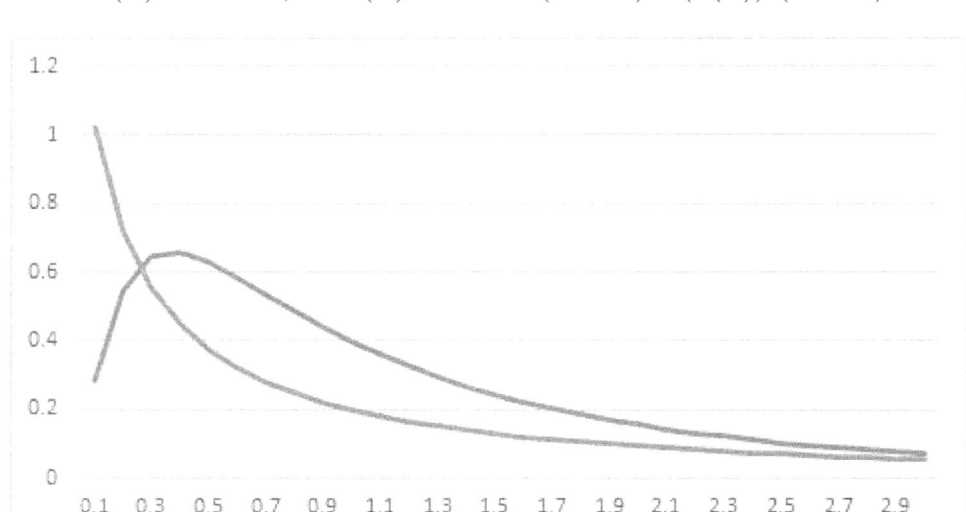

Lognormal Distribution $\sigma = 1$ and 2

Chapter 4, we will present a theoretical justification for usefulness of this distribution, using the Central Limit Theorem. Like other distributions, the lognormal distribution is used as a model for reasons other than the justification provided by the Central Limit Theorem. It arises when effects of many factors on a given response are not additive but multiplicative. Due to this fact, it has found a wide application in engineering-related problems. One common use of lognormal distribution is in modeling the lifetime of units whose failure modes are based on fatigue and stress. Since failure of mechanical devices are usually due to such factors, lognormal distribution is useful in assessing the lifetime of machines used in training, exercise, and sport-related activities.

Another area where lognormal distribution has found a wide application is the stock market. This is because the rate of growth for a single stock is sometimes modeled using normal distribution. Noting that for a continuous growth the value of an asset at time t and at the rate of r is

$$V(t) = V(0)e^{tr}$$

it follows that the distribution of $V(t)$ is lognormal.

EXAMPLE Suppose that a random variable $Y = lnX$ is distributed as $N(10,4)$. Then X has a lognormal distribution with mean and variance

$$E(X) = e^{10+(1/2)4} = e^{12}$$

and

$$\text{Var}(X) = e^{[2(10)+4]}(e^4 - 1) = e^{24}(e^4 - 1)$$

Also its mode and median are, respectively,

$$\text{Mode} = e^6 \simeq 403.43, \quad \text{Median} = e^{10} \simeq 22,026.$$

In order to calculate a specific probability, say $P(X \leq 1,000)$, we use the log transformation and calculate $P(lnX \leq ln1,000) = P(Y \leq ln1,000)$ instead. Using the given information, the required probability is then

$$P(Y \leq ln1,000) = P\left(z \leq \frac{ln1,000 - 10}{2}\right) = P(z \leq -1.55) = 0.0606.$$

A PRACTICE PROBLEM Suppose that a random variable Y has a lognormal distribution with mean e^3 and variance $e^6(e^4 - 1)$. Find the mean and variance of X where $Y = lnX$.

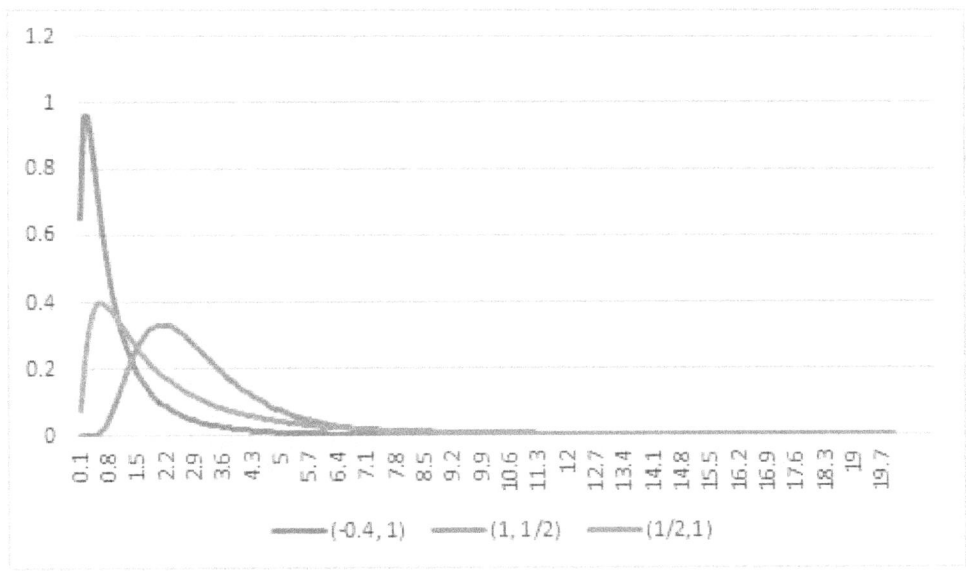

Lognormal Distribution (-0.4,1)(1,1/2)(1/2,1)

3.8.5 Gumbel Distribution

Gumbel distribution, also known as the Type I extreme value distribution, plays an important rule in analysis and modeling of the extreme values such the smallest or largest observations, for example, records. The extreme value theory deals with extreme values in place of typical or average values. It was originated from the needs of astronomers to be able to accept or reject extreme observations. Today it has found a wide range of applications since in many practical situations we are interested in or are concerned with the large or small values rather than average values. This is clearly the case in most sports where average performances are not of any interest, whereas extraordinary performance are of major interest.

In 1930, Gumbel studied the applications of the extreme value distributions. The first application was in modeling the old ages. The distribution then got the attention of engineers who applied that to model large floods.

The Gumbel's Type I extreme value distribution has two forms. The first form is based on the smallest extreme, or the minimum. The general form for the probability density function for the minimum is

$$f(x) = \frac{1}{\beta} e^{\frac{x-v}{\beta}} e^{-e^{\frac{x-v}{\beta}}}, \quad -\infty < x < \infty.$$

Consider the special case when $v = 0$ and $\beta = 1$. This is referred to as the standard Gumbel distribution. For this case,

$$f(x) = e^x e^{-e^x}, \quad -\infty < x < \infty$$

with the cumulative distribution function

$$F(x) = 1 - e^{-e^x}, \quad -\infty < x < \infty.$$

The cumulative distribution function for the minimum when $v \neq 0$ and $\beta \neq 1$ is

$$F(x) = 1 - e^{-e^{\frac{x-v}{\beta}}}, \quad -\infty < x < \infty.$$

Also

$$E(X) = v - 0.57772\beta, \quad \text{Mode} = v, \quad \text{Var}(X) = 1.64493\beta^2, \quad \text{Median} = v - 0.3665\beta.$$

or

$$E(X) = v - \gamma\beta, \quad \text{Mode} = v, \quad \text{Var}(X) = \pi^2/6\beta^2, \quad \text{Median} = v + ln(ln2)\beta$$

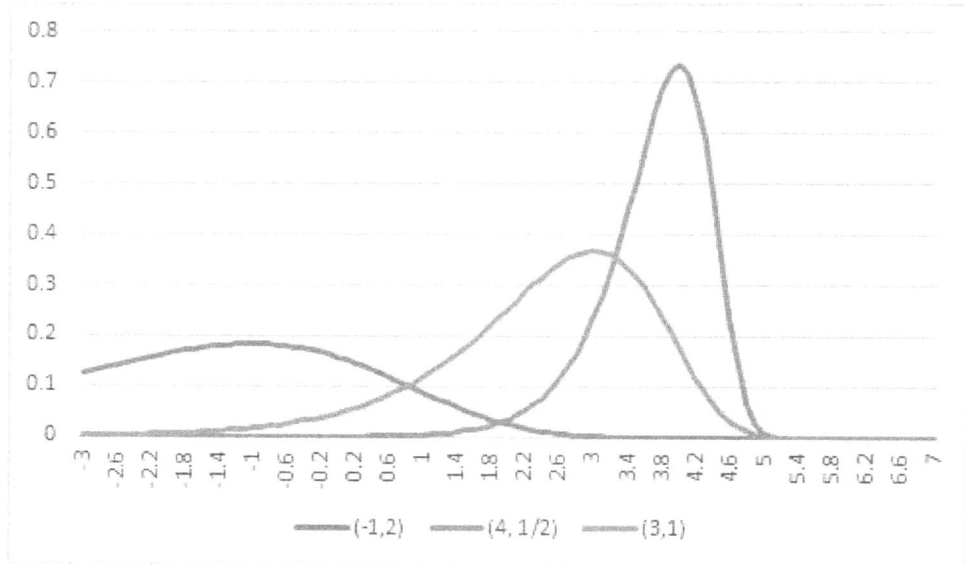

Extreme Value Distribution (-1,2)(4,1/2)(3,1)

Chapter 3: Some Special Distributions

EXAMPLE The annual mean maximum number of points scored in a basketball tournament in Philadelphia for a period of 40 years is given in the table below;

29.34	38.35	43.05	47.79
34.04	40.47	45.40	49.63
37.78	42.62	46.62	33.27
40.00	44.85	49.42	36.77
41.75	46.06	33.03	39.52
44.82	49.06	35.45	41.15
46.00	32.20	39.14	44.46
48.13	35.15	41.05	45.95
29.88	38.37	43.36	47.87
34.95	40.48	45.84	51.13

Suppose that the Gumbel's distribution is a good model for this data. Let X denote the annual mean maximum number of points scored in a given year. We have

$$F(x) = 1 - e^{-e^{\frac{x-v}{\beta}}}, \quad -\infty < x < \infty$$

where $v = 41.380$ and $\beta = 5.85$ are suggested values for v and β. Using the data, we want to answer the following questions:

(a) What is the probability of having a mean annual maximum greater than 41?
(b) What is the expected value and the variance of X?

SOLUTION:

(a)
$$P(X > 41.00) = 1 - P(X \leq 41.00) = 1 - F(41.00) = 1 - \exp[-e^{(41-41.38)/5.85}] = 0.608241.$$

(b)
$$E(X) = v - 0.57772\beta = 41.380 - 0.57772(5.85) = 38.0003,$$
$$\text{Var}(X) = 1.64493\beta^2 = 1.64493(5.85)^2 = 56.2936.$$

3.8.6 Weibull Distribution

The Weibull distribution takes its name from the Swedish physicist Waloddi Weibull who introduced the distribution in 1939. It is one of the three possible extreme value distributions (Type III) to be discussed further in Chapter 4. It is a widely applicable distribution that does have theoretical justification in some cases, but in many applications it simply provides a good fit to the observed data.

The Weibull distribution is used in reliability theory, life testing and in queuing theory where it is used to model interarrival and service times. Most commonly, the exponential distribution (which is the special case of Weibull distribution when $\beta = 1$) is used to model these phenomena, but if there are more large service times than the exponential distribution can account for, the Weibull distribution is considered to be a better model.

The Weibull distribution has also been used widely to model time to failure of multicomponent systems such as electronic circuits. It is not an appropriate model for systems subject to wear. In those cases, the normal distribution is usually used.

The random variable X has a Weibull distribution if its probability density function takes the form

$$f(x;\alpha,\beta,v) = \frac{\beta}{\alpha}\left(\frac{x-v}{\alpha}\right)^{\beta-1}\exp\left[-\left(\frac{x-v}{\alpha}\right)^{\beta}\right], \quad x \geq 0.$$

The distribution function is

$$F(x;\alpha,\beta,v) = 1 - \exp\left[-\left(\frac{x-v}{\alpha}\right)^{\beta}\right], \quad x \geq 0.$$

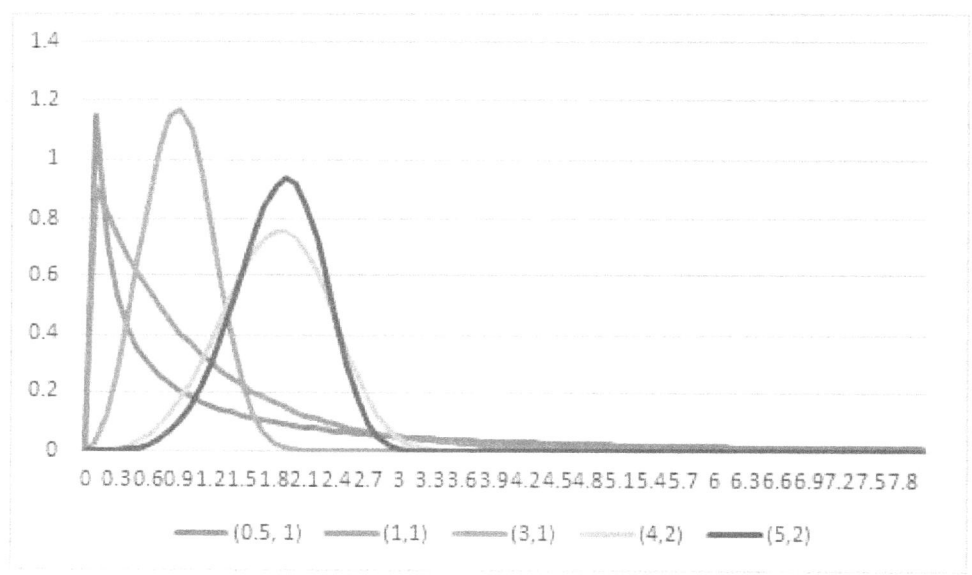

Weibull Distribution (0.5, 1), (1,1), (3,1), (4,2), (5,2)

The Weibull distribution has three parameters:

1. The location parameter v $(-\infty < v < \infty)$.
2. The scale parameter α $(\alpha > 0)$.
3. The shape parameter β $(\beta > 0)$.

The location parameter establishes the beginning of non-zero probability density on the *x*-axis. The scale parameter stretches or compresses the function along the *x*-axis. The shape parameter determines the location of the peak of the function on the *y*-axis, that is, the mode of the continuous distribution. Shown below are graphs of the Weibull probability density function for various values of the shape parameter with $v = 0$ and $\alpha = 1$. More general graphs are also presented.

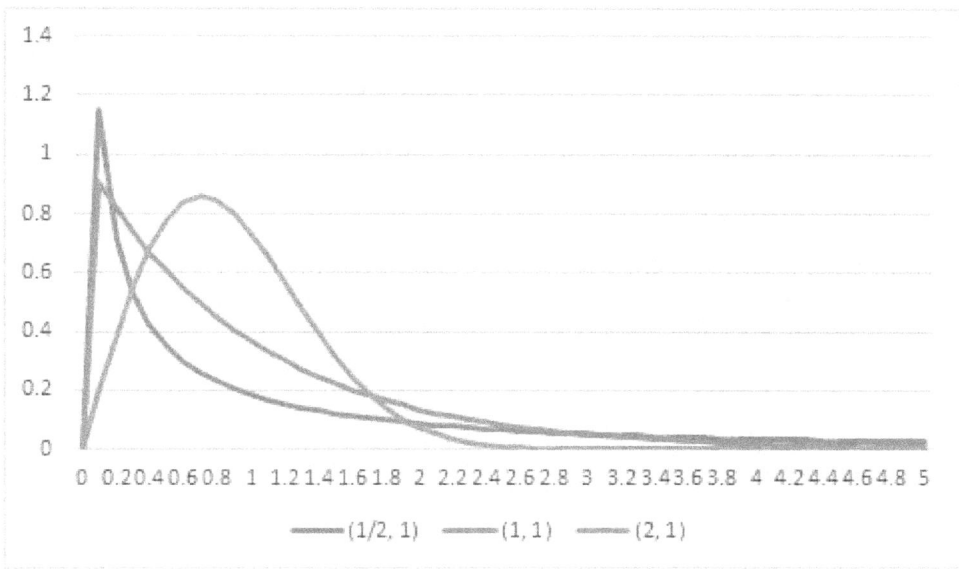

Weibull Distribution (1/2,1)(1,1)(2,1)

The Weibull distribution reduces to the exponential distribution when $v = 0$ and $\beta = 1$. The exponential distribution is a special case of both Gamma and Weibull distributions.
The expected value and variance of Weibull distribution are, respectively,

$$E(X) = v + \alpha \Gamma(1 + 2/\beta),$$
$$Var(X) = \alpha^2 \left(\Gamma(1 + 2/\beta) - (\Gamma(1 + 1/\beta))^2 \right).$$

Here the capital gamma refers to the gamma function, a continuous generalization of the factorial operation, as mentioned earlier, that is

$$\Gamma(\beta) = \int_0^\infty x^{\beta-1} e^{-x} dx \;,\; \Gamma(n) = (n-1)!$$

EXAMPLE Consider the following Weibull distribution.

$$f(x) = 5x\, e^{-2.5x^2} \;,\; x \leq 0.$$

Suppose that here X represents the length of the time (in years) a player can play before an injury occurs (survival time). Find the probability that (a) the player injures during the first 6 months, and (b) the player is able to play for longer than 1 year.

SOLUTION:

(a) $P(X \geq 0.5) = F(0.5) = 1 - e^{-2.5(0.5)^2} = 0.465.$
(b) $P(X > 1) = 1 - F(1) = e^{-2.5(1^2)} = 0.082.$

Also the mean and variance of X are (using $\Gamma(3/2) = \sqrt{\pi}/2$)

$$E(X) = 0.560499 \;,\; \text{Var}(X) = 0.085841.$$

We end this section by noting that the Weibull distribution has been successfully applied to many problems in sports. One interesting application has been in modeling the dart throws. Stern (1997) has demonstrated that this model fits the errors observed in throwing darts.

3.8.7 Fréchet Distribution

The Fréchett distribution is also referred to as Type II extreme value distribution. Other two types are Type I(Gumbel) and Type III (Weibull). Each of these distributions can also be written in two forms: one for the maximum values and one for the minimum values (left or right bound). Each form can be transformed into the other by replacing the fitted variable, x, by - x and by adjusting the location parameter.

For the maximum, the general form of the Frechet probability density function is

$$f(x) = \beta \left(\frac{x - v}{\alpha}\right)^{-(\beta+1)} \exp\left[-\left(\frac{x - v}{\alpha}\right)^{-\beta}\right], \quad x \geq v.$$

The parameter v is a location parameter and the parameter α is a scale parameter.

3.8.8 Maxwell Distribution

James Clerk Maxwell (1831–1879) used his knowledge of classical electromagnetism to come up with a set of Maxwell Equations. One of his more important contributions is the kinetic theory of gases, which resulted in the derivation of the Maxwell's distribution. He realized that thermal physics revolved around probability.

The Maxwell distribution can be used to calculate the fraction of molecules with velocities exceeding a critical value, say 1,500 m/s at 2,000°C. This might represent molecules that can escape from a planetary atmosphere or undergo a chemical reaction.

Imagine a container filled with gas. Maxwell made three probabilistic assumptions about the gas in this container:

1. The distribution of the molecules is uniform because they fill all space in the box and do not cluster in one area (ignoring the effects of gravity).
2. The three components of velocity are mutually independent.
3. The probability distribution is isotropic (identical). That is, all have the same form because none of the directions are more important than another.

Through using certain formulas developed by physicists, Maxwell derived the following distribution

$$f(x) = 4\pi n (m/2\pi kT)^{3/2} x^2 e^{-mx^2/2kT}$$

where n is the number of molecules per unit volume and X is the velocity of a molecule. Also T is the absolute temperature, m is the mass of a molecule, and k is the Boltzmann constant. This expression has the maximum velocity at $x = (2kT/m)^{1/2}$, which is called the most probable velocity.

3.8.9 Pareto Distribution

The Pareto distribution has applications in economics (e.g., in modeling the large incomes) and is used to model certain insurance loss amounts. It has a number of different equivalent formulations. The one we have chosen involves two parameters, α and β, and takes the form

$$f(x) = \frac{\alpha}{\beta}\left(\frac{\beta}{x}\right)^{\alpha+1}, \alpha > 2,\ x \geq \beta > 0.$$

For example, for $\alpha = 2.5$ and $\beta = 3$, the density function is

$$f(x) = \frac{2.5}{3}\left(\frac{3}{x}\right)^{3.5}, \text{ for } x \geq 3.$$

Note that here the density function starts at 3 and can represent, for example, an insurance policy with deductible of 3 units.

Since the Pareto distribution has a density which is a power function, $F(x)$ can be easily found. It is

$$F(x) = 1 - \left(\frac{\beta}{x}\right)^{\alpha}, \alpha > 2,\ x \geq \beta > 0.$$

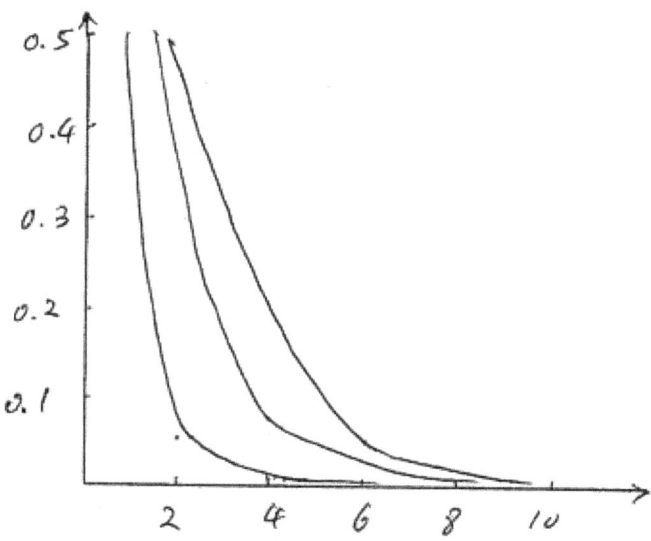

Pareto Distribution $(1/2, 2), (1, 1), (2, 1)$

Also the reliability or survival function is (see Section 3.10)

$$P(X > x) = R(x) = 1 - F(x) = (\beta/x)^{\alpha},\ \alpha > 2,\ x \geq \beta > 0.$$

EXAMPLE For the Pareto random variable with $\alpha = 2.5$ and $\beta = 3$, the cumulative distribution function is

$$F(x) = 1 - \left(\frac{3}{x}\right)^{2.5}, \text{ for } x \geq 3.$$

If the random variable X represents a possible loss amount, find the probability that the loss (a) is between 4 and 6 and, (b) is greater than 10.

SOLUTION:

(a) $P(4 \leq X \leq 6) = F(6) - F(4) = \left(\frac{3}{4}\right)^{2.5} - \left(\frac{3}{6}\right) = 0.3104$,

(b) $P(X > 10) = R(10) 1 - F(10) = \left(\frac{3}{10}\right)^{2.5} = 0.0493$.

The mean and variance of the Pareto distribution are

$$E(X) = \frac{\alpha\beta}{\alpha - 1}, \quad \text{Var}(X) = \frac{\alpha\beta^2}{\alpha - 2} - \left(\frac{\alpha\beta}{\alpha - 1}\right)^2.$$

For the Pareto distribution with $\alpha = 2.5$ and $\beta = 3$, the mean and variance are, respectively,

$$E(X) = \frac{2.5(3)}{2.5 - 1} = 5, \quad \text{Var}(X) = \frac{2.5(3)^2}{2.5 - 2} - \left(\frac{2.5(3)^2}{2.5 - 1}\right)^2 = 20.$$

Note that if we look at X as a loss amount in hundreds of dollars, then in this case the expected loss is $500. However, we have interpreted the insurance modeled as one for the loss less a deductible of $300. The random variable for the amount paid on a single claim is therefore $X - 3$. Thus, the expected amount of a single claim is

$$E(X - 3) = E(X) - 3 = 2.$$

3.8.10 Relationships Between Some Distributions

In sports, we are usually interested either in the best records of a given year (or season) or the performances that exceed a given threshold. Study of these types of data has led to two well-known family of distributions discussed below.

(a) Generalized Extreme Value Distribution

Let $X_1, X_2, \cdots$ be a sequence of independent and identically distributed random variables and X be the maximum of the first n, that is

$$X = \text{Max}(X_1, X_2, \cdots X_n).$$

It is shown that as $n \to \infty$ distribution of X tends to the generalized extreme value distribution that includes three families of distributions discussed in Sections 3.8.5, 3.8.6 and 3.8.7. These distributions are also known as Type I (Gumble), Type II (Frechet), and Type III (Weibull) distributions.

(b) Generalized Pareto Distribution

Generalized Pareto distribution arises when one is interested in values of a random variable above a large threshold. It describes and models the tail of the classical distributions discussed in this chapter.

Chapter 3: Some Special Distributions

The generalized Pareto distribution consists of three different families, Pareto(long-tail), Exponential(medium-tail) and Power(short-tail).

3.9 THE POISSON PROCESS

In Section 3.4, the Poisson distribution was introduced and some of its many applications in sport was mentioned. The Poisson distribution plays a very important role in science and engineering since it represents an appropriate probabilistic model for a large number of observational phenomena. It also arises in discussion of an extremely useful process known as Poisson process, which is related to other distributions such as the exponential, gamma, and uniform.

Consider a sporting event and the event of fans arriving at the stadium. Let X_t denote the number of fans arriving during a specified time period of length t before the game starts. The possible values of X_t are $0, 1, 2, \ldots$ Clearly, as t increase the chance that X_t takes a larger value is also increases. Thus X_t is a discrete random variable whose probability distribution depends on the value of t. In order to develop a suitable model for this situation, we need to make some assumptions about X_t. The plausibility of assumptions introduced below (recalling what X_t represents) is well substantiated. Indeed the empirical evidence obtained in practice strongly support the theoretical results that are derived below. As is well-known in deducting any mathematical results, we must accept some underlying postulates or axioms. Appropriate axioms are often set up to describe observational phenomena, and some axioms may be far more plausible (and less arbitrary) than others.

For the case introduced above, five assumptions will be made. These assumption are necessary for constructing a probabilistic model for the number of fans arriving at the stadium. Recall that the random variable X_t may assume the values $0, 1, 2, \ldots$. Let $P_n(t)$ represent the probability of observing exactly n arrivals in the time interval of length t, that is,

$$p_n(t) = P(X_t = n), \quad n = 0, 1, 2, \ldots.$$

ASSUMPTION 1: The number of arrivals during non overlapping time intervals are independent random variables.

ASSUMPTION 2: The distribution of the number of arrivals during any time interval depends only on the length of that interval and not on its endpoints. That is, if X_t is defined as above and if Y_t equals the number of arrivals during $[t, t+t)$, then for any $t > 0$, the random variables X_t and Y_t have the same probability distribution.

ASSUMPTION 3: Let Δt denote a small positive number, or equivalently a small time period. We assume that there is a positive constant λ such that when Δt is sufficiently small, $p_1(\Delta t)$ is approximately equal to $\lambda \Delta t$, that is, it is proportional to the length of interval. In this case, λ is called the rate. We denote this as $p_1(\Delta t) \sim \lambda \Delta t$. Throughout this section $g(\Delta t) \sim h(\Delta t)$ means that $g(\Delta t)/h(\Delta t) \to 1$ as $\Delta t \to 0$. This assumption states that if the interval is sufficiently small, the probability of obtaining exactly one arrival during that interval is directly proportional to the length of that interval. This also explains why the constant λ is called the rate.

ASSUMPTION 4: $\Sigma_{k=2}^{\infty} p_k(\Delta t) \sim 0$, which implies that $p_k(\Delta t) \sim 0$, $k \geq 2$. This assumption states that the probability of obtaining two or more arrivals during a sufficiently small interval is negligible.

ASSUMPTION 5: $X_0 = 0$, or equivalently $p_0(0) = 1$, that is, the probability of zero arrivals in the time interval of length zero is one. This is the initial condition for the model described below.

As we shall shortly demonstrate, the five assumptions listed above are sufficient for deriving an expression for $p_n(t) = P(X_t = n)$. Let us first present a theorem that states a number of conclusions drawn from the above assumptions.

THEOREM 3.9.1

(a) The random variables $X_{\Delta t}$ and $[X_{t+\Delta t} - X_t]$ are independent random variables with the same probability distribution (see figure below). Here 0 is used to represent a fixed starting point (initial starting time) from which the other times are measured.

```
|    |           |    |
0   Δt          t   t+Δt
```

(b)
$$p_0(\Delta t) \sim 1 - \lambda \Delta t. \tag{3.23}$$

(c)
$$p_0(t + \Delta t) \sim p_0(t)[1 - \lambda \Delta t],$$

and hence we have

$$\frac{p_0(t + \Delta t) - p_0(t)}{\Delta t} \sim -\lambda p_0(t).$$

PROOF: Assumptions 1 and 2 together imply conclusion (a). From Assumptions 3 and 4 we see that

$$p_0(\Delta t) = 1 - p_1(\Delta t) - \Sigma_{k=2}^{\infty} p_k(\Delta t) \sim 1 - \lambda \Delta t,$$

which leads to (b). Finally, combining (a) and (b) we see that

$$p_0(t + \Delta t) = P(X_t = 0 \text{ and } X_{t+\Delta t} - X_t = 0)$$
$$= p_0(t) p_0(\Delta t) \sim p_0(t)[1 - \lambda \Delta t],$$

and thus (c) is obtained.

Note that, using the last term, we have approximately

$$p_0(t + \Delta t) - p_0(t) = -\lambda p_0(t) \Delta t$$

or

$$\frac{p_0(t + \Delta t) - p_0(t)}{\Delta t} = -\lambda p_0(t).$$

Letting $\Delta t \to 0$, the left-hand side that represents the difference quotient of the function $p_0(t)$ approaches $p_0'(t)$. Thus, we have

$$p_0'(t) = -\lambda p_0(t) \Rightarrow \frac{p_0'(t)}{p_0(t)} = -\lambda.$$

Chapter 3: Some Special Distributions

Integrating both sides with respect to t yields $\ln p_0(t) = -\lambda t + C$, where C is a constant of integration. From Assumption 5 we find, by letting $t = 0$, that $C = 0$. Hence,

$$p_0(t) = e^{-\lambda t}. \tag{3.24}$$

Thus, these assumptions led to an expression for $P(X_t = 0)$. It is clear that $p_0(t) \to 0$ as $t \to \infty$, representing the common sense conclusion that if we wait long enough with a probability close to one fan will arrive. Using essentially the same approach, we shall now obtain $p_n(t)$ for $n \geq 1$. We start by the following theorem.

THEOREM 3.9.2 The number of arrivals during the time interval $[0, t)$ subject to the assumptions made above is a random variable having Poisson distribution with parameter λt, that is,

$$p_n(t) = e^{-\lambda t}(\lambda t)^n / n!, \quad n = 0, 1, 2, \ldots. \tag{3.25}$$

PROOF: Consider $p_n(t + \Delta t) = P(X_{t+\Delta t} = n)$. Clearly $X_{t+\Delta t} = n$ if only if $X_t = x$ and $[X_{t+\Delta T} - X_t] = n - x, x = 0, 1, 2, \ldots, n$. Using Assumptions 1 and 2, we have

$$p_n(t + \Delta t) = \Sigma_{x=0}^{n} p_x(t) p_{n-x}(\Delta t) - \Sigma_{x=0}^{n-2} p_n(t) p_{n-x}(\Delta t) + p_{n-1}(t) p_1(\Delta t) + p_n(t) p_0(\Delta t).$$

Using Assumptions 3 and 4 and also Equation (3.23), we obtain

$$p_n(t + \Delta t) \sim p_{n-1}(t) \lambda \Delta t + p_n(t)[1 - \lambda \Delta t].$$

and hence

$$\frac{p_n(t + \Delta t) - p_n(t)}{\Delta t} \sim \lambda p_{n-1}(t) - \lambda p_n(t).$$

Again letting $\Delta t \to 0$ and observing that the left-hand side represents the difference quotient of the function $p_n(t)$, we obtain

$$p_n'(t) = -\lambda p_n(t) + \lambda p_{n-1}(t), \quad n = 1, 2, \ldots.$$

This represents an infinite system of linear difference (differential) equations. If we define the function $q_n(t)$ by the relation $q_n(t) = e^{\lambda t} p_n(t)$, or $p_n(t) = e^{-\lambda t} q_n(t)$, then we have $p_n'(t) = -\lambda e^{-\lambda t} q_n(t) + e^{-\lambda t} q_n'(t)$. Substituting it in the above equation gives

$$q_n'(t) = \lambda q_{n-1}(t), \quad n = 1, 2, \ldots.$$

Since $p_0(t) = e^{-\lambda t}$, we have $q_0(t) = 1$. (Note that $q_n(0) = 0$ for $n > 0$.) Then we obtain, recursively,

$$q_1'(t) = \lambda, \quad \text{and hence} \quad q_1(t) = \lambda t$$

$$q_2'(t) = \lambda q_1(t) = \lambda^2 t, \quad \text{and hence} \quad q_2(t) = \frac{(\lambda t)^2}{2}.$$

In general, $q_n'(t) = \lambda q_{n-1}(t)$ and hence $q_n(t) = (\lambda t)^n / n!$ Recalling the definition of $q_n(t)$, we finally obtain

$$p_n(t) = e^{-\lambda t}(\lambda t)^n / n!, \quad n = 0, 1, 2, \ldots.$$

The theorem is thus proved.

3.9.1 Facts and Applications

(a) The following are two interesting facts about the Poisson Process:

  (i) Since $p_0(t) = e^{-\lambda t}$, it follows that the distribution of the time to the first arrival in a Poisson process is exponential. Also, it can be shown that the time to the r-th arrival has a Gamma distribution.

  (ii) If X_t and Y_t are number of arrivals (fans) for two competing teams (or from two different entrances) having Poisson distributions with parameters (rates) λ and μ, respectively, then the sum $X_t + Y_t$ has a Poisson distribution with parameter (rate) $\lambda + \mu$.

(b) It is important to realize that the Poisson distribution appeared as a consequence of certain assumptions we made. This means that whenever these assumptions are valid (or at least approximately so), the Poisson distribution should be used as an appropriate model. It turns out that there is a large class of phenomena for which the Poisson model is appropriate. Here are two examples:

  (i) Let X_t represent the number of telephone calls arriving at a ticket office of a sport organization during a time period of length t. The above assumptions are approximately satisfied, particularly during the "busy period" of the season. Hence X_t has a Poisson distribution.

  (ii) Let X_t represent the number of sport injuries in a given season. Again the above assumptions are appropriate, and hence, X_t has a Poisson distribution.

(c) The parameter λ originally appeared as a constant of proportionality in Assumption 3. The following interpretations of λ are worthwhile to note. If X_t represents the number of occurrences of some event during a time interval of length t, then $E(X_t) = \lambda t$, and hence $\lambda = [E(X_t)]/t$ represents the expected rate of occurrences.

(d) It is important to realize that the discussion presented in this section did not just deal with a random variable possessing a Poisson distribution. Rather, for every $t > 0$, we found that X_t had a Poisson distribution with a parameter dependent on t. Hence, we have an infinite collection of random variables $\{X_t, t > 0\}$. Such a collection of random variables is also known as a Poisson Process.

EXAMPLE 3.9.1 Fans arrive at a certain stadium according to a Poisson Process of rate $\lambda = 100$ per minute. Given that the stadium opens at 5:00 p.m., what is the probability that exactly 2,500 fans arrive by 5:30 p.m.

SOLUTION: Because of independence, the number of fans arriving in a half hour has a Poisson distribution with the rate of $(100)(30) = 3,000$. Thus we have

$$P(X_{30} = 2,500) = \frac{e^{-2,500}(2,500)^{30}}{30!}.$$

Chapter 3: Some Special Distributions

EXAMPLE 3.9.2 Suppose that students arrive at the university recreation center according to a Poisson Process with a rate of two students per minute. Let X_t be the number of students that arrive up to time t. Find

(a) $P(X_1 = 2)$,
(b) $P(X_1 = 2 \text{ and } X_3 = 6) = P(X_1 = 2 \cap X_3 = 6)$,
(c) $P(X_1 = 2 \mid X_3 = 6)$,
(d) $P(X_3 = 6 \mid X_1 = 2)$.

SOLUTION:

(a) $$P(X_1 = 2) = e^{-2} 2^2/2! = 2e^{-2} = 0.271.$$

(b) $$P(X_1 = 2 \text{ and } X_3 = 6) = P(X_1 = 2 \text{ and } X_3 - X_1 = 4) = P(X_1 = 2) \cdot P(X_2 = 4)$$
$$= e^{-2} 2^2/2! \cdot e^{-4} 4^4/4! = 0.053.$$

(c) $$P(X_1 = 2 \mid X_3 = 6) = \frac{P(X_1 = 2 \text{ and } X_3 = 6)}{P(X_3 = 6)}$$
$$= \frac{e^{-2} 2^2/2! \, e^{-4} 4^4/4!}{e^{-6} 6^6/6!}$$
$$= \frac{6!}{2! 4!} (2/6)^2 (4/6)^4$$
$$= \binom{6}{2} (2/6)^2 (4/6)^4 = 0.329.$$

(d) $$P(X_3 = 6 \mid X_1 = 2) = \frac{P(X_3 = 6 \text{ and } X_3 = 6)}{P(X_1 = 2)}$$
$$= \frac{e^{-2} 2^2/2! \, e^{-4} 4^4/4!}{e^{-2} 2^2/2!}$$
$$= e^{-4} 4^4/4! = P(X_2 = 4).$$

It is interesting to note (1) the relation of the conditional probability to the binomial distribution in part (c) and (2) the consequence of lack of memory property in part (d).

EXAMPLE 3.9.3 A complicated piece of exercise machinery, when running properly, can bring a profit of C dollars per hour ($C > 2$) to a firm. However, this machine has a tendency to break down at unexpected and unpredictable times. Suppose that the number of breakdowns during any period of length t hours is a random variable with a Poisson distribution with parameter t. If the machine breaks down X times during the t hours, the loss incurred (shutdown of machine plus repair) equals $(X^2 + X)$ dollars. Thus, the total profit P during any period of t hours equals $P = Ct - (X^2 + X)$, where X is the random variable representing the number of breakdowns of the machine. Therefore P is a random variable, and it might be of interest to choose t (which is at our disposal) in such a manner that the expected profit is maximized. We have

$$E(P) = Ct - E(X^2 + X).$$

Thus we find that $E(X) = t$ and $E(X^2) = t + t^2$. It then follows that $E(P) = Ct - 2t - t^2$. To find the value of t for which $E(P)$ is maximized, we differentiate $E(P)$ and set the resulting expression equal to zero. We obtain $C - 2 - 2t = 0$, yielding $t = \frac{1}{2}(C - 2)$ hours.

A PRACTICE PROBLEM During the year 2018, Major League Soccer (MLS), the following players were leading goal scoreres (by the minutes of play).

1. Josef Martinez from Atlanta, 1 goal every 93 min.
2. Zlatan Ibrahimovic from LA Galaxy, 1 goal every 97 min.
3. Bradley Wright-Phillips from New York Red Bulls, 1 goal every 126 min.
4. Mauro Manotas from Houston, 1 goal every 138 minutes.

Assuming that goals scored follows the Poisson Process find λ for each player. What are the probabilities that in the next 3 h of play each player scores one goal? How many goals do you expect to be scored in the next game (90 min) if all these players play for the same team? Find the probability distribution of the number of goals that will be scored by all these players in the next game (combined). What is the distribution of time to the next goal by Zlatan Ibrahimovic?

3.10 USING R

There are many existing functions in R for many known probability distributions. Besides ones we discussed in Chapter 2, we demonstrate examples of more calculations using R. Please note that the functions starting with a letter "d" is either a point mass function or probability density function; while functions starting with "p," for example, pbinom, are cumulative distribution functions.

EXAMPLE 3.10.1 Recall Example 3.2.1, the random variable X is a geometric random variable with $p = 0.6$. Therefore, we can calculate the probability $P(X = 3)$ as following:

```
# Syntax: dgeom(NUM_FAIL_BEFORE_FIRST_SUCCESS, PROB_OF_SUCCESS)
# So here we will use 3-1=2 as the first argument.
dgeom(2, 0.6)
```

EXAMPLE 3.10.2 Recall Example 3.2.4, X is a Negative Binomial random variable with parameters $k = 3$ and $p = 0.4$. To find $P(X = 10)$ one can use

```
# Syntax: dnbinom(NUM_FAIL, NUM_SUCCESS, PROB_OF_SUCCESS)
# So here we will use 10-3=7 as the first argument, 3 success.
dnbinom(7, 3, 0.4)
```

Chapter 3: Some Special Distributions

EXAMPLE 3.10.3 Recall Example 3.3.1, here X is hypergeometric random variable with $n = 8$, $k = 30$ and $N = 100$. To find $P(X = 3)$:

```
# Syntax: dhyper(x, k, N-k, n)
dhyper(3, 30, 70, 8)
```

EXAMPLE 3.10.4 Recall Example 3.4.1, we calculate $P(X = 4)$ when X is Poisson random variable with $\lambda = 5$.

```
dpois(4, 5)
```

EXAMPLE 3.10.5 In Example 3.6.3, we applied normal distribution to PPM variable X and assume $X \sim N(0.4263, 0.1159^2)$. Thus $P(X > 0.8291)$ is

```
1-pnorm(0.8291, 0.4263, 0.1159)
```

Please note we need to use standard deviation but not variance in the function.

Functions dchisq, pchisq, dgamma, pgamma, dbeta, pbeta, dexp, pexp, dlnorm, plnorm, dweibull, pweibull are prepared for Chi-square, Gamma, Beta, Exponential, Lognormal, Weibull distributions. Unfortunately there is no functions for Maxwell, Pareto, and Fréchet distributions, but there are available in other packages.

REFERENCES

Chatterjee, S., M. S. Handcock, and J. S. Simonoff. 1995. *A Casebook for a First Course in Statistics and Data Analysis*. New York: John Wiley.

Stern, Hal S. 1997. "Shooting Darts." *Chance* 10 (3):16–19.

CHAPTER 4

Limit Properties and Models

This chapter represents some limit properties and models that have proved extremely useful in applications. One such property is the celebrated Central Limit Theorem, often referred to as the most important theorem of statistics. This theorem is the foundation of a major part of statistical inference discussed in later chapters.

Models from limiting cases may be divided into three classes.

1. The model of sums: The Normal and Poisson Distributions.
2. The model of products: The Lognormal Distribution.
3. The model of extremes: The Extreme Value Distributions.

We start with a short introduction involving a special distribution discussed in chapter 3. Recall the Bernoulli random variable X_i with two possible values 0 and 1 and the corresponding probabilities

$$p = P(X_i = 1) \text{ and } q = 1 - p = P(X_i = 0).$$

Suppose that $X_i's$, $i = 1, 2, \cdots, n$ are independent and we are interested in the following sum:

$$X_1 + X_2 + \cdots + X_n.$$

It can be shown that the distribution of this sum is

(a) Binomial, if n is finite.
(b) Poisson, if $n \to \infty$, $p \to 0$ such that $np \to \lambda$ remains constant.
(c) Normal, if $n \to \infty$ and p is fixed.

This type of approximations are very useful, particularly when they take forms that are easier to manipulate. To clarify, consider the following example of approximating a hypergeometric probability by a binomial probability. Note that this approximation is not of practical interest but merely a demonstration.

EXAMPLE A production lot of 200 tennis shoes has 8 defective shoes. A random sample of 10 shoes is selected. Find the probability that the sample will contain exactly 1 defective shoe.

SOLUTION: The distribution of the number of defective shoes is hypergeometric and we have

$$P(X=1) = \binom{8}{1}\binom{192}{9} / \binom{200}{10} = 0.2878.$$

Since $n/N = 10/200 = 0.05 < 0.1$, it is possible to calculate this probability approximately using binomial distribution with $p = 8/200 = 0.04$. Then we have

$$P(X=1) \approx \binom{10}{1}(0.04)^1(0.06)^9 = 0.2770.$$

This approximation, of course, may not be satisfactory, but it serves as an example.

4.1 THE MODEL OF SUMS: NORMAL DISTRIBUTION

Many applications of statistics involve sums or averages of a set of measurements. For example, airlines have a baggage limit (weight or volume) and their concern is the total weight or volume of the baggage. In the stock market, daily changes are of less concern than moving averages. Most businesses are concerned with total sale rather than hourly or daily sales. In this section, we consider a model for sum and present the well-known Central Limit Theorem. This theorem plays a key role in statistics.

4.1.1 The Central Limit Theorem

Let $X_1, \ldots, X_n$ be n independent random variables with the same distribution (random sample), and consequently the same expected value μ and the same variance σ^2, and let

$$Y_n = X_1 + \ldots + X_n. \tag{4.1}$$

First recall that

$$E(Y_n) = E(X_1 + X_2 + \ldots + X_n) = E(X_1) + E(X_2) + \cdots + E(X_n) = n\mu$$
$$\text{Var}(Y_n) = \text{Var}(X_1 + X_2 + \ldots + X_n) = \text{Var}(X_1) + \text{Var}(X_2) + \cdots + \text{Var}(X_n)$$
$$= n\sigma^2.$$

If $X_i's$ are normally distributed random variables, then $Y_n \sim N(n\mu, n\sigma^2)$ and $(Y_n - n\mu)/\sigma\sqrt{n} \sim N(0,1)$ for any integer n.

If $X_i's$ are not normally distributed, it can be shown that for a large n

$$\frac{Y_n - n\mu}{\sigma\sqrt{n}} \tag{4.2}$$

is distributed approximately as the standard normal distribution $N(0,1)$. More precisely,

$$\lim_{n\to\infty} P(\frac{Y_n - n\mu}{\sigma\sqrt{n}} \leq y) = \Phi(y). \tag{4.3}$$

Chapter 4: Limit Properties and Models

This result is known as the Central Limit Theorem. It is a fundamental theorem and among other things demonstrates the importance of the normal distribution. It can be applied whenever n is sufficiently large. For practical purposes, we may replace

$$P(\frac{Y_n - n\mu}{\sigma\sqrt{n}} \leq y) \tag{4.4}$$

by its limit $\Phi(y)$ for $n \geq 10$. When $n > 30$, the approximation is very good.

We can also use a reformulation of the above result by introducing $\bar{X}$, the sample mean

$$\bar{X} = \frac{1}{n}(X_1 + \ldots + X_n). \tag{4.5}$$

For $\bar{X}$, instead of (4.2), we have the term

$$\frac{\bar{X} - \mu}{\sigma/\sqrt{n}}.$$

Again, for $n \geq 10$ we may use the following approximation:

$$P(\frac{\bar{X} - \mu}{\sigma/\sqrt{n}} \leq x) \approx \Phi(x). \tag{4.6}$$

We note once more that for normally distributed $X_1, \ldots, X_n$ this formula is exact. Most authors suggest using (4.6) when $n > 30$, although the approximation is quite good for a much smaller n, especially for symmetric parent distributions.

Many textbooks refer to distribution of $\bar{X}$ as the sampling distribution of the sample mean. In fact, as will be discussed further in Section 5.2, in many real-life situations one is not interested in distribution of a single measurement but the distribution of average of several measurements. This is partly because it is not reasonable to draw any conclusion based on a single observation. For example, when estimating population mean, the average of several measurements provides a much improved (standard deviation $\sigma/\sqrt{n}$ versus σ) estimate of the true mean.

EXAMPLE 4.1.1 Let X denote the number of points scored by a basketball player in a randomly selected game during the season. His last season statistics are 20 points per game with standard deviation 5. Let $\bar{X}$ denote the average number of points for this player in 64 games in the upcoming season. What is the probability that $\bar{X}$ will be less than 19? That is, what is the probability that his performance will be a little worse than the previous season. Assume that his minutes are the same as in previous seasons and that all other factors remain unchanged.

SOLUTION: Since 64 is greater than 30, $\bar{X}$ has approximately a normal distribution with mean of 20 and standard deviation of $5/\sqrt{64} = 0.625$. Hence, using (4.6) we have

$$P(\bar{X} < 19) \approx P(z < (19 - 20)/0.625) = \Phi(-1.6) = 0.0548.$$

EXAMPLE 4.1.2 In a small college 40 students want to try out for the basketball team. On the basis of years of experience, the coach knows that the time needed to evaluate a student is a random variable with an expected value of 6 minutes and a standard deviation of 6 minutes. If evaluation times are independent, and coach begins to evaluate at 7:50 a.m. and works continuously, what is the (approximate) probability that he is through evaluation before 12:00 p.m. when the sports news begins on TV? If the sports report begins at 12:10 p.m., what is the probability that he will miss part of the report if he waits until evaluation is done before turning on the TV?

SOLUTION: Let X_i denote the time needed to evaluate the ith student. Then the total time required for evaluation is
$$Y_n = X_1 + X_2 + \cdots + X_n.$$

Since $40 > 30$, Y_n has approximately a normal distribution with mean of $(40)(6) = 240$ and standard deviation of $6\sqrt{40} = 37.95$. Here we want to find $P(Y_n \leq 250)$ and $P(Y_n > 260)$, where 250 is the number of minutes from 7:50 a.m. to 12 p.m. We have

$$P(Y_n \leq 250) \approx P(z < (250 - 240)/37.95) = P(z \leq 0.26) = 0.6026,$$
$$P(Y_n > 260) \approx P(z > (260 - 240)/37.95) = P(z > 0.53) = 0.2981.$$

EXAMPLE 4.1.3 Consider the measurement errors due to the inaccuracy of an instrument, designed to measure the length. Suppose that we can divide the whole length to equal segments and measure each segment once or several times. If the errors are independent and have the same type of distribution, then the total error is the sum of many errors and according to the Central Limit Theorem, has a normal distribution.

A PRACTICE PROBLEM A manufacturer of soccer shoes claims that the distribution of the lengths of life of its best shoes has a mean of 48 months and a standard deviation of 6 months if it is used 3 times a week. Suppose a consumer group decides to check the claim by purchasing a sample of 49 of these shoes and subjecting them to tests that determine their life. Assuming that the manufacturer's claim is valid, describe the sampling distribution of the mean lifetime of a sample of 49 shoes (sample mean) and find the probability the consumer group's sample has a mean life of 46 or fewer months, that is, $P(\bar{X} \leq 46)$.

We end this section by noting that in many real-life situations, distribution of measurements are unknown. However, if we could come up with large samples, the distribution of sample average is normal and therefore inference regarding the sample mean is possible. This is a great help and shows the importance of the Central Limit Theorem to the researchers.

4.1.2 The Normal Approximation to the Binomial Distribution

Another limiting property that has proved useful in practice is the following: the binomial distribution with parameters n and p can be approximated by $N(np, npq)$, at least when $0.1 \leq p \leq q$ and $npq \geq 5$, or equivalently when np and nq are both greater than 5. Note that a binomial random variable is the sum of n independent Bernoulli random variables each assuming either 0 or 1. This shows the connection to the Central Limit Theorem. Note also that this approximation relates a discrete distribution to a continuous one.

Chapter 4: Limit Properties and Models

EXAMPLE 4.1.4 Suppose that two volleyball teams are to play a series of 40 games and suppose the value p, the probability for team A to win a randomly selected game, is 0.2. What is the probability that team A will win more than 10 games?

SOLUTION: Since $p = 0.2$, we have $q = 0.8$. The number of games team A wins in this series is a binomial random variable with $n = 40$ and $p = 0.02$. Let X denote this variable, then the required probability is

$$P(X > 10) = 1 - (p_0 + p_1 + p_2 + \ldots + p_{10})$$

where

$$p_i = \binom{40}{i}(0.2)^i(0.8)^{40-i}, \quad i = 0, 1, \ldots, 10.$$

A rather tedious calculation shows that this probability is approximately 0.1608.

Now, since $0.1 < p < q$ and $npq = 6.4 > 5$, we can use the normal distribution $N(np, npq)$ to approximate the required probability. In this case $np = 8$ and $npq = 6.4$. Note that for discrete random variables the probabilities $P(X > 10)$ and $P(X \geq 11)$ are equal. However, for continuous random variables they are different. To account for this, we consider the "middle value" and calculate $P(X > 10.5)$ when using normal approximation. This adjustment is called continuity correction. We then have

$$P(X > 10.5) \approx P(z > (10.5 - 8)/\sqrt{6.4}) = P(z > 0.99) = 0.1611$$

which is quite accurate.

As pointed out earlier, the above approximation is valid under certain conditions. What happens to the sum of Bernoulli random variables (binomial) if such conditions are not present, for example, when $p < 0.1$? The Section 4.2 presents a limiting theory that covers this case. Note that this situation arises when the probability of success is very small, that is, the event under consideration is rare. For this reason, the theorem dealing with this situation is often referred to as the Law of Rare Events.

A PRACTICE PROBLEM A fair coin is tossed 100 times. Find the probability of observing fewer than 45 heads.

4.2 POISSON LIMIT THEOREM

This useful limit property relates binomial distribution to the Poisson distribution. It is known as Poisson Limit Theorem and is applicable when $p \leq 0.1$ ($p \leq q$). This situation arises when the probability of success is small (rare event) and number of trials is large. In chapter 3, we presented several examples from soccer in the section where Poisson distribution was introduced. In soccer, possession of the ball is important, since whenever a team has the ball, it has an opportunity to attack and perhaps score. The probability p that any attack will result in a goal is small, but there are many such attacks in a game. If attacks are independent, then the distribution of the number of goals scored is binomial, but since p is usually small the Poisson approximation may be applied.

Theoretically, the Poisson Limit Theorem is valid when the value of p changes as n increases. In fact, it is more precise to use the notation p_n indicating that the value of p is dependent on n. The theorem states that if a constant $\lambda > 0$ exists such that

$$\lim_{n \to \infty} p_n = 0$$

and

$$\lim_{n \to \infty} np_n = \lambda$$

then the binomial distribution tends to the Poisson distribution with parameter λ. That is,

$$\lim_{n \to \infty} \binom{n}{k} p_n^k \cdot (1-p_n)^{n-k} = e^{-\lambda} \frac{\lambda^k}{k!}, \quad \text{where } \lambda = \lim_{n \to \infty} np. \tag{4.7}$$

Let us outline a proof for the Poisson Limit Theorem. First the left-hand side namely

$$\binom{n}{k} p_n^k (1-p_n)^{n-k}$$

can be written as

$$\frac{(np_n)^k}{k!} \cdot \frac{n(n-1)\cdots(n-k+1)}{n^k} (1-p_n)^{n-k}.$$

But

$$\frac{n(n-1)\cdots(n-k+1)}{n^k} = 1\left(1-\frac{1}{n}\right)\cdots\left(1-\frac{k-1}{n}\right) \to 1 \text{ as } n \to \infty,$$

$$\lim_{n \to \infty} \frac{(np_n)^k}{k!} = \frac{\lambda^k}{k!}$$

and

$$(1-p_n)^{n-k} = \left(1-\frac{np_n}{n}\right)^n \left(1-\frac{np_n}{n}\right)^{-k} \to e^{-\lambda} \cdot 1 = e^{-\lambda} \text{ as } n \to \infty.$$

Using these we obtain (4.7). As mentioned earlier, the Poisson Limit Theorem is also referred to as the Law of Rare Events.

For practical purposes when $n \geq 20$, $p \leq 0.1$ ($p \leq q$) one may, instead of the binomial term, use the corresponding Poisson term that is easier to calculate.

EXAMPLE 4.2.1 Suppose the chance of scoring a hole-in-one on a par three hole of a golf course is 1/500. Suppose that a certain player takes 100 shots, what is the probability that he will make no more than one hole-in-one?

SOLUTION: Let X denote the number of hole-in-one's he makes in 100 shots. Clearly X has a binomial distribution with $n = 100$ and $p = 1/500$. Hence, the required probability is

$$P(X \leq 1) = p_0 + p_1$$

where $p_0 = (499/500)^{100} = 0.819$ and $p_1 = (100!)/(1! \times 99!)(1/500)^1 (499/500)^{99} = 0.164$. Hence, $P(X \leq 1) = 0.819 + 0.164 = 0.983$. Now in this case $n = 100 > 20$, and $p = 1/500 < 0.1$. Thus, the

Poisson distribution with $\lambda = np = 0.2$ should approximate this probability quite well. Using the Poisson distribution we obtain

$$P(X \leq 1) = e^{-0.2}(1 + 0.2/1!) = 0.982.$$

We finish this section by noting that in sports the levels of excellence change constantly. This means that the probability of exceeding a level of excellence may change in fact, may decrease. This situation fits the Poisson Limit Theorem perfectly. In fact, this subject and its applications to sports are discussed in detail in Section 15.4 of this book.

PRACTICE PROBLEMS

1. In a city with 50,000 population, 1% play golf. If we choose 500 individuals randomly, what is the probability it includes 6 individuals who play golf?
2. Formula One, or so called F1, is the highest class of single-seater auto racing sanctioned by the Fdration Internationale de l'Automobile (FIA) and owned by the Formula One Group. US racer won twice in total 68 races from 1950 to 2018. What is the probability that a champion was selected and he came from United States?

4.3 NORMAL APPROXIMATION: A SUMMARY

This section summarizes some classical situations where normal approximation may be used. It includes the following cases:

- If X is a binomial random variable with parameters n and p, then for large n ($np \geq 5$, $n(1-p) \geq 5$), X is approximately normally distributed, with mean np and variance $np(1-p)$.
- If X is a Poisson random variable with parameter λ, then for large λ ($\lambda \geq 25$), X is approximately normally distributed, with mean λ and variance λ.
- If X is a hypergeometric random variable, then for large n and $N - n$ ($np \leq 5$, $n(1-p) \leq 5$, $(N-n)p \leq 5$, $(N-n)(1-p) \leq 5$), X is approximately normally distributed, with mean $nk/N = np$ and variance $nk(N-k)(N-n)/n^2(N-1)$.

It should be mentioned, once more, that when approximating a discrete distribution by a continuous one, it is necessary to apply a continuity correction. This is due to the fact that for discrete distributions, the probability that a random variable takes a particular value can be positive, but for a continuous distribution it is always equal to zero. Thus, for a discrete random variable, a continuity correction replaces the event $X = k$ with the event $k - 0.5 \leq X \leq k + 0.5$. The following are examples of the continuity correction.

$$P(X = 4) = P(3.5 \leq X \leq 4.5), \, P(4 \leq X \leq 7) = P(3.5 \leq X \leq 7.5),$$

$$P(4 < X < 7) = P(4.5 \leq X \leq 6.5), \, P(X > 7) = P(X > 7.5), \, P(X \geq 7) = P(X \geq 6.5).$$

Note that a continuity correction should only be applied when we are approximating discrete random variables with continuous ones.

EXAMPLE Suppose that the number of the goals scored during a season by a soccer team has a Poisson distribution with mean 100. Find the probability of fewer than 80 goals. Also the probability of more than 80, but less than 100 goals.

SOLUTION: First consider $P(X < 80)$. Using Poisson distribution this is equal to

$$\sum_{x=0}^{79} \frac{e^{-100} 100^x}{x!}$$

which is not easy to calculate. Since $\lambda > 25$, we can use normal approximation.

Here $\mu = 100, \sigma^2 = 100$ so that X, the number of goals, has a normal distribution with above mean and variance.

Using this we get

$$P(X < 80) \approx P(X \leq 79.5) = P\left(\frac{X-100}{10} \leq \frac{79.5-100}{10}\right) = P(z \leq -2.15) = 0.0158.$$

$$P(80 < X < 100) \approx P(80.5 \leq X \leq 99.5) = P\left(\frac{80.5-100}{10} \leq \frac{X-100}{10} \leq \frac{99.5-100}{10}\right)$$

$$= P(-0.95 \leq z \leq -0.05) = 0.4801 - 0.0256 = 0.4545.$$

To find the probability that X equals 100, we use

$$P(X = 100) \approx P(99.5 \leq X \leq 100.5) = P\left(\frac{99.5-100}{10} \leq \frac{X-100}{10} \leq \frac{105.5-100}{10}\right)$$

$$= P(-0.005 \leq z \leq 0.005) = 0.5199 - 0.4801 = 0.0398.$$

Note that the exact answer to this last calculation is

$$e^{-100} 100^{100}/100!$$

which is hard to compute with an ordinary calculator.

EXAMPLE Suppose that we are drawing 20 cards from a deck of cards and want to find the probability of drawing at least 8 spades. If we let X denote the number of spades, then since the required conditions are satisfied we have approximately

$$X \sim N\left(20\left(\frac{1}{4}\right), 20\frac{1}{4}\frac{3}{4}\frac{52-20}{52-1}\right) = N(5, 1.53^2)$$

Thus,

$$P(X \geq 8) = P(X \geq 7.5) = P\left(\frac{X-5}{1.53} \geq \frac{7.5-5}{1.53}\right) = P(z \geq 1.63) = 0.0516.$$

Note that to find this probability exactly, we need to use hypergeometric distribution and compute

$$1 - P(X \leq 7) = 1 - \sum_{k=0}^{7} \frac{\binom{13}{k}\binom{39}{20-k}}{\binom{52}{20}}.$$

A PRACTICE PROBLEM Suppose that in large universities with several sport programs the number of players who get injured in a given year has a Poisson distribution with mean 81. Show that the approximate probability for more than 100 players to get injured in a given year is 0.15.

FURTHER LARGE SAMPLE RESULTS

For later chapters of this book, we need to slightly generalize the results mentioned earlier. Suppose that we have a sample from a (possibly non-normal) distribution with mean μ and variance σ^2. Then for a large n,

$$\frac{\bar{X} - \mu}{\sigma/\sqrt{n}} \sim N(0,1), \quad \frac{\bar{X} - \mu}{s/\sqrt{n}} \sim N(0,1).$$

Similarly, if we have a sample from a Bernoulli distribution, then for large n such that $np > 5$, $n(1-p) > 5$ we have

$$\frac{\hat{p} - p}{\sqrt{p(1-p)/n}} \sim N(0,1), \quad \frac{\hat{p} - p}{\sqrt{\hat{p}(1-\hat{p})/n}} \sim N(0,1).$$

Note that in both cases, the first expression is derived from the Central Limit Theorem and the second formula just substitutes estimates of the standard deviation for the true standard deviation.

4.4 THE MODEL OF PRODUCTS: THE LOGNORMAL DISTRIBUTION

Multiplicative models While the normal distribution arises from the sum of many small effects, it is desirable also to consider the distribution of a phenomenon, which arises as the result of multiplicative mechanisms acting on a number of factors. An example of such a mechanism occurs in breakage processes, such as in the crushing of aggregate or transport of sediments in streams. The final size of a particle results from a number of collisions of particles varying in size traveling at different velocities. Each collision reduces the particle by a random proportion of its size at the time. Therefore, the size Y_n of a randomly chosen particle after the nth collision is the product of Y_{n-1} (its size prior to that collision and X_n the random reduction factor). Extending this argument back through previous collisions, we get

$$Y_n = Y_{n-1}X_n = Y_{n-2}X_{n-1}X_n = \cdots = Y_0 X_1 X_2 \cdots X_n.$$

It is also reasonable to think that in some sports, improvement of two different skills may have a multiplicative effect (rather than an additive) on performance of a player or a team. Also, risk factors for a person suffering from two conditions such as high blood pressure and high blood sugar may increase in multiplicative form rather than additive form.

In all these cases, the variable of interest Y is expressed as the product of a large number of variables, each of which is, in itself, difficult to study and describe. In many cases, however, something can be said of the distribution of Y. Take natural logarithms of both sides of any of the equations above. The result is of the form

$$\ln Y = \ln Y_0 + \ln X_1 + \ln X_2 + \cdots + \ln X_n.$$

Since X_i is a random variable, $\ln X_i$ is also a random variable. Calling upon the Central Limit Theorem, one may predict that the sum of a number of such variables will be approximately normally distributed. In this case, we expect $\ln Y$ to be normally distributed. Let

$$X = \ln Y$$

then, knowing that X is normally distributed, our problem is to determine the distribution of Y or

$$Y = e^X.$$

As was discussed in Chapter 3, a random variable Y whose logarithm is normally distributed has lognormal distribution. As discussed in Chapter 3, if X is normally distributed, that is

$$f_X(x) = \frac{1}{\sigma\sqrt{2\pi}} \exp\left[-\frac{1}{2}\left(\frac{x-\mu}{\sigma}\right)^2\right], \quad -\infty \leq x \leq \infty.$$

Then upon substituting, we find

$$f_Y(y) = \frac{1}{y\sigma\sqrt{2\pi}} \exp\left[-\frac{1}{2}\left(\frac{\ln y - \mu}{\sigma}\right)^2\right], \quad y \geq 0.$$

Thus the random variable Y is lognormally distributed, whereas its logarithm X is normally distributed. The range of Y is zero to infinity, whereas for X it is negative to positive infinity. If Y = 1, then X = 0, and if Y > 1, then X is positive. For $0 \leq Y \leq 1$, X ranges from minus infinity to zero, since the logarithm of a number between zero and one is negative. Y cannot have negative values, since the logarithm of a negative number is not defined.

4.5   THE MODEL OF EXTREMES: THE EXTREME VALUE DISTRIBUTIONS

In sports, concern often lies in the largest or smallest of a number of random variables (e.g., performance). Success or failure of an individual, team, or a system may rest solely on their ability to function under the maximum demand, pressure, or load to which they are subjected, not simply the typical values or conditions. Clearly in sports extreme performances are of interest, not average performances. If the key variable Y is the maximum of n random variables $X_1, X_2, \cdots, X_n$, then the probability distribution of Y may be obtained as

$$F_Y(y) = P[Y \leq y] = P(X_i \leq y, \text{for all } i) = P(X_1 \leq y, X_2 \leq y, \cdots, X_n \leq y).$$

If the X_i's are independent, then we have

$$F_Y(y) = P(X_1 \leq y)P(X_2 \leq y)\cdots P(X_n \leq y) = F_{X_1}(y)F_{X_2}(y)\cdots F_{X_n}(y).$$

In the special case where all the X_i's are identically distributed with distribution function $F_X(x)$, we have

$$F_Y(y) = (F_X(y))^n.$$

In a similar manner, one can obtain the distribution of the minimum Z using the fact that $Z > z$ occurs only if all X_i's are greater than Z. This leads to

$$F_Z(z) = 1 - (1 - F_X(z))^n.$$

If the X_i's in the last case are continuous random variables with common density function $f_X(x)$, then

$$f_Y(y) = \frac{d}{dy}F_Y(y) = n(F_X(y))^{n-1}f_X(y)$$

and

$$f_Z(z) = n(1 - F_X(z))^{n-1}f_X(z).$$

Recall that when studying reliability of a system, maximum and minimum play a critical role. This is because, for a series system, the system reliability equals the reliability of the weakest link (component), whereas for a parallel system it is equal to the reliability of the strongest component.

Clearly the above analysis applies only to a finite n. The question that arises is, what can be said about $F_Y(y)$ when $n \to \infty$. If the conditions of independence and common distribution hold among the X_i's, then in a number of cases of great practical importance, the shape of the distribution of Y is relatively insensitive to the exact shape of the distribution of the X_i. In these cases, limiting forms (as n grows) of the distribution of Y can be found, which can be expected to describe the behavior of the random variable even when the exact shape of the distribution of X_i is not known precisely. It is in this sense that this situation is not like those already encountered in Sections 4.3 and 4.4 on the normal and lognormal distributions. When dealing with extreme values, however, no single limiting distribution exists. The limiting distribution depends, obviously, on whether largest or smallest values are of interest, and also on tail behavior of the underlying distribution for X_i. Three specific cases that have been studied in some detail and are applied widely, will be mentioned here.

Suppose that in a given year there have been n best annual performances (e.g., highest number of points scored in soccer or basketball). The Extreme Value Theorem specifies the form of the limiting distribution of Y as $n \to \infty$. The three possible types of limiting distributions are

1. The Gumbel Distribution (Type I):

$$F_Y(y) = \exp(-e^{-y}), \quad \text{for } -\infty < y < \infty.$$

2. The Fréchet Distribution (Type II):

$$F_Y(y) = \begin{cases} 0, & \text{for } y \leq 0 \\ \exp(-y^k), & \text{for } y > 0 \ (k > 0). \end{cases}$$

3. The Weibull Distribution (Type III):

$$F_Y(y) = \begin{cases} \exp(-(-y^{-k})), & \text{for } y < 0 \ (k < 0) \\ 1, & \text{for } y \geq 0. \end{cases}$$

The shape parameter k reflects the weight of the tail of the parent distribution. The tail of the distribution $F_Y(y)$ is either declining exponentially (Type I) or by a power (Type II) or is finite (Type III). The three types may be combined into Generalized Extreme Value Distribution as

$$F_Y(y) = \begin{cases} \exp[-(1-k(y-\alpha)/\sigma)]^{1/k} & k \neq 0, \\ \exp[-\exp(-(y-\alpha)/\sigma)] & k = 0. \end{cases}$$

EXAMPLE 4.5.1 First recall that the general form of Type I extreme value distribution for maxima is

$$F_Y(y) = e^{-e^{-(y-\alpha)/\beta}}, \quad -\infty < y < \infty.$$

For this distribution we have

$$E(Y) = \alpha + 0.57772\beta, \quad \text{Mode} = \alpha,$$

$$\text{Var}(Y) = 1.64493\beta^2, \quad \text{Median} = \alpha + 0.3665\beta.$$

Now, consider the following problem. The yearly maximum flow discharge measured in cubic meters per second at a given location of a river where sailing competitions are held during the last 60 years has provided the following estimates:

$$\hat{\alpha} = 42.89, \quad \hat{\beta} = 9.47.$$

To find the probability of a flow discharge greater than 50 m³/s we proceed as follows:

$$P(Y > 50) = 1 - P(Y \leq 50) = 1 - F(50) = 1 - e^{-e^{-(50-42.89)/9.47}} = 0.376.$$

Also,

$$E(Y) = 42.89 + 0.57772(9.47) = 48.36,$$

$$\text{Var}(Y) = 1.64493(9.47)^2 = 147.519.$$

This information may be used when planning for the next competitions.

A PRACTICE PROBLEM Research the champion score point of F1 car race from 1950 to 2018, the average score points was 108.4, and variance was 10,843.3. What is the probability that champion score points will be larger than 100 in the year 2019?

4.6 EXCEEDANCES

In many situations in sports, instead of records we are interested in the events associated with the exceedances of certain performance measures of the variable under study. In sports such as long jump, one needs to jump certain length to qualify for a competition. This leads to deal with the frequencies (number of athletes) instead of the values on the random variable itself. For instance, we may like to know how many baseball players will have more than 50 home runs next season. Or how many of the players will surpass the performance of the third best player in the history of a given sport. Many authors have discussed exceedances and their applications. This study has followed Castillo's presentation although using sport examples.

DEFINITION: Let Y be a random variable representing the performance in a given sport and x a real number representing the level of excellence. We say that the event $Y = y$ is an exceedance of the level x if $y > x$. One important practical problem is the following: Assuming independent and identically distributed trials, determine the probability of r exceedances in the next n trials. For example, in the next season consisting of n games, how many times a player will score (exceed) more than his third best record of the last season.

Checking the assumptions, it becomes clear that here we are dealing with a Bernoulli experiment (only two outcomes are possible: "exceedance" or "not exceedance") repeated n times. Thus, the number of exceedances is a binomial variable with parameters n and $p(x)$, where $p(x)$ is the probability of exceedance of the level x. Note that $p(x) = P(X > x) = 1 - F(x)$ is similar to survival function.

The probability of r exceedances of level x in the next n trials is then

$$P[e_n(x) = r] = \binom{n}{r} [1 - F(x)]^r F^{n-r}(x), \quad 0 \leq r \leq n \tag{4.8}$$

where $e_n(x)$ is the number of exceedances.

Now, suppose that, rather than a fixed level, we make the level x dependent on n, x_n say. For example, as number of athletes increases or number of the years competition are held in increased, we increase the bar or change the level of excellence. If we do this in a way that the following condition is satisfied:

$$\lim_{n \to \infty} n[1 - F(x_n)] = \tau; \quad 0 \leq \tau \leq \infty.$$

Then, the probabilities in (4.8) can be approximated by those of a Poisson distribution (process). This leads to the following theorem, which is essentially Poisson approximation to the binomial probabilities discussed earlier.

THEOREM (EXCEEDANCES AS A POISSON PROCESS): Let $\{X_n\}$ be a sequence of independent and identically distributed random variables with distribution function $F(x)$. Assume that the sequences $\{x_n\}$ of real numbers satisfies the condition

$$\lim_{n \to \infty} n[1 - F(x_n)] = \tau; \quad 0 \leq \tau \leq \infty \tag{4.9}$$

then

$$\lim_{n \to \infty} P[e_n(x_n) = r] = \frac{e^{-\tau}\tau^r}{r!}; \quad r \geq 0 \qquad (4.10)$$

where $e_n(x_n)$ is the number of $X_i (i=1,,2,...,n)$ exceeding x_n and the right-hand side of (4.9) must be taken as 1 or 0 depending on whether $\tau = 0$ or $\tau = \infty$, respectively.

As pointed out, the proof of this theorem is just the well-known Poisson limit (discussed in Section 4.2) for the binomial distribution if $0 < \tau < \infty$ and some trivial considerations for the two extreme cases $\tau = 0$ and $\tau = \infty$ are satisfied. The practical importance of this theorem lies in the fact that exceedances approximately follow a Poisson distribution.

As an example, suppose that the level x_n is selected such that $1 - F(x_n) = 1/n$, then $\tau = 1$ and we have

$$\lim_{n \to \infty} P[e_n(x_n) = r] = 1/r!e.$$

Thus, the probability of two exceedances in the next n trials is $1/2e = 0.184$.

The following is another interesting problem related to the exceedance. Suppose that for independent and identically distributed trials, we want to determine the probabilities of r exceedances, in the next N trials, of the m-th largest observation in the past n trials. This is useful when choosing the member of a team based on past performances prior to a competition.

In this case, if p_m is the probability of exceedance of the m-th largest observation in the past n trials, the probability of r exceedances in the N future trials can be calculated from the following binomial distribution:

$$\binom{N}{r} p_m^r [1-p_m]^{N-r}. \qquad (4.11)$$

It can also be shown that p_m has a Beta distribution with density function

$$f(p_m) = \frac{p_m^{m-1}[1-p_m]^{n-m}}{B(m, n-m+1)}, \quad 0 \leq p_m \leq 1. \qquad (4.12)$$

The mean number of exceedances, $\bar{r}(n, m, N)$, taking into account that the mean of the binomial variable in (4.10) is Np_m, and using the total probability rule together with the fact that the order statistics of size n from any distribution divides, on average, the area under the density function to $n+1$ equal parts is

$$\bar{r}(n, m, N) = \int_0^1 Np_m f(p_m) dp_m = \frac{Nm}{n+1}. \qquad (4.13)$$

Also the variance of the number of exceedances is given by

$$\sigma^2(n, m, N) = \frac{Nm(n-m+1)(N+n+1)}{(n+1)^2(n+2)}. \qquad (4.14)$$

The variance takes a minimum value for $m = (n+1)/2$. However, the coefficient of variation decreases with m as should be expected. Note that these results are distribution free in the sense that they are true for any $F(x)$.

EXAMPLE 4.6.1 The yearly maximum number of points scored in high school basketball competitions in a certain district in New York during the last 40 years was 102. Determine the mean value

Chapter 4: Limit Properties and Models

and variance of the number of exceedances of 102 of the yearly maximum number of the points during the next 30 years. If the second largest number of the points scored was 97, what are the mean value and variance of the number of exceedances of this value in the next 20 years.

SOLUTION: From (4.13) and (4.14), we get respectively

$$\bar{r}(40, 1, 30) = 30/41 = 0.732$$
$$\sigma^2(40, 1, 30) = (30)(40)(71)/(41)^2(42) = 1.207$$

and

$$\bar{r}(40, 2, 20) = (20)(2)/41 = 40/41 = 0.976$$
$$\sigma^2(40, 2, 20) = (20)(2)(39)(61)/(41)^2(42) = 1.348.$$

EXAMPLE 4.6.2 Consider the yearly maximum number of home runs (or for continuity home runs per at bat) in baseball during the last 60 years (seasons). As a standard of excellence, we want to choose a value in order to have a mean value of 4 exceedances in the next 20 years.

SOLUTION: According to (4.12), we have

$$\bar{r}(60, m, 20) = \frac{20m}{61} \approx 4 \Rightarrow m \approx 12$$

which shows that the value to be chosen is the 12th largest order statistic in the data.

4.6.1 Return Periods

Most of us have heard terms such as 50-year floods or 100-year earthquakes. These terms refer to return periods or average time it takes to observe such events. Consider an event (breaking a record, winning the championship, etc.) such that its probability of occurrence in a unit period of time (normally 1 year) is p. Assume that occurrences of such an event in different periods are independent. Then, as time passes, we have a sequence of equally likely Bernoulli experiments (only two outcomes: (1) occurrence or (2) not occurrence, are possible). Thus, the time (measured in unit periods) to the first occurrence is a geometric random variable with parameter p and mean value $1/p$. This motivates the following definition.

DEFINITION: Let A be an event, and T the random time between consecutive occurrences of A. The mean value, τ, of the random variable T is called the return period of the event A.

Note that if $F(x)$ is the distribution function of the yearly maximum of a random variable, the return period of that random variable to exceed the value x is $1/[1 - F(x)]$ years. Similarly, if $F(x)$ is the distribution function of the yearly minimum of a random variable, the return period of the variable to go below the value x is $1/F(x)$ years.

Also note that if a given engineering work fails when, and only when, the event A occurs, its mean lifetime coincides with the return period of A. The importance of return periods in engineering is due to the fact that many design criteria are defined in terms of return periods.

The probability of occurrence of the event A before the return period is (see the geometric distribution)
$$F(\tau) = 1 - (1-p)^\tau = 1 - (1-p)^{1/p}$$
which for $\tau \to \infty (p \to 0)$ tends to the value 0.63212.

EXAMPLE 4.6.3 One very interesting example for this is the return period for a player like Michael Jordan. In Section 3.6, it was shown the distribution of points per minute for guards who played in 92-93 season was normal with mean 0.4236 and standard deviation 0.1159. Using this we found that for Jordan we have
$$P(PPM > 0.8291) \approx 1/5,000.$$

This means that the return period for a player who would perform like or better than Jordan is 5,000 players. So, if 100 guards are added to the NBA each year, on average, it takes about 50 years to produce such a player.

EXAMPLE 4.6.4 Suppose that the distribution function of the yearly maximum number of points scored in basketball in a given region is given by
$$F(x) = \exp\left[-\exp\left(-\frac{x - 38.5}{7.8}\right)\right].$$

Then the return periods of yearly maximum score of 60 and 70 points are, respectively
$$\tau_{60} = \frac{1}{1 - F(60)} = 16.25 \text{ years,}$$
$$\tau_{70} = \frac{1}{1 - F(70)} = 57.24 \text{ years.}$$

This means that the maximum score of 60 and 70 points occur, on average, once every 16.25 ($\approx 16 - 17$) and 57.24 ($\approx 57 - 58$) years, respectively.

EXAMPLE 4.6.5 Suppose that in a certain tournament, an exceptional performance is defined as that value with a return period of 50 years, and the yearly maximum number of the points scored is known, from past experience, to have a limiting Gumbel distribution with the distribution function
$$F(x) = \exp\left[-\exp\left(-\frac{x - 15}{4}\right)\right].$$

Then the number of the points h, must satisfy the equation
$$\frac{1}{1 - F(h)} = 50$$
from which we get $h = 30.61 \approx 31$ points.

Chapter 4: Limit Properties and Models

4.6.2 Exceedances and English Premier League

Astrophysicists at the University of Warwick studying the extreme variability in X-rays emitted from matter falling into black holes have discovered that their research methods also show that the world's top division soccer matches have an unusually large proportion of high scoring games—so much that international soccer actually shows a pattern of "extreme events" similar to that seen in the large bursts of X-rays from the accretion discs of black holes. However, analysis of just the English premier soccer league and cup games showed that English top division soccer is in fact 30 times less likely to have high scoring games than the rest of the world taken as a whole, and could thus be seen by some people as 30 times more boring.

"Extermal events," that is, large events that are more likely than would be expected from a random process, can be a signature of complexity in nature. In the case of matter moving in accretion disks around black holes it tells us about the turbulent flow in the accreting matter.

When comparing this distribution of events with other patterns in the world around us, two University of Warwick postgraduate physics students (John Greenhough and Paul Birch), with their supervisor Sandra Chapman and colleague George Rowlands, looked at the number of goals scored by the home and away teams in over 135,000 games in 169 countries since 1999 and found that the results followed the pattern of "extremal statistics."

However, they also compared their data with an analysis they made of the scores of 13,000 English top division games and 5,000 FA Cup matches between the 1970/71 and 2000/01 seasons. They found that these scores contained far less high-scoring games than the world as a whole and rather than fitting an extremal statistics pattern the English games more closely fitted either Poisson or negative binomial distributions.

In summary, their analysis revealed that a total score over 100 goals in any one game occurs approximately only once in every 10,000 English top division matches (once every 30 years) but in top division matches world-wide, such a score is seen once in 300 games (about once every day).

4.6.3 Characteristic Values

If the event A consists of the exceedances of the level x of a random variable X (we can write $A(x)$ instead of A), then p becomes $p(x) = 1 - F(x)$ and

$$\tau(x) = \frac{1}{1 - F(x)}.$$

In this case, the expected value of the number of events $A(x)$ in n periods is given by (see the binomial random variable)

$$n[1 - F(x)].$$

Of special interest is the value of the level x leading to an expected value of 1. This value is called the characteristic largest value for that period. More precisely, we have the following definition:

DEFINITION: A certain value x of a random variable X is said to be the characteristic largest value for a period of duration n units if the mean value of the number of exceedances of that value in such a period is unity. Because of its dependence on n, the characteristic largest value will be denoted by u_n.

Change of X into—X allows us to define, in a similar manner, the characteristic smallest value, v_n.

Note that the characteristic largest value satisfies the equation

$$n[1 - F(u_n)] = 1 \Rightarrow F(u_n) = 1 - \frac{1}{n}. \tag{4.15}$$

Similarly, the characteristic smallest value satisfies

$$nF(v_n) = 1 \Rightarrow F(v_n) = \frac{1}{n}. \tag{4.16}$$

Note the symmetry of Expressions (4.15) and (4.16).

The probability of exceeding the characteristic largest value in the period is

$$1 - F^n(u_n) = 1 - \left(1 - \frac{1}{n}\right)^n$$

which for large n tends to $1 - \exp(-1) = 0.6321$.

EXAMPLE 4.6.6 Recall the Example 4.6.4. The distribution function of the yearly maximum number of points scored was

$$F(x) = \exp\left[-\exp\left(-\frac{x - 38.5}{7.8}\right)\right].$$

Then, according to (4.15), the characteristic largest value is the solution of the following equation:

$$F(u_5) = 1 - \frac{1}{5} = 0.8$$

from which we get $u_5 = 50.2$ points.

We stop the discussion of probability concepts here and proceed to present some methods of statistical inference in the Chapter 5.

4.7 USING R

In this chapter, instead of discussing the use of existing functions in R to calculate probabilities, we would like to demonstrate how to create your own function in R. It is important one can define desired functions for proposed algorithms.

We will need to name the defined function in a similar manner as naming variables. A keyword *function* will be used to declare the function, and list of parameters should be provided also. The syntax of declaring an R function is

```
FUN_NAME = function(PARA_LIST){
FUN_BODY
}.
```

Chapter 4: Limit Properties and Models

Let's solve the practice problem in Section 4.5 here as an example. We will calculate the desired probability using Type I extreme value distribution for maxima. Therefore, we need to declare the following formula in R:

$$F(y) = P(Y < y) = e^{-e^{-(y-\alpha)/\beta}}, \quad -\inf < y < \inf$$

as following:

```
F = function(y, a, b){
    exp( -exp( -(y-a)/b ) )
}
```

Next, we need to calculate the value of α and β using known information. Because the variance of all 68 years scores is $10,843.3$, thus $\beta = \sqrt{10,843.3/1.64493} = 81.19$, and with known average score 108.4 points we have $\alpha = 108.4 - 0.57772 \times \beta = 61.49$. We were looking for the probability of $P(Y > 100) = 1 - F(100)$. We can now calculate this probability using our defined function above.

```
beta  = sqrt(10843.3/1.64493)
alpha = 108.4 - 0.57772*beta
1 - F(100, alpha, beta)
```

We end this section by noting that user-defined functions can be used in exactly same way as all built-in functions in R, which means users can find integral or run simulations with customized functions. However, user-defined functions are temporary stored in user's workspace. Only when user saved workspace, functions will be retained for next use. Otherwise, users please keep function definitions in your R code and reload them next you want to use.

CHAPTER 5

Estimation

Statistical inference constitutes the major part of statistical science. It has two major components: estimation and testing a hypothesis. Estimation deals with problems such as prediction, and testing a hypothesis (a theory) deals with problems such as verifying a claim. For example, we may want to estimate the number of people who watch a certain sport, or predict the result of the next soccer match. For testing, we may want to investigate a claim that a certain type of exercise improves the performance, or to test a claim that more people play tennis than volleyball. Since estimation and testing usually are carried out based on a limited information (data), we require techniques that utilize such information efficiently.

Problems involving estimation arise in decision-making and planning. As such they have become an accepted part of scientific methodology and also everyday life. Estimates are often referred to as statistics. For example, in year 2003, in honor of Valentine's Day, the United States Census Bureau compiled the following statistics about what takes place in the United States. According to estimates there are 2.3 million marriages annually (6,400 per day); median ages at first marriage are 25.1 and 26.8 for women and men, respectively; the age women marry is up 4.3 years since 1970; the age for men has climbed 3.6 years since 1970; 52 and 57 are the percentages of American women and men, respectively, who are 15 and over and married but not separated; 13.6 million individual, ages 25 to 34 have never been married (34% of all individuals in this age group; 4.9 million is the number of opposite-sex unmarried partner living together.

Problems involving estimation may also arise as a result of measuring a continuous variable. In earlier chapters, the differences between discrete and continuous random variables were discussed. In practice, most variables we consider or encounter are continuous. However, once we measure them, the resulting values are expressed in discrete form since we do not yet have tools to measure continuous variables to a degree of accuracy. Due to this fact, most of the values attributed to continuous variables are nothing but estimates of the true values. For example, when you say you weigh 158 pounds, this is not your exact weight but only an estimate. An important question that arises is how can we improve such estimates, and more importantly, if we find a way to do this, how do we specify (or quantify) the improvement achieved?

In this chapter, we will provide an answer to this question and present some classical methods for estimation of the parameters of interest such as population mean or population variance.

5.1 THE PROBLEM DESCRIPTION

Consider a sample space Ω and a chance experiment E, that generates a random variable X on Ω in the way described earlier in Section 2.1. Assume that X has a distribution that depends on a parameter θ such as population mean μ, or probability of success p. For example, μ may be the average income of college graduates in a certain year, and p may be the percentage of high school students who play baseball. We are interested in estimating θ. As was pointed out earlier, a survey of the whole population to find the true value of θ may be too tedious and unnecessary and it may even be impossible. So we draw a sample from the population and use the information in the sample to estimate θ. If we use a criterion and estimate θ by a single value, $\hat{\theta}$, then $\hat{\theta}$ is called a point estimator and its particular value is called a point estimate. For example, we may, based on a telephone survey, estimate the percentage of the people who watch college basketball using a single number such as 30% or 35%.

DEFINITION: An estimator is a rule that tells us how to calculate an estimate of a parameter based on the observed sample.

DEFINITION: A point estimate is a single value estimating the true value of the population parameter. For example, the sample mean is a point estimate for the population mean. A point estimator uses the sample to produce a single number to estimate a population parameter.

EXAMPLE 5.1.1 Suppose that new body fitting equipment is introduced to the market to help the users lose weight. Here the population Ω is the possible weights lost by all individuals who may use this equipment. We may ask how many pounds a person is expected to lose by using this equipment regularly for a certain time period. Although not essential, here we assume that weight losses are normally distributed with a known σ^2. Then using the data attempt to estimate μ, the true population mean (the average or expected weight loss). By taking a random sample, we will be using a chosen procedure to get an estimate $\hat{\mu}$ for μ. For instance, we may use the sample mean $\bar{X}$ as $\hat{\mu}$, or perhaps the sample median. Other possibilities are trimmed mean (mean after dropping the extreme values) or the sample mode (value with the highest frequency).

Because there will always be an error in estimating a parameter θ (due to the lack of complete information) we may also be interested in providing two numbers, an upper and a lower limit such as $\hat{\theta}_1 \leq \theta \leq \hat{\theta}_2$. This is called an interval estimate or after adding a confidence level, a confidence interval. If we can set up such an interval, then we may be able to say that we are, for example, 90% sure that θ would fall in this interval or range. The percentage such as 90% is called the confidence level. This subject will be discussed in detail in Section 5.5.

DEFINITION: An interval estimate is a range of values estimating the true value of the population parameter. The probability that this range contains the parameter is called the confidence level.

Sometimes the value of the parameter of interest is known based on past information and we want to know whether this value has changed. In other words, we are not interested in estimating but only confronted with a yes–no decision. For example, the decision may be related to the following question is the true value of θ greater than (less than or not equal to) some chosen value

θ_0 or not? As a more general example, one may just be interested in finding out if more people like basketball than baseball and use some data to decide whether such a claim should be rejected or not. In such cases, rather than estimating, one may pose a hypothesis, say the claim is not true, and try to disprove it using the information in the sample regarded as the support or evidence. These two problems, estimation and testing hypotheses, are related, and as mentioned earlier, constitute the main body of inferential statistics. We will discuss the testing problem in Chapter 6.

When estimating a parameter, whatever the method used, we may want to find out how "good" our estimate is. Suppose, we know that the average height of adult male in a certain area is 71 inches with standard deviation of 6 inches. Clearly, one factor that affects our estimate directly is variability of height measured by standard deviation. The smaller the standard deviation (less variability) the higher the chance of obtaining a "good" estimate. Another factor having an effect on our estimate is the sample size. Clearly, larger samples contain more information and thus result in "better" estimates. To clarify this, suppose that you pick an adult male randomly, measure his height and ask me to guess his height (of course you are not going to show him to me). What would be my best guess or estimate? Clearly 71 inches, since my chance of being correct is the highest. Next you pick two adult males, measure their heights and average. Then ask me to guess the value of the average. What would be my best guess? Clearly 71 inches again. But in which case my guess or estimate is closer to the true value? Clearly in the second case and this becomes more evident as you pick more people, do the same, and ask me to guess. The next section discusses this issue and some related problems.

5.2 SAMPLING DISTRIBUTION

The idea of sampling distribution stems from knowledge gained long ago. People simply realized that they could not rely on one measurement, and that average of several measurements provides a much "better" estimate (e.g., for population mean or any other parameter). Let us, in line with the last paragraph of the previous section, consider a real-life example. As we all know, insurance is based on the premise that individuals (such as players of a certain sport) faced with large and unpredictable losses can reduce the financial effects of such losses by forming a group and sharing the losses incurred by the group as a whole. This principle can be justified mathematically using the Law of Large Numbers, as follows:

Suppose that $X_1, \cdots, X_n$ are respective losses faced by n different individuals over the coming period. If each individual agrees to pay $(X_1 + \cdots + X_n)/n$ toward the aggregate loss $X_1 + \cdots + X_n$, then according to the Law of Large Numbers, the amount that each person will pay becomes more predictable as the size of the group increases, provided that the X_j's are independent and have the same distribution. That is, the distribution of $(X_1 + \cdots + X_N)/N$ becomes more concentrated around μ, where μ is the loss that each individual expects to incur in the absence of insurance.

Of course, this does not mean that a person in a risk-sharing arrangement will never have to pay an unexpected large amount. Indeed, any loss that an individual faces before entering into a risk-sharing arrangement is still possible after entering into such an arrangement. However, the probability that an individual will have to pay an unexpected large amount is greatly reduced when the individual participates in a risk-sharing arrangement. The reason is that the distribution of $\overline{X}$ becomes concentrated around μ as $N \to \infty$.

We always assume that the sample we draw from Ω is a random sample of size N. That is, each of the N members of the sample has the same distribution as X and they are independent. In this way, the sample generates N replicas of X, denoted by $X_1, \ldots, X_N$, all with the same distribution as X and also independent. Hence, as we know from the Central Limit Theorem, the random variable

$$\bar{X} = \frac{1}{N}(X_1 + \ldots + X_N) \tag{5.1}$$

will be normally distributed—at least approximately even if X is not. Usually, by drawing a sample, one obtains a realization of $\bar{X}$, which in this case is simply the sample mean $\bar{x}$ discussed in Section 1.1. It is important to note that the value $\bar{x}$ is a product of chance. Other samples will produce other values of $\bar{x}$. That is, $\bar{x}$ will vary from sample to sample and hence is a realization of a random variable $\bar{X}$. For example, if you randomly select 40 male college students and find the average of their heights, the value of the sample mean will be different from the one obtained by your friend. In any case, for a sufficiently large N, the distribution of $\bar{X}$ is approximately normal even if X is not a normally distributed random variable itself (Central Limit Theorem). When this is the case, we say that the sampling distribution of the sample mean is normal. The sampling distribution is needed for reliability assessment of the estimate. This type of assessment helps us get an idea of how close or far we may be from the true value of the parameter.

EXAMPLE 5.2.1 Consider the following winning times in the Men's 100-m Freestyle Olympic swimming competitions from 1980 to 2016.

Year	1980	1984	1988	1992	1996	2000	2004	2008	2012	2016
Time (s)	50.4	49.8	48.6	49.0	48.7	48.3	48.2	47.2	47.5	47.6

If we draw a random sample of size 4 from this data with a replacement (one number is drawn at a time, and the selected number will be put back before the next drawing), then it is possible to get a sample such as $\{49.0, 49.8, 47.2, 47.2\}$. Notice that in this sample the number 47.2 appears twice. This is possible because we are sampling with replacement, which ensures that each drawing represents the same distribution. For this sample, the value of $\bar{x}$ is

$$(49.0 + 49.8 + 47.2 + 47.2)/4 = 48.3.$$

Clearly, it is also possible to draw a sample like $\{50.4, 48.6, 48.7, 47.5\}$. For this sample the value of $\bar{x}$ is

$$(50.4 + 48.6 + 48.7 + 47.5)/4 = 48.8.$$

So the value of $\bar{x}$, the realization of the random variable $\bar{X}$, depends on the sample and is a product of chance (it is a random variable). We should emphasize once more that knowing the sampling distribution of $\bar{X}$ will enable us to assess the reliability of the estimate and calculate the margin of error. If we consider all 10 winning times as our population, then we have $\mu = 48.53$ and therefore the first sample leads to an underestimation, whereas the second sample leads to an overestimation. As we shall discuss our goal is to assess the performance of the estimation procedure, in this case, use of sample mean as an estimate for population mean. Note that a more interesting problem will be to the prediction of times for future Olympics.

Chapter 5: Estimation

Let us now consider an example and through that summarize properties of the sampling distribution of the sample mean and also the Central Limit Theorem and its applications. Throughout we will use results presented in Chapters 2 and 3.

EXAMPLE 5.2.2 A sport organization has an airplane to transport players, coaches, and other staff to the away games. It has room for 64 passengers and a total baggage limit of 4,000 lb. Based on data collected in the past several years, the total weight of the baggage checked by a member traveling is a random variable with mean of 60 lb and standard deviation of 16 lb. If 64 members board the flight, what is the approximate probability that the total weight of their baggage will exceed the limit?

SOLUTION: Let X denote the weight of baggage for a randomly selected member of the team. We need to calculate the following probability.

$$P(X_1 + X_2 + ... + X_{64} > 4,000) = P(\bar{X} > 4,000/64) = P(\bar{X} > 62.5).$$

For this, we may adopt the procedure summarized below. The four steps are usually referred to as Rules 1, 2, 3, and 4.

X (Individual)	$\bar{X} = (X_1 + X_2 + ... + X_{64})/64$ (Group Average)
1. $E(X) = 60$	$E(\bar{X}) = (E(X_1) + E(X_2) + ... + E(X_{64}))/64 = 60.$
2. $SD(X) = 16$	$SD(\bar{X}) = 16/\sqrt{64} = 2.$
3. Is $X \sim N(60, 16)$?	No. Nothing can be said about the distribution of $\bar{X}$ unless N is large.
4. Is $N > 30$?	Yes. Applying the Central Limit Theorem we have approximately $\bar{X} \sim N(60, 2)$.

Thus

$$P(\bar{X} > 61.5) = P(z > \frac{62.5 - 60}{2}) = P(z > 1.25) = 0.1056.$$

Note that, if the answer to step (question) 3 were yes, then the sampling distribution of $\bar{X}$ would be exactly normal with mean 60 and standard deviation 2. When this were the case, step 4 would not be needed.

Now that we are introduced to the point and interval estimates, the next step will be to discuss well-known classical methods used for estimation and to study their statistical properties. We start with point estimators.

5.3 POINT ESTIMATOR

Suppose that the random variable X has a density function $f(x; \theta)$ or a probability mass function $p_k(\theta)$, that depend on an unknown parameter θ. Suppose that we wish to estimate θ. Let $X_1, \ldots, X_N$ be a random sample. Any function of $X_1, \ldots, X_N$ such as $\hat{\theta} = g(X_1, \ldots, X_N)$ is called a point estimator of θ (also known as statistics) and any realization of it, that is $g(x_1, ., x_N)$ is called a point estimate. For example, suppose that $f(x; \theta)$ represents an exponential distribution with density function

$$f(x; \theta) = \theta e^{-\theta x}, \quad x > 0.$$

Since, for exponential random variable X, we have $E(X) = 1/\theta$, we may replace the sample mean $\bar{X}$ for population mean $E(X)$ and obtain a point estimate for θ as

$$\hat{\theta} = 1/\bar{X} = N/(X_1 + X_2 + \cdots + X_n).$$

Because $\hat{\theta}$ is a random variable, it has some distribution and it makes sense to speak about $E(\hat{\theta})$, $\text{Var}(\hat{\theta})$, and so on. Since an unknown parameter could be estimated using different functions (different $g(\cdot)$), one may like to know which of the several possible estimates is the "best." To answer this question, we need to talk about properties of a "good" estimator. Although different estimators may be considered "good" in different situations, it is generally accepted that such estimators should have one or more of the properties introduced below.

An estimator $\hat{\theta}$ of θ is called

- **Unbiased** if $E(\hat{\theta}) = \theta$ (i.e., if the expected value of the estimator is the true value of the parameter).
- **Efficient** if $\text{Var}(\hat{\theta})$ is minimal. That is, it has the least variation possible (i.e., the square of the deviation from the true value has the minimal expected value).
- **Consistent** if $P(|\hat{\theta} - \theta| < \varepsilon) \to 1$ for any given $\varepsilon > 0$ as $N \to \infty$. (i.e., with increasing sample size it gets more and more sure that the estimate is close to the true value).

To clarify, we first cite some well-known estimators with all the properties mentioned above and then present a demonstrating example. Let $X_1, \ldots, X_N$ be a random sample representing N duplicates of X and consider the estimators

$$\bar{X} = \frac{1}{N} \sum_{i=1}^{N} X_i \quad \text{and} \quad S^2 = \frac{1}{N-1} \sum_{i=1}^{N} (X_i - \bar{X})^2. \tag{5.2}$$

- If X is distributed as $N(\mu, \sigma^2)$, then $\bar{X}$ and S^2 are unbiased, efficient, and consistent estimators of μ and σ^2, respectively.
- If X has a Poisson distribution with parameter λ, then $\bar{X}$ is an unbiased, efficient, and consistent estimator of λ. Note that here λ is the mean value of the Poisson distribution.
- If X is distributed as $B(N, p)$, then $\bar{X}$ defined as the number of successes divided by N, is an unbiased, efficient, and consistent estimator of p.

Thus, if in Example 5.1.1, the sample mean $\bar{X}$ is used as a point estimator of the average weight loss of the population, then $\bar{X}$ is an unbiased, efficient, and consistent estimator of μ (the true mean weight loss).

DEMONSTRATION

To demonstrate the above properties, suppose that the estimation of the mean of a distribution based on a random sample, $X_1, X_2, \ldots, X_N$ of size N is desired. Consider, as an estimator, a linear combination of the sample values. That is

$$\hat{\mu} = \alpha_1 X_1 + \alpha_2 X_2 + \ldots + \alpha_N X_N = \sum_{i=1}^{N} \alpha_i X_i$$

where $\alpha_1, \alpha_2, \ldots, \alpha_N$ are constants. Note that here $g(x_1, x_2, \cdots, x_N) = \alpha_1 x_1 + \alpha_2 x_2 + \cdots + \alpha_N x_N$. Recall, for example, that if all the α_i's are same and equal to $1/N$, then $\hat{\mu} = \bar{X}$. Other classical estimators are the sample median, the sample trimmed mean, and so on. Assuming that $E(X_i) = \mu$ and Var$(X_i) = \sigma^2$ we have

$$\begin{aligned} E(\hat{\mu}) &= E(\alpha_1 X_1 + \alpha_2 X_2 + \ldots + \alpha_N X_N) \\ &= E(\alpha_1 X_1) + E(\alpha_2 X_2) + \ldots + E(\alpha_N X_N) \\ &= \alpha_1 E(X_1) + \alpha_2 E(X_2) + \ldots + \alpha_N E(X_N) \\ &= \alpha_1 \mu + \alpha_2 \mu + \ldots + \alpha_N \mu = \mu \sum_{i=1}^{N} \alpha_i. \end{aligned}$$

For $\hat{\mu}$ to be unbiased, we should have $\sum_{i=1}^{N} \alpha_i = 1$. This means that the sample mean for which all α_i's are equal to $1/N$, and the sample median for which α corresponding to the middle value (for ordered data) is 1 and the rest of the α_i's are 0 (for N odd) are both unbiased. Of course, there are infinitely many sets of α_i''s that provide us with unbiased estimators. All we need is to make sure that sum of the α-values equals 1. Therefore, $\hat{\mu}$ presents a weighted average. Which of these infinitely many unbiased estimation is "better" or "'best"'? Using the properties mentioned above to find such an estimate we need to find out which of the unbiased estimators has the smallest variance. Since the X_i's are independent we have

$$\begin{aligned} \text{Var}(\hat{\mu}) &= \alpha_1^2 \text{Var}(X_1) + \alpha_2^2 \text{Var}(X_2) + \ldots + \alpha_N^2 \text{Var}(X_N) \\ &= \sigma^2(\alpha_1^2 + \alpha_2^2 + \ldots + \alpha_N^2) = \sigma^2 \sum_{i=1}^{N} \alpha_i^2. \end{aligned}$$

To minimize this, we need to minimize the sum of the squares of α_i's. It can be shown that $\sum_{i=1}^{N} \alpha_i^2$ with the constraint $\sum_{i=1}^{n} \alpha_i = 1$ (for being unbiased) is minimum when all the α_i's are equal to $1/N$, that is

$$\alpha_1 = \alpha_2 = \ldots = \alpha_N = 1/N.$$

Note that for this choice, we have
$$\hat{\mu} = \bar{X}.$$

Thus, the sample mean as an estimate of the population mean is also efficient. In fact, among all unbiased estimators formed as linear combinations of sample values, $\bar{X}$ has the Minimum Variance equal to σ^2/N. For this reason, sample mean is often referred to as the Minimum Variance Unbiased Estimator (MVUE) of a population mean. This means there is no other unbiased linear estimator (of the form discussed above) that would have a variance less than that for $\bar{X}$. Note that the value σ^2/N was obtained by replacing $1/N$ for α_i's in the expression for Var$(\hat{\mu})$. We hope students can show that any other choice values for α_i's will lead to a variance greater than σ^2/N.

5.4 MAXIMUM LIKELIHOOD PRINCIPLE

There are several methods for finding an estimator for the parameters of interest. Among these, the maximum likelihood method is considered to be the most powerful and the most appealing.

The maximum likelihood method was introduced by R. A. Fisher and since then has become a major method used for estimation. Of course, being the most powerful method, application of this method requires more information than other classical methods.

To obtain the maximum likelihood estimator (MLE) based on a sample of size N, we first use the density function $f(x,\theta)$ or the probability mass function $p(x,\theta)$ to construct the likelihood function below;

$$L(\theta) = L(x_1,\ldots,x_N;\theta) = \begin{cases} f(x_1,\theta) \cdot f(x_2,\theta)\ldots f(x_N,\theta) & \text{(continuous case)} \\ p(x_1,\theta) \cdot p(x_2,\theta)\ldots p(x_N,\theta) & \text{(discrete case)} \end{cases} \quad (5.3)$$

Roughly speaking, this presents the likelihood of observing a particular sample, for example, the one already observed. Here we think of θ as a real variable in parameter space denoted by Θ rather than the true value of the parameter.

The MLE, based on a realization $x_1,\ldots,x_N$, is the value of θ for which the likelihood function L, assumes a maximum. The philosophy behind this principle is that, events that are more likely to happen, that is, happen more often. Here, $L = L(x_1,\ldots,x_N;\theta)$ is the probability of observing $(x_1,\ldots,x_N)$ as a sample realization. Thus maximizing L as a function of θ means seeking the value of θ for which it is most likely to observe just this realization $(x_1,\ldots,x_N)$. In other words, among all possible values of θ, we choose the one under which observing the realization $(x_1,\ldots,x_N)$ has the highest or the maximum probability.

It should be noted that in some cases and for certain observed samples, the maximum value of the likelihood function may not actually be attained for any value of the parameter. In such a case, a MLE does not exist. Also for other certain observed samples, the maximum value may actually be attained at more than one value. In such a case, the MLE is not uniquely defined, and any one of these values can be chosen as the estimate.

We shall now illustrate the method of maximum likelihood and the possibilities mentioned by considering a few examples. In each example, we shall attempt to determine the MLEs.

EXAMPLE 5.4.1 Consider a box that contains a number of black and white balls. Suppose that it is known that the ratio of the numbers is 3:1 but it is not known whether the black or the white balls are more numerous. This means that the probability of drawing a black ball is either 1/4 or 3/4. If N balls are drawn from the box with replacement, the probability distribution (probability mass function) of X (the number of black balls) is binomial. That is,

$$f(x;p) = \binom{N}{x} p^x q^{n-x} \quad x = 0,1,2,\cdots,N, \quad p = 1/4, 3/4$$

where $q = 1-p$ and p is the probability of drawing a black ball.

Suppose that we have decided to draw three balls, with replacement and attempt to estimate the unknown parameter p of the distribution. The estimation problem is particularly simple in this case because we only have to choose between two possible values 0.25 and 0.75 for p. Let us anticipate the results of the drawing. The possible outcomes and their probabilities are given below.

Chapter 5: Estimation

x	0	1	2	3
$f(x;3/4)$	1/64	9/64	27/64	27/64
$f(x;1/4)$	37/64	27/64	9/64	1/64

In the present example, if we found $X = 0$ (no black ball) in a sample of three, the estimate 0.25 for p would be preferred over 0.75 because the probability 27/64 is greater than 1/64. Therefore, a sample with $X = 0$ is more likely to arise from a population with $p = 0.25$ than from the one with $p = 0.75$. In general, we should estimate p by 0.25 when $X = 0$ or 1, and by 0.75 when $X = 2$ or 3. Thus, the estimator may be defined as

$$\hat{p} = \begin{cases} 0.25, & x = 0, 1, \\ 0.75, & x = 2, 3. \end{cases}$$

The estimator selected for every X is the value say $\hat{p}$, such that

$$f(x;\hat{p}) > f(x;p')$$

where p' is the alternative value of p.

More generally, if several alternative values of p were possible, we obtain an estimate in the same manner. For example, if we found $X = 6$ in a sample of 25 from a binomial population, we should substitute all possible values of p in the expression (the likelihood of observing 6 black balls)

$$f(6;p) = \binom{25}{6} p^6 (1-p)^{19}, \ 0 \leq p \leq 1$$

and choose as an estimate of p the value, which maximizes $f(6;p)$. For the above case, one would find $\hat{p} = 6/25$ provided that this value is among the possible values of p. In general, the position of the maximum can be found by putting the derivative of the function defined above with respect to p equal to 0 and solving the resulting equation for p. Those who had calculus remember that this is how we find the maximum of a function when it exists. Thus,

$$\frac{d}{dp}f(6;p) = \binom{25}{6} p^5 (1-p)^{18} [6(1-p) - 19p]$$

and by putting this equal to 0 and solving for p, we find that $p = 0, 1, 6/25$ are the possible roots. The first two roots give a minimum, and so the required estimate is $\hat{p} = 6/25$. This estimate has the property that for any alternative p'

$$f(6;\hat{p}) > f(6;p').$$

EXAMPLE 5.4.2 Considered two volleyball teams, Team One and Team Two. Suppose we do not have any information about their past records but wish to estimate the value of p, the chance that Team One defeats Team Two in a randomly selected game. We let them play 20 games against each other. Suppose that in the first 10 games Team One wins 6 out of 10, and on the second 10 games 7 out of 10. Based on these, it is natural to guess that the value of p should be about 13/20. Why? We will show below that the MLE points to the same natural guess.

Using the likelihood function (5.3) and the formula for binomial distribution we have

$$L(6,7;p) = \binom{10}{6} p^6 \cdot (1-p)^4 \binom{10}{7} p^7 \cdot (1-p)^3$$

$$= \binom{10}{6}\binom{10}{7} p^{13} \cdot (1-p)^7.$$

We can now test different values for p ($0 < p < 1$) to find the maximum. We can also find the derivative of $L(6,7;p)$ with respect to p to get

$$L'(6,7;p) = \binom{10}{6}\binom{10}{7} p^{12} \cdot (1-p)^6 (13 - 20p).$$

Solving the equation

$$L'(6,7;p) = 0$$

we obtain the maximum likelihood estimate $\hat{p} = 13/20$. Note that the other two values $p = 0$ and $p = 1$ are not acceptable. They provide minimum values.

For this example, it was also possible to combine the information and use the following likelihood function:

$$L(13,p) = \binom{20}{13} p^{13}(1-p)^7$$

in the first place. This will lead to the same answer. We hope students will check this.

EXAMPLE 5.4.3 Suppose that your insurance company is giving a medical test for a certain disease. The test is 90% reliable in the following sense: if a person has the disease, there is a probability of 0.9 that the test will give a positive response, whereas, if a person does not have the disease, there is a probability of only 0.1 that the test will give a positive response. We shall let X stand for the result of the test, where $X = 1$ means that the test is positive and $X = 0$ means that the test is negative. Note that $p = 0.1$ means that the person tested does not have the disease, and $p = 0.9$ means that the person does have the disease. This parameter space was chosen so that, given p, X has a Bernoulli Distribution with parameter p. The likelihood function is

$$f(x;p) = p^x(1-p)^{1-x}.$$

If $X = 0$ is observed, then

$$f(0,p) = \begin{cases} 0.9, & \text{if } p = 0.1, \\ 0.1, & \text{if } p = 0.9. \end{cases}$$

Clearly, $p = 0.1$ maximizes the likelihood when $X = 0$ is observed. If $X = 1$ is observed, then

$$f(1;p) = \begin{cases} 0.1, & \text{if } p = 0.1, \\ 0.9, & \text{if } p = 0.9. \end{cases}$$

Chapter 5: Estimation

Clearly, $p = 0.9$ maximizes the likelihood when $X = 1$ is observed. Hence, combining these, what we have is

$$\hat{p} = \begin{cases} 0.1, & \text{if } X = 0, \\ 0.9, & \text{if } X = 1. \end{cases}$$

EXAMPLE 5.4.4 A statistically minded basketball player wants to estimate his free-throw percentage, p. He keeps records on how many shots he has to try before making one. In a particular day, he makes a shot on the second, fourth, first, third, second, third, third, and fourth try. Assuming that p is constant in that day, how do you think he should estimate p?

SOLUTION: We can approach this problem in two different ways.

(1) Using binomial distribution we have

$$L(8,p) = \binom{22}{8} p^8 (1-p)^{14} \rightarrow \hat{p} = 8/22 = 4/11.$$

(2) Using geometric distribution we have

$$L(p) = (1-p)p(1-p)^3 p\, p(1-p)^2 p\, (1-p)p\, (1-p)^2 p\, (1-p)^2 p\, (1-p)^3 p$$
$$= p^8 (1-p)^{14} \rightarrow \hat{p} = 8/22 = 4/11.$$

EXAMPLE 5.4.5 (ESTIMATION OF POPULATION SIZE) The following example, known as capture recapture problem due to William Feller, is a good illustration of the likelihood principle. It has been used, for example, to estimate the size of the population of fishes in a given lake. Suppose that a manager of a company producing tennis shoes wants to know the demand for their shoes and thus is interested in knowing the population size in the region of interest. This manager conducts a survey and interviews 500 clients at random. During the interview, each of these clients is issued an indentification code. After several months, the manager interviews another round of clients at random, and finds that 50 clients in the second round were also interviewed in the first round. What conclusion can he draw regarding the size of the population in this region?

SOLUTION: Let

N = Population size to be estimated
n_1 = Number of clients interviewed in the first round
n_2 = Number of clients interviewed in the second round
r = Number of clients in the second round who were identified in the first round
$L_r(N)$ = Probability that the second round interview contains exactly r clients who were interviewed in the first round

Using the hypergeometric distribution, we have

$$L_r(N) = \frac{\binom{n_1}{r}\binom{N-n_1}{n_2-r}}{\binom{N}{n_2}}.$$

For any given particular set of n_1, n_2, and r, we may want to find the value of N that maximizes the probability $L_r(N)$. In other words, for a given set of data, the value $\hat{N}$ is among all possible values of N, which is most consistent with the given set of data in terms of maximizing the likelihood. To compute the maximum likelihood, consider the ratio

$$\frac{L_r(N)}{L_r(N-1)} = \frac{(N-n_1)(N-n_2)}{(N-n_1-n_2-r)N}.$$

It is easy to see that this is greater than 1 if $Nr < n_1 n_2$ and is less than 1 if $Nr > n_1 n_2$. This means that as N increases, the sequence $L_r(N)$ first increases and then decreases. It reaches its maximum when N is the largest integer before the number $n_1 n_2 / r$. Thus, $\hat{N} \sim n_1 n_2 / r$. In this example, the maximum likelihood estimate of the population size is $\hat{N} = 5,000$.

PRACTICE PROBLEMS

1. Half of the year is off-season for a soccer team. However, coach wants to have practice every Saturday unless a player on the team has a birthday. Assume that birthdays are evenly distributed during year. How many player coach should recruit to make the number of practice days maximal. Note that, for example, two players practicing 15 Saturdays is better than one player practicing 26 Saturdays.
   Hint: Let n denote the number of players and show that the likelihood function is $L(n) = 26n(1-1/26)^n$. Then follow the method used in Example 5.4.5.

2. Let $x_1, \cdots, x_n$ be an i.i.d. sample from a Poisson distribution with parameter λ with follow probability mass function

$$P(X=x) = \frac{\lambda^x e^{-\lambda}}{x!}$$

please express the MLE of λ in terms of $x_1, \cdots, x_n$.

EXAMPLE 5.4.6 Suppose $x_1, x_2, \cdots x_N$ is a sample of size N from a population with density function

$$f(x;\lambda) = \lambda e^{-\lambda x}, \quad x \geq 0.$$

Estimate λ using the maximum likelihood method.

SOLUTION:

$$L(\lambda) = L(x_1, x_2, \cdots x_N; \lambda) = (\lambda e^{-\lambda x_1})(\lambda e^{-\lambda x_2})\cdots(\lambda e^{-\lambda x_N})$$
$$= \lambda^N e^{-\lambda(x_1+x_2+\cdots x_N)} = \lambda^N e^{-\lambda N \bar{x}}.$$

Chapter 5: Estimation

To find the maximum likelihood estimate, we first find the log-likelihood function by taking logarithm of both sides. That is
$$logL(\lambda) = Nlog\lambda - \lambda N\bar{x}.$$

Note that this simplifies the calculation. Also since log transformation preserves the position of the maximum, we can work with the log-likelihood function in place of the likelihood function. Next, we take derivative and set that equal to zero. This leads to
$$N/\lambda - N\bar{x} = 0 \Rightarrow \hat{\lambda} = 1/\bar{x}.$$

PRACTICE PROBLEMS

1. Suppose $x_1, x_2, \cdots, x_N$ is a sample of size N from population with density function
$$f(x;\lambda) = (1+\lambda)x^\lambda, \quad 0 \le x \le 1.$$
Estimate λ using the maximum likelihood method.

2. Let $x_1, \cdots, x_n$ be an i.i.d. sample from following distribution with parameter θ
$$f(x;\theta) = \frac{1}{2}e^{-|x-\theta|}, x \in R$$
please find MLE of θ.

5.4.1 Method of Moments

We finish this section by noting that in the statistical literature, one finds other frequently used estimating methods. One such method, referred to as the method of moments, estimates population parameters by their sample counterparts, for example, population mean and variance by sample mean and sample variance. This method is not as powerful as the maximum likelihood method, but it also does not require the knowledge of the probability density function or probability distribution.

EXAMPLE Recall the Example 5.4.6. To find and estimate for λ using the method of moments, we first find sample mean $\bar{x}$. For exponential density function the population mean is $E(X) = 1/\lambda$. This leads to $1/\lambda = \bar{x} \rightarrow \tilde{\lambda} = 1/\bar{x}$, which is the same as the maximum likelihood estimate.

EXAMPLE 5.4.7 Suppose $x_1, x_2, \cdots, x_N$ is a sample of size N from a population with density function
$$f(x, \theta) = \begin{cases} 1/\theta, & 0 \le x \le \theta \\ 0, & \text{elsewhere.} \end{cases}$$

Estimate θ using both the maximum likelihood method and also the method of moments.

SOLUTION:
$$L(\theta) = f(x_1, \theta)f(x_2, \theta)\cdots f(x_N, \theta) = 1/\theta^N.$$

Clearly $L(\theta)$ is maximum for the minimum possible value of θ. But θ cannot be smaller than the observed maximum value, that is, max $(x_1, x_2, \cdots, x_N)$. Thus the maximum likelihood estimate of θ is the largest observed value, that max $(x_1, x_2, ... x_N)$.

Now for the method of moments, we need to find $E(X)$. It is easy to see that it is $\theta/2$. The sample counterpart of this is $\bar{x}$. Thus, the method of moment estimate of θ is $2\bar{x}$.

A PRACTICE PROBLEM Consider the last practice problem. Apply method of moments and compare the results with the maximum likelihood estimate.

5.5 INTERVAL ESTIMATION AND CONFIDENCE INTERVALS

Although point estimators are useful, in many instances they do not provide a completely satisfactory answer to many of the estimation problems. For example, when studying the blood pressure of a healthy person, it is not reasonable to call a person healthy only if his or her blood pressure is 120. Similarly, it is not reasonable to call a child normal if his or her height equals an fixed prespecified value. In sports too, it is not reasonable to try to predict the exact value of the next record. For these examples, it would be more sensible to think of two numbers (an interval) consisting of a lower value and an upper value. This will provide a range of possible values, for example, 70 to 120 for blood pressure of a healthy person. A great advantage of this approach is that it is possible to include a measure of confidence for the estimation or prediction. This section discusses this approach and presents many different situations that arise in application.

Let us start with an example. Suppose that X is distributed as $N(\mu, \sigma^2)$. Then as was discussed in Section 3.6, the random variable $\frac{\bar{X}-\mu}{\sigma/\sqrt{N}}$ is distributed as $N(0,1)$. This means that 95% of the values of this random variable will fall within two standard deviations of mean. To be exact, 1.96 standard deviations of mean (see Table B3). That is, we have

$$P(-1.96 \leq \frac{\bar{X}-\mu}{\sigma/\sqrt{N}} \leq 1.96) = 0.95.$$

Rearranging the terms, we get

$$P(\bar{X}-1.96\frac{\sigma}{\sqrt{N}} \leq \mu \leq \bar{X}+1.96\frac{\sigma}{\sqrt{N}}) = 0.95.$$

This means that the random interval $[\bar{X}-1.96\frac{\sigma}{\sqrt{N}}, \bar{X}+1.96\frac{\sigma}{\sqrt{N}}]$ contains the true value of μ with 95% probability. This interval is called a 95% confidence interval for population mean, μ. The "95% probability" is called the confidence level. Note that, for each realization or sample, there will be a different confidence interval. In other words, if two people independently draw a sample of size N and construct a confidence interval, they will not produce the same interval. One meaning of a 95% confidence interval is that, in 5% of the cases, this interval may not include μ. Putting this differently, if 100 people take a random sample of size N and construct a 95% confidence interval, 95 of these intervals are expected to contain the true value of μ.

Chapter 5: Estimation

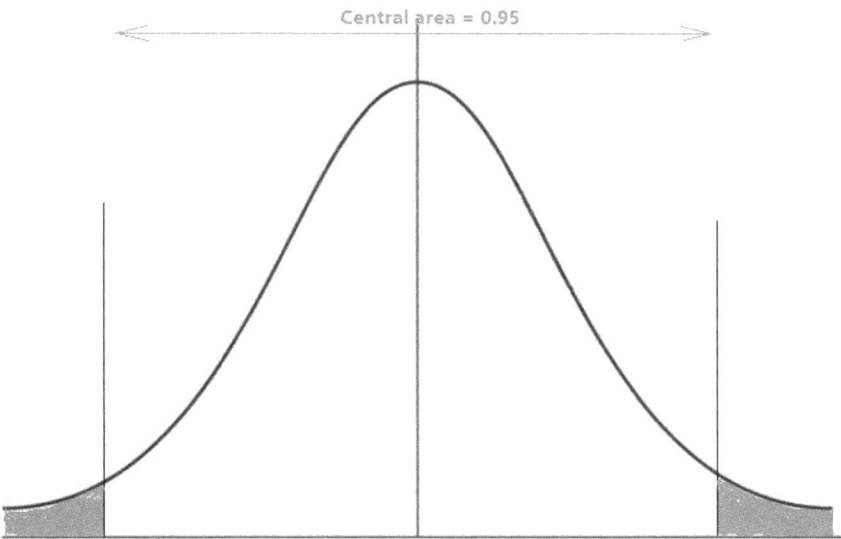

FIGURE 5.5.1 z-Values for 95% Level of Confidence
Confidence Interval for μ When σ Is Known

Rather than using 95%, we can consider a general confidence level $100(1-\alpha)\%$ and construct a confidence interval. Steps for constructing confidence intervals are presented in the remainder of this section. A formal definition of confidence interval for population mean is as follows:

DEFINITION: A $100(1-\alpha)\%$ confidence interval for a population mean, μ, is an interval such that μ lies within $100(1-\alpha)\%$ of such intervals if repeated samples of the same size were formed and confidence intervals were constructed or interval estimates were made.

MARGIN OF ERROR: For a 95% confidence interval, the term $1.96\sigma/\sqrt{N}$ is called the margin of error. In other words, a 95% confidence interval is found by adding and subtracting the margin of error to the sample mean. Such error depends on the chosen confidence level (increases as confidence level is increased), population standard deviation (increases as standard deviation is increased), and the sample size (decreases as sample size is increased). Since high confidence levels and narrow intervals are desirable (wide intervals provides little information and hence are not considered good estimates), one needs a large sample to satisfy such requirements (Figure 5.5.1).

In general, a $100(1-\alpha)\%$ confidence interval for μ, the mean of a normal population, when σ is known is given by

$$(\bar{X} - z_{1-\alpha/2}\frac{\sigma}{\sqrt{N}}, \ \bar{X} + z_{1-\alpha/2}\frac{\sigma}{\sqrt{N}}) \tag{5.4}$$

where $z_{1-\alpha/2}$ presents a percentile point of $N(0,1)$. See Section 3.6 for further discussion concerning $z_{1-\alpha/2}$. The values of $z_{1-\alpha/2}$ can be obtained from the table of normal distribution (Table B3). Here $z_{1-\alpha/2}\frac{\sigma}{\sqrt{N}}$ is the margin of error.

EXAMPLE 5.5.1 The attendance of football games at a certain stadium is normally distributed with a standard deviation of 3,000. Data were collected on the attendance of 16 randomly selected recent games and it was found that the average attendance of these 16 games was 40,000. Use this data to develop a 90% confidence interval for the expected number, or average attendance, μ of football games at this stadium.

SOLUTION: For this example, $1 - \alpha = 0.90, \alpha = 0.10, 1 - \alpha/2 = 0.95$ and hence $z_{1-\alpha/2} = z_{0.95} = 1.65$ (see Table 3.6.1 or Table B3). Using (5.4), the corresponding confidence interval is

$$(40,000 - 1.65 \frac{3,000}{\sqrt{16}}, 40,000 + 1.65 \frac{3,000}{\sqrt{16}})$$

which is

$$(38762.5, 41237.5).$$

Thus, the probability that the true value of average attendance lies somewhere between 38,762.5, and 41,237.5 is 90%. Or, we are 90% confident that the true average falls in this interval. Note that here the confidence interval represents an estimate for the expected number of attendance for the future games.

If, aside from you, 99 people take a random sample size of 16 from the population of attendances and construct a 90% confidence interval for μ, we expect 90 of these intervals to cover the true value of μ.

NOTE: If the distribution of X is not normal but N is sufficiently large, then according to the Central Limit Theorem the distribution of $\bar{X}$ is approximately normal and therefore $\frac{(\bar{X}-\mu)}{\sigma/\sqrt{N}}$ is approximately distributed as $N(0,1)$. Hence, we can still apply (5.4) to construct a confidence interval for μ. Moreover, for a large N, we can replace sample standard deviation s for σ (when it is unknown) and construct an approximate confidence interval. We will come back to this point later.

EXAMPLE To estimate the mean temperature μ at which a particular type of metallic glass becomes brittle, 36 pieces of this metallic glass were randomly sampled from a recent production run. The temperature at which brittleness first appeared was recorded for each piece in the sample. The following results were obtained: $\bar{x} = 480°F, s = 11°F$. Use a 90% confidence interval to estimate μ.

SOLUTION: Let X denote the temperature at which a randomly selected piece of metallic glass becomes brittle. Here is the questions; in what temperature we should expect this to happen? That is $E(X) = ?$ $\bar{x} = 480°$ $s = 11°F$. We need to estimate $\mu = E(X)$. First we ask the following questions:
Is distribution of X normal? No.,   Is $N > 30$? Yes.,   Is σ known? No.

Then an approximate 90% confidence interval for the true mean temperature at which this type of metallic glass becomes brittle is

$$\bar{x} \pm (z_{1-\alpha/2}) \frac{s}{\sqrt{N}} = 480 \pm (1.645)(11/\sqrt{36}) \rightarrow (476.98, 483.02).$$

A PRACTICE PROBLEM The manager of a recreation center has recorded the times 64 users spent on a treadmill. The sample mean and sample deviations are 35 and 10 min, respectively. Construct a 99% confidence interval for the true average time users spend on a treadmill.

Chapter 5: Estimation

CHOOSING THE SAMPLE SIZE

From expression (5.4), it is clear that as we increase the confidence level the length of the interval, $2z_{1-\alpha/2}\frac{\sigma}{\sqrt{N}}$ increases. In other words, the interval becomes wider. This is because we are seeking an interval to cover the true value of μ with a higher probability or with a higher confidence. In practice, the wide intervals are neither desirable nor informative. What is the point of being 100% confident that the average blood pressure of a healthy person is between 0 and 500? In order to decrease the length of the interval, our only reasonable option is to increase the sample size. This is, in fact, good news since unlike the variation, this is a factor we can control. The only problem is the higher cost and whether it is possible to have a large sample.

When σ is known, the 95% confidence interval for μ was constructed based on the fact that, for approximately 95% of all random samples, $\bar{x}$ will be within $1.96\sigma/\sqrt{N}$ of μ. The quantity $1.96\sigma/\sqrt{N}$ (one-half of the length of confidence interval) is the margin of error discussed earlier. It is sometimes called the bound on the error of estimation associated with a 95% confidence level. This means that with 95% confidence, the point estimate $\bar{x}$ will be no further than $1.96\sigma/\sqrt{N}$ away from μ. Before collecting any data, an investigator may wish to determine the size of a sample for which a particular value of this bound is achieved. For example, with μ representing the average pulse rate for an adult male of a certain age, the objective of the research may be to estimate μ to within 1 beat with 95% confidence. The value of N necessary to achieve this is obtained by equating 1 to $1.96\sigma/\sqrt{N}$ and solving for N.

More generally, suppose that it is desired to estimate μ to within an amount B (the specified error of estimation) with 95% confidence. Finding the necessary sample size requires solving the equation $B = 1.96\sigma/\sqrt{N}$ for N. This leads to

$$N \geq \left[\frac{1.96\sigma}{B}\right]^2.$$

Note that a large value of σ or a small value of B results in a large value of N.

Use of the sample size formula requires σ to be known. In practice, this is rarely the case. To overcome this problem, one possibility is to carry out a preliminary study and use the resulting sample standard deviation (or a somewhat larger value, to be conservative) to determine N for the main study. Another possibility is to simply make an educated guess about the value of σ and use that when calculating N. For a population distribution that is not too skewed, dividing the range (the difference between the largest and smallest measured values) by 4 often provides a rough estimate for the value of the standard deviation. This is particularly recommended for normal distribution, since 95% of the measurements fall within two standard deviations of the mean.

In general, the sample size required to estimate a population mean μ to within a margin of error B or an amount B with $100(1-\alpha)\%$ confidence should satisfy the following inequally:

$$N \geq \left[\frac{z_{1-\alpha}\sigma}{B}\right]^2 = \left[\frac{(z-Critical)(\text{standard deviation})}{\text{margin of error}}\right]^2.$$

If σ is unknown, it may be estimated based on previous information or, by using (range)/4 if the population is not too skewed.

EXAMPLE Suppose that we wish to estimate the mean cost of participating in a sport (shoes, clothing, etc.), per semester, for students at a particular university. In order for the estimate to be useful, we want it to within $20 of the true average cost (population mean). How large of a sample should be used in order to be 95% confident of achieving this level of accuracy?

SOLUTION: In order to determine the required sample size, we need a value for σ. An expert is pretty sure that the amount spent varies widely, with most values being between $50 and $450. Using this, a reasonable estimate of σ is

$$\frac{\text{Range}}{4} = \frac{450 - 50}{4} = \frac{400}{4} = 100.$$

The required sample size is then

$$N \geq \left[\frac{1.96\sigma}{B}\right]^2 = \left[\frac{(1.96)(100)}{2}\right]^2 = [9.8]^2 = 96.04 \rightarrow N \geq 97.$$

Note that, since a fraction of a sample is not possible, we always round the resulting value, upward.

PRACTICE PROBLEMS

1. Find the sample size necessary for estimating the average height of the male students to within 1 inch. Assume that the student's height varies between 60 and 88 inches.
2. By researching F1 champions' ages data since 1950, we found their ages ranging from 23 to 46. If I would like to construct a 90% confidence interval for champion's average age achieving an accuracy of 3 years old, what is the proper sample size I should use?

CONFIDENCE INTERVAL FOR μ WHEN σ IS UNKNOWN (NORMAL DISTRIBUTION)

The confidence intervals discussed so far are valid only when σ is known. In practice, this is hardly the case. In fact, if μ is unknown, it is not reasonable to assume that σ is known. When σ is unknown, it is replaced by its sample counterpart s. In this case, the sampling distribution of $\frac{\bar{X} - \mu}{s/\sqrt{N}}$ is no more $N(0,1)$ but t (student t) with $(N-1)$ degrees of freedom (see Chapter 3), because $\frac{\bar{X}-\mu}{s/\sqrt{N}} = \frac{(\bar{X}-\mu)/(\sigma/\sqrt{N})}{s/\sigma}$ with the numerator being $N(0,1)$ and the denominator having a $\chi^2_{(N-1)}$ distribution. Using this, a $100(1-\alpha)\%$ confidence interval for μ when σ is unknown, is constructed as

$$(\bar{X} - t_{1-\alpha/2,(N-1)}\frac{s}{\sqrt{N}}, \ \bar{X} + t_{1-\alpha/2,(N-1)}\frac{s}{\sqrt{N}}). \tag{5.5}$$

Here, the sample standard deviation s (see Section 5.3) is instead of the unknown σ, and the t-distribution is used instead of the $N(0,1)$. $t_{1-\alpha/2,(N-1)}$ is the notation for the $1-\alpha/2$ percentile of the t-distribution, with a degree of freedom $N-1$. This means that if a variable T has the t-distribution with $N-1$ degrees of freedom, then

$$P(T \leq t_{1-\alpha/2,(N-1)}) = 1 - \alpha/2.$$

Chapter 5: Estimation

Recall that if N is sufficiently large, the values of $t_{1-\alpha/2,(N-1)}$ and $z_{1-\alpha/2}$ are very close to each other and we can use the latter to develop an approximate confidence interval.

Before considering an example, we should note that when population variance is unknown, one may wish to construct a confidence interval for it too. For this, we need to use chi-squared distribution (discussed in Chapter 3).

EXAMPLE Pulse rate is an important measure of the fitness of a person's cardiovascular system. A random sample of 21 adult males who jog at least 15 miles per week had a mean pulse rate of 52.6 beats per minute and a standard deviation of 3.22 beats per minute. Find a 95% confidence interval for the mean pulse rate of all adult males who jog at least 15 miles per week. Assume that the distribution of pulse rate is normal.

SOLUTION: Let X denote the pulse rate of a randomly selected adult male who jogs at least 15 miles per week. We need to estimate $\mu = E(X)$. First we ask the following questions: Is $N > 30$? No. Is σ known? No. Is the distribution of X normal? Yes. Then a 95% confidence interval for μ, the true mean pulse rate of all adult males who jog at least 15 miles per week, is

$$\bar{x} \pm t_{1-\alpha/2,(N-1)}\, s/\sqrt{N} = 52.6 \pm 2.09(3.22/\sqrt{21})$$
$$= 52.6 \pm 1.47 \rightarrow (51.13, 54.07).$$

Note that if no information about the distribution of the measurements is provided, we have to assume normality since otherwise we cannot construct a confidence interval.

A PRACTICE PROBLEM Exercise time for 16 students had a mean value of 5 h/week with a standard deviation of 2 h/week. Construct a 99% confidence interval for population mean exercise time/week for all students.

CONFIDENCE INTERVAL FOR σ^2

A $100(1-\alpha)\%$ confidence interval for σ^2, when X is distributed as $N(\mu, \sigma^2)$, is given by

$$\left(\frac{(N-1)s^2}{\chi^2_{1-\alpha/2,(N-1)}}, \frac{(N-1)s^2}{\chi^2_{\alpha/2,(N-1)}} \right). \tag{5.6}$$

Here, $\chi^2_{\alpha/2,(N-1)}$ represents a percentile point of the χ^2 distribution with $N-1$ degrees of freedom (see Section 3.5). This means that if X has an $\chi^2_{(N-1)}$ distribution and p is a number between 0 and 1, then

$$P(X \leq \chi^2_{p,(N-1)}) = p.$$

As was mentioned earlier, in evaluation of individual stocks, standard deviation is often used as a quantity for measuring the risk. In fact, there are many circumstances where estimation of variance or standard deviation is more useful than estimation of the mean. In sports, we may use standard deviation to measure consistency of players.

EXAMPLE 5.5.2 In Example 5.5.1, we assumed that the standard deviation is known to be 3,000. In practice, however, the value of standard deviation is often unknown. This is when we use formula (5.5) instead of (5.4). Let us assume that the attendance of football games at this stadium is normally distributed, but both the mean and standard deviation are unknown. The attendance of 16 recent games has shown a sample average of 40,000 and a sample standard deviation (calculated from these 16 attendances) 3,000. Use this data to develop a 90% confidence interval for the mean attendance and a 90% confidence interval for the standard deviation of attendance of the football games at this stadium.

SOLUTION: The sample size in this case is $N = 16$. Hence the degree of freedom is $N - 1 = 15$. The value of $\alpha = 0.1$ and the value of $t_{1-\alpha/2,(N-1)}$, obtained from table for t-distribution (Table B4) is 1.73. Using formula (5.5), the corresponding confidence interval for the population mean is

$$\left(40,000 - 1.73 \frac{3,000}{\sqrt{16}}, 40,000 + 1.73 \frac{3,000}{\sqrt{16}}\right)$$

or

$$(38,702.5, 41,297.5).$$

Note that, as expected, this interval is wider than the one obtained before (known variance case).

For a 90% confidence interval for the standard deviation σ, we first construct a 90% confidence interval for the variance σ^2 using formula (5.6). Here $1 - \alpha = 0.90$ and therefore $\alpha = 0.10$ and $\alpha/2 = 0.05$. The values of $\chi^2_{1-\alpha/2,(N-1)} = \chi^2_{.95,(15)}$ and $\chi^2_{\alpha/2,(N-1)} = \chi^2_{.05,(15)}$ are 25 and 7.26 respectively (Table B5). Hence, a 90% confidence interval for σ^2 is

$$\left(\frac{15 \times 3,000^2}{25}, \frac{15 \times 3,000^2}{7.26}\right)$$

or

$$(5,400,000, 18,595,000).$$

Taking the square root of these two numbers, we get a 90% confidence interval for the standard deviation σ as

$$(2,323.79, 4,312.19).$$

A PRACTICE PROBLEM In sports such as basketball players may be judged based on their average scores per game. Although this is a reasonable measure for performance some coaches prefer to use a measure reflecting the players' consistency. Suppose that consistency is measured by standard deviation. If we construct confidence intervals (for standard deviation) for two different players, we can conclude that one is more consistent than the other only if the two intervals do not overlap. Select two players of your choice (and sport of your choice) and set up confidence intervals to see if you can establish whether one is more consistent than the other.

CONFIDENCE INTERVALS BASED ON CHEBYSHEV'S RULE

As is clear from what is covered so far, the normal distribution plays the key role in estimation. This is to a large extent due to the Central Limit Theorem, which is applicable when the sample is large. There is, however, a situation where construction of a confidence interval involving a large sample is not carried out using this approach. This occurs when sample comes from a relatively

Chapter 5: Estimation

small population that is not normally distributed. It is based on the Chebyshev's rule discussed in Chapter 1. Chebyshev's rule implies that at least $(1 - \frac{1}{k^2})$ of the sample means must be within k standard errors (SEs) of the mean of all the samples (within $k \cdot SE$ of μ) when applying this approach once the equation

$$1 - \frac{1}{k^2} = (1 - \alpha)$$

is solved for k and SE is determined, the endpoints of the desired confidence interval for μ can be located at $\bar{x} \pm k \cdot SE$.

EXAMPLE About 250 professional tennis players are believed to live and work in the immediate vicinity of New York City. A random sample of 50 of these players yielded a mean annual income of $\bar{x} = \$180,000$ and an estimated standard deviation of $s = \$20,000$. Use the information to determine an 80% confidence interval for the mean annual income of the population of all 250 players.

SOLUTION: Although this is a large sample, it comes from a relatively small population. The sample includes 20% of the population. In practice, the large sample theory based on the Central Limit Theorem is applied when sample size is less than 5% of the population. Here, the sample size is not less than 5% of the population. Also, because of the unusual nature of player's income, assumption of normal distribution does not seem reasonable. Therefore, Chebyshev's rule should be used to construct the desired confidence interval. To construct an 80% confidence interval, we first solve

$$1 - \frac{1}{k^2} = 80\%$$

for k. This yields $k = 2.24$. Also, here we have

$$SE = \frac{s}{\sqrt{N}} = \frac{20,000}{\sqrt{50}} = 2,828.43.$$

Therefore, the endpoints of the confidence interval are at $\bar{x} \pm k \cdot SE$, or $180,000 \pm 2.24 \cdot 2828.43$. This gives the following approximate 80% confidence interval

$$(17,3664.32, 18,6335.68).$$

A PRACTICE PROBLEM Research information regarding similiar sports (golf, etc.) and construct appropriate confidence intervals.

CONFIDENCE INTERVAL FOR PROPORTION (p)

Suppose now that we are interested in estimating p, the percentage (or proportion) of individuals with a certain characteristic. For example, we may want to estimate the proportion of the people who support a local team or have a certain political party affiliation. We take a random sample of size N and calculate

$$\hat{p} = \frac{\text{Number of individuals with that characteristic}}{\text{Sample size}} = \frac{X}{N}$$

where X represents the number with the characteristic of interest. To determine the reliability of the estimator $\hat{p}$, we need to know its sampling distribution. We know that X has a binomial distribution with $E(X) = Np$ and $\text{Var}(X) = Npq$, which for a large N, is approximately normally distributed random variable with the same mean and variance. Thus, for large N, $\hat{p}$ has an approximate normal distribution with $E(\hat{p}) = \frac{1}{N}E(X) = p$ and $\text{Var}(\hat{p}) = 1/N^2 \text{Var}(X) = pq/N$ where $q = 1-p$. That is

$$\hat{p} \sim N(p, \sqrt{\frac{pq}{N}}).$$

Here, N is considered large if the interval, $\hat{p} \pm 3\sigma_{\hat{p}}$, does not include 0 or 1. An alternative condition for normal approximation (the one mentioned in Section 4.1) is the following. We should have $0.1 \le p \le q$ and $npq \ge 5$, or $n\hat{p}$ and $n\hat{q}$ should both be bigger than 5. The latter is used in most of the exercises in this book.

When all the conditions are satisfied, a large sample $100(1-\alpha)\%$ confidence interval for p is

$$\hat{p} \pm z_{1-\alpha/2} \sigma_{\hat{p}} \approx \hat{p} \pm z_{1-\alpha/2} \sqrt{\frac{\hat{p}\hat{q}}{N}} \qquad (5.7)$$

or equivalently

$$X/N \pm z_{1-\alpha/2} \sqrt{\frac{(X/N)(1-X/N)}{N}}.$$

The length (width) of this interval is

$$L = 2B = (\hat{p} + z_{1-\alpha/2}\sqrt{\frac{\hat{p}\hat{q}}{N}}) - (\hat{p} - z_{1-\alpha/2}\sqrt{\frac{\hat{p}\hat{q}}{N}}) = 2z_{1-\alpha/2}\sqrt{\frac{\hat{p}\hat{q}}{N}}.$$

EXAMPLE 5.5.3 In a university with a population of about 20,000 students, 1,000 students were interviewed and asked how much they knew about the sport "Cricket." Among them 400 said "not much." Construct a 90% confidence interval for p, the population proportion of the student body in the university who do not know much about Cricket.

SOLUTION: Here $N = 400$, $\hat{p} = 400/1,000$, and $\sigma_{\hat{p}}$ is approximately

$$\sqrt{\frac{\hat{p}\hat{q}}{N}} = 0.0155.$$

Clearly the interval $\hat{p} \pm 3\sigma_{\hat{p}}$ does not include 0 or 1, and N is large enough for application of (5.7). Thus a 90% confidence interval for p is

$$(0.4 - 1.65\,(0.0155),\ 0.4 + 1.65\,(0.0155)) = (0.3744,\ 0.4256).$$

Note that it is easy to construct a one-sided confidence interval, if one is interested only on upper or lower end values.

A PRACTICE PROBLEM It seems that chewing tobacco and snuff are part of the game of baseball. In fact, these two types of smokeless tobacco are the most commonly used in the United States. A study published in the *Journal of the American Medical Association*, addressed several issues

Chapter 5: Estimation

associated with the use of smokeless tobacco. Major and minor league baseball players were selected for this study because of their high rate of smokeless tobacco use. Among the 1,094 people studied, 493 were nonuser of smokeless tobacco, 138 were former users, and 463 were current users. The researchers found the presence of oral leukoplakia (white, opaque, leathery-appearing plaque) in 196 of the players who had used smokeless tobacco within the previous week. It was found in only 7 of the players who quit the use of smokeless tobacco. Construct a 90% confidence interval for population proportions.

As mentioned, in order to apply (5.7), a different but equivalent condition is used in many textbooks is to calculate $N\hat{p}$ and $N(1-\hat{p})$. If both values are greater than 5, then (5.7) is a valid confidence interval. In the above example, $N\hat{p} = 160$ and $N(1-\hat{p}) = 240$, both are greater than 5.

Sample size determination is very similar to that for mean. We illustrate this using an example.

EXAMPLE To demonstrate the calculation of sample size, consider the Example 5.5.3 and the following question. How large of a sample would be necessary to estimate the population proportion with a sampling error of $B \leq 0.05$ and a 95% confidence level? Or what sample size is required to estimate p to within 0.05 of the true value?

SOLUTION: Here we have

$$0.05 = (1.96)\sqrt{(0.4)(0.6)/N} \Rightarrow N \geq \frac{(1.96)^2(0.4)(0.6)}{(0.05)^2} = 368.79 \approx 369.$$

Note that if $\hat{p}$ is not known one needs to be conservative and consider the worst case scenario, that is use 0.5. This is because the maximum value of pq when $p+q=1$, occurs at $p=q=0.5$, and these values produce the largest possible value for $\sigma_{\hat{p}}$. Applying this to the example above, we get

$$N \geq \frac{(1.96)^2(0.5)(0.5)}{(0.05)^2} = 384.16 \approx 385.$$

Summarizing these, to determine the sample size the general formula is

$$N \geq \frac{\hat{p}\hat{q}z_{1-\alpha/2}^2}{B^2} = \hat{p}(1-\hat{p})\left(\frac{z_{1-\alpha/2}}{B}\right)^2.$$

where B is the margin of error. The question of sample size is discussed in more detail in Section 6.7.1.

PRACTICE PROBLEMS

1. To estimate the next president's final approval rating, how many people should be sampled so that the margin of error is 3% with 95% confidence?
2. A sample of 100 students were asked if they play any sports. Twenty-five stated that they do. If we want to estimate the true proportion of students who play sports with a sampling error of 0.10 or less at a 90% confidence level, how many students should we include in our sample?

CONFIDENCE INTERVAL FOR $\mu_1 - \mu_2$ WHEN σ_1^2 AND σ_2^2 ARE KNOWN

Suppose we are interested in comparing two populations with regard to their means. For example, we may believe that people spend more time watching or playing sport A than sport B and wish to estimate the difference between the average times they spend watching or playing these two sports. We can do this by constructing a confidence interval for $\mu_A - \mu_B$.

From many situations where confidence intervals can be constructed for two population cases, we mention only the interval estimation for the difference between the means of the two normal distributions when the populations variances are known. The reason for this will be explained after presenting the procedure and an example. Confidence intervals for other cases, including the difference between the means when population variances are unknown, difference between population proportions, and the ratio of population variances will be discussed in Chapter 6.

Let $X^{(1)}$ be $N(\mu_1, \sigma_1^2)$, and $X^{(2)}$ be $N(\mu_2, \sigma_2^2)$ with known standard deviations σ_1, σ_2, respectively. The sample sizes are N_1 and N_2, respectively. Then the random variable

$$Y = \frac{\bar{X}^{(1)} - \bar{X}^{(2)} - (\mu_1 - \mu_2)}{\sqrt{\sigma_1^2/N_1 + \sigma_2^2/N_2}} \tag{5.8}$$

is distributed as $N(0,1)$. The confidence limits are then easily constructed using

$$P(|Y| \leq z_{1-\alpha/2}) = 100(1-\alpha)\%. \tag{5.9}$$

In fact, a $100(1-\alpha)\%$ confidence interval for $\mu_1 - \mu_2$ is given by

$$(\bar{X}^{(1)} - \bar{X}^{(2)}) \pm z_{1-\alpha/2} \sqrt{\sigma_1^2/N_1 + \sigma_2^2/N_2}. \tag{5.10}$$

EXAMPLE 5.5.4 To study the effect of carbohydrate on fuel substrates during exercise in women, a group of female students who participated in regular aerobics exercise is compared with a group who did not. The data on percent body fat obtained by skin folds measured at seven sites is summarized in the following table. The standard deviation of the two populations is also listed in the table. Estimate the true mean difference in percent body fat between the two populations using a 90% confidence interval.

Group	Sample Size	Mean Percent Body Fat	Population Standard Deviation
Aerobic Exercise	80	20.1	4.0
No Aerobic	60	20.7	3.5

SOLUTION: Using (5.10) the required confidence interval is

$$(20.1 - 20.7) \pm z_{0.95} \sqrt{4.0^2/80 + 3.5^2/60}$$

that is

$$-0.6 \pm 1.645 \times 0.636 \Rightarrow (-1.646, 0.446).$$

Chapter 5: Estimation

Just like single population cases, if distribution of X_1 and X_2 are not normal but sample sizes are large, we can, based on the Central Limit Theorem, still use (5.10). Moreover if σ_1^2 and σ_2^2 are not known, they are replaced by their sample counterparts to provide an approximate confidence interval for $\mu_1 - \mu_2$.

We end this section by explaining the reason for stopping the discussion of confidence intervals and proceeding to consider the testing problems. Consider the interval obtained for the difference between the percent body fats in Example 5.5.4. Since the interval covers both negative and positive percentages, we cannot draw a conclusion that female students in one group have an average percent body fat less than the other group. This is because the true difference could fall anywhere in the interval. In other words, we do not have sufficient evidence to support the claim (hypothesis) that $\mu_1 \neq \mu_2$. Thus estimating the difference between the mean ($\mu_1 - \mu_2$) is not necessary unless we first establish that $\mu_1 \neq \mu_2$. It is therefore more appropriate to study the tests of hypotheses such as $\mu_1 = \mu_2$ versus $\mu_1 \neq \mu_2$ first and then, if needed, estimate $\mu_1 - \mu_2$.

A PRACTICE PROBLEM In order to estimate the difference between the average number of points scored in two different basketball tournaments the scores of 36 games from each tournament was recorded. The means and standard deviations are, respectively, 82, 71, and 10 and 20. Construct a 95% confidence interval for the difference between the mean number of points scored in these two tournaments.

FINITE POPULATION CORRECTION ERROR

Except for one case, the populations we have sampled so far have been very large or assumed to be infinite. What if the sampled population is not infinite, or not even very large? In such instances, we make some adjustments in the way we compute the SE of the sample means and the SE of the sample proportions. A finite population can be rather small; it could be all the players of one team. It can also be very large, such as all of the tennis fans in London.

For a finite population, where the total number of objects is n and the size of the sample is N, an adjustment is made to the SEs of the sample means and the sample proportions by multiplying them by a factor

$$\sqrt{\frac{n-N}{n-1}}.$$

This adjustment is called the finite population correction factor. Why is it necessary to apply a factor, and what is its effect? Logically, if the sample is a substantial percentage of the population, then estimates are more precise than those for a smaller sample. To understand this we can study the effect of the term $\frac{n-N}{n-1}$. Suppose that population size is 1,000 and the sample size is 100. Then this ratio is $(1{,}000 - 100)/(1{,}000 - 1)$, or 900/999. Taking the square root gives a correction factor, 0.9492. Multiplying by the standard error reduces the error by about 5% ($1 - 0.9492 \cong 0.05$). This reduction in the size of the SE yields a smaller range of values in estimating the population mean. If the sample size is 200, the correction factor is 0.8949, meaning that the SE has been reduced by more than 10%.

EXAMPLE 5.5.5 There are 250 families in a small town in Pennsylvania. A poll of 40 families revealed the mean annual contribution to district sports programs is $450 with a standard deviation of $75. Construct a 95% confidence interval for the mean annual contribution per family.

SOLUTION: First note the population is finite. That means there is a limit to the number of people in the town. Second, the sample constitutes more than 5% of the population. That is, $N/n = 40/250 = 0.16$. Hence, the finite population correction factor must be used. The 95% confidence interval is constructed as follows:

$$\bar{x} \pm z_{1-\alpha/2} \frac{s}{\sqrt{n}} \left(\sqrt{\frac{n-N}{n-1}} \right) = 450 \pm 1.96 \frac{75}{\sqrt{40}} \left(\sqrt{\frac{250-40}{250-1}} \right)$$

$$= 450 \pm 23.243(\sqrt{0.8434}) = 450 \pm 21.35 \to (428.65, 471.35).$$

SUMMARY: CONFIDENCE INTERVALS AND SAMPLE SIZES

Point Estimate A single value, computed from a sample, that is used to estimate a population parameter.

Confidence Interval A range of values constructed from a sample so that the parameter of interest falls within the range with a specified probability. The specified probability is called the level of confidence or confidence level.

1. The factors that determine a confidence interval for a mean are:

   - The number of observations in a sample N.
   - The variability in a population, usually estimated by the sample standard deviation s.
   - The level of confidence $1 - \alpha$, which determines the needed z-value (z-critical).

   A confidence interval for the mean is then either

   $$\bar{x} \pm z_{1-\alpha/2} \frac{s}{\sqrt{N}}$$

   or

   $$\bar{x} \pm t_{1-\alpha/2,(N-1)} \frac{s}{\sqrt{N}}.$$

2. The factors that determine a confidence interval for a proportion are

   - The number of observations in a sample.
   - The value of $\hat{p}$, computed by dividing the number of successes in the sample X by the number observations in the sample, N.
   - The level of confidence $1 - \alpha$, which determines the z-value.

   A confidence interval for proportion is then

   $$\hat{p} \pm z_{1-\alpha/2} \sqrt{\frac{\hat{p}(1-\hat{p})}{N}}.$$

3. The required size of a sample can be determined for both mean and proportion. The factors that determine the size of a sample for a mean are:

Chapter 5: Estimation

- The desired level of confidence to determine the z-value.
- The maximum allowable error B.
- For mean, the value of σ or its estimate. If no value is available use range/4.
- An estimate of the population proportion. If no estimate is available, then 0.50 is used.

The formula for sample size for mean and proportion are, respectively,

$$N \geq \left(\frac{z_{1-\alpha/2}(\text{Standard Deviation})}{B}\right)^2, \quad N \geq \hat{p}(1-\hat{p})\left(\frac{z_{1-\alpha/2}}{B}\right)^2.$$

4. The finite population correction factor is applied to a confidence interval if $\frac{N}{n}$ (where n is the population size) is more than 0.05. The correction factor is

$$\sqrt{\frac{n-N}{n-1}}$$

and affects only the second term. Thus we have

$$\bar{x} \pm z_{1-\alpha/2} \frac{s}{\sqrt{N}} \left(\sqrt{\frac{n-N}{n-1}}\right).$$

The finite population correction factor for selected sample sizes when the population size is 1,000 is given below,

Sample Size	Fraction of Population	Correction Factor
10	0.010	0.9955
25	0.025	0.9879
50	0.050	0.9752
100	0.100	0.9492
200	0.20	0.8949
500	0.50	0.7075

5. A confidence interval for the variance is

$$\left(\frac{(N-1)s^2}{\chi^2_{1-\alpha/2,(N-1)}}, \frac{(N-1)s^2}{\chi^2_{\alpha/2,(N-1)}}\right).$$

5.6 BOOTSTRAP CONFIDENCE INTERVAL

The procedures described in the previous sections for constructing confidence intervals are applicable when certain assumptions are met or certain conditions are satisfied. In practice, it is neither always easy to justify the assumptions made nor is easy to check to see if required conditions are satisfied. In recent years, there have been many attempts to look for methods that require fewer assumptions or conditions.

One of the most important developments in statistics in the last two decades has been the appearance of statistical methods designed to take advantage of the availability of high-speed computing. These methods require few assumptions for their valid applications. They provide a line of attack on problems previously regarded as intractable. A good example of this is the bootstrap technique introduced and developed by Bradley Efran. In general, the bootstrap method consists of estimating a characteristic of an unknown population by simulating the characteristic when the true population is replaced by the estimated one. The appeal of this approach is its wide applicability to complex data structures in both parametric and nonparametric (distribution free) problems. Several authors have discussed the consequences of this method in various statistical inference problems. The basic premise is that, a random sample can be expected to mirror the population from which it was drawn. Bootstrap goes a step further and uses the actual sample observations themselves as a surrogate for the underlying population whose true characteristics remain unknown.

The bootstrap method is especially useful when the underlying statistic whose distribution is sought as intrinsically complicated and hence not amenable to theoretical approximations. Furthermore, the method is automatic and exact. It is automatic because no special calculations are needed for specific problems and it is exact because, theoretically, given an infinite number of simulated samples, we can compute the distribution of the underlying statistic exactly. The bootstrap method has caught the imagination of the statistical community because of its elegance, mathematical subtlety, and ease of use. For a mathematical treatment of bootstrap, confidence intervals and a discussion of their desired properties under quite general conditions, refer to Diciccio and Romano (1988) and reference therein.

5.6.1 An Example of Bootstrapping

We present here a simple illustration of bootstrapping to a problem in sport.

Let $R_1, R_2, \ldots R_n$ be a random sample presenting records for some sport, such as men's long jump. Suppose that we wish to find a point estimate and a confidence interval for R_b, that is the next best record. Consider, for example, a magnitude $\hat{R}_b$, that exceeds the best record in the sample by an amount equal to the difference between the best and the second best records in the sample. Let's consider this value as an estimate for R_b.

A confidence interval (95%, for example) for R_b can be found by determining the distribution through repeated samples of size N (with replacement) taken from the underlying distribution and finding a value $\hat{R}_b^*$ such that

$$P\left(R < R_b < \hat{R}_b^*\right) = 0.95$$

Here R is the best record in the sample. The problem that remains to be addressed is that, the distribution of $\hat{R}_b^*$ depends on the underlying distribution for R_i's, which is unknown. This problem may be attacked using the bootstrap method.

The bootstrap paradigm dictates that we assume that the sample $S = \{R_1, R_2, \ldots R_n\}$ itself constitutes the underlying distribution. Then, by resampling from S many times and computing

Chapter 5: Estimation

$\hat{R}^*_{jb}$ for each resample (jth resample), we get a bootstrap distribution for R_b which approximates the actual distribution of $\hat{R}_b$.

Suppose that the following sample is given that represents records (in meters) for long jump in a certain district.

$$\Omega = \{2.9, 3.0, .3.2, 3.6, 4.1, 4.5, 4.9, 5.0, 5.4, 5.9\}.$$

This is a sample of size 10 for which $\hat{R}_b$ is 6.4. That is, $5.9 + (5.9 - 5.4)$. To find a confidence interval using the bootstrap method we need to take the following four steps:

Step 1: Resampling using a pseudorandom number generator. Draw a sample of 10 values, with replacement, from Ω and thus obtain the bootstrap resample such as

$$\Omega^*_1 = \{2.9, 4.9, 2.9, 3.6, 5.0, 5.9, 5.9, 2.9, 3.0, 2.9\}.$$

Note that this is like writing each of the above 10 values onto a million slips of paper, placing them into a large box, and drawing 10 slips. Observe that some of the original sample values appear more than once and others not at all. It is important that Ω^* be of the same size as Ω.

Step 2: Calculation of the bootstrap estimate. The value of $\hat{R}^*_{1b}$ obtained from Ω^*_1 is 6.8. That is, the largest value, 5.9, plus 0.9, which is the difference between 5.9 and 5.0, the second largest value in Ω^*_1.

Step 3: Repeat steps 1 and 2 a large number of times (say, $M = 200$) to obtain a total of 200 bootstrap estimates, $\hat{R}^*_{1b}, \hat{R}^*_{2b}, \ldots, \hat{R}^*_{200b}$.

Step 4: Construct the estimated population by sorting the bootstrap estimates into an increasing order

$$\hat{R}^*_{1b} \leq \hat{R}^*_{2b} \leq \ldots \leq \hat{R}^*_{200b}, \text{ taking } M = 200.$$

For a $100(1-\alpha)\%$ confidence interval, let L = integer part of $M \times \alpha/2$ and $U = M - L + 1$. Then $\hat{R}^*_{Ub}$ gives the upper confidence limit. For example, if $\alpha = 0.05$ and $M = 200$, we have $L = 10$ and $U = 191$, $\hat{R}^*_{191b}$ is the upper limit. Suppose that in our example, the eleven largest values in the set of 200 bootstrap are

$$\ldots, 6.8, 6.8, 6.8, 6.9, 6.9, 6.9, 7.2, 7.2, 7.3, 7.3, 7.7.$$

Then $\hat{R}^*_{191b} = 6.8$ is taken as the 95% upper confidence limit for R_b.

A PRACTICE PROBLEM The number of points scored in the final games of a basketball tournament in the last 15 years are 85, 66, 72, 81, 85, 82, 79, 67, 92, 74, 89, 90, 91, 67, 77. Using the bootstrap method find a 90% confidence interval for

(a) The population mean number of points
(b) The population maximum number of points (the possible upper bound)
(c) Obtain a 90% confidence interval for population mean using t-distribution and compare that with the result in part (a)

Final Word

In this chapter, we discussed both point and interval estimates. The two ways of reporting an estimate of, for example, a population mean (the point and the interval estimates), are related. If the half-width of a $100(1-\alpha)\%$ confidence interval for μ is equal to a number B, then the probability that a point estimate will deviate from μ by no more than B is $(1-\alpha)$, and vice versa. For example, if the half-width of a 95% confidence interval is equal to B, that is, if

$$B = 1.96\sigma/\sqrt{N}$$

then whenever the confidence interval encloses μ, $\bar{x}$ will deviate from μ by no more than B. This will occur 95% of the time. Therefore, 95% of the time, the error in estimating μ will be less than B.

When a typical newspaper reports an estimate, they usually report a bound (a limit) on the sampling error, that is, the error of estimation. That is, they provide an indication that their estimate is correct to within plus or minus some number, say B. This means that the sampling error will be less than B with a high (usually 0.95) probability. In other words, they provide a margin of error or a measure for reliability of the estimate.

5.7 USING R

In this chapter, we demonstrate how to construct proper confidence interval using known distributions, such as normal distribution, and bootstrap method.

5.7.1 Using Known Distributions

There are built-in functions whose names starts with **q** are used to obtain the proper critical value according to a desired probability, for example qnorm, qbeta, qt, qchisq, qexp, and so on.

EXAMPLE 5.7.1 Suppose in a survey of car racer in a city shows in every 5,000 residence we surveyed, there exists 10 car racers. Please construct a 95% confidence interval of car racer proportion in this city.

SOLUTION: According to our discussion in this chapter, we will construct the interval using normal distribution.

```
# Find point estimate
phat = 10/5000
# 95% confidence interval, qnorm(1-alpha/2)
# here 1-alpha/2=1-0.05/2=0.975
LL = phat - qnorm(0.975)*sqrt(phat*(1-phat)/5000)
UL = phat + qnorm(0.975)*sqrt(phat*(1-phat)/5000)
```

You will find point estimate $\hat{p} = 0.2\%$, while the 95% confidence interval is $[0.08\%, 0.32\%]$.

EXAMPLE 5.7.2 Suppose we have following sample in our F1 race study

$$23, 28, 29, 30, 30, 31, 31, 31, 34, 34, 36, 40$$

Chapter 5: Estimation

Please construct 95% confidence interval of average age using bootstrap approach.

SOLUTION: Computers in fact generate random values using technology called pseudorandom generator. Therefore, if we fixed a number named random seed, we all will obtain exactly same numbers. This will be helpful if one need to represent same results from others.

In this example, we will also demonstrate how to make a loop in R thus we can repeat sampling step. The function **sample()** is used for sampling a new sample from given sample. Its syntax is **sample(GIVEN_SAMPLE, SIZE, replace=TRUE/FALSE)**.

```
# Set a randoom seed as 123
set.seed(123)
age_sample = c(23, 28, 29, 30, 30, 31, 31, 31, 34, 34, 36, 40)
# Varaible bs_avg will be used to store bootstrap averages
bs_avg = c()
# Bootstrap here, here we use 1000 bootstraps.
for (i in 1:1000){
        bs_sample = sample(age_sample, length(age_sample), replace=TRUE)
        bs_avg=c(bs_avg, mean(bs_sample))
}
# Find confidence interval now
# Again, LL from alpha/2=0.05/2=0.025
# and UL from 1-alpha/2=1-0.05/2=0.975
quantile(bs_avg, c(0.025, 0.975))
```

Now you will see the 95% bootstrap confidence interval is [29.25, 33.58].

REFERENCE

Diciccio, Thomas J., and Joseph P. Romano. 1988. "A Review of Bootstrap Confidence Intervals." *Journal of the Royal Statistics Society: Series B* 50(3):338–54.

CHAPTER 6

Statistical Testing

This chapter is concerned with statistical testing which, as pointed out earlier, is an integral part of inferential statistics. The problems considered under this title include testing a hypothesis about the unknown parameter or parameters of a distribution (population), testing claims, and testing scientific theories. The decision (rejecting or not rejecting the hypothesis) is made by using the data (sample) as evidence, and by considering how much error one is willing to tolerate when making such a decision. Testing hypotheses also includes topics such as establishing relationships among variables of interest as well as detecting differences among the groups. It is important to remember that, the major part of science and scientific investigations is based on certain accepted or agreed upon theories or hypotheses. Examples include evolution theory, big bang theory, tectonic theory, and so on. These theories are used as long as no plausible alternative supported by data is introduced.

Let us start with a real-life example. It is known that companies frequently conduct surveys to help them make decisions concerning the effectiveness of their advertising. Suppose that a company that produces running shoes and presently enjoys 10% of the market share launches a new advertising campaign. At completion of the campaign the company wants to know whether the campaign was successful in raising its market share. A random sample of purchasers of the shoes will provide evidence upon which a conclusion may be drawn (an answer may be found). Here the company should test $p = 0.10$ versus $p > 0.10$ where p, measured as percentage, presents the market share after the campaign. These values represent two alternative hypotheses namely, advertising was not effective (noting changed) and advertising was effective. These two hypotheses are referred to as the "Null hypothesis" and the "Alternative hypothesis," respectively. A simpler example is what happens in a court of law. There, not guilty is null and guilty is the alternative hypotheses, respectively.

To understand the decision-making process, suppose that we are given a coin and asked to determine whether or not it is a fair coin. One reasonable approach is to flip the coin 20 times and make a decision based on the outcome. Note that the number of times we flip the coin is a decision in itself. However, we are not going to discuss this problem at this point.

Next we need to define a decision criterion. For example, we may call the coin fair if we observe between 7 and 13 heads, or be a little bit more conservative and call it fair if between 8 and 12 heads are observed. The question that arises at this point is how do we decide which criterion is "better"? To find an answer, let us assume that the coin is really a fair coin. If this is the case, what is the probability of making a wrong decision and declaring the coin "not fair"? Let us call this error type I error. Note that for a fair coin, the expected number of heads is 10. Using the binomial distribution, the probabilities related to these two decision criterion are, respectively, as

$$1 - P(7 \leq X \leq 13) = 1 - \sum_{x=7}^{13} \binom{20}{x} (1/2)^x = 0.116,$$

and

$$1 - P(8 \leq X \leq 12) = 1 - \sum_{x=8}^{12} \binom{20}{x} (1/2)^{20} = 0.264.$$

This shows that by requiring stronger evidence (a number closer to 10), one would face a higher risk of rejecting a true hypothesis (coin is fair). Stating this in a reverse order, if one can tolerate a higher risk of rejecting a true hypothesis, then a deviation of observed from the expected outcome (in this case 10 heads for a fair coin) will provide stronger evidence against the hypothesis (coin is fair).

On the other hand, suppose that the coin is not fair and the chance of observing a head is, for example, 60%. Then the probability of declaring the coin fair when it is not is

$$P(7 \leq X \leq 13) = \sum_{x=7}^{13} \binom{20}{x} (0.6)^x (0.4)^{20-x} = 0.744$$

for the first criterion and

$$P(8 \leq X \leq 12) = \sum_{x=8}^{12} \binom{20}{x} (0.6)^x (0.4)^{20-x} = 0.563$$

for the second one. Let us call this error type II error. If, rather than 60%, probability of observing a head is 80%, then respective probabilities are

$$P(7 \leq X \leq 13) = \sum_{x=7}^{13} \binom{20}{x} (0.6)^x (0.4)^{20-x} = 0.087,$$

and

$$P(8 \leq X \leq 12) = \sum_{x=8}^{12} \binom{20}{x} (0.8)^x (0.2)^{20-x} = 0.032.$$

So, what is the conclusion? Well, we have to think about two factors; which type of error (type I or type II) is more important for the problem at hand, and the level of the error that we can tolerate. The next step is to use these to define a decision criterion. For example, rejecting a true null hypothesis may be considered more serious than accepting a false alternative. That is, type I error is more important than type II error. We will elaborate on this point in Sections 6.2 through 6.17.

We end the introduction by providing a clarification based on the "rare event rule." According to this rule, if under a given hypothesis, the probability of a particular event is "small" (e.g., less than 5%) and yet such an event actually occurs, we may conclude that the hypothesis is probably not correct. For example, if we roll a fair die 100 times, we expect to observe (approximately) an even number 50 times. For a fair die, the probability of observing an even number 30, 40, and 47 times are, respectively, 0.00004, 0.028, and 0.31. We see that, except for the last case, the probabilities are small (less than 0.05), that is, events are rare. Thus, if any of the first two events occur, we can safely conclude (at 5% level of significance) that the die is biased. In summary, occurrences of rare events (events not expected to occur under the assumed hypothesis) are used as evidence against the hypothesis.

6.1 STATISTICAL HYPOTHESIS

A statistical test is a method to decide whether or not to reject a statistical hypothesis by inspecting a sample. A very simple example is when we want to know whether a given observation belongs to a known population (Example 6.1.1). As a sport example, suppose that a claim is made that a specific drug can increase endurance of basketball players. By testing the drug on several players, we collect data and use that to make a decision to reject or to not reject the claim.

Let us fix a population Ω and a random variable X. In general, a statistical hypothesis H is a statement concerning the distribution of X. We may start with the simplest case in which we assume that X is distributed normally ($X \sim N(\mu, \sigma^2)$), and H is a statement about the population mean μ or population variance σ^2. The tests for such cases are called parameter or parametric tests. This is because we are assuming that the distribution of X is known except for some parameter, for example, the mean μ or the variance σ^2. To illustrate, let Ω denote the population of individuals who work out using a certain body fitting equipment and let X represent the amount (in pounds) of weight lost after using the equipment for 1 month. Assume that X is a normally distributed random variable $N(\mu, \sigma^2)$ with known σ^2. We would like to confirm (if we can) that the equipment actually has an effect ($\mu > 0$). We state the hypothesis H_0 (null hypothesis) as $\mu = 0$ (no effect), hoping that, based on the available data, H_0 will be rejected. H_0 is called a null hypothesis because if H_0 is not rejected all we have done will be "null and void." The reason for formulating H_0 in a negative way will become clear shortly.

EXAMPLE 6.1.1 Suppose that in a community the distribution of income is normal with mean $50,000 and standard deviation $10,000. You meet someone whose income is $84,000. You wonder whether this person is a member of this community. This could be the null hypothesis. The hypothesis that the person is not a member of this population, could be the alternative hypothesis. Let X denote the income of a randomly selected member of this community. To test the hypothesis of interest, we may ask the following question: what is the probability that a randomly selected member of this community makes $84,000 or more?

This is easy to answer. We have

$$P(X > 84,000) = P(z > \frac{84,000 - 50,000}{10,000}) = P(z > 3.4) = 1 - 0.9997 = 0.0003.$$

This means that, in this community, only 3 people out of 10,000 have this type of income. Thus, either this person is one of the very few people with high income, or else he or she is not a member of this community. But which of these two possibilities is more likely? In other words, our observation (person making $84,000 or more) provides stronger evidence in favor of which hypothesis "is" or "is not" a member of this community? The statistical method for testing hypotheses considers the probability as small as 0.0003 a strong evidence in favor of "is not" a member of this community, simply because the probability of such an event occurring is very small. In other words, the *p*-value (probable value) is small and we do not expect this event to occur. So the occurrence of such (rare) event provides evidence against the null hypothesis, (is).

EXAMPLE 6.1.2 A coin was flipped 10 times and 8 heads were observed. The probability of such an event is less than 5% (in fact, 0.044 using binomial distribution with $p = 1/2$) for a fair coin. This, therefore, provides evidence against the null hypothesis that the coin is fair. Again this is because the probability of such an event happening (*p*-value) is very small (a rare event). For further discussion concerning *p*-values, see Section 6.5.

PRACTICE PROBLEMS

1. A die is rolled 18 times and 9 times a number greater than 4 was observed. Do we have evidence against the null hypothesis that the die is fair? Find the *p*-value.
2. A racer would like to replace the engine in his race car, and high-speed engine manufactured by Honda is preferred. However, he suspected the store provided him a fake Honda engine. It is known from Honda that the average frequency of rotation (RPM) to obtain highest torque is around 6,000. Now he took his race car to a professional testing center for a test, and found his engine got RPM of 6,300 for peak torque. Do you think his engine is fake?

In what follows, we will discuss several different situations that occur frequently in practice.

6.2 TESTS

Assume that X is distributed as $N(\mu, \sigma^2)$ and σ^2 is known. The null hypothesis H_0 to be tested is: population mean equals a specific value μ_0, $\mu = \mu_0$.

If $X_1, \ldots, X_N$ is a random sample of size N, then the random variable

$$T = T(X_1, \ldots, X_N) = \frac{\bar{X} - \mu_0}{\sigma/\sqrt{N}}, \quad (\bar{X} = \frac{1}{N}\sum_{i=1}^{N} X_i) \tag{6.1}$$

is distributed as $N(0,1)$, given that H_0 is true. Using the $100(1-\alpha/2)\%$ percentile point of $N(0,1)$ distribution, we have

$$P(T < -z_{\alpha/2}) + P(T > z_{1-\alpha/2}) = P(|T| > z_{1-\alpha/2}) = \alpha \tag{6.2}$$

for some chosen level α, say 5%. This shows that if H_0 is true, then there is only a small chance, in this case 5%, of obtaining a realization of T (x_1, ..., x_N replaced for X_1, ..., X_N to obtain $T(x_1, \ldots, x_N) = \frac{\bar{x} - \mu_0}{\sigma/\sqrt{N}}$) with an absolute value greater than $z_{0.975} = 1.96$. Using the reasoning in contradictory manner, if we get a sample realization leading to $|T| > 1.96$, then there is a 95% probability that H_0 is not true. So when $|T| > 1.96$, we reject H_0 at the 5% significance level. In the other words, the assumed null hypothesis may not be true. So, although it is possible to obtain such a realization by chance (perhaps bad luck), we use its occurrence as a "strong" evidence against the null hypothesis. When the null hypothesis H_0 is rejected, an appropriate alternative hypothesis denoted by H_1 may be considered and be accepted.

T is called test quantity. Its distribution, in this case $N(0,1)$, is called test statistic, and α is the level of significance of the test. The region $|T| > z_{1-\alpha/2}$ is called the critical region or rejection region, and $z_{1-\alpha/2}$ is known as the critical value of the test. It should be pointed out that although test quantity and test statistics are not the same in principle, in this book they are often used interchangeably.

The situation described above is similar to what happens in a court of law, where H_0 is "not guilty" and H_1 is "guilty." Most of the time a correct decision is made. However, it is possible to declare the person guilty when he (she) is, in fact, not guilty (Type I error) or the opposite (Type II error) that is declaring the person not guilty when he (she) is guilty. The probability of a Type I error, is the probability of rejecting a null hypothesis when it is true, is the same as α discussed above. It is called the significance level of the test or level of significance. In most cases, a Type I error is fixed in advance. However, there are cases where a Type II error is of prime importance and may be fixed in advance. For example, for most courts the Type I error (declaring an innocent person guilty) is of prime concern. But in some medical tests false negative (Type II error) may be more critical than false positive (Type I error). For example, in preventive medicine declaring a person diabetic while he or she is not is less risky than the opposite and in the long run less costly. The same is true when testing for HIV since a false negative leads to a more severe consequence, through infecting other people.

SUMMARY

A **Statistical Hypothesis** is a claim or a conjecture about a population parameter. This conjecture may or may not be true.

The **Null Hypothesis**, symbolized by H_0, is a statistical hypothesis that states that there is no difference between a parameter and a specific value, or states that there is no difference between two parameters. H_0 is initially assumed to be true.

The **Alternative Hypothesis**, symbolized by H_1, is a statistical hypothesis that states that there is a difference between a parameter and a specific value, or states that there is a difference between two parameters. It usually contains a statement of inequality such as $>$, $<$, or $\neq$.

A **Type I Error** occurs if a null hypothesis is rejected when it is actually true.

A **Type II Error** occurs if a null hypothesis is not rejected when it is actually false.

Description of Possible Errors

Decision	Truth	
	H_0	H_1
Reject H_0	Type I Error	Correct Decision
Do Not Reject H_0	Correct Decision	Type II Error

Note that decreasing the probability of making one type of error will lead to an increase in the probability of making the other type of error.

Students sometimes ask whether Type I or Type II errors should be identified with the diagnosis of illness or well-being. They also often ask which error is more important. We already discussed this briefly. Consider a hypothetical scenario like this one: suppose the local government imposes mandatory HIV blood tests on all citizens. Assume the tests have a 0.01 false-positive rate and a 0.01 false-negative rate. Unknown to the testers, 500 out of 170,000 citizens are HIV-positive. As a result of this, we should expect 5 false negatives and 1,695 false positives out of 170,000 tests.

Which is the more serious error? Is it 5 undetected HIV carriers or 1,695 people who are falsely believed to be HIV-positive? This leads to catch onto the point that the rarity of a disorder or disease cannot only make the diagnosticity of a test problematic but can also alter our perceptions of which error is the worse one. We also start to see some of the difficulties that arise from using imperfect diagnostic tests on nonclinical populations. Note that under different circumstances one error may be perceived as more serious than the other, and so we need to worry about both.

We now turn to specific tests and describe the necessary steps.

6.3 z-TEST

The z-test, as is clear from the name, is based on the normal distribution. To see how it is applied, we summarize the procedure described in Section 6.2 as follows:

1) Assume that X is distributed as $N(\mu, \sigma^2)$ and σ^2 is known.
2) Null hypothesis to be tested is $H_0: \mu = \mu_0$. We may introduce $H_1: \mu \neq \mu_0$ as an alternative hypothesis.
3) Test quantity is (value of the evidence)

$$T(x_1, \ldots, x_N) = \frac{\bar{x} - \mu_0}{\sigma/\sqrt{N}}. \tag{6.3}$$

4) The distribution of $\frac{\bar{X} - \mu_0}{\sigma/\sqrt{N}}$ is $N(0,1)$.
5) The critical value corresponding to the level α is $z_{1-\alpha/2}$.
6) Decision rule is to reject H_0, if $|T| > z_{1-\alpha/2}$. If $|T| \leq z_{1-\alpha/2}$, then H_0 cannot be rejected.

Note that by applying the above steps we control the Type I error, namely the probability of rejecting H_0 when it is actually true. However, on the other hand, we pay no attention to the Type II error, namely the probability of not rejecting H_0 despite being false. The Type II error may well happen in 50% of the cases. That is, the reason to formulate H_0 "negatively." We would like to verify at a high significance level that H_0 is false. If we reject H_0 when H_0 is actually true, we have committed an error less than or equal to α. If, on the other hand, we want to verify that H_0 is true and that the test does not reject H_0, then there may be a 50% probability that H_0 is false. The "proof" or "disproof" of H_0 then is just like flipping a coin. As mentioned earlier, one interesting question that arises in testing a hypothesis is which error is more important and should be controlled. For example, most courts try to avoid the Type I error that is, declaring an innocent person guilty. However, in other cases such medical tests where we could have false-positive or false-negative

Chapter 6: Statistical Testing

situations may be different. As an example, when testing HIV clearly a false negative is more important and should be avoided or minimized, if possible.

NOTE: If the distribution of X is not normal but N is sufficiently large, then according to the Central Limit Theorem the distribution of $\bar{X}$ is approximately normal and z-test may still be applied. Moreover for large samples, we can replace sample standard deviation s for σ (when it is unknown) and follow the steps described earlier. This, of course, will provide us with an approximate procedure.

EXAMPLE 6.3.1 Data were collected on the golf ball driving distances by professional golfers in a recent tournament. The data on the first drives of 36 randomly selected golfers yielded a sample mean distance of 247.5 yards. Assuming the standard deviation of the driving distances on the first drive is 10 yards, carry out a z–test on $H_0{:}\mu = 250$ versus $H_1{:}\mu \neq 250$ at the 10% significance level.

SOLUTION: Let X denote the driving distance of a randomly selected golf ball.

ASSUMPTION: $X \sim N(\mu, 10^2)$. Using (6.3) we have

$$T = z = \frac{247.5 - 250}{10/\sqrt{36}} = -1.5.$$

The critical value is $z_{0.95} = 1.65$. Since $|T| < 1.65$, H_0 cannot be rejected. That is, although we have some evidence against H_0, but it is not "strong" enough (at 10% significance level) to justify its rejection. Note that, the same evidence may be considered "strong" for a different (in this case higher) level of significance. That is, if we can tolerate a larger error, we may consider the evidence strong (in that level) and reject the null hypothesis.

EXAMPLE 6.3.2 In recent years college basketball in United States has become increasingly popular since it creates types of excitement and competition that are not present in professional sports. One rule of these games is the 35-second shot clock, which fans have some concern about. They think that it slows the games and makes them less exciting. However, NCAA officials claim that this rule has not affected the game in any way fans think. Data collected show that the average time that a college team takes to set up a shot was 20 s in the past and is still 20 s. To test this, a random sample of games was selected. The average time taken to set up a shot based on 625 opportunities is found to be 19.1 with $s = 7.2$. Does data support the claim made by the officials? Use $\alpha = 0.05$.

SOLUTION: Since we have a large sample the hypotheses

$$H_0{:}\mu = 20 \quad vs \quad H_1{:}\mu < 20$$

may be tested using

$$T = z = \frac{19.1 - 20}{7.2/\sqrt{625}} = -3.125.$$

The rejection region is $T < -1.645$. Thus, we have sufficient evidence to conclude that the true average time taken to set up a shot is even less than 20 s.

PRACTICE PROBLEMS

1. A new diet and exercise program claims that participants will lose on average at least 5 pounds during the first week of the program. A random sample of 49 people participating in the program showed a sample mean weight loss of 4 pounds. The sample standard deviation was 2.5 pounds.

    (a) Test the claim made by the diet program at 10% significance level?
    (b) What is the p-value? (p-value = $p(z \leq -2.8)$)

2. A physician claimed that average heart beat rate of a healthy person will be 75 bpm after 10-min walking. A random sample of 40 healthy volunteers take 10-min walk and have their heart rate monitored. We found the average heart rate is 63 bpm, and standard deviation is 15 bpm. Please test the physician's claim at 5% significance level.

Although other types of alternatives and tests for proportions are not yet discussed, here we present Figure 6.3.1, which illustrates the way they work. We also present a different summary of the terms and procedure used for hypothesis testing below.

Terms and Procedure for Hypothesis Testing

HYPOTHESIS A statement about a population parameter developed for the purpose of testing.

HYPOTHESIS TESTING A procedure based on sample evidence and probability theory to determine whether the hypothesis is a reasonable statement.

NULL HYPOTHESIS A statement about the value of a population parameter.

ALTERNATIVE HYPOTHESIS A statement that is accepted if the sample data provide enough evidence that the null hypothesis is false (is rejected).

LEVEL OF SIGNIFICANCE The probability of rejecting the null hypothesis when it is true.

TYPE I ERROR Rejecting the null hypothesis, H_0, when it is true.

TYPE II ERROR Not rejecting the null hypothesis when it is false.

TEST QUANTITY A value, determined from sample information, used to determine whether to reject or not to reject the null hypothesis. A value that quantifies the evidence against the null hypothesis.

TEST STATISTICS Distribution of the test quality

CRITICAL VALUE The dividing point between the region where the null hypothesis is rejected and the region where it is not rejected.

	Researcher	
Null Hypothesis	Do Not Reject H_0	Rejects H_0
H_0 is True	Correct Decision	Type I Error $= \alpha$
H_0 is False	Type II Error $= \beta$	Correct Decision

Step 1	Step 2	Step 3	Step 4	Step 5	
State null and alternate hypotheses	Select a level of significance	Identify the test statistic	Formulate a decision rule	Take a sample, arrive at decision	Do not reject H_0 or Reject H_0 and accept H_1

FIGURE 6.3.1 Alternative Hypothesis, Rejection Region for z-Test (a) Upper-Tailed, (b) Lower-Tailed, and (c) Two-Tailed

Requirements for Tests

To conduct a test, we need the following information:

1. The value claimed for the population parameter.
2. The value of the sample counterpart of the population parameter (e.g., $\bar{x}$ for μ or $\hat{p}$ for population proportion p).
3. The sample size.
4. If the population parameter of interest is μ, then the value of the sample standard deviation is S.

The decision regarding the null hypothesis is made based on the probability of observing the sample statistics found in the study (p-value), assuming that the null hypothesis is true. If sample statistics is approximately normally distributed, the z value may be calculated as

$$\frac{\text{Observed Statistics} - \text{Null Hypothesis Parameter Value}}{\text{Standard Deviation of Statistics}}$$

and the corresponding probability of observing a z value so far (or further) from zero. If the alternative hypothesis is two-tailed, we double this probability to obtain the two-tailed p-value. If population standard deviation is unknown, it is replaced by its sample counterpart and the p-value is determined from the t distribution with $N-1$ degrees of freedom.

6.3.1 Testing Using Confidence Interval

The confidence interval discussed in Chapter 5 and testing presented in this chapter are related. To test $H_0: \mu = \mu_0$ against $H_1, \mu \neq \mu_0$ at significance level of α, it is possible to construct a $100(1-\alpha)\%$ confidence interval and make a decision depending on whether this interval includes or excludes μ_0. If it does, we do not reject H_0. Otherwise, we do. In Example 6.3.1, a 90% confidence interval for μ is $247.5 \pm 1.645(10/\sqrt{36}) \rightarrow (244.76, 250.24)$. Since this interval includes 250, we cannot reject

H_0. Again choosing a different α may lead to a confidence interval that may exclude μ_0 and hence lead to the rejection of H_0.

A PRACTICE PROBLEM Recall the Example 6.3.2. Construct a 90% confidence interval for μ and use that to test $H_0: \mu = 20$ verses $H_1: \mu < 20$.

EXAMPLE 6.3.3 Which one is more effective, diet or exercise? Recent studies reveal that losing weight has become an issue of great concern in modern societies. Wood et al. (1998) published a study in *The New England Journal of Medicine* comparing two methods of losing weight: diet alone and exercise alone. Eighty-nine sedentary men were studied over a 1-year period. Forty-two men dieted and lost an average of 7.8 kg over the year, with a standard deviation of 5.8 kg. Forty-seven men who exercised lost an average of 4.6 kg, with a standard deviation of 5.5 kg. They then concluded that diet was more effective. Use confidence intervals to confirm their finding.

SOLUTION: Since sample sizes are greater than 30, we have large samples. Thus, for example, 90% confidence intervals for the means are, respectively

$$7.8 \pm 1.645 \frac{5.8}{\sqrt{42}} \quad \text{and} \quad 4.6 \pm 1.645 \frac{5.5}{\sqrt{47}}$$

or

$$(6.33, 9.27) \quad \text{and} \quad (3.28, 5.92).$$

Since these intervals do not overlap, we can conclude that the population means are different. This indicates that the true mean weight loss from exercising may be less than that from dieting. Note that if two intervals overlap, one cannot conclude that the true means are different. A formal test for the situation presented in this problem is described in Section 6.4.

A PRACTICE PROBLEM During the presidential election, a random sample of voters were asked who they did vote for. The data led to the following 95% confidence intervals for percentages voted Democratics and Republicans respectively

$$(0.31, 0.35) \quad \text{and} \quad (0.34, 0.38)$$

Can we conclude that more people voted Republican? What happens if we replace 95% with a lower confidence level?

6.4 TWO-SAMPLE z-TEST

This test is useful when the hypothesis or claim involves comparison of two population means or two population proportions. The following describes the situation for the former.

1) Assume that $X^{(1)}$ and $X^{(2)}$ are distributed as $N(\mu_1, \sigma_1^2)$ and $N(\mu_2, \sigma_2^2)$ respectively, and σ_1^2 and σ_2^2 are known and we have samples of sizes N_1 and N_2 on $X^{(1)}$ and $X^{(2)}$, respectively.
2) The null hypothesis to be tested is $H_0: \mu_1 = \mu_2$ or $\mu_1 - \mu_2 = 0$. We may introduce the alternative as $H_1: \mu_1 - \mu_2 \neq 0$.

Chapter 6: Statistical Testing

3) Test quantity is

$$T(x_1^{(1)},\ldots,x_{N_1}^{(1)};x_1^{(2)},\ldots,x_{N_2}^{(2)}) = (\bar{x}^{(1)} - \bar{x}^{(2)})/\sqrt{\frac{\sigma_1^2}{N_1} + \frac{\sigma_2^2}{N_2}}. \quad (6.4)$$

4) This is the distribution of $(\bar{X}^{(1)} - \bar{X}^{(2)})/\sqrt{\frac{\sigma_1^2}{N_1} + \frac{\sigma_2^2}{N_2}}$ is $N(0,1)$.
5) Critical value is $z_{1-\alpha/2}$ corresponding to level of significance α.
6) Decision rule is to reject H_0 if $|T| > z_{1-\alpha/2}$. Otherwise, H_0 cannot be rejected.

Note that if distributions of $X^{(1)}$ and $X^{(2)}$ are not normal but N_1 and N_2 are sufficiently large then, according to the Central Limit Theorem, the distribution of $(\bar{X}_1 - \bar{X}_2)/\sqrt{\sigma_1^2/N_1 + \sigma_2^2/N_2}$ is approximately normal. Hence, we can still apply the procedure described above. Moreover, for large samples the procedure may be applied approximately by replacing s_1^2 and s_2^2 for σ_1^2 and σ_2^2, respectively, when variances are unknown.

EXAMPLE 6.4.1 Let $X^{(1)}$ and $X^{(2)}$ represent the amount of weight loss under two different exercise programs. Suppose that hundreds of people ($N_1 = 100$ and $N_2 = 100$) were treated by each of these programs and the measured values of weight losses, in pounds, are $x_1^{(1)},\ldots,x_{N_1}^{(1)}$ and $x_1^{(2)},\ldots,x_{N_2}^{(2)}$, respectively. Suppose that from this data we get $\bar{x}^{(1)} = 10$ and $\bar{x}^{(2)} = 8.5$. Also assume that $\sigma_1 = 4$ and $\sigma_2 = 3$. Are these two programs equally effective or not? Carry out a two-sample z-test using a significance level 5%.

SOLUTION: Since $N_1 = N_2 = 100 > 30$, we have large samples. Therefore using (6.4) we find

$$T = (10 - 8.5)/\sqrt{16/100 + 9/100} = 3.$$

From Section 3.6 (or Table B3) we have $z_{1-\alpha/2} = z_{0.95} = 1.96$. Since $|T| > 1.96$ we reject the null hypothesis and conclude that the two programs are not equally effective. Note that it is also possible to find out which program is more effective. This is discussed later in this chapter, where rather than $\mu_1 \neq \mu_2$ alternatives such as $\mu_1 > \mu_2$ or $\mu_1 < \mu_2$ are considered.

A PRACTICE PROBLEM Do male athletes outperform female athletes in sports like soccer? To find an answer, a test measuring the basic skills was administered. A summary of the test scores is displayed in the table below

	Males	Females
N	175	175
$\bar{x}$	48.9	48.4
S	4.1	3.75

Is there evidence of a difference between the true mean test scores for two population at 5% level of significance?

6.5 OBSERVED SIGNIFICANCE LEVEL, *p*-VALUES

Recall the Example 6.3.1 where based on the available data, we did not reject $H_0: \mu = 250$ at the 10% significance level. As pointed out earlier, if we could tolerate a larger error for rejecting H_0 when it is actually true, we may end up reversing our decision and reject H_0. To clarify this suppose that, rather than 10%, we choose a significance level of 20%. Since $z_{0.90} = 1.28$ and $|T| = 1.50$, our evidence would be "strong" and therefore we reject H_0. So there is a range of values for α that lead to the rejection of H_0. We then may ask the following: what is the smallest value of α for which H_0 will be rejected? This question can be answered using a concept known as probable value or simply *p*-value. We already learned a few things about this concept and its role in hypothesis testing. In fact, rather than deciding about level of significance and repeating the decision process for different levels, we can calculate the value of α that separates the rejection region from the "do not reject" region. This is a useful concept since we may not be able to make a decision about the Type I error in advance. Moreover, when using software there is no need to carry the test for different values of α. In fact, we can use the *p*-value for making a decision about a range of values for α.

DEFINITION: The observed significance level, or *p*-value, for a statistical test is the probability (assuming H_0 is true) of observing a value of the test statistic that is at least as contradictory to the H_0 and supportive of the H_1 as the one computed from the sample. Putting this differently, the *p*-value for a hypothesis, expressed as a claim about a population parameter, is the probability of selecting a sample at least as extreme as the observed sample assuming that the (null) hypothesis is true. It measures how unlikely the value of the observed test statistics is, assuming the null hypothesis is true. The smaller the *p*-value is, the greater the weight of evidence favoring rejection of the null hypothesis and acceptance of the alternative hypothesis.

To clarify this, consider once more the test statistics T and the two different cases, namely, the one- and two-tailed alternatives. First, consider the one-tailed tests where the alternative only considers the values that are either less than or greater than the specific value presented by the null hypothesis. (These types of alternatives are discussed in more detail in Section 6.10). That is,

$$\begin{cases} H_0: \mu = \mu_0, & H_1: \mu < \mu_0, \\ H_0: \mu = \mu_0, & H_1: \mu > \mu_0. \end{cases}$$

The Figure 6.5.1 and 6.5.2 illustrate how the *p*-value is found for the one-tailed and a two-tailed tests.

EXAMPLE Suppose we have
$$H_0: \mu = 3{,}000 \quad vs \quad H_1: \mu < 3{,}000$$
and $T = -2$. Then the *p*-value is

$$p\text{-value} = p(z < -2) = 0.0228.$$

If α is bigger than 0.0228, the test statistics falls in the rejection region and H_0 is rejected. Note that, in general, the small *p*-values will lead to the rejection of H_0.

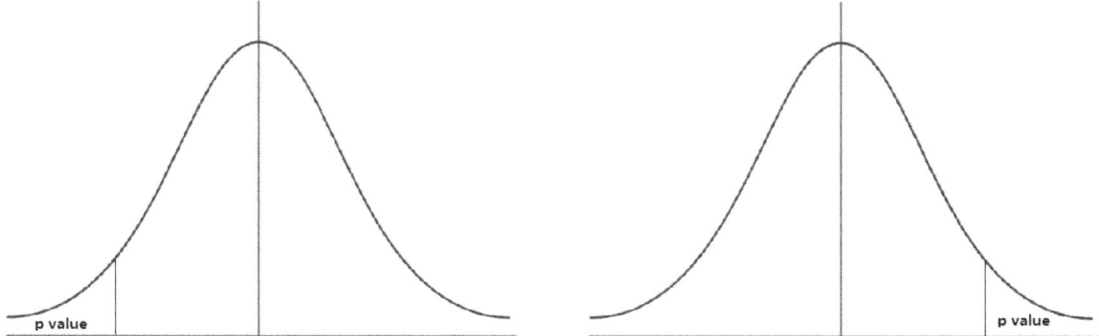

FIGURE 6.5.1 (a) Lower Tailed, $H_1 : \mu < \mu_0$ (b) Upper Tailed $H_1 : \mu > \mu_0$
Demonstration of the *p*-value for a One-Tailed Test

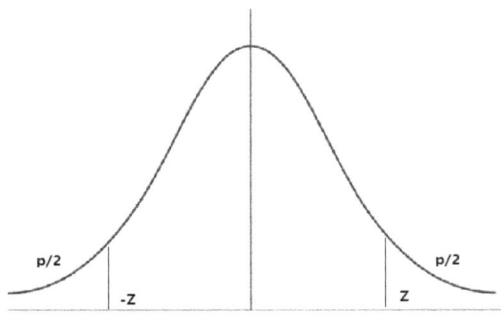

FIGURE 6.5.2 Next consider the two-tailed case, that is, $H_0 : \mu = \mu_0 \quad H_1 : \mu \neq \mu_0$
Demonstration of *p*-value for a Two-Tailed Test

EXAMPLE Suppose we have

$$H_0 : \mu = 3{,}000 \text{ vs } H_1 : \mu > 3{,}000$$

and $T = 1.75$. Then the *p*-value is

$$p\text{-value} = p(z > 1.75) = 1 - 0.9599 = 0.0401$$

If α is bigger than 0.0401 the test statistics falls in the rejection region and H_0 is rejected.

EXAMPLE Suppose we have

$$H_0 : \mu = 3{,}000 \text{ vs } H_1 : \mu \neq 3{,}000$$

and $T = 0.95$. Then the *p*-value is

$$p\text{-value} = 2p(z > 0.95) = 2p(z < -0.95) = 0.3422.$$

Since the *p*-value is large, we cannot reject H_0.

SUMMARY: Making a Decision About H_0 Based on the *p*-value

1. Choose the maximum value of α (Type I error) that you are willing to tolerate.
2. If the observed significance level (p-value) of the test is less than the chosen value of α, reject H_0.

Therefore, if $\alpha = 0.05$, and the p-value equals 0.0228 as in the first case, then H_0 will be rejected. However, when the p-value equals 0.3422 as in the second case, H_0 will not be rejected.

A PRACTICE PROBLEM Suppose we have $H_0: \mu = 100$ and three different alternatives namely; $\mu < 100, \mu > 100$, and $\mu \neq 100$. If the values of test statistics are, respectively, $T = -0.125, T = 2.10$, and $T = 1.12$, calculate the P-value for each case.

6.6 *t*-TEST

6.6.1 Test for a Population Mean

In Sections 6.3 and 6.4, we assumed that the variances for the normal populations were known. This is not usually the case in practice and therefore the z-test cannot always be used. In such cases we should apply a test known as the *t*-test, which is based on the *t*-distribution described in Section 3.7. A student of mine used to say, "we must leave the z-world and enter the *t*-world." The following summarizes the steps.

1) Assume that X is distributed as $N(\mu, \sigma^2)$ and σ^2 is unknown. Suppose we have a sample of size N and let $s^2 = \frac{1}{N-1} \sum_{i=1}^{N}(x_i - \bar{x})^2$.
2) Null hypothesis to be tested is $H_0: \mu = \mu_0$. We may introduce $H_1: \mu \neq \mu_0$ as an alternative hypothesis.
3) Test quantity is

$$T(x_1, \ldots, x_N) = \frac{\bar{x} - \mu_0}{s/\sqrt{N}}. \quad (6.5)$$

4) The distribution of $\frac{\bar{X} - \mu_0}{S/\sqrt{N}}$ is $t_{(N-1)}$ (t-distribution with N-1 degrees of freedom).
5) Critical value is $t_{1-\alpha/2,(N-1)}$ corresponding to the level of significance α.
6) Decision rule is to reject H_0 if $|T| > t_{1-\alpha/2,(N-1)}$. Otherwise, H_0 cannot be rejected.

Note that when $N \geq 30$, then practically there is a little difference between $t_{(N-1)}$ and $N(0,1)$ distributions. Thus, one could use the z–test as an approximate procedure.

EXAMPLE 6.6.1 The NBA started allowing teams to use a zone defense in the 2001-2002 season. Many coaches believe that this will lead to a lower average points per game in future seasons. To get a preliminary idea we looked at the data for Philadelphia 76's during the seasons 2000-2001 and 2001-2002. Their average points per game for the 2000-2001 season was 98.29. A random sample of 25 Sixers (76ers) games during the season 2001-2002 had a mean of 88.44 and a standard deviation of 14.17. Is there sufficient evidence to indicate that the mean points per game for this team during the season 2001-2002 was less than that of the 2000-2001 NBA season? Use a 10% level of significance.

Chapter 6: Statistical Testing

SOLUTION: Let X denote the points scored by the 76ers during a randomly selected game in 2001-2002 season. Let $\mu = E(X)$. Is the distribution of X normal? Here we have to assume that it is. Then

$$H_0: \mu = 98.29 \quad vs \quad H_1: \mu < 98.29.$$

$$T = \frac{\bar{x} - \mu}{s/\sqrt{n}} = \frac{88.44 - 98.29}{14.17/\sqrt{25}} = -3.48.$$

Rejection Region: $T < -t_{0.90,(24)} = -1.32$, Decision: Reject H_0

Conclusion: There is sufficient evidence to indicate that for the Sixers (76ers) the average points per game in the 2000-2001 season was greater than the average points per game in the 2001-2002 season at the 10% significance level.

EXAMPLE 6.6.2 The following data show the total yardage accumulated during an NCAA college football season for a sample of 10 receivers.

$$744, \quad 652, \quad 576, \quad 596, \quad 852, \quad 809, \quad 820, \quad 975, \quad 941, \quad 907.$$

Let X denote the total yardage and μ denote the mean value of the total yardage accumulated during this season by all receivers. Carry out a t–test for $H_0: \mu = 800$ versus $H_1: \mu \neq 800$ at the 10% significance level, assuming that the total yardage is normally distributed.

SOLUTION: For this data, $\bar{x} = 787.2$ and $s = 141.55$. Using (6.5) we have

$$T = \frac{787.2 - 800}{141.55/\sqrt{5}} = 0.202.$$

The critical value $t_{0.95,(9)}$ in this case is 1.83. Since $|T| < 1.83$, H_0 cannot be rejected. There is not sufficient evidence at the 10% significance level that the population mean of total yardage differs from 800.

PRACTICE PROBLEMS

1. A manufacturer of a sports car is testing a new engine to determine whether it meets new air-pollution standards. The mean emission μ of all engines of this type must be less than 20 parts per million of carbon. Ten engines are manufactured for testing purposes, and the emission levels were determined. The mean and standard deviation were respectively 17.17 and 2.98. Can we, at 10% level of significance, conclude that this type of engine meets the pollution standard? What assumptions do we need to make?
2. Remington, a firearm manufacture, advertised their famous Remington 700 bolt action rifle will achieve an average accuracy of 0.5 MOA at 100 yards with 5 rounds group. To verify that, 20 Ramington 700 rifles were shot at 100 yards for a 5-round group test. The average accuracy was 0.8 MOA, while the standard deviation was 0.3 MOA. Would you please tell whether Remington 700 is accurate enough as claimed with 10% significance level?

6.6.2 Test for a Population Correlation Coefficient

As discussed in Chapter 1, in some situations decision maker are concerned with the extent to which two variables are related. For this, a statistical technique known as correlation analysis is used to determine the strength of the (linear) relationship between the two variables. The linear association is measured by a number between -1 and $+1$, referred to as the sample correlation coefficient.

The sample correlation coefficient, r, is often used to determine whether the population correlation, ρ, is significantly different from zero. That is, if data supports the existence of a linear relationship. This is done by performing a hypothesis test. A hypothesis test for ρ can be one-tailed or two-tailed. The null and alternative hypotheses are as follows:

$$H_0{:}\rho = 0 \ vs \ H_1{:}\rho > 0, \ \text{or} \ \rho < 0, \ \text{or} \ \rho \neq 0.$$

A t-test is used to see whether the correlation between the variables of interest is significant. The test quantity takes the following form:

$$T = \frac{r}{\sqrt{\frac{1-r^2}{N-2}}}.$$

It is shown that the test statistics, that is, the sampling distribution of T, is a t-distribution with $N-2$ degrees of freedom, where N is the number of observed or measured pairs.

Note that if the hypothesis test shows a significant linear correlation, the next step will be to find equation of the line that best describes the relationship. This line, called the regression line, can be used to predict the values of one variable (response) for the given values of other variable (predictor) (see Chapters 1 and 8).

EXAMPLE 6.6.3 Suppose that using eight pairs of data we find $r \approx 0.913$. We want to test the significance of the correlation coefficient at $\alpha = 0.05$.

SOLUTION: The null and alternative hypotheses are, respectively,

$$H_0{:}\rho = 0 \ \text{(no correlation)} \ \text{and} \ H_1{:}\rho \neq 0 \ \text{(significant correlation)}.$$

Because there are eight pairs of data in the sample, the t-distribution has $8 - 2 = 6$ degrees of freedom. The test is two-tailed, $\alpha = 0.05$, and the rejection regions are $T < -2.447$ and $T > 2.447$ ($|T| > 2.447$). The test quantity is

$$T = \frac{r}{\sqrt{(1-r)^2/(n-2)}} = \frac{0.913}{\sqrt{\frac{1-(0.913)^2}{8-2}}} \approx 5.482.$$

Because T is in the rejection region, we reject the null hypothesis. There is sufficient evidence to indicate that there is a significant correlation at 5% level of significance.

EXAMPLE 6.6.4 A coach of a professional basketball team is concerned about the performance of their star player. He wants him to play around 35 min per game but is not sure if there is any relationship between his performance and the way this time is distributed over intervals or segments during the game. The data below represents results of 7 similar games with 35 min being divided over a different number of segments. For example, 2 segments means that he would play for a certain amount of time, rest for a while, and play again to complete the playing time of 35 min. Test, using $\alpha = 0.05$, to see whether there is any positive correlation between the number of segments and the number of points scored by the star player.

Game:	1	2	3	4	5	6	7
Number of Segments:	2	5	1	3	4	1	5
Scores:	24	28	22	26	25	24	26

SOLUTION:

$$H_0: \rho = 0 \quad vs \quad H_1: \rho > 0.$$

For this data

$$r = \frac{\Sigma x_i y_i - (\Sigma x_i)(\Sigma x_i)/N}{\sqrt{\Sigma x_i^2 - (\Sigma y_i)^2/N}\sqrt{\Sigma y_i^2 - (\Sigma y_i)^2/N}} = 0.854.$$

The test (quantity) statistics is

$$T = r/\sqrt{\frac{1-r^2}{N-2}} = 0.854/\sqrt{\frac{1-(0.854)^2}{7-2}} = 0.854/0.233 = 3.67.$$

Rejection regions is $T > t_{0.95,(5)} = 2.02$. Since $T > 2.02$, we reject H_0 and conclude that the number of segments and scores are positively linearly related at 5% level of significance.

PRACTICE PROBLEMS

1. Five couples have reported the age at which they have stopped playing sports. Using $\alpha = 0.05$, test to see whether there is a significant correlation (linear association) between the ages stopped playing sports for male and female partners

$$(33,36), (31,28), (32,42), (20,19), (24,25).$$

2. Use the data in Example 6.6.4 to find the equation of regression line. Predict the star player's score for 6 segments.

6.7 LARGE-SAMPLE TEST OF HYPOTHESIS FOR POPULATION PROPORTION

In Sections 6.3, 6.4, and 6.6, tests for population mean were discussed. Population mean is an important parameter that carries a great deal of information about an attribute of interest. In this section, testing and estimation of population proportion is considered. As one may anticipate, the statistical inference for mean and proportion is related. Recall that for binomial distribution, we have $\mu = Np$, and since N is known, we can consider parameter p, rather than μ, and perform a test. Proportion is also a useful parameter when dealing with qualitative data. For example, we may collect data by posing a question such as do you play any sports? For this case responses are in the form of yes or no. We may then want to use the data and estimate the true proportion or percentage of individuals who play a sport, or test a hypothesis about it. To set up a test regarding population proportion we may follow the steps described below.

1) Check to make sure that a normal approximation to binomial is applicable. This requires that $p_0 \pm 3\sigma_{\hat{p}}$ do not include 0 or 1, or equivalently $N\hat{p}$ and $N(1-\hat{p})$ should both be greater than 5, where $\hat{p} = \frac{X}{N}$.
2) Null hypothesis to be tested is $H_0: p = p_0$. We may introduce $H_1: p \neq p_0$ as an alternative hypothesis.
3) Test quantity is

$$T(x_1, \ldots, x_N) = \frac{\hat{p} - p_0}{\sqrt{\frac{p_0 q_0}{N}}}$$

where $q_0 = 1 - p_0$.
4) The distribution of $\frac{X/N - p_0}{\sqrt{p_0 q_0/N}} = \frac{\hat{p} - p_0}{\sqrt{p_0 q_0/N}}$ is $N(0, 1)$.
5) Critical value is $z_{1-\alpha/2}$ corresponding to the level α.
6) Decision rule is to reject H_0 if $|T| > z_{1-\alpha/2}$. If $|T| \leq z_{1-\alpha/2}$ then H_0 cannot be rejected.

The rejection regions, for different alternatives are summarized in the following table:

H_0	H_1	Rejection Region
$p = p_0$	$p > p_0$	$T > z_{1-\alpha}$
$p = p_0$	$p < p_0$	$T < -z_{1-\alpha}$
$p = p_0$	$p \neq p_0$	$T < -z_{1-\alpha/2}$ or $T > z_{1-\alpha/2}$

EXAMPLE 6.7.1 Soccer fans think that most goals are scored during the first and the last 5 min of each half. In the summer of 1998, France hosted the World Cup of men's soccer. In 64 games, a total of 171 goals were scored of which 60 came during the first or the last 5 min of the first and second halves. Does this indicate that the proportion of total number of goals scored in these time periods is higher than 20/90 (the expected proportion for uniform distribution). Use $\alpha = 0.01$.

It is interesting to note that in the World Cup 2002 in Japan/Korea many goals were scored in the first or the last 5 min of the halves. In particular South Korea (host nation) scored many goals in the last 5 min. One notable case was the game between Turkey and South Korea (for the third position) where the first goal was scored by Turkey just after 10 s (World Cup record) and a goal was scored by South Korea in the last minute of injury time. Perhaps, you would like to examine the data for World Cup 2002 or for Euro 2004, and repeat the analysis presented below.

SOLUTION: Let X denote the number of goals scored (out of 171) within the first and last 5 min of the first and second halves of a randomly selected game. Here we may use p and $\hat{p}$ to represent

Chapter 6: Statistical Testing

the true and estimated proportion of the goals scored in the first or the last 5 min of the halves. Recall that

$$E\left(\frac{X}{N}\right) = \frac{1}{N}E(X) = \frac{1}{N}(Np) = p$$

1. $\hat{p} = 60/171 = 0.35$, $N\hat{p} = 171(0.35) = 60 > 5$, $N(1-\hat{p}) = 171(0.65) = 111 > 5$.
2. $H_0: p = 2/9 = 0.22$ vs $H_1: p > 2/9$ (one-tailed).
3. Test statistics

$$T = \frac{\hat{p} - p_0}{\sqrt{\frac{p_0(1-p_0)}{N}}} = \frac{0.35 - 0.22}{\sqrt{\frac{(0.22)(0.78)}{121}}} = 4.104.$$

4. Rejection region

$$T > z_{0.99} = 2.58.$$

5. Decision: reject H_0.
6. Conclusion: There is sufficient evidence to indicate that the percentage of goals scored in the first and last 5 min of each half of a World Cup 98 soccer game is greater than 22% at the 5% significance level. Thus, the fans' observation is supported.

EXAMPLE 6.7.2 A mail-order company that sells sporting goods has observed that 10% of its customers have placed an additional order within 6 months of their original order. However, the records for a random sample of 1,000 recent customers indicate that only 80 customers placed an additional order within 6 months of their original order. Is there evidence (at $\alpha = 0.10$) that the proportion of customers who place additional orders has changed?

SOLUTION: First we need to make sure that normal approximation to binomial is applicable. We have $\hat{p} = 80/1,000 = 0.08$ and

$$N\hat{p} = 1,000(0.08) = 80 > 5 \text{ and } N(1-\hat{p}) = 1,000(0.92) = 920 > 5.$$

In this case,
$$H_0: p = 0.10 \text{ vs } H_1: p \neq 0.10$$

and $\hat{p} = 0.08$. Using the formula for test quantity we have

$$T = \frac{\hat{p} - p_0}{\sqrt{p_0 q_0/N}} = \frac{0.08 - 0.10}{\sqrt{(0.10)(0.90)/1,000}} = -2.11.$$

Since $|-2.11| > z_{0.95} = 1.645$, we reject H_0. At 10% level of significance, the proportion of such customers has changed. Now that the null hypothesis is rejected, we may want to know what is the new percentage. For this, we consider a confidence interval. For example, a 90% confidence interval for p is

$$\hat{p} \pm z_{1-\alpha/2}\sqrt{\hat{p}\hat{q}/N} = 0.08 \pm 1.645\sqrt{(0.08)(0.92)/1,000} = 0.08 \pm 0.014 \Rightarrow (0.066, 0.094).$$

Note that $p = 0.10$ does not fall in this interval indicating that H_0 should be rejected. In fact, according to this interval, the new percentage is less than 10%.

Finally we could have made our decision based on the p-value. Here we have

$$p\text{-value} = 2P(z < -2.11) = 2(0.0174) = 0.0348.$$

Since this is less than 0.10, H_0 is rejected.

EXAMPLE 6.7.3 Table 6.7.1 present Chris Mullin's (a professional basketball player) free-throw percentage for the years 1988 to 1998. Assume that Mullin's free-throw shooting is simply a series of Bernoulli trials. We estimate the fixed probability for the trials by his overall percentage $p = 0.851$. The expected value for his shooting percentage after any number of trials is $p_0 = 0.851$. The standard deviation for N trials is $p_0(1-p_0)/\sqrt{N} = 0.361/\sqrt{N}$. The last column shows the number of standard deviations from the mean for each year. The large numbers, 3.13 in 97-98 and -4.07 in 93-94 suggest that the variation in Mullin's performance is not simply due to randomness. In other words, the values would be highly unlikely to result from Bernoulli trials with fixed $p_0 = 0.851$. The probability that Mullins' percentage would be as low as 0.753 (as it was in 93-94) is less than 0.00003 (p-value) if the Bernoulli model with $p_0 = 0.851$ is assumed. This means that a null hypothesis such as $H_0: p = p_0 = 0.851$ could easily be rejected based on data for the season 93-94. The results may suggest that Mullin had a "hot" year in 97-98 and a "cold" year in 93-94 for reasons other than random variation (see Section 2.7). In other words, we have evidence to conclude that his free-throw percentage varies and cannot be considered fixed over the years.

Also a 95% confidence interval for long-term (true percentage) performance of Chris Mullin is

$$\hat{p} \pm z_{1-\alpha/2}\sqrt{\hat{p}\hat{q}/N} = 0.851 \pm 0.0123 \Rightarrow (0.839, 0.863).$$

Again we note that many of the yearly percentages are outside this interval providing further evidence against the assumption of constant p_0.

EXAMPLE 6.7.4 Consider the data in Table 6.7.1. Suppose now that we want to compare Mullin's free-throw percentages for the last two seasons, 97-98 and 96-97. The question of interest is the following: which of these two seasons was he a better free-throw shooter?

SOLUTION: Clearly Mullin performed better in the 97-98 season (0.939 compared to 0.864). But remember that such a difference could be due to the chance. In order to compare his potential performances in these two seasons, we can set up a 95% confidence interval for these two seasons. Thus, we have

$$97\text{-}98: \quad 0.939 \pm 1.96\sqrt{\frac{(0.939)(0.061)}{164}} = 0.939 \pm 0.0366 \rightarrow (0.9024, 0.9756)$$

$$96\text{-}97: \quad 0.864 \pm 1.96\sqrt{\frac{(0.864)(0.136)}{213}} = 0.864 \pm 0.0235 \rightarrow (0.8405, 0.8875)$$

From these intervals we are 95% confident that Mullin's true free-throw percentages were between 0.9024 and 0.9756 for the 97-98 season and between 0.8405 and 0.8875 for the 96-97 season. These two intervals do not overlap. Even for the worst scenario the percentage for the 97-98 season was higher than the best case in season 96-97 with 95% confidence. We can therefore draw a conclusion that he performed better in the 97-98 season. If on the other hand, we compare, for example, the 96-97 season to the 95-96 season, we find that intervals overlap and therefore we

Chapter 6: Statistical Testing

TABLE 6.7.1 Chris Mullin's Free-Throw Percentage 1988–1998

Year	Made/Attempted x/N	Free Throw % $\tilde{p}$	Standard Deviation $\sqrt{0.361/N}$	z-Score $(\tilde{p} - 0.851)/\sqrt{0.361/N}$
1997–1998	154/164	0.939	0.0281	3.13
1996–1997	184/213	0.864	0.0247	0.526
1995–1996	137/160	0.856	0.0285	0.17
1994–1995	184/213	0.864	0.0247	0.526
1993–1994	165/219	0.753	0.0244	−4.017
1992–1993	183/226	0.809	0.0240	−1.75
1991–1992	350/420	0.833	0.0176	−1.02
1990–1991	513/580	0.884	0.0150	2.20
1989–1990	505/568	0.889	0.0151	2.50
1988–1989	493/553	0.892	0.0154	2.67
Total	2,823/3,316	0.851		

cannot draw such a conclusion. Note that a different approach for handling this problem would be to test $H_0: p_{97} - p_{96} = 0$ versus $H_1: p_{97} - p_{96} > 0$. This approach will be discussed in Section 6.11.6.

PRACTICE PROBLEMS

1. In a random sample of 900 tennis players 105 are left-handed. Would this data support the claim that 10% of tennis players are left-handed? Find a 95% confidence interval for the true proportion of left-handed players.
2. In the Soccer World Cup 2002, Germany scored 14 goals, of which 8 of them were scored by head (in air). Would this provide sufficient evidence (at the 5% significant level) to conclude that this team scores more than half of its goals by head?
3. By researching F1 car race champions list, I found there are 15 champions chose Ferrari engines in all 69 seasons between 1950 and 2018. Meanwhile, I found 13 champions used Ford engines. Can you make a conclusion whether Ford engine is as popular as Ferrari engine using significance level of 0.1.

RESEARCH PROBLEMS

1. Soccer experts and fans believe that 60% of the goals in international soccer leagues are scored from set pieces such as penalty kicks, corner kicks, and so on. Collect data for a tournament of your choice (e.g., World Cup) and test the theory.
2. Are there more male Democrats for female Democrats in your college? Collect data and apply the procedure described in Example 6.7.4.

6.7.1 Determination of Sample Size

We now discuss determination of the sample size necessary to make an inference about a binomial parameter p. Recall that this subject was also covered in Chapter 5. Once the null hypothesis is rejected, we may want to estimate p with a certain degree of accuracy. Referring to Example 6.7.2, it is easy to see that

$$L = 2z_{1-\alpha/2}\sqrt{\hat{p}\hat{q}/N} = 2(1.645)\sqrt{(0.08)(0.92)/1,000} = 0.028$$

This denotes the length (width) of the confidence interval for p, and one-half of that (0.014) is the margin of error. Suppose we want to be 95% confident that the estimate for the true proportion of customers who placed an additional order within 6 months of their original order, is within 0.01 of the true value of p. In other words, we want the length of the confidence interval to be 0.02, or the margin of error to be 0.01. How many recent customers should we sample?

To answer this question, we need to solve the following equation for N:

$$0.02 = 2z_{1-\alpha/2}\sqrt{\hat{p}\hat{q}/N.} = 2(1.645)\sqrt{(0.08)(0.92)/N}.$$

Simplifying this we find that, this can be accomplished if $N \geq 2{,}827.4$, or $N \geq 2{,}828$, since N has to be an integer.

In general, we have the following formula for the determination of the sample size:

$$N \geq \frac{(z_{1-\alpha/2})^2(pq)}{B^2}, \quad L = 2B$$

where B is the margin of error. Since the value of the product pq is unknown, it may be replaced by its estimate $\hat{p}\hat{q}$ from a prior sample.

EXAMPLE 6.7.5 Suppose that a national chain wishes to estimate p, the proportion of the customers who purchase sporting goods. If they want to be 90% confident that their estimate is within 0.02 of the true value of p, how many customers should they include in their sample?

SOLUTION: Suppose from past data, it is known that this proportion is 0.15. Using this we have

$$B = 0.02, \quad z_{0.95} = 1.645, \quad \hat{p} = 0.15, \quad \hat{q} = 0.85$$

and hence

$$N \geq (1.645)^2 \left(\frac{(0.15)(0.85)}{(0.02)^2}\right) = 862.5$$

or $N \geq 863$. If we had no idea about the value of p, then we should consider the worst scenario or a conservatively large N by approximating p. For instance, we could consider $p = 0.30$ or even $p = 0.50$. These values will lead to a much bigger N. In fact, $p = 0.50$ corresponds to the maximum value of pq and therefore the largest value of N. So if nothing is known about the value of p, to be on the safe side, we may use $p = 0.50$.

A PRACTICE PROBLEM Suppose that the athletic director of a university wishes to estimate p, the percentage of freshman who are physically active. If she wants to be 99% confident that her estimate is within 0.03 of the true value of p, how many freshman should she include in her sample. Assume that the last year's percentage was 0.25.

6.8 χ^2 TEST FOR VARIANCE (STANDARD DEVIATION)

In sports, like many other areas of application, there are cases where a hypothesis regarding the variation of a given attribute (random variable) is of prime interest. For example, the amount by which records are broken these days appears to be smaller than those of 50 years ago. If this is the case, then variation in data such as the best annual records of a given sport should be smaller for the past 50 years than that for annual records of more than 50 but less than 100 years ago. Also, when buying sporting goods or equipment, customers like to see that they are closely identical. This means that they do not expect to see a substantial variation in performance measures or lifetimes. Another example of interest is the consistency of the players. This can be assessed by variance or standard deviation, since smaller variation in performance measures means a higher level of consistency.

When inference regarding the variance is of interest, we can apply a test known as the χ^2-test, which is based on the χ^2-distribution described in Section 3.7. Steps for this test are described below. It should be noted that unlike normal and t distributions the χ^2 distribution is not symmetric and takes only positive values. This means that to find rejection region we need to specify two critical values, not just one.

1) Assume that X is distributed as $N(\mu, \sigma^2)$ and both μ and σ^2 are unknown. Suppose that we have a sample of size N.
2) The null hypothesis to be tested is $H_0: \sigma^2 = \sigma_0^2$. We may introduce $H_1: \sigma^2 \neq \sigma_0^2$ as an alternative hypothesis.
3) Test quantity is

$$T(x_1, \ldots, x_N) = (N-1)\frac{s^2}{\sigma_0^2}. \tag{6.6}$$

4) The distribution of $(N-1)\frac{s^2}{\sigma_0^2}$ is $\chi^2_{(N-1)}$.
5) Critical values are $\chi^2_{\alpha/2,(N-1)}$ and $\chi^2_{1-\alpha/2,(N-1)}$.
6) Decision rule is to reject H_0 if $T < \chi^2_{\alpha/2,(N-1)}$ or $T > \chi^2_{1-\alpha/2,(N-1)}$. Otherwise H_0 cannot be rejected.

Note that the alternative hypothesis can be just greater (less) than a specified value σ_0^2. The following table provides the rejection regions for different alternatives.

H_0	H_1	Rejection Region
$\sigma^2 = \sigma_0^2$	$\sigma^2 > \sigma_0^2$	$T > \chi^2_{1-\alpha,(N-1)}$
$\sigma^2 = \sigma_0^2$	$\sigma^2 < \sigma_0^2$	$T < \chi^2_{\alpha,(N-1)}$
$\sigma^2 = \sigma_0^2$	$\sigma^2 \neq \sigma_0^2$	$T < \chi^2_{\alpha/2,(N-1)}$ or $T > \chi^2_{1-\alpha/2,(N-1)}$

EXAMPLE 6.8.1 Referring to Example 6.6.2, let σ^2 denote the variance of the total yardage accumulated during the NCAA college football season by all receivers. Also assume that this population has a normal distribution. Carry out a χ^2-test for $H_0: \sigma^2 = 18,000$ versus $H_1: \sigma^2 \neq 18,000$ at a 5% significance level.

SOLUTION: For this data $s^2 = 20,037$. From (6.6) we have

$$T = 4 \cdot \frac{20,037}{18,000} = 4.45.$$

Using Table B5 for χ^2, the critical values are $\chi^2_{\alpha/2,(N-1)} = \chi^2_{0.025,(9)} = 0.48$ and $\chi^2_{1-\alpha/2,(N-1)} = \chi^2_{0.975,(9)} = 11.14$. Since T is between these two values, H_0 cannot be rejected. There is not sufficient evidence, at 5% significant level, that σ^2 is different from 18,000.

EXAMPLE 6.8.2 The mean lifetime of items such as light bulbs and transistors is an important characteristic, but equally important is the variation of lifetime. If σ^2, the variance of, for example, a bulb's lifetime is large, some bulbs will last for a very short time and others very long. Suppose regulatory agencies specify that the standard deviation of the lifetime should be less than 100 h. To see whether the products meet this specification, we take a random sample of size 51 and find that $s = 150$. Does this data provide sufficient evidence, at 5% significance level, to indicate that the variability is as small as desired? Assume that lifetime has a normal distribution.

SOLUTION: To answer this question, we use the procedure described above. Here we should test

$$H_0: \sigma^2 = 100^2 \quad vs \quad H_1: \sigma^2 > 100^2.$$

The test statistics is

$$T = (51-1)\frac{150^2}{100^2} = 112.5$$

and degrees of freedom is $(N-1) = 50$. Thus $\chi^2_{0.95,(50)} = 67.5048$. Since $T > \chi^2_{0.95,(50)}$, we reject H_0, and accept H_1. This means the product under consideration does not meet the specifications. In other words, there is sufficient evidence, at 5% significance level that $\sigma^2 > 10,000$.

Since the null hypothesis is rejected, we may want to have some idea about the true variation of the lifetime. For this we can provide an estimate (confidence interval) for the variation. Here a $100(1-\alpha)\%$ confidence interval for σ^2 is constructed using

$$\frac{(N-1)s^2}{\chi^2_{1-\alpha/2,(N-1)}} < \sigma^2 < \frac{(N-1)s^2}{\chi^2_{\alpha/2,(N-1)}}.$$

For the above example, a 95% confidence interval for σ^2 is

$$\frac{50(150)^2}{71.42} < \sigma^2 < \frac{50(150)^2}{32.26}$$

or

$$(15,752 \quad 34,768)$$

Chapter 6: Statistical Testing

Note that this interval does not include 10,000, indicating that the specification is not met.

If needed we can take the square roots of the above numbers and obtain a 95% confidence interval for the standard deviation. For the above example we get (125.5, 186.5) as a 95% confidence interval for standard deviation.

EXAMPLE 6.8.3 In recent years investing in the stock market has become a part of most people's life. Suppose that you have some money and wish to invest. How do you decide between two different stocks? One way to do this is to measure the "risk" of each stock and use that to make up your mind. But, how do people measure the risk of a given stock? One simple and intuitively appealing measure is the variance. The bigger the variance, the greater the risk. Thus you can look at the past performance and, using that, set up a confidence interval for variances. If the intervals do not overlap, then you will know which stock is more risky. If you are a risk-taker you should go for the risky stock. After all if you like reward you need to take risk. On the other hand, if you are conservative and cannot tolerate to lose you should go for the less risky stock. In fact, your investment is a reflection of your personality.

A PRACTICE PROBLEM A random sample of size 64 produced a standard deviation $s = 2$. Does this data provide sufficient evidence, at 10% significance level, that the variability is greater than 1.8?

6.9 *F*-TEST

In Section 6.3, comparison of population centers based on sample means was discussed. Here we have a similar situation where two are compared, but now with respect to their variability based on sample variances. For example, we may want to compare two similar products to find out which one provides a more consistent performance, that is, which one exhibits less variation in, for example, it's lifetime or in any other measure of performance. In other words in which case items are more identical. Or we may want to compare two different stocks to see which one is more risky when risk is measured by variation. In sports we may want to compare two players to see which one is more consistent. Or, as discussed in Section 6.8, we may want to compare the variation of best annual records for two different time segments (past and recent) to confirm that the amount by which records are broken now is smaller than that in the past. The following summarizes the steps required for such comparisons.

1) Assume that $X^{(1)}$ and $X^{(2)}$ are distributed as $N(\mu_1, \sigma_1^2)$ and $N(\mu_2, \sigma_2^2)$, respectively, and μ_1, μ_2, σ_1^2, and σ_2^2 are all unknown. We have samples of sizes N_1 and N_2 on $X^{(1)}$ and $X^{(2)}$ and therefore sample variances s_1^2, and s_2^2 can be calculated.
2) The null hypothesis to be tested is $H_0: \sigma_1^2 = \sigma_2^2$. We may introduce the alternative as $H_1: \sigma_1^2 \neq \sigma_2^2$.
3) Test quantity is

$$T(x_1^{(1)}, \ldots, x_{N_1}^{(1)}; x_1^{(2)}, \ldots, x_{N_2}^{(2)}) = \frac{s_1^2}{s_2^2}. \tag{6.7}$$

4) The distribution of $\frac{s_1^2}{s_2^2}$ is $F_{(N_1-1, N_2-1)}$.

5) Critical values corresponding to the level of significance α are $F_{\alpha/2,(N_1-1,N_2-1)}$ and $F_{1-\alpha/2,(N_1-1,N_2-1)}$, respectively.
6) Decision rule is to reject H_0 if $T < F_{\alpha/2,(N_1-1,N_2-1)}$ or $T > F_{1-\alpha/2,(N_1-1,N_2-1)}$. Otherwise, H_0 cannot be rejected.

EXAMPLE 6.9.1 Suppose we want to compare two different drugs used to lower the blood pressure. Let $X^{(1)}$ and $X^{(2)}$ represent, respectively, the amount by which the blood pressure is lowered using these drugs. Assume that in both cases random variables of interest have normal distribution. Suppose that $N_1 = 10$ and $N_2 = 21$ persons were treated by these drugs and the measured values are $x_1^{(1)}, \ldots, x_{N_1}^{(1)}$ and $x_1^{(2)}, \ldots, x_{N_2}^{(2)}$, respectively. Suppose from these data we get $s_1^2 = 89$ and $s_2^2 = 100$. Carry out the F-test for $H_0: \sigma_1^2 = \sigma_2^2$ versus $H_1: \sigma_1^2 \neq \sigma_2^2$ using a significance level of 10%.

SOLUTION: Using (6.7) we get
$$T = 89/100 = 0.89.$$

The critical values are $F_{\alpha/2,(N_1-1,N_2-1)} = F_{0.05,(9,20)} = 0.34$ and $F_{1-\alpha/2,(N_1-1,N_2-1)} = F_{0.95,(9,20)} = 2.39$. Since T is between these two values, we cannot reject H_0. In other words, the variation of blood pressures is not significantly different for these two drugs at 10% level of significance.

In the above example, it happens that $s_1^2 < s_2^2$ and thus $T < 1$. In practice, it may happen that $s_1^2 \geq s_2^2$. For instance, if in the above example we switch the notations for the two samples, then we would have $N_1 = 21$, $N_2 = 10$, $s_1^2 = 100$, $s_2^2 = 89$, and $T = 100/89$. The critical values will now be $F_{\alpha/2,(20,9)} = F_{0.05,(20,9)} = 0.42$ and $F_{1-\alpha/2,(20,9)} = F_{0.95,(20,9)} = 2.94$. These values also lead to the conclusion that H_0 cannot be rejected. Since the result of the test is not changed, most users set up the problem so that $T > 1$. That is, always name the sample with a larger sample variance as coming from population 1. As is explained below this simplifies the testing procedure.

One property of the F distribution that is helpful for determination and simplification of the rejection region is the relationship regarding the values of the F distribution on its lower and upper tails. We have
$$F_{1-\alpha,(v_1,v_2)} = 1/F_{\alpha,(v_2,v_1)}.$$

For instance
$$F_{0.05,(20,9)} = 1/F_{0.95,(9,20)} = 1/2.39 = 0.42.$$

This was used in Example 6.9.1. This implies that, when $T > 1$ we only need to check if the test quantity falls in the rejection region in the upper tail.

EXAMPLE 6.9.2 Performances of two basketball players are compared to see which one is a more consistent player. The measure and data used here is the number of points scored per game. Assume that both players have played the same number of minutes per game. The players are compared for consistency based on variability of their performance summarized in the following table:

Chapter 6: Statistical Testing

	Player A	Player B
$\bar{x}$	18.42	20.75
s	5.50	3.60
N	13	10

Note that other than testing for consistency, we may want to compare their performances based on average number of points scored per game and select one player for certain recognition. As was discussed earlier, this type of comparison can be done using a *t*-test. Since the samples are small, in order to perform the *t*-test for the difference between the means, we need to assume that the population variances are equal. If we have doubts about the validity of this assumption, we can first apply the *F*-test to check its validity of such assumption. Use the data given above to perform such a test using a significance level of $\alpha = 0.10$.

SOLUTION: We wish to test the hypothesis

$$H_0: \sigma_1^2 = \sigma_2^2 \quad vs \quad H_1: \sigma_1^2 \neq \sigma_2^2$$

where σ_1^2 and σ_2^2 are the population variances of the points per game by two players, respectively. The test requires the following assumptions:

(i) Both populations are normal.
(ii) The samples are obtained randomly and independently from the two populations.

The rejection region consists of values of the *F* statistic for which $T > F_{0.95,(12,9)} = 3.07$. The value of the test statistic (test quantity) is

$$T = \frac{\text{Larger sample variance}}{\text{Smaller sample variance}} = \frac{(5.50)^2}{(3.60)^2} = 2.33.$$

The computed value of the test statistic does not lie within the rejection region. There is insufficient evidence to conclude that the population variances are significantly different at $\alpha = 0.10$. This means that, at 10% significant levels, we cannot declare one player more consistent than the other, using the variance as a measure of consistency.

A PRACTICE PROBLEM Test equality of the population means for data in Example 6.9.2. Use $\alpha = 0.10$.

As pointed out that when the hypothesis of equal variances is not rejected, it is possible to pool the samples and use that to estimate the (equal) population variance. This is particularly useful when N_1 and N_2 are small. For Example 6.9.2, the hypothesis of equal variances is not rejected. Thus, we may combine the samples and estimate the (equal) population variance as if all the data have come from the same population (pooled estimate). Of course, if we consider a different α we may end up rejecting H_0. For example, for $\alpha = 0.20$, we have $F_{0.90,(12,9)} = 2.21$ and hence the hypothesis of equal variance is rejected. When equality of variance is rejected, we may set up a confidence interval for the ratio of the population variances. For the above example a 90% confidence interval for the ratio of two population variances is constructed as

$$\left(\frac{s_1^2}{s_2^2}\right)\left(\frac{1}{F_{0.95,(12,9)}}\right) < \frac{\sigma_1^2}{\sigma_2^2} < \left(\frac{s_1^2}{s_2^2}\right) F_{0.95,(9,12)}.$$

Recall that,

$$s_1^2 = (5.50)^2 = 30.25, \quad F_{0.95,(12,9)} = 3.07,$$
$$s_2^2 = (3.60)^2 = 12.96, \quad F_{0.95,(9,12)} = 2.80.$$

Substitution of these values into the general expression yields

$$\left(\frac{30.25}{12.96}\right)\left(\frac{1}{3.07}\right) < \frac{\sigma_1^2}{\sigma_2^2} < \left(\frac{30.25}{12.96}\right)(2.80)$$

or

$$0.760 < \frac{\sigma_1^2}{\sigma_2^2} < 6.535.$$

Once more, since equal variance means that the ratio should be equal to 1 and that 1 is included in this interval, we cannot reject the hypothesis of equal variances, at least at 10% significant level.

A PRACTICE PROBLEM A study to compare the athletic achievement test scores of male and female students has produced information summarized in the table below. The researchers hypothesized that the distribution of test scores for male is more variable than the corresponding distribution for females. Test this using $\alpha = 0.01$.

	Males	Females
$\bar{x}$	49	48
S	13	12
N	18	17

6.10 OTHER ALTERNATIVE HYPOTHESIS

So far, except in a few cases, we considered a single hypothesis, the null hypothesis H_0, and this hypothesis was tested against its opposite. In other words, H_0 was tested against the alternative that H_0 is not true. As was demonstrated in some of the examples discussed earlier, in many cases H_1, the alternative hypothesis is not exactly the opposite of H_0 and we want to test whether our sample (evidence) favors H_0 or H_1. For example, we may have $H_0: \mu = \mu_0$ and $H_1: \mu = \mu_1 (\neq \mu_0)$ (two-sided or two-tailed) or $H_0: \mu = \mu_0$ and $H_1: \mu > \mu_0$ (one-sided, right-tailed or one-tailed), and so on. The latter happens when one is testing to see whether something is better than or larger than something else, not just different. All the tests mentioned apply mutatis mutandis. Clearly, the critical values (step 5) and the critical region (steps 5 and 6) need to be changed accordingly. For example, consider

$$H_0: \mu = \mu_0 \quad vs \quad H_1: \mu < \mu_0$$

where μ is the mean of a normal distribution with known σ^2. To discuss this in context, suppose that a firm produces material for a certain type of sport shoe, usually with a mean tensile strength of μ_0 (target value). As a part of quality control program, they test their product periodically. Based on the last investigation, the manager thinks that the production has been changed (is off-target) and suspects that the materials produced are "weaker" now and he wants to examine this. To do this, we proceed as follows:

1) Assume that X is distributed as $N(\mu, \sigma^2)$ and σ^2 is known. Suppose a sample of size N is available.
2) The null and the alternative hypothesis are $H_0: \mu = \mu_0$ and $H_1: \mu < \mu_0$, respectively.
3) Test quantity is

$$T(x_1,\ldots,x_N) = \frac{\bar{x} - \mu_0}{\sigma/\sqrt{N}}. \qquad (6.8)$$

4) Assuming that the null hypothesis is true, the distribution of $\frac{\bar{X}-\mu_0}{\sigma/\sqrt{N}}$ is $N(0,1)$.
5) Critical value is $z_\alpha = -z_{1-\alpha}$ corresponding the level α.
6) Decision rule is to reject H_0 and accept H_1 if $T < z_\alpha = -z_{1-\alpha}$. Otherwise H_0 cannot be rejected and H_1 cannot be accepted.

If we replace $\mu < \mu_0$ with $\mu > \mu_0$, then step 5 should be replaced by $z_{1-\alpha} = -z_\alpha$ and step 6 by $T > z_{1-\alpha} = -z_\alpha$.

Note that, for the example described above, μ is the true mean tensile strength for the materials that were produced recently and H_0 states that nothing is changed (on-target), whereas H_1 says that the materials' average strength is less (off-target)than what it was.

EXAMPLE 6.10.1 Data were collected on the golf ball driving distances by professional golfers in a recent tournament. The data on the first drives of 36 randomly selected golfers yielded a sample mean distance of 247.5 yards. Assuming the standard deviation of the driving distances on the first drive is 10 yards, carry out a z–test on $H_0: \mu = 250$ versus $H_1: \mu < 250$ at the 10% significance level.

SOLUTION: First note that X represents the driving distance for a randomly selected golf ball and we would like to find out if the mean driving distance is the same as before or has decreased. We have

$$T = \frac{247.5 - 250}{10/\sqrt{36}} = -1.5.$$

The critical value equals -1.28. Since $T < -1.28$, H_0 should be rejected. This leads to the conclusion that the mean driving distance is less than 250 yards (the average target driving distance) at a 10% significance level. Recall that the last part of this sentence means that the Type I error, that is, the probability of rejecting H_0 when it is true is 10%.

Note that we could alternately have made our decision based on the *p*-value. Here we have

$$p\text{-value} = P(z < -1.5) = 0.0668.$$

Since this is less than 10%, H_0 should be rejected.

We finish this part by mentioning that in all the tests discussed in this chapter, decisions are made based on incomplete information. As such, we may end up rejecting a true null hypothesis or accepting a false alternative hypothesis. That is, we may commit Type I and Type II errors. Let us check the probability of making Type I error, that is, rejecting H_0 when it is true. If H_0 is true, then T is distributed as $N(0,1)$ and the probability of rejecting H_0 is given by

$$P(T < z_\alpha = -z_{1-\alpha}).$$

This is equal to α by definition of the α percentile. Recall that α is the probability of making a Type I error. In Example 6.10.1, this means blaming the firm for weakness of the materials (rejecting H_0 and accepting H_1), when the firm is on target (H_0 is true). This situation only occurs with probability equal to α.

Not rejecting H_0 when H_1 is true is the Type II error and its probability is usually denoted by β. The value $1 - \beta$, that is, the probability of not rejecting H_0 when H_1 is true, is known as power of the test.

Finally, we should once more mention that if distribution of X is not normal but N is sufficiently large, we can still apply the above procedure. In fact, this is exactly what we did in Example 6.10.1.

Summary of Null and Alternative Hypothesis

The **Null Hypothesis**, or H_0, gives a specific value for a population parameter. Thus, it has the form that includes equality

$$H_0 \text{:Population Parameter} = \text{Claimed Value}.$$

The **Alternative Hypothesis**, or H_1, has one of the following forms:

$$\text{(Left-tailed) } H_1 \text{:Population Parameter} < \text{Claimed Value},$$
$$\text{(Right-tailed) } H_1 \text{:Population Parameter} > \text{Claimed Value},$$
$$\text{(Two-tailed) } H_1 \text{:Population Parameter} \neq \text{Claimed Value}.$$

Remember that the hypotheses must be formulated before sample data are analyzed, and one should never test a hypothesis using the same data that originally suggested the hypothesis.

6.11 MORE ON INFERENCE ABOUT TWO POPULATIONS

In practice, there are many situations that result in comparison of two populations. Also depending on what parameter or parameters of the populations of interest are compared, investigators design appropriate experiments in order to collect proper data. This section includes many subsections presenting further discussion concerning tests applied to two population cases.

6.11.1 Large-Sample Inference About the Difference Between Two Population Means

In this section, we consider large-sample inference about the difference between two population means using independent samples. We start with an example.

EXAMPLE 6.11.1 Fitness club A claims that due to the programs they offer, the average amount of time their members spend in the club is greater than that of fitness club B. In order to test this claim, a local newspaper conducted a survey that included independent random samples of 100 members of club A and 150 members of club B. A summary of information regarding the hours spent in the club during the past 6 months was recorded as follows:

Sample 1: Club A	Sample 2: Club B
$\bar{x}_1 = 219.7$ h	$\bar{x}_2 = 213.4$ h
$s_1 = 12.3$ h	$s_2 = 9.9$ h
$N_1 = 100$ members	$N_2 = 150$ members

Let X_1 and X_2 represent, respectively, the amount of time randomly selected members of club A and club B spend in the club. Also let μ_1 and μ_2 be the true average amount of time spent in club A and club B, respectively. We would like to solve two problems. First, test $H_0: \mu_1 - \mu_2 = 0$ (i.e., $\mu_1 = \mu_2$), no difference between population means verses $H_1: \mu_1 > \mu_2$ (i.e. $\mu_1 - \mu_2 > 0$) at a significance level $\alpha = 0.05$. Second, if H_0 is rejected, find a 90% confidence interval for $\mu_1 - \mu_2$ (the difference). Clearly, a natural point estimate for $\mu_1 - \mu_2$ is $\bar{x}_1 - \bar{x}_2$. To do inference, we need to know the sampling distribution of $\bar{X}_1 - \bar{X}_2$. Since we have large samples that are independent, we have approximately (using the Central Limit Theorem)

$$\bar{X}_1 - \bar{X}_2 \sim N\left(\mu_1 - \mu_2, \frac{\sigma_1^2}{N_1} + \frac{\sigma_2^2}{N_2}\right)$$

where σ_1^2 and σ_2^2 are the variances of the two populations, respectively. Thus a large-sample ($N_1 \geq 30$, $N_2 \geq 30$) confidence interval for $\mu_1 - \mu_2$ may be constructed as

$$\bar{X}_1 - \bar{X}_2 \pm z_{1-\alpha/2} \sqrt{\frac{\sigma_1^2}{N_1} + \frac{\sigma_2^2}{N_2}} \approx \bar{X}_1 - \bar{X}_2 \pm z_{1-\alpha/2} \sqrt{\frac{s_1^2}{N_1} + \frac{s_2^2}{N_2}}.$$

Note that when σ_1^2 and σ_2^2 are unknown and are replaced by their sample counterparts s_1^2 and s_2^2, $t_{1-\alpha/2}$ may be used in place of $z_{1-\alpha/2}$. However, as discussed earlier for large samples, the difference is not usually significant.

Next, we consider the large-sample test of hypotheses for $\mu_1 - \mu_2$. Here we present all forms of alternatives and therefore follow the pattern of the general case introduced in the Section 6.10.

$$\text{Test Quantity: } T = \frac{\bar{x}_1 - \bar{x}_2 - D_0}{\sqrt{\frac{\sigma_1^2}{N_1} + \frac{\sigma_2^2}{N_2}}} \approx \frac{\bar{x}_1 - \bar{x}_2 - D_0}{\sqrt{\frac{s_1^2}{N_1} + \frac{s_2^2}{N_2}}}$$

where $D_0 = \mu_1 - \mu_2$ is the hypothesized difference. The rejection regions will depend on the alternatives and are given in the following table:

H_0	H_1	Rejection Region
$\mu_1 - \mu_2 = D_0$	$\mu_1 - \mu_2 > D_0$	$T > z_{1-\alpha}$
$\mu_1 - \mu_2 = D_0$	$\mu_1 - \mu_2 < D_0$	$T < -z_{1-\alpha} = z_\alpha$
$\mu_1 - \mu_2 = D_0$	$\mu_1 - \mu_2 \neq D_0$	$T > z_{1-\alpha/2}$ or $T < -z_{1-\alpha/2}$

Note that this test is also valid for small samples provided that σ_1^2 and σ_2^2 are known, and distributions are normal.

SOLUTION: Example 6.11.1

$$H_0 : \mu_1 - \mu_2 = D_0 = 0 \quad vs \quad H_1 : \mu_1 - \mu_2 > D_0 = 0.$$

Here we have $\alpha = 0.05$, $z_{0.95} = 1.645$, and

$$T = \frac{\bar{x}_1 - \bar{x}_2 - 0}{\sqrt{\frac{\sigma_1^2}{N_1} + \frac{\sigma_2^2}{N_2}}} \approx \frac{\bar{x}_1 - \bar{x}_2 - 0}{\sqrt{\frac{s_1^2}{N_1} + \frac{s_2^2}{N_2}}} = \frac{219.7 - 213.4}{\sqrt{\frac{12.3^2}{100} + \frac{9.9^2}{150}}} = 4.28.$$

Since $T > z_{0.95} = 1.645$, we reject H_0 and conclude that the average amount of time spent by members of club A in the club is significantly greater than that of club B, at a 5% level of significance. The p-value $= P(z > 4.28) \approx 0$, leading to the same conclusion.

Next we find a 90% confidence interval for $\mu_1 - \mu_2$. We have

$$\bar{X}_1 - \bar{X}_2 \pm z_{1-\alpha/2}\sqrt{\frac{\sigma_1^2}{N_1} + \frac{\sigma_2^2}{N_2}} \approx \bar{X}_1 - \bar{X}_2 \pm z_{1-\alpha/2}\sqrt{\frac{s_1^2}{N_1} + \frac{s_2^2}{N_2}}.$$

Here $z_{1-\alpha/2} = z_{0.95} = 1.645$. Hence, the required interval is

$$219.7 - 213.4 \pm 1.645\sqrt{\frac{12.3^2}{100} + \frac{9.9^2}{150}} = 6.30 \pm 18.9 = (4.31, 8.19).$$

Note that this interval does not include zero, leading to the rejection of the null hypothesis of equal population means.

A PRACTICE PROBLEM A dietitian has developed a diet that is low in fats and carbohydrates. Two random samples of 100 people who are physically active are selected. One group is placed on this diet and the other group in a regular diet. For each person, the amount of weight lost (or gained) in a months period is recorded and is summarized in the table below. Test an appropriate hypothesis and construct a confidence interval. Use $\alpha = 0.05$ and a confidence level of 90%.

	Lowfat	Regular
$\bar{x}$	10	8
s	5	1
N	100	100

Chapter 6: Statistical Testing

6.11.2 Small-Sample Inference About the Difference Between Two Population Means (Normal Populations)

Next, we consider the small-sample inferences about the difference between two population means. Again, we present an example to demonstrate the procedure.

EXAMPLE 6.11.2 Suppose that an educational researcher wishes to determine whether there is a difference in the development of athletic ability between boys and girls. Six second-grade girls and four second-grade boys were given a 20-point test to evaluate their athletic talents. The means and standard deviations of the scores are as follows:

Girls	Boys
$\bar{x}_1 = 14.82$	$\bar{x}_2 = 16.80$
$s_1 = 1.12$	$s_2 = 1.46$
$N_1 = 6$	$N_2 = 4$
μ_1: true mean	μ_2: true mean

Is there evidence (at $\alpha = 0.05$) that second-grade boys have developed a greater athletic ability, than second-grade girls as measured by this test?

Since the samples are small, the test must be based on a two-sample t statistics. The following assumptions are required for the valid applications of this procedure.

1. Both sampled populations have distributions that are approximately normal.
2. The population variances are equal. This is helpful since we can combine the data and estimate the pooled variance. If we have doubts about equality of variances, we can perform an F-test to see if we can confirm this.
3. The samples are randomly and independently selected from the populations.

Under these assumptions, the confidence interval for $\mu_1 - \mu_2$ is given by

$$\bar{X}_1 - \bar{X}_2 \pm t_{1-\alpha/2, (N_1+N_2-2)} \sqrt{S_p^2 \left(\frac{1}{N_1} + \frac{1}{N_2}\right)},$$

where

$$S_p^2 = \frac{\sum(X_1 - \bar{X}_1)^2 + \sum(X_2 - \bar{X}_2)^2}{N_1 + N_2 - 2}$$

$$= \frac{(N_1 - 1)S_1^2 + (N_2 - 1)S_2^2}{N_1 + N_2 - 2}$$

is a pooled sample estimator of σ^2.

The following is the small-sample test of hypotheses for $\mu_1 - \mu_2$.

$$\text{Test Quantity:} \quad T = \frac{\bar{x}_1 - \bar{x}_2 - D_0}{\sqrt{S_p^2 \left(\frac{1}{N_1} + \frac{1}{N_2}\right)}}$$

where $D_0 = \mu_1 - \mu_2$ is the hypothesized difference between the population means. The rejection region is given in the following table where for all cases the degrees of freedom is $N_1 + N_2 - 2$.

H_0	H_1	Rejection Region
$\mu_1 - \mu_2 = D_0$	$\mu_1 - \mu_2 > D_0$	$T > t_{1-\alpha}$
$\mu_1 - \mu_2 = D_0$	$\mu_1 - \mu_2 < D_0$	$T < -t_{1-\alpha} = t_\alpha$
$\mu_1 - \mu_2 = D_0$	$\mu_1 - \mu_2 \neq D_0$	$T > t_{1-\alpha/2}$ or $T < -t_{1-\alpha/2} = t_{\alpha/2}$

SOLUTION: (Example 6.11.2) Let X_1 and X_2 be random variables representing the scores of a randomly selected girl and boy, respectively. Let μ_1 and μ_2 be their respective means. We like to test the following:

$$H_0: \mu_1 - \mu_2 = 0 \quad vs \quad H_1: \mu_1 - \mu_2 < 0.$$

The test quantity is

$$T = \frac{\bar{x}_1 - \bar{x}_2 - 0}{\sqrt{s_p^2(\frac{1}{N_1} + \frac{1}{N_2})}}.$$

Since

$$s_p^2 = \frac{(N_1 - 1)s_1^2 + (N_2 - 1)s_2^2}{N_1 + N_2 - 2} = \frac{(6-1)(1.12)^2 + (4-1)(1.46)^2}{6 + 4 - 2} = 1.58335,$$

we have

$$T = \frac{14.82 - 16.80}{\sqrt{1.58335(1/6 + 1/4)}} = \frac{-1.98}{0.8122} = -2.44.$$

For $\alpha = 0.05$ and degrees of freedom $6 + 4 - 2 = 8$, we have $t_{0.95(8)} = 1.860$. Since $T < -1.860$, we reject H_0 and conclude that the mean test score for second-grade boys is higher than that for girls at a 5% significance level. Furthermore, a 90% confidence interval for $\mu_1 - \mu_2$ is

$$\bar{x}_1 - \bar{x}_2 \pm t_{0.95,(N_1+N_2-2)}\sqrt{s_p^2(\frac{1}{N_1} + \frac{1}{N_2})}$$

$$= 14.82 - 16.8 \pm 1.86\sqrt{1.58335(\frac{1}{6} + \frac{1}{4})}$$

$$= -1.98 \pm 1.86(0.8122) \to (-3.4907, -0.4693).$$

Note that this interval does not include 0, indicating that H_0 should be rejected.

EXAMPLE 6.11.3 Recent data collected on soccer players indicate more injuries to females than males although female players may be more flexible. Consider a test that is conducted to compare the low-back flexibility of selected male and female soccer players. The following measurements, in centimeters, were obtained using the sit-and-reach test:

Chapter 6: Statistical Testing

Male Soccer Players	Female Soccer Players
$\bar{x}_1 = 22.5$	$\bar{x}_2 = 25.6$
$s_1 = 2.5$	$s_2 = 3.0$
$N_1 = 10$	$N_2 = 8$

Here we have

$$s_p^2 = \frac{(10-1)(2.5)^2 + (8-1)(3.0)^2}{10+8-2} = 7.453.$$

The test quantity is

$$T = \sqrt{7.453(1/10 + 1/8)} = -2.40.$$

Since $t_{0.95,(16)} = 1.75$ and $-2.40 < -1.75$, $H_0{:}\mu_1 - \mu_2 = 0$ is rejected and $H_1{:}\mu_1 - \mu_2 < 0$ is accepted. Based on this data, there is sufficient evidence to indicate that female players are more flexible than male players in the hip and low-back joints as measured by the sit-and-reach test, at a 5% significant level.

PRACTICE PROBLEMS

1. Use the data in Example 6.11.3 to construct a 90% confidence interval for $\mu_1 - \mu_2$. Use the interval to test $H_0{:}\mu_1 - \mu_2 = 0$ versus $H_1{:}\mu_1 - \mu_2 < 0$.
2. Use the data in Example 6.11.1 assuming that now $N_1 = 10$ and $N_2 = 15$. Test the same hypothesis and draw appropriate conclusions. Also construct a confidence interval.

The Magnitude of the Difference (Size of Effect)

Statistical tests help us to decide whether differences are statistically significant. A difference that is, or is not statistically significant, can be practically insignificant or significant. Declaring the test statistics to be significant at 5%, or some similar level, only provides evidence that the differences are real and that they did not occur merely by chance. This is statistical significance. In many situations, we must also consider what is of practical significance. If the samples are large enough, if the standard deviations are small enough, and especially if the experiment can be repeated, statistically significant differences may be found between means that are quite close together in value. This small, but statistically significant difference may not be large enough to be of much use in a practical application. How important is the size of the mean difference?

Some statisticians suggest the use of the omega squared value (ω^2) to determine the importance, or usefulness, of the mean difference. Omega squared is an estimate of the percentage of the total variance (the difference between the means) that can be explained by the influence of the independent variable (the treatment). For a t-test, the formula for omega squared is given by

$$\omega^2 = \frac{t^2 - 1}{t^2 + N_1 + N_2 - 1}.$$

Applying this to the data from Example 6.11.3, which compared male and female players for their low-back flexibility yields

$$\omega^2 = \frac{(-2.4)^2 - 1}{(-2.4)^2 + 10 + 8 - 1} = 0.21.$$

In this case, 21% of the differences between male and female soccer players in hip and low-back flexibility can be attributed to gender. The remaining 79% of the variance is due to individual differences among subjects, other unidentified factors, and measurement errors. Note that this is similar to decomposition of variation in regression analysis (discussed in Chapter 8) and one-way analysis of variance (discussed in Chapter 7).

An important question that arises is the following: how large must omega squared be before it is considered important? The answer to this question is not statistically based. Each investigator or consumer of the research must determine the importance of omega squared. In this example, it is meaningful to know that 21% of the variance can be explained by gender difference. But, gender clearly is not the only variable that affects hip and low-back flexibility. Other unidentified factors are at work in this study.

6.11.3 Inference About the Difference Between Two Population Means (Unequal Variances)

In Section 6.11.2, we considered methods for comparing means when we are unable to detect any difference in the population variances. We now turn our attention to the problem of comparing means when σ_1^2 and σ_2^2 are not equal.

If we have convincing evidence that $\sigma_1^2 \neq \sigma_2^2$, it makes no sense to pool the data. There is no common population variance to be estimated. Instead, we simply substitute these unknown variances with their unbiased estimators S_1^2 and S_2^2 to obtain the test quantity

$$\frac{(\bar{X}_1 - \bar{X}_2) - (\mu_1 - \mu_2)}{\sqrt{(S_1^2/N_1) + (S_2^2/N_2)}}.$$

It has been found that this random variable has approximately a t distribution with degrees of freedom equal to

$$v = \frac{[(S_1^2/N_1) + (S_2^2/N_2)]^2}{[(S_1^2/N_1)^2/(N_1 - 1)] + [(S_2^2/N_2)^2/(N_2 - 1)]}.$$

Confidence intervals and test of hypotheses on $\mu_1 - \mu_2$ can be carried out in the usual manner. In particular, a $100(1 - \alpha)\%$ confidence interval for $\mu_1 - \mu_2$ is given by

$$(\bar{X} - \bar{X}_2) \pm t_{1-\alpha/2,(v)} \sqrt{(S_1^2/N_1) + (S_2^2/N_2)}.$$

Chapter 6: Statistical Testing

Hypothesis tests are performed using the test quantity

$$\frac{(\bar{X}_1 - \bar{X}_2) - D_0}{\sqrt{(S_1^2/N_1) + (S_2^2/N_2)}}$$

which has a t distribution with v degrees of freedom.

The formula given for computing the degrees of freedom has been widely accepted. However, one should be aware of the fact that other formulas do exist. The issue of how to best estimate the number of degrees of freedom associated with the random variable

$$\frac{\bar{X}_1 - \bar{X}_2}{\sqrt{(S_1^2/N_1) + (S_2^2/N_2)}}$$

has not been resolved. One should also realize that it is unusual for v to be an integer. Since the number of degrees of freedom associated with a t distribution must be a positive integer, one suggestion is to round the calculated value of v down to the nearest positive integer.

EXAMPLE 6.11.4 A manager of a store that sells sporting goods is interested in determining the hours that will be most profitable. Two different schedules are under consideration, early morning and late afternoon. There is no preconceived notion as to which schedule is better. The problem is to estimate the difference in mean sales for the two schedules via a 95% confidence interval. The following data are obtained on the number of people that came to the center daily in the last 25 days.

$$\text{Morning}: N_1 = 25, \bar{x}_1 = 570, s_1^2 = 1,600,$$
$$\text{Afternoon}: N_2 = 25, \bar{x}_2 = 600, s_2^2 = 625.$$

SOLUTION: Since $s_1^2/s_2^2 = 1,600/625 = 2.56$, we do have evidence that the population variances are different. We use the above procedure to compute the number of degrees of freedom as

$$v = \frac{[(s_1^2/N_1) + (s_2^2/N_2)]^2}{[(s_1^2/N_1)^2/N_1 - 1)] + [(s_2^2/n_2)^2/N_2 - 1)]}$$

$$= \frac{[(1,600/25) + (625/25)]^2}{[(1,600/25)^2/24] + [(625/25)^2/24]} = 40.27 \approx 40.$$

A 95% confidence interval for $\mu_1 - \mu_2$ is then given by

$$(57 - 600) \pm 2.021 \sqrt{\frac{1,600}{25} + \frac{625}{25}} \Rightarrow (-49.07, -10.93).$$

Since 0 does not lie in this interval, we can conclude that the afternoon schedule produces a higher average sale than the morning schedule.

EXAMPLE 6.11.5 The age at which players in a certain sport turn professional has decreased in recent years. To see whether this is the case, 25 players between the ages of 30 and 40 are randomly selected and the age at which they turned professional is determined. It is found that $\bar{x}_1 = 22.3$ and $s_1^2 = 4.52$. Similar statistics for 25 players between the ages of 16 and 30 reveal $\bar{x}_2 = 18.7$ and $s_2^2 = 2.00$. Let us test

$$H_0: \mu_1 - \mu_2 = 0 \quad vs \quad H_1: \mu_1 - \mu_2 > 0.$$

To do so, we first consider the ratio $s_1^2/s_2^2 = 4.52/2.00 = 2.26$. This shows that $\sigma_1^2 \neq \sigma_2^2$. Thus we will not pool variances. The number of degrees of freedom associated with T is

$$v = \frac{[(4.52/25) + (2.00/25)]^2}{[(4.52/25)^2/24] + [(2.00/25)^2/24]} = 42.5 \approx 42,$$

$$T = \frac{(22.3 - 18.7)}{\sqrt{452/25) + (2/25)}} = 7.05 > t_{0.99,(42)} = 2.418.$$

We reject H_0 and conclude that players today do tend to turn professional at an earlier age than in the past at 1% level of significance.

As pointed out, the procedure described here for comparing population means is somewhat controversial and differences of opinion do exist.

A RESEARCH PROBLEM In recent years, more and more basketball players skip college and join the NBA. Collect data and follow the steps of the Example 6.11.5 to draw an appropriate conclusion.

6.11.4 Inference About the Difference Between Two Population Means: Paired Difference Experiment

In Section 6.11.2 and 6.11.3, we assumed that there was no correlation between measurements from the two groups involved (independent samples). In other words, data were randomly selected from the populations and were independent of each other. In many situations such as a pre-post (before and after) tests are applied to a group of subjects twice, and therefore the measurements are no longer independent. Dependent samples assume that there is a correlation between the measurements. When subjects are measured twice, or when to reduce variability subjects are matched before experimentation, the collected data from dependent samples.

The next example, presents inference about the difference between two population means using the paired difference experiments. One advantage of pair difference experiment is that, since the same subjects are used twice, indirectly factors other than the one investigated are indirectly controlled.

EXAMPLE 6.11.6 To compare the popularity of two sports channels, five sport fans were interviewed. The average amount of time per month they spend watching each of these two channels are listed in the following table:

Fan	Channel A	Channel B	Difference x_D
1	14	12.3	1.7
2	20	17.8	2.2
3	21.4	20.9	0.5
4	27.1	25.7	1.4
5	35.6	33.0	2.6

$$\bar{x}_D = 1.68$$
$$s_D = 0.8044$$
$$N_D = 5$$

Can we conclude (at $\alpha = 0.01$) that the mean time spent watching Channel A is greater than that of Channel B?

First note that, unlike the previous example, these samples is not independent. This is because we are using the same people (fans) and the time they spend watching either channel is not independent. Clearly, while watching Channel A, fans are not watching Channel B. Thus, the procedure based on the assumption of independent samples cannot be used. For this case, a valid test may be carried out using a single sample t-test described below. Let

$$\mu_D = \mu_1 - \mu_2 = \text{Population mean of differences,}$$
$$\bar{x}_D = \text{Sample mean of differences,}$$
$$s_D = \text{Sample standard deviation of differences,}$$
$$N_D = \text{Number of pairs.}$$

We need to make the following assumptions:

1. The population of differences is approximately normal.
2. The sample differences are randomly selected from the population of differences. Here we do not need to assume $\sigma_1^2 = \sigma_2^2$.

With these assumptions, like a single sample case, we use the test quantity

$$T = \frac{\bar{x}_D - \mu_D}{s_D/\sqrt{N_D}}$$

and construct a paired difference confidence interval for $\mu_D = \mu_1 - \mu_2$ as

$$\bar{X}_D \pm t_{1-\alpha/2,(N_D-1)}(S_D/\sqrt{N_D}).$$

The rejection regions for the different alternative hypotheses are given below. In all cases, the degrees of freedom is $N_D - 1$.

H_0	H_1	Rejection Region
$\mu_D = \mu_1 - \mu_2 = D_0$	$\mu_D = \mu_1 - \mu_2 > D_0$	$T > t_{1-\alpha}$
$\mu_D = \mu_1 - \mu_2 = D_0$	$\mu_D = \mu_1 - \mu_2 < D_0$	$T < -t_{1-\alpha} = t_\alpha$
$\mu_D = \mu_1 - \mu_2 = D_0$	$\mu_D = \mu_1 - \mu_2 \neq D_0$	$T > t_{1-\alpha/2}$ or $T < -t_{1-\alpha/2} = t_{\alpha/2}$

SOLUTION: (Example 6.11.6)

$$H_0: \mu_D = 0 \quad vs \quad H_1: \mu_D > 0$$

and the test quantity is

$$T = \frac{\bar{x}_D - 0}{s_D/\sqrt{N_D}} = \frac{1.68}{0.8044/\sqrt{5}} = 4.67.$$

The degrees of freedom is $N_D - 1 = 4$, $\alpha = 0.01$, and hence $t_{0.99,(4)} = 3.747$. Since $T > 3,747$, we reject H_0. Thus, we conclude that the mean time spent watching Channel A is greater than that for Channel B at $\alpha = 0.01$. In other words, Channel A has been more popular.

A 95% confidence interval for $\mu_D = \mu_1 - \mu_2$, the difference between the average times spent watching Channels A and B, is

$$\bar{x}_D \pm t_{0.975,(4)}(s_D/\sqrt{N_D}) = 1.68 \pm 2.776(0.8044/\sqrt{5}) = 1.68 \pm 0.9986 \rightarrow (0.681, 2.679).$$

Note that this interval does not include 0 leading to the same conclusion.

EXAMPLE 6.11.7 In a certain country there has been some discussion among psychologists that the athletic ability of a first-born child tends to be slightly higher than that of the second-born child. To test the theory, observations are obtained on 200 siblings using series of tests that measure athletic ability. The data obtained, led to $\bar{x}_D = 4.3$ and $s_D = 2$. Test the appropriate hypotheses and report your findings.

SOLUTION: $H_0: \mu_D = 0$, vs $H_1: \mu_D > 0$.
The test quantity is

$$T = \frac{\bar{x}_D - 0}{s_D/\sqrt{N_D}} = \frac{4.3 - 0}{2/\sqrt{200}} = 30.406.$$

Since this is larger than $t_{0.99,(199)} = 2.326$ we reject H_0. There is sufficient evidence to indicate that first-born children tend to be more athletic than second-born children at 1% level (in fact at any level) of significance (p-value is zero).

EXAMPLE 6.11.8 Home court advantage is considered to be an important factor for the success of NBA teams. Both experts and fans believe in this theory. In recent years, there has been some concern among the fans that officials are intimidated by the home crowd and tend to call more fouls on the visitors than on the home teams. To provide support for this point of view, 20 games are randomly selected and the number of fouls called on the home teams and the visitors are recorded. The data are summarized in the following table:

Home	Visitor	Home	Visitor
15	18	10	9
21	24	9	13
12	14	7	9
15	15	18	17
19	24	16	16
8	17	21	22
23	20	23	26
20	20	19	18
30	27	15	23
32	35	6	10

Do these data support the contention that officials tend to call more fouls on the visitors than on the home teams? Use $\alpha = 0.01$.

SOLUTION:

$$H_0 : \mu_D = 0 \quad vs \quad H_1 : \mu_D < 0.$$

The test quantity is ($\bar{x}_D = -2$, $s_D = 3.2$)

$$T = \frac{-2 - 0}{3.2/\sqrt{20}} = -2.795.$$

This is less than $-t_{0.99,(19)} = -2.539$. Thus, we reject H_0 and conclude that the average number of fouls called on the home teams is less than that for the visitors at 0.01 level of significance.

A PRACTICE PROBLEM Two procedures are used to manufacture golf balls. Method A utilizes a liquid center and method B, a solid center. To compare the distances, 12 golfers were allowed to drive a ball of each type, and the length of the drives (in yards) were measured. The following results were obtained:

Type A: 188.0 215.8 140.6 182.7 193.8 100.2 195.2 117.6 199.0 179.5 122.3 106.7
Type B: 172.7 202.5 128.1 173.9 180.7   88.7 188.9 108.8 186.5 175.9 112.7   99.8

Show that a 90% confidence interval on mean difference is (7.9, 11.14). Is there reason to believe that one ball tends to yield greater distances on the drive? Test this using $\alpha = 0.05$.

A RESEARCH PROBLEM Test the home advantage theory for a sport of your choice. For example, you may collect NBA playoff data.

6.11.5 Fitness Test

One of the interesting questions faced by exercise physiologists is how to measure fitness. It is well-known that the rate of a person's heartbeat and the time required for that rate to return to normal upon cessation of exercise are partly determined by the person's physical condition. A person who is physically well-conditioned will also be less affected by a given amount of exercise than a poorly conditioned person. The fitter the person is, the faster his or her pulse rate will return to normal after exercise. Since it is very easy to measure person's pulse rate, many fitness tests have been developed, which are based on person's pulse rate recovery after exercise.

The pulse ratio tests, as they are called, involve stepping up and down at a specified rate off a box of a specified height. One such test is the Harvard Step Test, originally designed for use with male students at Harvard University. It is a very straightforward test to administer and consequently it has been adapted for groups other than male students. The values used for the height of the box and rate are dependent upon the age and sex of the person since values suitable for an adult would be inappropriate for a child.

The test for adult males is as follows: The subject steps up and down off a box 20 inches high at a rate of 30 times per minute. When the subject steps up onto the box, he must attain a position in which the body is erect; crouching is not permitted. The stepping procedure involves four stages: left foot is placed on the box, the right foot is placed on the box, then the left foot is placed on the floor, and the right foot is placed on the floor. The person is permitted to change the order of the feet provided that the order of the four stages and the rate of stepping are maintained.

The stepping continues for 5 min unless exhaustion is reached previously. In either case, the duration of stepping is recorded. Immediately after completion the person sits down in a chair and three pulse counts are taken (at the wrist) at the following times: 1 to $1\frac{1}{2}$, 2 to $2\frac{1}{2}$, and 3 to $3\frac{1}{2}$ min after stepping ceased.

The person's Harvard Index is then obtained from the formula

$$\text{Harvard Index} = \frac{\text{Duration of exercise in seconds} \times 100}{2 \times (\text{Sum of three pulse counts during recovery})}.$$

The physical condition, or fitness, of the subject is then determined according to the following scale. If the Harvard Index is greater than 90, level of fitness is excellent, 80-89 is good, 55-79 is average, and less than 55 is poor. As an example, suppose that a Harvard step test using a highly trained marathon runner as a subject gave the following three pulse counts: 50, 47, and 45. The Harvard Index is therefore

$$\frac{300 \times 100}{2(50+47+45)} = 105.63$$

so we conclude that the subject was extremely fit. Marathon runners can be expected to score high values of the Harvard Index since the index is a measure of recovery from prolonged exercise and the marathon is certainly a prolonged exercise.

EXAMPLE The manager of a basketball team wants to know how fit his players are. He records their indicates prior to, and on completion of a 1-month period of pre-season training. He thinks that training should have improved the players' fitness. For the 31 players on the team he records the difference between the indices after and before training and finds that

$$\bar{x}_D = 20.99, \quad s_D = 11.69.$$

He wants to test

$$H_0: \mu_D = 0 \ vs \ H_1: \mu_D > 0.$$

The test quantity is

$$T = \frac{20.99 \times \sqrt{31}}{11.69} = 9.997.$$

The two levels of significance commonly used are the 5% and 1%, which in the case of a one-tailed test and 30 degrees of freedom, give the critical values of t as 1.697 and 2.457, respectively (note that we can use z criticals as approximation).

The difference is thus highly significant. As anticipated there is a definite improvement in the fitness of the players as a result of their training program. The manager can therefore conclude that he has a worthwhile training program.

6.11.6 Inference About the Difference Between Two Population Proportions

Finally consider the inference about the difference between population proportions based on two independent binomial experiments. Again we do this through an example.

EXAMPLE 6.11.9 Suppose that a sample of male students (M) and a sample of female students (F) were polled on an issue regarding a change on policy regarding student athletes. Of 200 female students, 90 were in favor of the change and 110 were against it. Of 100 male students, 58 were for change and 42 were not. Can we conclude that more male students are in favor of the change than female students at a 1% level of significance?

Let X_1 and X_2 denote, respectively, the number of male and female students who are in favor of change. We have the following information:

Sample proportion for M:$\hat{p}_1 = X_1/N_1 = 58/100$;

Population proportion for M:p_1, which is unknown; Sample proportion for F:$\hat{p}_2 = X_2/N_2 = 90/200$; Population proportion for F:p_2, which is unknown; Hypothesis: H_0:$p_1 - p_2 = 0$ versus H_1:$p_1 - p_2 > 0$.

Since the required conditions, namely $N_1\hat{p}_1 = 58 > 5, N\hat{p}_2 = 90 > 5$, are satisfied we know that we can use the normal approximation to the binomial distribution. Thus, we have, $\hat{p}_1 \sim N(p_1, \sqrt{p_1 q_1/N_1})$ and $\hat{p}_2 \sim N(p_2, \sqrt{p_2 q_2/N_2})$. Here $q_i = 1 - p_i$ for $i = 1, 2$. Note that an equivalent condition, often referred to as condition for large sample, is that $\hat{p}_i \pm 3\sigma_{\hat{p}_i}$ does not include 0 or 1, for $i = 1, 2$. Checking the equivalent conditions is left as an exercise. Since the samples are independent, it follows that

$$\hat{p}_1 - \hat{p}_2 \sim N(p_1 - p_2, \sqrt{p_1 q_1/N_1 + p_2 q_2/N_2}).$$

Hence a large-sample $100(1 - \alpha)\%$ confidence interval for $p_1 - p_2$ (for two independent, samples form binomial distributions) can be constructed as

$$(\hat{p}_1 - \hat{p}_2) \pm z_{1-\alpha/2}\sqrt{p_1 q_1/N_1 + p_2 q_2/N_2} \approx (\hat{p}_1 - \hat{p}_2) \pm z_{1-\alpha/2}\sqrt{\hat{p}_1 \hat{q}_1/N_1 + \hat{p}_2 \hat{q}_2/N_2}.$$

Since $p_1 q_1$ and $p_2 q_2$ are unknown, we replace them with their estimates $\hat{p}_1 \hat{q}_1$ and $\hat{p}_2 \hat{q}_2$. Fortunately pq is relatively insensitive to the value chosen to approximate p. Referring to the above example, the point estimate of $p_1 - p_2$ is $\hat{p}_1 - \hat{p}_2 = 116/200 - 90/200 = 26/200 = 0.13$. A 95% confidence interval is then

$$0.13 \pm 1.96\sqrt{\frac{(0.58)(0.42)}{100} + \frac{(0.45)(0.55)}{200}} \rightarrow (0.122, 0.138).$$

Note that this interval does not include zero indicating that H_0 cannot be rejected. In what follows we will confirm this by testing an appropriate hypothesis.

The large-sample test of hypothesis for two independent samples from two binomial distributions can be carried out using the following steps:

Test quantity is

$$T = \frac{\hat{p}_1 - \hat{p}_2}{\sqrt{p_1 q_1/N_1 + p_2 q_2/N_2}} \approx \frac{\hat{p}_1 - \hat{p}_2}{\sqrt{\hat{p}\hat{q}(1/N_1 + 1/N_2)}}$$

where $\hat{p} = (X_1 + X_2)/(N_1 + N_2) = (N_1\hat{p}_1 + N_2\hat{p}_2)/(N_1 + N_2)$ is a pooled estimate of p (the common value of $p_1 = p_2 = p$ assuming that H_0 is true, that is $p_1 = p_2$). The rejection regions are given in the following table:

H_0	H_1	Rejection Region
$p_1 - p_2 = 0$	$p_1 - p_2 > 0$	$T > z_{1-\alpha}$
$p_1 - p_2 = 0$	$p_1 - p_2 < 0$	$T < -z_{1-\alpha} = z_\alpha$
$p_1 - p_2 = 0$	$p_1 - p_2 \neq 0$	$T > z_{1-\alpha/2}$ or $T < -z_{1-\alpha/2} = z_{\alpha/2}$

Note that the test can be adopted for $p_1 - p_2 = D_0$, but in practice, usually $D_0 = 0$.

SOLUTION: (Example 6.11.9)

$$H_0: p_1 - p_2 = 0 \ vs \ H_1: p_1 - p_2 > 0$$

with

$$\hat{p}_1 = x_1/N_1 = 58/100 = 0.58$$

$$\hat{p}_2 = x_2/N_2 = 90/200 = 0.45$$

and the pooled estimate

$$\hat{p} = (x_1 + x_2)/(N_1 + N_2) = (58 + 90)/(100 + 200) = 0.493.$$

Hence,

$$T = \frac{0.58 - 0.45}{\sqrt{(0.493)(0.507)(1/100 + 1/200)}} = 2.12.$$

The critical value is $z_{0.99} = 2.33$. Since $T < z_{0.99}$ we cannot reject H_0. There is not sufficient evidence to indicate that more male students (than female students) favor the change at the 1% level of significance. For this example, p-value $= P(z > 2.12) = P(z < -2.12) = 0.017$. Since this is greater than 0.01, H_0 cannot be rejected.

EXAMPLE 6.11.10 In the past two decades, there have been intensive anti-smoking campaigns sponsored by both federal and private agencies. Suppose that we randomly sampled 1,500 adults in 1992 and then sampled 2,000 adults in 2002 to determine whether there was evidence that the percentage of smokers had decreased. The results of the two-sample surveys are $x_1 = 576$ and $x_2 = 652$, respectively. Does this data indicate that the fraction of smokers decreased over this 10-year period? Use $\alpha = 0.05$.

Chapter 6: Statistical Testing

SOLUTION: If we define p_1 and p_2 as the true proportions of adult smokers in 1992 and 2002 the elements of the test are:

$$H_0: p_1 - p_2 = 0 \text{ vs } H_1: p_1 - p_2 > 0.$$

The test quantity is

$$T = (\hat{p}_1 - \hat{p}_2) / \sqrt{\hat{p}\hat{q}\left(\frac{1}{N_1} + \frac{1}{N_2}\right)}$$

and the rejection region is $T > z_{0.95} = 1.645$.

We can now calculate the sample proportions of smokers

$$\hat{p}_1 = \frac{576}{1,500} = 384, \quad \hat{p}_2 = \frac{652}{2,000} = 0.326, \quad \hat{p} = \frac{x_1 + x_2}{n_1 + n_2} = \frac{576 + 652}{1,500 + 2,000} = 0.351.$$

Thus

$$T = \frac{0.384 - 0.326}{\sqrt{(0.351)(0.649)\left(\frac{1}{1,500} + \frac{1}{2,000}\right)}} = \frac{0.058}{0.0164} = 3.56.$$

This is greater than 1.645. There is sufficient evidence to indicate that the proportion of adults who smoke has decreased over the 1992-2002 period at a 5% significance level. We can construct a confidence interval for $p_1 - p_2$ if we are interested in estimating the extent of the decrease. We leave this for students to try.

We end this section by noting that to estimate $\mu_1 - \mu_2$ and $p_1 - p_2$ within a given bound B with $100(1-\alpha)\%$ confidence, we can use the following formula when sample sizes are equal:

$$N_1 = N_2 \geq (z_{1-\alpha/2})^2 (\sigma_1^2 + \sigma_2^2)/B^2,$$

$$N_1 = N_2 \geq (z_{1-\alpha/2})^2 (p_1 q_1 + p_2 q_2)/B^2.$$

PRACTICE PROBLEMS

1. College basketball games (NCAA) in the United States have become extremely popular. One concern of fans is officiating. It has been claimed that officials are getting better each year. To substantiate this claim, 50 games of each of the two successive years were randomly selected and their films were carefully examined to find out officiating errors. Six errors were found in year 1 and four errors in year 2. Test the claim at $\alpha = 0.05$ and draw a conclusion. Set up a 90% confidence interval for the difference.

2. In a certain university, there has been some concern about academic performance of students who play for varsity teams. The athletic director claims that not only is there no ground for concern but, in fact, failure rate, p_1, among varsity athletes is lower than rate p_2 for other students. She asks a statistics professor to help her gain statistical support for this. This collaboration results in the following data $\hat{p}_1 = 30/150 = 0.2$, $\hat{p}_2 = 43/200 = 0.215$. Set up an appropriate hypothesis and test to see whether the director's claim can be supported. Use $\alpha = 0.05$.

6.11.7   Case Study: 2018-2019 NBA MVP

As of today when this section is written, the two leading candidates for the honor are the Milwaukee Bucks' Giannis Antetokounmpo and the Houston Rockets' James Harden. Let's analyze their performance and make a prediction of winner.

Through statistical methods, one can determine which one of the two players mentioned above deserved the honor. Here (as measures of performance) we have chosen five criteria from which we will base our decision. These are:

1. Average number of points scored per game and the consistency.
2. Average number of rebounds per game and the consistency.
3. Average number of assists per game and the consistency.
4. Average number of blocks per game and the consistency.
5. Average number of steals per game and the consistency.

Data are obtained from the official NBA website, www.nba.com. We have used a 0.05 level of significance for all tests. Note that here we have data for all the games played. Therefore, all we are doing here is to decide whether the differences between performance measures are significant assuming that data represent sample of the games that could have been played. For the tests involving population means, we have large samples that makes application of Central Limit Theorem possible. For test involving population variances, we are assuming that data come from normal populations and are independent.

Variables

N_1 = The number of games played by Giannis.
N_2 = The number of games played by Jamesn.
X_{P_1} = The number of points Giannis scored in a randomly selected game.
X_{P_2} = The number of points James scored in a randomly selected game.
X_{R_1} = The number of rebounds Giannis had in a randomly selected game.
X_{R_2} = The number of rebounds James had in a randomly selected game.
X_{A_1} = The number of assists Giannis had in a randomly selected game.
X_{A_2} = The number of assists James had in a randomly selected game.
X_{B_1} = The number of blocks Giannis had in a randomly selected game.
X_{B_2} = The number of blocks James had in a randomly selected game.
X_{S_1} = The number of steals Giannis had in a randomly selected game.
X_{S_2} = The number of steals James had in a randomly selected game.

Population parameters are defined accordingly. For example, $E(X_{P_1}) = \mu_{P_1}$ or $E(X_{S_2}) = \mu_{S_2}$, and so on.

First Performance Measure: Points

Giannis: $N_1 = 54$ Mean = 27.22 Standard Deviation = 7.76
James: $N_2 = 55$ Mean = 36.47 Standard Deviation = 9.84

$H_0: \mu_{P_1} - \mu_{P_2} = 0$ $H_1: \mu_{P_1} - \mu_{P_2} < 0$ Test Statistic: $T = -5.454$ p-value < 0.0001
Rejection Region: Reject H_0 if $T < -1.66$
Decision: Reject H_0
Conclusion: At 0.05 level of significance, there was sufficient evidence to indicate that $\mu_{P_1} < \mu_{P_2}$, that is, James's point average was higher.

$H_0: \sigma_{P_1}^2 / \sigma_{P_2}^2 = 1$ $H_1: \sigma_{P_1}^2 / \sigma_{P_2}^2 < 1$
Test Statistic: $T = 0.62$
Rejection Region: Reject H_0 if $T < 0.64 = F_{0.05,(53,54)}$
Decision: Reject H_0
Conclusion: At 0.05 level of significance, there was sufficient evidence to indicate that Giannis was a more consistent scorer.

Second Performance Measure: Rebounds

Giannis: $N_1 = 54$ Mean = 12.70 Standard Deviation = 3.94
James: $N_2 = 55$ Mean = 6.65 Standard Deviation = 3.15

$H_0: \mu_{R_1} - \mu_{R_2} = 0$ $H_1: \mu_{R_1} - \mu_{R_2} > 0$
Test Statistic: $T = 8.86$ p-value < 0.0001
Rejection Region: Reject H_0 if $T > 1.66$
Decision: Reject H_0
Conclusion: At 0.05 level of significance, there was sufficient evidence to indicate that Giannis's rebound average was higher.

$H_0: \sigma_{R_1}^2 / \sigma_{R_2}^2 = 1$ $H_1: \sigma_{R_1}^2 / \sigma_{R_2}^2 > 1$
Test Statistic: $T = 1.569$
Rejection region: Reject H_0 if $T > 1.573 = F_{0.95,(53,54)}$
Decision: Do not eject H_0
Conclusion: At 0.05 level of significance, there was not sufficient evidence to indicate that James was a more consistent rebounder.

Third Performance Measure: Assists

Giannis: $N_1 = 54$ Mean = 5.98 Standard Deviation = 2.62
James: $N_2 = 55$ Mean = 7.69 Standard Deviation = 3.66

$H_0: \mu_{A_1} - \mu_{A_2} = 0$ $H_1: \mu_{A_1} - \mu_{A_2} < 0$
Test Statistic: $T = -2.81$ p-value = 0.003
Rejection Region: Reject H_0 if $T < -1.66$
Decision: Reject H_0

Conclusion: At 0.05 level of significance, there was sufficient evidence to indicate that James's assist average was higher.

$H_0: \sigma_{A_1}^2 / \sigma_{A_2}^2 = 1 \quad H_1: \sigma_{A_1}^2 / \sigma_{A_2}^2 < 1$
Test Statistic: $T = 0.52$
Rejection Region: Reject H_0 if $T < 0.64 = F_{0.05, (53, 54)}$
Decision: Reject H_0
Conclusion: At 0.05 level of significance, there was sufficient evidence to indicate that Giannis was a more consistent assist provider.

Fourth Performance Measure: Blocks

Giannis: $N_1 = 54$ Mean = 1.447   Standard Deviation = 1.11
James: $N_2 = 55$ Mean = 0.76   Standard Deviation = 0.84

$H_0: \mu_{B_1} - \mu_{B_2} = 0 \quad H_1: \mu_{B_1} - \mu_{B_2} > 0$
Test Statistic: $T = 3.61$ p-value = 0.0002
Rejection Region: Reject H_0 if $T > 1.66$
Decision: Reject H_0
Conclusion: At 0.05 level of significance, there was sufficient evidence to indicate that Giannis's block average was higher.

$H_0: \sigma_{B_1}^2 / \sigma_{B_2}^2 = 1 \quad H_1: \sigma_{B_1}^2 / \sigma_{B_2}^2 > 1$
Test Statistic: $T = 1.76$
Rejection Region: Reject H_0 if $T > 1.573 = F_{0.95, (53, 54)}$
Decision: Reject H_0
Conclusion: At 0.05 level of significance, there was sufficient evidence to indicate that James was a more consistent blocker.

Fifth Performance Measure: Steals

Giannis: $N_1 = 54$ Mean = 1.41   Standard Deviation = 1.6
James: $N_2 = 55$ Mean = 2.15   Standard Deviation = 2.27

$H_0: \mu_{S_1} - \mu_{S_2} = 0 \quad H_1: \mu_{S_1} - \mu_{S_2} < 0$
Test Statistic: $T = -2.76$ p-value = 0.0034
Rejection Region: Reject H_0 if $T < -1.66$
Decision: Reject H_0
Conclusion: At 0.05 level of significance, there was sufficient evidence to indicate that James's steal average was higher.

$H_0: \sigma_{S_1}^2 / \sigma_{S_2}^2 = 1 \quad H_1: \sigma_{S_1}^2 / \sigma_{S_2}^2 < 1$
Test Statistic: $T = 0.705$
Rejection Region: Reject H_0 if $T < 0.64 = F_{0.05, (53, 54)}$
Decision: Do not reject H_0
Conclusion: At 0.05 level of significance, there was not sufficient evidence to indicate that Giannis was a more consistent steal maker.

Chapter 6: Statistical Testing

Summary: Who Did Better in Each Category

Criteria	Average	Consistency
Points	James	Giannis
Rebounds	Giannis	–
Assists	James	Giannis
Blocks	Giannis	James
Steals	James	Giannis

Giannis won in 5 of the 10 categories. James won in 4 of the 10 categories. There was no significant difference in 1 of the 10 categories.

Based on these results, Giannis will be a better candidate for the MVP of the 2018-2019 NBA regular season.

A PRACTICE PROBLEM For all the cases tested in Section 6.11.7, construct 90% confidence intervals for the difference between the means and the ratio of two variances. Use these intervals to confirm the test results.

6.12 DISTRIBUTION FREE χ^2-TEST

Most of the tests presented so far were developed for numerical or quantitative data. This section and Section 6.13 are devoted to development of two statistical techniques for categorical or qualitative data. The first is a goodness-of-fit test, which is a generalization of a binomial experiment. It is applicable to data produced by a multinomial experiment. The second is applicable to data arranged in a table called contingency table based on two attributes. Its purpose is to determine whether or not two classifications of a population of qualitative data are statistically independent. The sampling distribution of the test statistics in both cases is the chi-square, discussed in Chapter 3.

Up to now we always assumed that X is a normally distributed random variable (or we have large samples) and considered parameter or parametric tests. Tests for checking hypotheses on general distributions with nothing assumed about their form are called non-parametric or distribution free. These methods do not rely on the form of a special distribution such as $N(\mu, \sigma^2)$ and hence results obtained are valid for all possible distributions. The most important of these tests is the χ^2-test, discussed below. Non-parametric statistical inference will be discussed in more detail in Chapter 10.

Let H (or H_0) represent the hypothesis that X has a specified density function $f(x)$. For example, it may be hypothesized that X has a uniform distribution. Our objective is to find, through testing, if available data support this claim. To apply this test we decompose the range of the values of X into finitely many, say M, segments $I_1, \ldots, I_M$. The numbers

$$p_i = \int_{I_i} f(x)dx, \quad i = 1, \ldots, M \tag{6.9}$$

are by definition the (theoretical) probabilities of the events, "X takes a value in I_i." In the case where X is discrete, we can (essentially) use the probability mass function of X to define the p_i's. If $X_1, \ldots, X_N$ is a random sample on X then, based on the relative frequency definition of probability, Np_i values in the sample are expected to fall in I_i. In this test the theoretical frequencies Np_i are compared with the observed or empirical (experimental) frequencies $J_1, \ldots, J_M$, found in the sample. Suppose that J_i is the number of the values among $X_1, \ldots, X_N$, which lie in the interval I_i. If H is true, then we expect the discrepancies between the observed and theoretical frequencies $J_i - Np_i$ to be small. It can be shown that as $N \to \infty$ the test statistic

$$T(X_1,\ldots,X_N) = \sum_{i=1}^{M} \frac{(J_i - Np_i)^2}{Np_i}$$

$$= \sum \frac{(\text{Observed Frequency} - \text{Expected Frequency})^2}{\text{Expected Frequency}} \qquad (6.10)$$

is distributed approximately as $\chi^2_{(M-1)}$. For practical purposes it suffices to choose N and I_i in such a way that the theoretical frequencies Np_i are not smaller than 5. So, we can work with $\chi^2_{(M-1)}$ as the test statistic. The critical values of $\chi^2_{1-\alpha,(M-1)}$ define the critical region. In other words, $T > \chi^2_{1-\alpha,(M-1)}$ defines the region for rejecting H, where α is the level of significance of the test. The probability of rejecting H despite H being true is then

$$P(T > \chi^2_{1-\alpha,(M-1)}) \approx \alpha. \qquad (6.11)$$

EXAMPLE 6.12.1 How are the basketball players' ages distributed? Consider all players who played in the 2018-2019 NBA season. We want to test a hypothesis that the age is uniformly distributed between 19 and 42. That is, we almost have the same number of players in segment(subsegments) of equal length. We will use the data given below and a significance level of 10%.

Age	19	20	21	22	23	24	25	26	27	28	29	30
Number of Players	10	23	35	42	54	49	44	50	30	37	26	32

Age	31	32	33	34	35	36	37	38	39	40	41	42
Number of Players	16	19	16	8	5	3	3	3	0	1	0	1

SOLUTION: Let X denote the age of a randomly selected player. We want to test if X is uniformly distributed over the interval [19,43) using the given data. For this, we decompose the interval into 24 segments [19,20), [20,21), ..., [42,43). Assuming that the hypothesis of uniform distribution is true, p_i defined in (6.9) equals 0.042 for all $i = 1, 2, \ldots, 24$. Also J_i, $i = 1, 2, \ldots, 24$, are equal to the number of players in the above table, and the sample size N equals the sum of J_i's, that is, $N = 507$. If the distribution of age is uniform then, we expect to have 21.125 players in each segment. Note

Chapter 6: Statistical Testing

that this is a theoretical expectation, or average, although practically not possible. The test statistic using (6.10) is then

$$T = \sum_{i=1}^{24} \frac{(J_i - 507 \times 0.042)^2}{507 \times 0.042} = \frac{(10 - 21.125)^2}{21.125} + \cdots + \frac{(1 - 21.125)^2}{21.125} = 7620.6/21.125 = 360.74.$$

Also the critical value is $\chi^2_{0.90,(24-1)} = 32.01$. Since the value of T is much larger than 32.01, we reject the hypothesis that X has a uniform distribution over the interval $[19, 43)$ at 10% significant level.

EXAMPLE 6.12.2 The sports director of a small high school knows from the past data that 20% of male students like basketball, 25% tennis, 30% soccer, 15% football, and 10% baseball. From 100 freshmen entering the school 18 like basketball, 21 tennis, 27 soccer, 19 football, and 17 baseball. Are this year's freshmen different from past students in their sports preference?

SOLUTION: We would like to test

H_0: Distribution of this year's freshmen liking of sports is the same as previous years

H_1: H_0 is not true

	Basketball	Tennis	Soccer	Football	Baseball
Expected Frequencies	20	25	30	15	10
Observed Frequencies	18	21	27	19	17

$$T = \frac{(18-20)^2}{20} + \cdots + \frac{(17-10)^2}{10} = 7.11.$$

The critical value is $\chi^2_{0.90,(4)} = 7.78$. So H_0 cannot be rejected. There is not sufficient evidence, at the 10% significant level, to indicate that the distribution of this year's freshmen liking of sports differs from previous years.

EXAMPLE 6.12.3 Kim was a 75% free-throw shooter last year. After a summer of rest she is back to practice. She wants to know whether her free-throw percentage is still the same. She tries 100 free throws in groups of 4 (25 groups) and records the number of hits. Suppose that the counts are

$$2, 1, 4, 2, 3, 3, 3, 2, 1, 2, 2, 1, 3, 4, 0, 3, 3, 2, 2, 2, 4, 3, 2, 3, 1.$$

Test an appropriate hypothesis and draw conclusions at 5% level of significance.

SOLUTION: We would like to test

H_0: Number of hits has a binomial distribution with $p = 0.75$ *versus*
H_1: H_0 is not true

The frequencies are given below

Expected	3	3	3	3	3	3	3	3	3	3	3	3	3	3	3	3	3	3	3	3	3	3	3	3	3
Observed	2	1	4	2	2	2	2	2	1	2	2	1	3	4	0	3	3	2	2	2	4	3	2	3	1

$$T = \frac{(3-2)^2}{3} + \frac{(3-1)^2}{3} + \cdots + \frac{(3-1)^2}{3} = \frac{37}{3} = 12.33.$$

The rejection region is $T > \chi^2_{0.95,(24)} = 36.415$. Since $12.33 < 36.45$, H_0 cannot be rejected. This means that she does not have sufficient evidence to indicate that here free-throw percentage has changed at 5% level of significance.

A PRACTICE PROBLEM A basketball coach claims that the starters in his team are equally good free-throw shooters. During the practice, each of the starters tried 40 free-throws and made, respectively, 28, 25, 22, 31, and 29 shots. Apply the chi-square test and draw a conclusion using $\alpha = 0.05$.

6.13 TEST OF INDEPENDENCE, CONTINGENCY TABLES

In this section, we will, for the first time, use bivariate statistics, which is a special case of multivariate statistics. To clarify what this means, consider a random variable X, which assigns to each member of the population a pair of numbers rather than only one number. For example, for each soccer game the number of shots intended and the number of goals scored are recorded, or for each basketball player the height and weight are measured. In order to distinguish this from univariate random variable we present the bivariate random variable as $X = (X^{(1)}, X^{(2)})$.

There are many interesting results regarding the bivariate random variables. In this section, the question of interest is to determine whether $X^{(1)}$ and $X^{(2)}$ are independent random variables. Recall that independence means that the multiplication of the probabilities should hold. That is, we should have

$$\begin{aligned} p_{ik} &= P(X^{(1)} = i, X^{(2)} = k) \\ &= P(X^{(1)} = i) \cdot P(X^{(2)} = k) \\ &= p_{i.}p_{.k} \end{aligned} \quad (6.12)$$

where $p_{i.} = P(X^{(1)} = i)$ and $p_{.k} = P(X^{(2)} = k)$. To test the hypothesis of "independence" we use the χ^2–test.

As in Section 6.12, first the range of the random variables $X^{(1)}$ and $X^{(2)}$ are decomposed into R and S classes or segments, respectively (see Table 6.13.1), assuming that $X^{(1)}$ and $X^{(2)}$ are continuous. Suppose from a realization of a sample on $(X_1^{(1)}, X_1^{(2)}), \ldots, (X_N^{(1)}, X_N^{(2)})$ we get N_{ik} sample points $(X_j^{(1)}, X_j^{(2)})$ falling into the i-th class with respect to $X^{(1)}$ and into the k-th class with respect to $X^{(2)}$. Let $N_{i.} = \sum_{k=1}^{S} N_{ik}$ and $N_{.k} = \sum_{i=1}^{R} N_{ik}$ represent, respectively, the sums of rows and columns in the rectangular scheme, called a contingency table. If independence of $X^{(1)}$ and $X^{(2)}$ holds, then we

Chapter 6: Statistical Testing

TABLE 6.13.1 Contingency Table

i \ k	1	2	...	S	
1	N_{11}	N_{12}	...	N_{1S}	$N_{1.}$
2	N_{21}	N_{22}	...	N_{2S}	$N_{2.}$
.	.	.	.	.	.
.	.	.	.	.	.
R	N_{R1}	N_{R2}	...	N_{RS}	$N_{R.}$
	$N_{.1}$	$N_{.2}$	...	$N_{.S}$	N

should have $p_{ik} = p_{i.} p_{.k}$. Using the relative frequencies $\frac{N_{ik}}{N}$, $\frac{N_{i.}}{N}$, $\frac{N_{.k}}{N}$ as estimates of the probabilities p_{ik}, $p_{i.}$, and $p_{.k}$, respectively, we should then expect to have

$$\frac{N_{ik}}{N} \approx \frac{N_{i.}}{N} \cdot \frac{N_{.k}}{N}. \tag{6.13}$$

The test statistics

$$T = N \sum_{i=1}^{R} \sum_{k=1}^{S} \frac{(\frac{N_{ik}}{N} - \frac{N_{i.}}{N} \cdot \frac{N_{.k}}{N})^2}{\frac{N_{i.}}{N} \cdot \frac{N_{.k}}{N}} = \sum_{i,k} \frac{(N_{ik} - N_{i.} N_{.k}/N)^2}{N_{i.} N_{.k}/N} \tag{6.14}$$

is distributed approximately as $\chi^2_{(f)}$ with $f = (R-1)(S-1)$ degrees of freedom. Therefore, the rejection region for the testing of the hypothesis of independency of $X^{(1)}$, $X^{(2)}$ is

$$T > \chi^2_{1-\alpha,(f)}.$$

EXAMPLE 6.13.1 Of 24 football players 25 years of age or over, 4 had an injury last year. Of 36 football players under 25 years of age, 16 had injuries in the same period. Is there any relationship between age and number of injuries ($\alpha = 0.05$)? In other words, are these two classifications dependent?

	Injury	No Injury	Row Total
Under 25	16	20	36
25 or older	4	20	24
Column Total	20	40	60

SOLUTION: First note that here we have

$$N_{11} = 16, \ N_{12} = 20, \ N_{21} = 4, \ N_{22} = 20,$$

$$N_{1.} N_{.1}/N = (36)(20)/60 = 12, \ N_{1.} N_{.2}/N = (36)(40)/60 = 24,$$

$$N_{2.} N_{.1}/N = 8, \ N_{2.} N_{.2}/N = 16.$$

Hence, the test statistics

$$T = \sum \frac{(\text{Observed Frequency} - \text{Expected Frequency})^2}{\text{Expected Frequency}}$$

$$= \frac{(16-12)^2}{12} + \frac{(20-24)^2}{24} + \frac{(4-8)^2}{8} + \frac{(20-16)^2}{16} = 5.$$

Here the degrees of freedom is equal to $(2-1)(2-1) = 1$, and $\chi^2_{0.95,(1)} = 3.84$. Since $\chi^2 > 3.84$, we reject H_0 (age and number of injuries are independent) and conclude that age and number of injuries are dependent. In other words, the hypothesis of "two classifications are independent" is rejected at 5% significance level.

EXAMPLE 6.13.2 The management of a certain major league baseball team wants to find out whether there are differences in support for the team among various age groups. The information below was collected during interviews with fans selected at random. Based on this data, can we conclude that there is a relationship between age and number of games attended per year, or are the two classifications independent? Use a significance level of 5%.

		Number of Games Attended Per Year			
		1 or 2	3 to 5	Over 5	Total
	Under 20	78	107	17	202
	21 – 30	147	87	13	247
Fans' ages	31 – 40	129	86	19	234
	41 – 55	55	103	40	198
	Over 55	23	74	22	119
	Total	432	457	111	1,000

SOLUTION: Let $X^{(1)}$ and $X^{(2)}$ denote the age of a randomly selected fan and the number of games he or she attended per year, respectively. Let H_0 be the hypothesis that $X^{(1)}$ and $X^{(2)}$ are independent and H_1 the hypothesis that H_0 is not true. If the above data lead to a rejection of H_0 at the given significance level 5%, then we can conclude that there is a relationship (dependency) between age and the number of games attended per year.

The range of $X^{(1)}$ is decomposed into five segments: under 20, 21–30, 31–40, 41–55, and over 55. The range of $X^{(2)}$ is divided into three segments: 1–2, 3–5, and over 5. Hence, we have $R = 5$, $S = 3$, $N = 1,000$, $N_{11} = 78$, and so on. Using (6.14) we compute the test statistic T as

$$T = \sum_{i=1}^{5} \sum_{k=1}^{3} \frac{(N_{ik} - N_{i.}N_{.k}/1,000)^2}{N_{i.}N_{.k}/1,000} = 103.08.$$

The degrees of freedom is $f = (5-1)(3-1) = 8$. Thus, the critical value is $\chi^2_{0.95,(8)} = 15.51$. The value of T is way above the critical value, so we reject H_0. That is, there is a dependency between age and the number of games attended per year.

A PRACTICE PROBLEM In many sports people refer to "home field or home court advantage" as being a reliable predictor for games. In other words, it is thought that the team who is playing at home will have a better chance of winning. The following data focus on the home court advantage

Chapter 6: Statistical Testing

for the NBA team Atlanta Hawks. The table summarizes the win/loss data for the Atlanta Hawks from the 1990/1991 seasons through the 2003/2004 seasons and is split by home and away games. (http://www.bballsports.com).

	Win	Loss	Total
Home	340	208	548
Away	205	353	558
Total	545	561	1,106

Show that the value of test statistics is 70.82 and test the dependency of the two classifications at a 5% level of significance.

6.13.1 Further Discussions on "Hot Hand" in Sports

In Section 2.7, we used basic probability concepts to study the "Hot Hand" phenomenon in sports, and particularly in basketball. The belief in "Hot Hand" is largely based on the assumption that "success breeds success." In Section 2.7, however, we expressed our doubts about this assumption, and presented data showing that this assumption may or may not be valid. In this section, we continue the discussion by using the contingency table. This will help us to test whether future and past successes could be considered independent. The data presenting the performance of nine members of the Boston Celtics described in Tables 6.13.3 and 6.13.4 are used for the analysis.

We conduct this study in two steps. First, we show that instead of χ^2 distribution, we can use the standard normal distribution to perform the test like the one presented in an article Wardrop (1998) on basketball appeared in the book "Statistics in Sports" edited by Jay Bennett, Arnold and Oxford University Press, New York. Next we will use χ^2 distribution to conduct a test similar to those introduced in this chapter. The analysis presented here is essentially due to Wardrop (1995).

To begin with, consider once more the question of "Hot Hand" in basketball. We analyze the pattern of free-throws to find out whether the outcomes of the first and the second free throws are independent or dependent. We have a contingency table of the form

The null hypothesis is, "the outcome of the second shot is statistically independent of the outcome of the first shot," and the alternative is that the null hypothesis is not true. It is interesting to note that if the null hypothesis is true, then conditional on the values of the row and column totals, it can be shown that the cell count N_{11} has a hypergeometric distribution with

$$E(N_{11}) = N_{1.}N_{0.1}/N, \quad Var(N_{11}) = \frac{N_{1.}N_{2.}N_{0.1}N_{0.2}}{(N-1)N^2}.$$

Also, under the null hypothesis the distribution of

$$z = \frac{N_{11} - E(N_{11})}{\sqrt{Var(N_{11})}}$$

can be approximated by the standard normal curve. Now for Larry Bird, $N_{11} = 251$, $E(N_{11}) = 252.12$, and $Var(N_{11}) = 4.575$. Hence his z-score is -0.52. This leads to a p-value equal to 0.3015. Thus, the null hypothesis cannot be rejected at any reasonable significance level.

Table 6.13.3 lists the outcomes of successive free throws of nine members of the Boston Celtics. The last column shows the z-scores for each player. From this table, Wardrop has made the following observations:

1. McHale provides the strongest evidence in support of the "Hot Hand" theory.
2. Carr provides the strongest evidence in support of an inverse relationship between the outcomes of the two shots (note that the z- score is negative).
3. Four players shot better after a hit ($z > 0$).
4. Five players shot better after a miss ($z < 0$).
5. To summarize, separate analysis of individual players performed by Wardrop indicate that four players shot better after a hit and five players shot better after a miss, but none of the individual players' pattern is convincing. By contrast, the analysis of the collapsed data gives statistically significant evidence in support of the hot hand phenomenon.

Let us now summarize the observed frequencies for the free-throw data collapsed over the nine Celtics players (Table 6.13.4). This provides a contingency table similar to Table 6.13.2. We can now

TABLE 6.13.2 Contingency Table for Free Throws

First Shot	Second Shot		Row Total
	Hit	Miss	
Hit	N_{11}	N_{12}	$N_{1.}$
Miss	N_{21}	N_{22}	$N_{2.}$
Column Total	$N_{0.1}$	$N_{0.2}$	N

TABLE 6.13.3 Outcomes of Successive Shots for Nine Members of the Boston Celtics (Wardrop)

Player	N_{11}	$E(N_{11})$	$\text{Var}(N_{11})$	z
Kevin McHale	93	88.23	7.633	1.73
Cedric Maxwell	245	240.20	14.667	1.25
Robert Parish	164	160.75	13.061	0.90
Nate Archibald	203	202.26	8.38	0.26
Rick Robey	54	54.81	10.257	-0.25
Gerald Henderson	77	77.58	4.858	-0.26
Larry Bird	251	252.12	4.575	-0.52
Chris Ford	36	37.03	3.1	-0.58
M. L. Carr	39	41.20	3.62	-1.16
Collapsed Data	1,162	1,143.03	72,015	2.24

Chapter 6: Statistical Testing

TABLE 6.13.4 Collapsed Data

First Shot	Second Shot		
	Hit	Miss	Row Total
Hit	1,162	311	1,473
Miss	428	148	576
Column Total	1,590	459	2,049

conduct a χ^2 test for independence, and see whether we can confirm the evidence collapsed data provided in favor of hot hand.

Here we have

$$N_{11} = 1,162, \quad N_{12} = 311, \quad N_{21} = 428, \quad N_{22} = 148,$$
$$N_1.N_{0.1}/N = 1,143, \quad N_1.N_{0.2}/N = 330,$$
$$N_2.N_{0.1}/N = 447, \quad N_2.N_{0.2}/N = 129.$$

Hence

$$T = \frac{(1,162-1,143)^2}{1,143} + \frac{(311-330)^2}{330} + \frac{(428-447)^2}{447} + \frac{(148-129)^2}{129} = 5.016.$$

Here the degrees of freedom equals 1, and $\chi^2_{0.95,(1)} = 3.84$. Since $\chi^2 > 3.84$, we reject H_0 (outcomes of first and second shorts are independent) and arrive at a conclusion that supports the hot hand phenomenon.

A RESEARCH PROBLEM Select a team from NCAA or National Basketball Association (NBA). Research the statistics for the players in the team for a chosen season. Apply the method described in Section 6.13.1 and draw a conclusion regarding the hot hand.

6.13.2 Case study: Comparison of Men and Women Professional Basketball Players

Do female professional basketball players shoot as well as male basketball players? In what follows an attempt is made to provide an answer to this question. Since the best players in the United States play either in the WNBA and in the NBA or in NCAA Division I College Basketball, we use data from these games. We present this section as a report. We note that in recent years, women athletes have closed the gap in their performances with men in many different sports.

(1) **NCAA** This part considers NCAA Division I College Basketball for comparison of male and female players. Basketball is arguably the most prominent co-ed sport played in the United States today, with media attention being given to both Men's Division I games and Women's too. Extra media time is even given to the two sexes respective post-season tournaments.

Watching men's and women's basketball games, though, one begins to notice that the final score of women's games tend to be somewhat lower than those of the men's games. Thus, one may wonder whether gender is a factor in the final scoring average of a Division I basketball team.

In order to test this we looked at the offensive scoring averages of the top 5 men's and women's teams in the Division I for the year 2001. For data, we visited ESPN.com and looked up the necessary statistics under the men's and women's college basketball headings. The results for the women's teams include the top 5 teams being, in order, Eastern Kentucky, Connecticut, Howard, Oklahoma, and Tennessee. The results for the men's include top 5 teams being in order, TCU, Duke, Maryland, Virginia, and McNeese State. These rankings are for a period prior to March 2001.

Having gathered the necessary statistics, we posed the following question,"Is gender a factor in determining the scoring average of NCAA team at a 5% significance level?" We then formed a contingency table based on the statistics from ESPN.com, shown in the table below. The table includes gender, actual scoring averages for the top 5 teams, row and column totals, and the expected scores for the top 5 teams.

From these we formed the null hypothesis that the two classifications are independent of each other, and an alternative hypothesis that the null hypothesis is not true. In order to find out whether or not we had to reject the null hypothesis, we performed the chi-squared test. Here, the test statistic is the sum of the actual value minus the expected value quantity squared, divided by the expected value. This resulted in a test statistic equal to 0.0704.

Next we proceeded to draw up a rejection region. In order to find the rejection region, we had to determine the number of degrees of freedom by multiplying the number of rows minus 1 and the number of columns minus 1. We found the degrees of freedom to be 4. From Table B5, the rejection region includes all the values of test statistic that are greater than 9.49.

The test statistic, 0.0704, did not fall in the rejection region, so we did not reject the null hypothesis, drawing the conclusion that there was not sufficient evidence to indicate that the classifications were not independent at a 5% significance level. This implied that shooting and gender can be considered dependent. Summarizing this, we have the followings:

Question: Is gender a factor in determining the scoring average of a NCAA team at 5% significance level?

Data for NCAA

Gender	Team 1	Team 2	Team 3	Team 4	Team 5	Total
Male	93.6	91.8	85.7	85.0	83.5	439.6
(Expected)	(92.42)	(91.30)	(85.82)	(85.46)	(84.60)	
Female	88.4	88.0	83.3	83.3	83.1	426.1
(Expected)	(89.58)	(88.50)	(83.18)	(82.84)	(82.00)	
Total	182	179	169	168.3	166.6	865.7

1. H_0: The two classifications are independent, H_1: H_0 is not true.

2. Test Quantity (Statistics)

$$T = \frac{(93.6-92.42)^2}{92.46} + \cdots + \frac{(83.1-82)^2}{82} = 0.0704.$$

3. Rejection Region

$$T > \chi^2_{0.95,(4)} = 9.49.$$

4. Decision: Do not reject H_0.

5. Conclusion: There is not sufficient evidence to indicate that the classifications are not independent at a 5% significance level.

(2) **NBA** Like the previous part, the purpose of analyzing the data was to determine whether or not there was a significant difference between the shooting percentages of the leaders of the NBA compared with the leaders of the (women's) *WNBA*. The result is somewhat surprising. Most fans are under the impression that the men, most likely, would be able to out shoot the women of professional basketball. The analysis of data did not support this. This is despite the fact that there are much more players in the NBA than WNBA, noting that the WNBA is relatively new.

Again we used ESPN.com, NBA.com, and WNBA.com to obtain the statistics for leader's Field Goal Percentage (FG%). There were two reasons for the slight difference in the sample sizes of the men compared to the sample sizes of the women. The FG% of the top 33 players in the NBA had a low value of 0.471 or 47.1%, as did the FG% of the top 20 players in the WNBA. Also, the ESPN.com and NBA.com listed the FG% for the top 50 male players, whereas ESPN.com and WNBA.com listed the FG% only for the top 20 female players. None of the websites listed more than the top 20 leaders for women. This probably is due to the fact that there are much less woman players than men, so, taking a larger sample for men may be justified. After obtaining the statistics we set up a table and calculated the totals and the expected frequencies along with the observed frequencies and plugged them into the formula to determine value of the test statistics χ^2. Finally, we determined the degrees of freedom and used the χ^2 table to find the critical value at a 5% significant level. As mentioned earlier, we were a little surprised at the results. This study determined that there is not enough evidence to indicate that men have a better shooting percentage than women, which although it should not have been, was a little surprising. As mentioned earlier, one may think that for some reason men should have better percentages than women. We don't know whether the results would change with a larger sample size. If we run into more statistical information on this subject, we should definitely like to try this study again to see whether the results indeed do stay the same. Summarizing these we have the following:

Question: Can women shoot a basketball as well as men? A sample including players with the top field goal percentage, the top 33 NBA players and the top 20 WNBA players were taken. A chi-square test was applied in an attempt to provide an answer to this question. The results of this test are as follows:

Observed Frequencies for NBA and WNBA

FG%	47.1–49.5	49.6–51.8	51.9–54.1	54.2–56.4	56.5–59.0	Total
Men	24	6	2	0	1	33
Women	10	6	2	1	1	20
Totals	34	12	4	1	2	53

H_0: Gender and field goal percentages are independent
H_1: H_0 is not true

Expected Frequencies for NBA and WNBA

FG%	47.1–49.5	49.6–51.8	51.9–54.1	54.2–56.4	56.5–59.0
Men	21.17	7.472	2.491	0.623	1.245
Women	12.83	4.528	1.509	0.377	0.755

Test Statistics: $T = \sum \frac{(\text{Observed Frequency} - \text{Expected Frequency})^2}{\text{Expected Frequency}} = 3.185$

Rejection Region: $T > 9.48773 = \chi^2_{0.95,(4)}$.

Decision: Do not reject H_0. There is not sufficient evidence to indicate that the two classifications are independent at a 5% significant level.

We end this section by recalling the problem of possible relationship of the greatness and the height of the US presidents discussed in Section 1.7. The following table presented in Sommers (2002), *The College Mathematics Journal*, Vol. 33, No. 1, shows the relationship between these two attributes. By dividing all the presidents through Reagan into four groups (see the table), the calculated test statistic is 10.065. The probability that χ^2 variable will be as large as this value is only 0.0015. The conclusion is that greatness does depend on height.

Contingency Table Relating Presidential Greatness and Height, Washington Through Reagan

Rating	Height	
	6' or Shorter	Taller Than 6'
"Great" or "Near great"	3	5
"Above Average,"		
"Average," "Below Average," or "Failure"	26	3

PRACTICE PROBLEMS

1. People interested in track and field know that the athletes running sprints are usually heavy and muscular, whereas distance runners are usually small and thin. Then one question of interest might be to see whether weight and distance have a dependency for the sprinters. The following table provides statistics from 2,000 Olympics for men. Use $\alpha = 0.05$ to test such a hypothesis.

Weight	100 m	200 m	400 m	800 m	Row Total:
145–160	3	2	3	5	13
161–175	2	1	3	2	8
176–190	2	4	1	0	7
>190	0	1	1	0	2
Column Total	7	8	8	7	30

2. In an article authored by Copper et al. (1992) appeared in *Chance*, Vol. 5, the question of home court (ice) advantage and its dependency on the type of sport is discussed. Using the following data test to see whether home court advantage is dependent on the type of sport. Use $\alpha = 0.01$.

Chapter 6: Statistical Testing

	Football	Hockey	Basketball	Baseball
Number of home wins	57	50	127	53
Number of away wins	42	53	71	47

6.14 KOLMOGOROV-SMIRNOV TEST

Like chi-square test, the Kolmogorov-Smirnov test is nonparametric or distribution-free. It is, in general, more powerful than the chi-square test. The test is based on the comparison of the theoretical (cumulative) distribution function—given by the hypothesis H(or H_0), and the cumulative relative frequency distribution (Chapter 1), based on data. Again, the range of X is decomposed into segments (cells) and the sample data, ordered by magnitude, are used to construct the cumulative relative frequency distribution.

Let F_i denote the cumulative relative frequency calculated from the theoretical (hypothesized) distribution and F_i^* denote the corresponding cumulative relative frequency of the sample or empirical distribution, where i is the number of the segment. Then,

$$T = \max_i |F_i - F_i^*| \tag{6.15}$$

is used as the test quantity (statistics). The test statistic has the well-known Kolmogorov-Smirnov distribution which, like many classical distribution, is tabulated. From its critical values the critical regions for rejection of the hypothesis "X has the given distribution function" are constructed as usual. To demonstrate the application we now apply this test to the data given in Example 6.12.1, where the goal was to test the hypothesis that X has a uniform distribution.

EXAMPLE 6.14.1 Consider the data regarding the age distribution of the NBA players in 2018–2019 season.

Age	19	20	21	22	23	24	25	26	27	28	29	30
Number of Players	10	23	35	42	54	49	44	50	30	37	26	32

Age	31	32	33	34	35	36	37	38	39	40	41	42
Number of Players	16	19	16	8	5	3	3	3	0	1	0	1

Test the hypothesis that data come from a uniform distribution on [19, 43).

SOLUTION: We have

Age(x)	$F_i(x)$	$F_i^*(x)$	$|F_i(x) - F_i^*(x)|$
19	1/24	10/507	0.022
20	2/24	33/507	0.023
21	3/24	68/507	0.092
22	4/24	1,102/507	0.175

(Continued)

Age(x)	$F_i(x)$	$F_i^*(x)$	$\|F_i(x) - F_i^*(x)\|$
23	5/24	164/507	0.282
24	6/24	213/507	0.378
25	7/24	257/507	0.465
26	8/24	307/507	0.564
27	9/24	337/507	0.623
28	10/24	374/507	0.696
29	11/24	400/507	0.747
30	12/24	432/507	0.810
31	13/24	448/507	0.842
32	14/24	467/507	0.879
33	15/24	483/507	0.911
34	16/24	491/507	0.927
35	17/24	496/507	0.937
36	18/24	499/507	0.943
37	19/24	502/507	0.948
38	20/24	505/507	0.954
39	21/24	505/507	0.954
40	22/24	506/507	0.956
41	23/24	506/507	0.956
42	24/24	507/507	0.958

resulting in test quantity equal to $T = 0.958$. For $\alpha = 0.10$ and $N = 24$ the KS table below yields a critical value of 0.242. Since $T > 0.242$, we reject the null hypothesis that distribution of age is uniform at a 10% significance level. This confirms the conclusion of the test performed earlier.

KS Tables: Critical Values, One-Sample Kolmogorov-Smirnov Test

Sample Size	0.10	0.05	0.01
1	0.950	0.975	0.995
2	0.776	0.842	0.929
3	0.636	0.708	0.829
4	0.565	0.624	0.734
5	0.509	0.563	0.669
6	0.468	0.519	0.617
7	0.436	0.483	0.576
8	0.410	0.454	0.542
9	0.387	0.430	0.513

Sample Size	0.10	0.05	0.01
10	0.369	0.409	0.489
11	0.352	0.391	0.468
12	0.338	0.375	0.449
13	0.325	0.361	0.432
14	0.314	0.349	0.418
15	0.304	0.338	0.404
16	0.295	0.327	0.392
17	0.286	0.318	0.381
18	0.279	0.309	0.371
19	0.271	0.301	0.361
20	0.265	0.294	0.352
21	0.259	0.287	0.344
22	0.253	0.281	0.337
23	0.253	0.281	0.337
24	0.242	0.269	0.323
25	0.238	0.264	0.317
26	0.233	0.259	0.311
27	0.229	0.254	0.305
28	0.225	0.250	0.300
29	0.221	0.246	0.295
30	0.218	0.242	0.290
31	0.214	0.238	0.285
32	0.211	0.234	0.281
33	0.208	0.231	0.277
34	0.205	0.227	0.273
35	0.202	0.224	0.269
36	0.199	0.221	0.265
37	0.196	0.218	0.262
38	0.194	0.215	0.258
39	0.191	0.213	0.255
40	0.189	0.210	0.252
Approximation for $N > 40$	$\frac{1.22}{\sqrt{n}}$	$\frac{1.36}{\sqrt{n}}$	$\frac{1.63}{\sqrt{n}}$

PRACTICE PROBLEMS

1. The following presents the winning scores in 36 college basketball competition (NCAA). Test the null hypothesis that these data come from a normally distributed population with a mean of 85 and a standard deviation of 15. Use $\alpha = 0.10$.

$$\begin{array}{ccccccccc}
58 & 78 & 84 & 90 & 97 & 70 & 90 & 86 & 82 \\
59 & 90 & 70 & 74 & 83 & 90 & 76 & 88 & 84 \\
68 & 93 & 70 & 94 & 70 & 110 & 67 & 68 & 75 \\
80 & 68 & 82 & 104 & 92 & 112 & 84 & 98 & 80
\end{array}$$

Note that here some scores are repeated. To make the calculations shorter first complete the following tables:

Score (x)	Frequency	Cumulative Frequency	$F_i^*(x)$
58	1	1	1/36 = 0.0278
59	1	2	2/36 = 0.0556
67	1	3	3/36 = 0.0833
68	3	6	6/46 = 0.1667
70	4	10	10/36 = 0.2778
⋮	⋮	⋮	⋮

Score(x)	z-Score	$F_i(x)$
58	−1.80	$0.0359 = P(z < -1.80)$
59	−1.73	$0.0418 = P(z < -1.73)$
67	−1.20	$0.1151 = P(z < -1.20)$
68	−1.13	$0.1292 = P(z < -1.13)$
70	−1.00	$0.1587 = P(z < -1.00)$
⋮	⋮	⋮

Show that the value of test statistics is 0.1485.

2. Test the hypothesis that the following data

$$0, 0, 0.1, 0.1, 0.1, 0.2, 0.2, 0.3, 0.3, 0.4, 0.45, 0.45, 0.5, 0.5, 0.6, 0.9.$$

comes from a population with distribution function

$$F(x) = 1 - (1 - 1.111x)^{1.7095}, \quad 0 \leq x \leq 0.9.$$

Show that the value of test statistics is 0.0714.

Chapter 6: Statistical Testing

6.15 SIGN TEST

The sign test is yet another distribution-free test. It is easy to use but is not very powerful. This test is based on two "matched samples" of the continuous random variables X and Y. The hypothesis H (or H_0) we want to test states that X and Y have the same distribution.

Let $(X_1, Y_1), \ldots, (X_N, Y_N)$ be the paired samples and put $S_i = 1$ if $X_i > Y_i$ and $S_i = 0$ if $X_i \leq Y_i$. The value of $T = S_1 + \ldots + S_N$ is expected to be approximately equal to $N/2$ if H is true. If N is not too small, say it is greater than or equal to 30, then the critical values are approximately equal to $s_\alpha^\pm = \frac{1}{2}(N \pm z_{1-\alpha/2}\sqrt{N})$, where $z_{1-\alpha/2}$ is the $100(1 - \alpha/2)$ percentile of $N(0,1)$ distribution. Hence, the hypothesis of same distribution is rejected at the significance level α, if $S > s_\alpha^+$ or $S < s_\alpha^-$.

EXAMPLE 6.15.1 Let X and Y be the points per game (PPG) made by the top NBA basketball players in 1997–1998 and 1998–1999 regular seasons, respectively. Although these are discrete we will use them to demonstrate the procedure. We want to fin2d out whether top players' performance were the same or different in these two seasons. So we test the hypothesis H_0:X and seasons. So we test the hypothesis H_0:X and Y have the same distribution, using the sign test at 5% level of significance. The point per game of the top 30 players of each year as well as the value of S_i are listed in the Table 6.15.1. From the table we see that $N = 30$ and $T = S_1 + \ldots + S_{30} = 27$. With $\alpha = 0.05$ we have

$$s_\alpha^\pm = s_{0.05}^\pm = \frac{1}{2}(30 \pm 1.96\sqrt{30}) = 15 \pm 5.4.$$

Clearly $T > s_\alpha^+ = 20.4$. Hence we reject H_0. That is, based on the above measure of performance, we cannot conclude that the top players performed equally well in these two seasons or distributions are the same.

A PRACTICE PROBLEM Generate 100 numbers from the normal distribution N(0,1). Compare the first 50 with the last 50. Apply the sign test to see whether it confirms that these two sets of data have an identical distribution.

TABLE 6.15.1 Point Per Game for the Top 30 Players, 1997–1999

1997–1998	28.7	28.3	27.0	23.2	22.4	22.3	22.3	22.0	21.9	21.6
1998–1999	26.8	26.3	23.8	23.0	21.8	21.7	21.7	21.3	21.2	21.1
S_i	1	1	1	1	1	1	1	1	1	1
1997–1998	21.5	21.1	21.1	20.1	19.7	19.6	19.5	19.5	19.2	19.2
1998–1999	20.8	20.5	20.2	20.1	19.9	19.7	18.9	18.8	18.4	18.4
S_i	1	1	1	0	0	0	1	1	1	1
1997–1998	18.9	18.5	18.4	18.4	18.0	17.8	17.7	17.7	17.3	17.2
1998–1999	18.3	17.4	17.1	17.0	16.9	16.8	16.5	16.5	16.4	16.3
S_i	1	1	1	1	1	1	1	1	1	1

6.16 FINAL WORDS

As pointed in the beginning of this chapter, hypothesis testing constitutes the major part of statistical inference and, as such, is widely used in applications. Looking at the literature, one finds numerous books and professional journals filled with interesting tests and theories. Some of these tests that are either very useful or very controversial find their way into media and popular newspapers. The term "proved" though not appropriate, is used frequently and easily by referring to statistics or an experiment. Average citizens usually have a hard time separating facts and useful information from misleading and wrong ones. The only term that can explain this type of misuse is "garbage in, garbage out," which reminds us of the fact that a beautifully executed test will be ruined by biased sample data. Clearly a selective reporting of results is potentially one of the most serious abuse of hypothesis tests. To illustrate this, consider a "researcher" who wants to show "proof" that a newly developed pill brings your blood pressure down. He picks 50 people, measures their blood pressures before and after using the pill. If the test result is not significant in the direction he wants, he tries another 50 and repeats this until he finds the one that is statistically significant. Then writes a long report and claims that statistics "prove" that $\cdots$. Some people refer to this as data mining or fishing expedition. Most statisticians know very well that if only worthless theories are examined, the "researcher" will eventually stumble on a worthless theory that seems to be supported by the data. Consider the following case: Steve Carroll, an accountant in San Francisco, had introduced a theory known as the J theory. Observing that in the 1980s, the Super Bowls were won by teams that had a quarterback whose first name began with letter J and had three letters, he suggested that the two teams, 49ers and Redskins, should come on top since they follow the pattern of that decade. Now, although this theory is supported by the observation that the New York Giants with Phil Simms as their quarterback were the winner of 1987 Super Bowl, it makes absolutely no sense to think that somebody's name will determine the outcome of a game.

Another issue is the level of significance. We already noted that a hypothesis can be rejected in one level of significance but not rejected in another (p-value). When newspapers report results of a test they hardly report the level of significance or margin of error.

Finally, a theory rejected based on one sample may be accepted based on another. In other words, a model that is good or best for one set of data may not be the best or good for other data taken in a different time or place. So understanding the basics of a discipline from which data come is essential when testing or modeling. In fact, many models developed purely based on data and statistical methods does not make sense physically. In short, statistical results may need validation based on criterion other than statistical goodness of fit.

6.17 USING R

We have learned many hypothesis tests in this chapter, therefore we demonstrated R programming about that in this section.

6.17.1 Hypothesis test of mean values

In practice, we would like to use t-test all the time because it will very close to z-test when the conditions of Central Limit Theorem are satisfied.

Chapter 6: Statistical Testing

EXAMPLE 6.17.1 Generate a sequence of random values, x, from normal distribution N(3, 1) in this example. Let's test whether the mean value of population that your sample x drawn is 5.0, that is, $H_0: \mu = 5.0$ and $H_1: \mu \neq 5.0$.

```
# Generate the sample x, size 50
x = rnorm(50, 3, 1)
# Run the hypothesis test
t.test(x, mu=5, alternative=''two.sided'')
```

You will get

One Sample t-test

```
data: x
t = -13.625, df = 49, p-value < 2.2e-16
alternative hypothesis: true mean is not equal to 5
95 percent confidence interval:
2.574487 3.198023
sample estimates:
mean of x
2.886255
```

therefore we reject H_0.

EXAMPLE 6.17.2 Now we generate another sequence of random values, y, from normal distribution N(10, 4). Let's compare the population mean to the one of previous example: $H_0: \mu_1 - \mu_2 = 0$, $H_1: \mu_1 - \mu_2 < 0$.

```
# Generate the sample y, size 30
y = rnorm(30, 10, 2)
# Run the hypothesis test
t.test(x, y, alternative=''less'')
```

The output:

Welch Two Sample t-test

```
data: x and y
t = -18.301, df = 40.059, p-value < 2.2e-16
alternative hypothesis: true difference in means is less than 0
95 percent confidence interval:
-Inf -6.488154
sample estimates:
mean of x mean of y
2.886255 10.031858
```

You can see we reject H_0, therefore population of y has larger mean value.

6.17.2 Hypothesis test of variances

EXAMPLE 6.17.3 We still focus on x and y here. The variances of x and y are 1.2 and 3.9 respectively. Now we would like to see $H_0 : \sigma_1^2/\sigma_2^2 = 1$, $H_1 : \sigma_1^2/\sigma_2^2 < 1$.

```
# Calculate variances
var(x)
var(y)
# Run the test
var.test(x, y, alternative=''less'')
```

You will obtain the output as:

```
        F test to compare two variances

data: x and y
F = 0.31245, num df = 49, denom df = 29, p-value = 0.0001617
alternative hypothesis: true ratio of variances is less than 1
95 percent confidence interval:
 0.0000000 0.5308114
sample estimates:
ratio of variances
 0.3124478
```

Therefore we conclude that the population of y has larger variance.

6.17.3 χ^2-Test

EXAMPLE 6.17.4 (Contingency table) According to Agresti (2007) p. 39, we summarized counts of members of parties in following table:

Gender	Democrat	Independent	Republican
M	762	327	468
F	484	239	477

Are the counts of members uniformly distributed?

```
# Prepare data
M <- as.table(rbind(c(762, 327, 468), c(484, 239, 477)))
dimnames(M) <- list(gender = c(''F,'' ''M''),
  party = c(''Democrat,'' ''Independent,'' ''Republican''))
# Run the test
chisq.test(M)
```

The p-value less than 0.0001, therefore, gender and choices of party were not independent. You will see there were more female Democrats and male Republicans than the hypothesis of independence predicts.

Chapter 6: Statistical Testing

EXAMPLE 6.17.5 (Goodness-of-fit) Suppose in a sample, we obtain the frequencies: A–20, B–15, and C–25. Is that possible to claim A, B, and C are equally likely?

```
# Prepare data
a = c(A=20, B=15, C=25)
# Run the test
chisq.test(a)
```

We got a *p*-value of 0.2865, therefore we can say A, B, and C were equally likely. Alternatively, you can specify the probabilities as following.

```
# Prepare data
a = c(A=20, B=15, C=25)
prob = rep(1/3, 3)
# Run the test
chisq.test(a, p = prob)
```

6.17.4 Kolmogorov-Smirnov test

EXAMPLE 6.17.6 Recall the random values in your sample x, let's test whether x was drawn from a normal distribution?

```
# K–S test
ks.test(x, pnorm)
```

We obtained a small *p*-value! As we know x dose come from a normal distribution, why we have this contradiction? **pnorm** here indicates the CDF you would like to test with, so you can change it to other distributions, such as chisq, t, or F. However, pnorm here means standard normal distribution, that is, $N(0,1)$. If you calculate sample mean $\bar{x} = 2.89$, you will find it was not close to 0. So the proper way is

```
# K–S test
ks.test(x, pnorm, 3)
```

The parameter **3** here indicating mean of normal distribution. We "guess" from $\bar{x}$ that μ will be close to 3.

EXAMPLE 6.17.7 Whether two sequences x and y come from a same distribution?

```
# K–S test
ks.test(x, y)
```

The *p*-value is less than 0.0001, which indicates x and y come from different populations.

REFERENCE

Wardrop. 1995. "Simpson's Paradox and the Hot Hand in Basketball." *The American Statistician*, 49 (1):24–8.

CHAPTER 7

Analysis of Variance and Experimental Design

In many places in this book, we talked about experimental sciences. Obviously, experimental sciences did not begin with statistics, but with experimental design, as we now use the term. Experimental design does not refer to laboratory work or to the design of instruments to measure experimental responses. It refers to a plan for generating data (collecting and analyzing) so that the investigator can apply methods, which are well-developed, tested, tried, and their validity is already established. In this chapter, the experimental design together with an established method known as the analysis of variance (ANOVA), are introduced and are discussed. Some of their applications to sport-related problems are also presented.

7.1 COMPONENTS OF VARIANCE

In Chapter 6, tests of hypotheses regarding population means were discussed. The discussion included both single and two population cases. This chapter considers a more general case that involves several populations and presents a powerful technique known as the ANOVA for testing the equality of their means. Ironically, this technique analyzes the sample variance in order to test and estimate means, and that is the reason for the name used. To start, let us present some examples where such an analysis may be applied.

(1) A chain store executive needs to determine whether or not sales of new running shoes are effected by the aisle in which the shoes are displayed or stored. If there are several aisles in the store, she may place the shoes in different aisles during different periods of time and record the weekly sales. Then, she may test to find out whether or not the mean weekly sales differ significantly.

(2) A farm products manufacturer wants to determine whether the yield of a crop is different when the soil is treated with various fertilizers. Similar plots of land are planted with the same type of seed but are fertilized differently. At the end of the growing season, the mean yields from the sample plots can be compared. This type of experiment was one of the first to use the ANOVA, and the terminology of the original experiment is still used. Note that in both examples, the test is designed to determine whether significant differences exist among the treatment means.

To make the discussion relevant to sports, suppose that we want to compare four different methods for treatment of a certain common injury in a contact sport such as soccer, football, or basketball. If we decide to apply what we have learned so far, we can consider a pairwise comparison discussed in Chapter 6. That is, we compare Treatment 1 with 2, 1 with 3, and so on, but as we will demonstrate shortly, this is cumbersome and inadequate. In fact, a pairwise comparison leads to a relatively large Type I error. To clarify this, suppose that there are only three treatments. A pairwise comparison will involve $\binom{3}{2} = 3$ comparison (1-2, 1-3, and 2-3). For a 5% level of significance, the probability of making a correct decision in all 3 comparisons is $(0.95)^3$. Thus the probability of making at least one wrong decision is $1 - (0.95)^3 = 0.143$. Clearly, as the number of populations is increased, this error will increase and will tend to 1. That is, it becomes almost sure that at least one wrong decision will be made. The frame for handling such problems is the ANOVA, a powerful and extremely useful technique discussed below.

Let the population Ω be decomposed (say, by different treatments) into J subpopulations $\Omega_1, \ldots, \Omega_J$. From each of the subpopulations Ω_j a sample of size N_j is drawn. Let y_{ij} denote the ith observed value in the jth sample. First, note that this value can be expressed as the sum of the mean μ_j (of the subpopulation Ω_j) and the deviation from the mean, usually referred to as the error term ε_{ij}. That is,

$$y_{ij} = \mu_j + \varepsilon_{ij} \quad i=1,\ldots,N_j, \quad j=1,\ldots,J \tag{7.1}$$

In other words, every observation is made up of two components, one due to the treatment (μ_j) and one due to all of the other factors (ε_{ij} referred to as noise). Since it is reasonable to assume that positive and negative deviations are equally likely, and small errors are more likely than large errors, the errors are assumed to be independent and identically distributed as $N(0, \sigma^2)$. The random variations of all observed $N = N_1 + \ldots + N_J$ values y_{ij} from their "grand mean" μ (the average of μ_j's) is measured by the term

$$SSTotal = \sum_{j=1}^{J} \sum_{i=1}^{N_j} (y_{ij} - \mu)^2. \tag{7.2}$$

Here SS stands for the sum of squares and (7.2) is the total SS regardless of the treatments.

Before proceeding further, let us consider an example. Suppose that there are 40 boys participating in a training program. There are four different training methods and the goal is to decide which one works better or is more "effective." To do this, we divide the boys into four groups of 10 each and administrate one method to each group. In the end of the period, we give them the same test. Here, y_{ij} is the score of boy number i in the group number j, μ_j is the population mean for group j, and μ is the overall mean. Here i takes the values 1 to 10 and j takes values from 1 to 4.

Keeping j fixed, we get in the same way $\sum_{i=1}^{N_j}(y_{ij} - \mu_j)^2$ as a measure of the deviation of the measurements from their mean μ_j within the j-th sample or group. Summing up over all samples we get a measure for within variation denoted by $SSWithin(SSError)$ as,

$$SSWithin = \sum_{j=1}^{J} \sum_{i=1}^{N_j} (y_{ij} - \mu_j)^2 = (N_1 - 1)s_1^2 + (N_2 - 1) + \cdots + (N_J - 1)s_J^2. \tag{7.3}$$

Chapter 7: Analysis of Variance and Experimental Design

Finally, $\sum_{j=1}^{J}(\mu_j - \mu)^2$ measures the deviation of the sample means μ_j from their mean, which is the same as the grand mean μ. Weighted by the sample sizes N_j we define *SSBetween(SSTreatment)* as

$$SSBetween = \sum_{j=1}^{J} N_j(\mu_j - \mu)^2. \quad (7.4)$$

The basis of ANOVA is the splitting the total variation into the components of the variance as follows:

$$SSTotal = SSBetween + SSWithin.$$

To see how this was obtained, we write $y_{ij} - \mu$ as

$$y_{ij} - \mu = (y_{ij} - \mu_j) + (\mu_j - \mu).$$

Upon squaring both sides and summing we get

$$\sum_{j=1}^{J}\sum_{i=1}^{N_j}(y_{ij} - \mu)^2 = \sum_{j=1}^{J}\sum_{i=1}^{N_j}(y_{ij} - \mu_j)^2 + \sum_{j=1}^{J}\sum_{i=1}^{N_j}(\mu_j - \mu)^2$$

$$+ 2\sum_{j=1}^{J}\sum_{i=1}^{N_j}(y_{ij} - \mu_j)(\mu_j - \mu). \quad (7.5)$$

But the last term can be written as $2\sum_{j=1}^{J}(\mu_j - \mu)\sum_{i=1}^{N_j}(y_{ij} - \mu_j)$ and is equal to 0 since $\sum_{j=1}^{J}(\mu_j - \mu) = \sum_{j=1}^{J}\mu_j - N\mu = N\mu - N\mu = 0$. Also the second term can be written as $\sum_{j=1}^{J}N_j(\mu_j - \mu)^2$. It is interesting to note the similarity of the relation (7.5) to Pathagours theorem in geometry (after replacing zero for the third term).

Summarizing these, the ANOVA make use of the fact that the total variation can be written as sum of two variations, between treatments and within treatments. The former is due to the difference between the treatments and the latter is due to the difference between individuals within the group or treatment. Putting it differently, the term ANOVA refers to the idea of analyzing variability in the data to see how much of it can be attributed to differences in the μ'_js and how much is due to variability in the individual populations. If differences between the sample means can be explained by within-sample variability, there is no compelling reason for rejecting the null hypothesis of equal means. This situation arises when the *SSBetween* is small relative to SS *Within*. For this reason, it is sensible to base the test statistic on the ratio of these two terms.

To clarify this further, suppose that we are testing a drug that is used to reduce blood pressure. For the individuals using this drug, the reduction in blood pressure can be considered significant only if it is much more than natural changes in their blood pressure without using the drug. So if someone's blood pressure varies a great deal during the day, a small reduction in his or her blood pressure after taking the drug cannot be attributed to the drug. It can be just a within-sample variability.

EXAMPLE 7.1.1 Let Ω denote the PPGs (points per game) made by the top National Basketball Association (NBA) basketball players in the 3-year periods 2015–2018. Suppose that we are interested in seeing whether the top player's performance was the "same" in each of these 3 years, or performances of at least two of these years were significantly different from each other. To do this, we decompose Ω into three subpopulations: the PPG in seasons 2015–2016, 2016–2017, and 2017–2018. The following table gives the PPG of the top 20 players for each of the three seasons. The total number of observations equals 60, $J = 3$, $N_j = 20$ for $j = 1, 2, 3$, and y_{ij}'s are the data in the table. When the values of μ and μ_j's are known, *SSWithin*, *SSBetween*, and *SSTotal* are computed using the formulas presented above. On the other hand, when these values are unknown they are estimated from the samples, and this obviously changes the statistical nature of the problem. We will discuss this issue in Section 7.2 and the results presented there provide an answer to the question posed here and in Example 7.3.1 in Section 7.3.

Performance of Top 20 National Basketball Association (NBA) Players (1st to 10th and 11th to 20th)

2015–2016	30.1	29.0	28.2	26.9	25.3	25.1	24.3	23.5	23.5	23.1
2016–2017	31.6	29.1	28.9	28.0	27.3	27.0	27.0	26.4	25.5	25.3
2017–2018	30.4	28.1	27.5	26.9	26.9	26.4	25.4	24.4	23.1	23.1
2015–2016	22.2	22.1	21.8	21.2	21.2	20.9	20.9	20.8	20.7	20.6
2016–2017	25.2	25.1	25.1	23.9	23.7	23.6	23.2	23.1	23.1	23.0
2017–2018	23.0	22.9	22.6	22.6	22.2	22.1	21.9	21.4	21.4	21.3

EXAMPLE 7.1.2 Suppose that we want to see if the offense for the 3 league leaders of the American Baseball League were similar by using the data for Toronto, Minnesota, and Seattle, the leaders of the AL East, AL Central, and AL West in 2001. The following is their number of runs from 5 of their games in the 2001 season. Does this information provide evidence against the claim that the three teams have similar offenses by their runs scored, at a 5% significance level? Information from: www.sports.lycos.com

Toronto	Minnesota	Seattle
7	7	4
8	7	2
12	1	2
1	8	2
2	1	4
Mean: 6	6.2	3.6
Variance: 20.52	8.70	2.79

Again, we can answer this question using the method discussed in Section 7.2 below.

7.2 ONE-WAY CLASSIFICATION

The analysis of experimental data, as one should expect, depends on the design of the experiment, which refers to the way the data were collected. As will be discussed later, a very useful and relatively simple design called the completely randomized design is one in which random samples are independently selected from each of the populations. This design results in observations that are classified only according to the population from which they came; hence the designation is as a one-way classification. Note that the populations under consideration may just exist in concept. For example, in sports, one may want to determine the effect of various procedures (e.g., strategies or coaching), usually referred to as treatments, on a variable of interest, for example, performance. This section describes how the ANOVA may be applied when only one factor is under consideration.

We start by noting that in practice the mean of population j, μ_j is usually unknown and is replaced by its sample counterpart

$$\hat{\mu}_j = \bar{y}_{.j} = \frac{1}{N_j} \sum_{i=1}^{N_j} y_{ij}.$$

Similarly μ, the grand or overall mean, is replaced by

$$\hat{\mu} = \bar{y} = \frac{1}{N} \sum_{j=1}^{J} \sum_{i=1}^{N_j} y_{ij}.$$

Now replacing $\hat{\mu}_j$ for μ_j in (7.5) leads to an expression similar to a sample variance. In fact, the first summation on the right-hand side is now $\sum_{i=1}^{N_j} (y_{ij} - \bar{y}_{.j})^2$ and, like sample variance, has $N_j - 1$ degrees of freedom. Since this sum involves J summations of this form, its total degrees of freedom is $(N_1 - 1) + (N_2 - 1) + \ldots + (N_j - 1) = N - J$. Note that from the SS we can find the so-called mean squares (denoted by MS) as SS divided by the degrees of freedom (d.f.), which is the number of terms in SS minus one (see Chapter 1). Thus the degrees of freedom for SSTotal is $N - 1 = N_1 + N_2 + \ldots + N_j - 1$, for SSBetween (treatment SS) is $J - 1$ and for SSWithin (error SS) is $(N_1 - 1) + (N_2 - 1) + \ldots + (N_j - 1) = N - J$, respectively. These values are listed in Table 7.2.1.

TABLE 7.2.1 One-Way Analysis of Variance Table

Source of Variation	Sum of Squares	Degrees of Freedom (d.f)	Mean Squares	F-Ratio
Between Samples	$SSB = \sum_{j=1}^{J} N_j (\hat{\mu}_j - \hat{\mu})^2$	$J - 1$	$MSB = \frac{SSB}{J-1}$	$F = \frac{MSB}{MSW}$
Within Samples	$SSW = \sum_{j=1}^{J} \sum_{i=1}^{N_j} (y_{ij} - \hat{\mu}_j)^2$	$N - J$	$MSW = \frac{SSW}{N-J}$	
Total	$SST = \sum_{j=1}^{J} \sum_{i=1}^{N_j} (y_{ij} - \hat{\mu})^2$	$N - 1$		

In the last column, we have added the F–ratio defined as

$$\frac{MSB}{MSW} = \frac{\text{Mean Square Between}}{\text{Mean Square Within}} \tag{7.6}$$

Under the assumption, we made regarding the independence and normality of the error terms ε_{ij}, both numerator and denominator are distributed as χ^2 and, by definition, the ratio has an F-distribution with $(J-1, N-J)$ degrees of freedom.

The aim of the ANOVA is primarily testing the hypothesis that

$$H_0 : \mu_1 = \mu_2 = \ldots = \mu_J = \mu$$

versus the alternative that at least two population means differ. Under H_0, there is no significant differences between the subpopulations. Thus, if we want to establish that there are actually differences between the means of subpopulations, we have to look for evidence to reject H_0. Clearly, for H_0 to be supported the F ratio has to be "small"; in fact, it must be a number close to 1. This is because, under H_0, the between and within variations are expected to be very close to each other. To make a decision concerning the rejection of H_0 or otherwise, we construct the critical (rejection) region using the F-distribution. The rejection region is given by

$$F > F_{1-\alpha, (J-1, N-J)}.$$

If the F–ratio in the ANOVA table is greater than the $100(1-\alpha)$ percentile of the F distribution, then we conclude, at significance level α, that there is a difference between at least two means of subpopulations. Note that for this procedure to be valid, the errors need not be exactly normally distributed. In fact, we can apply ANOVA if they are only approximately normal. Also, the variance of the errors need not to be exactly equal. However, they should be at least of the same order of magnitude. If the expected values of the errors are not exactly or approximately zero, or if they are not independent from one another, then this will change the situation seriously and the results of ANOVA may not be valid.

Computational Formulas

When performing calculations for ANOVA, it is easier to use the computational formulas given below

$$SST = \sum_{j=1}^{J} \sum_{i=1}^{N_j} y_{ij}^2 - \frac{1}{N} (\sum_{j=1}^{J} \sum_{i=1}^{N_j} y_{ij})^2$$

$$SSB = \sum_{j=1}^{J} \frac{1}{N_j} (\sum_{i=1}^{N_j} y_{ij})^2 - \frac{1}{N} (\sum_{j=1}^{J} \sum_{i=1}^{N_j} y_{ij})^2$$

$$SSW = SST - SSB \quad (7.7)$$

As a calculation procedure we may use the scheme described in Table 7.2.2 where $Z_j = \sum_{i=1}^{N_j} y_{ij}$.

Before presenting an example, we note that for one-way classification we have the following model:

$$y_{ij} = \mu_j + \varepsilon_{ij} = \mu + \alpha_j + \varepsilon_{ij}$$

where μ is the overall mean and $\alpha_j = \mu_j - \mu$ is the jth treatment effect.

Chapter 7: Analysis of Variance and Experimental Design

TABLE 7.2.2 Calculation Scheme for Analysis of Variance (ANOVA)

	y	y^2	Z	Z^2	N	Z^2/N
Sample 1	y_{11} $\vdots$ y_{N_11}	y_{11}^2 $\vdots$ $y_{N_11}^2$	Z_1	Z_1^2	N_1	Z_1^2/N_1
Sample 2	y_{12} $\vdots$ y_{N_22}	y_{12}^2 $\vdots$ $y_{N_22}^2$	Z_2	Z_2^2	N_2	Z_2^2/N_2
$\vdots$	$\vdots$	$\vdots$	$\vdots$	$\vdots$	$\vdots$	$\vdots$
Sample j	y_{1j} $\vdots$ y_{N_jj}	y_{1j}^2 $\vdots$ $y_{N_jj}^2$	Z_j	Z_j^2	N_j	Z_j^2/N_j
	$\sum y$	$\sum y^2$			N	$\sum Z_j^2/N_j$

EXAMPLE 7.2.1 Consider the amount by which the blood pressure may be lowered using four different treatment programs. Each program combines a certain amount of exercise, diet, and medicine. Suppose that 16 individuals are treated using each of these programs (group of 16 people for each program), and the measured values are, respectively, $y_{1,1},\ldots,y_{16,1}$, $y_{1,2},\ldots,y_{16,2}$, $y_{1,3},\ldots,y_{16,3}$, and $y_{1,4},\ldots,y_{16,4}$. Suppose also that for this data we have

$$\sum_{i=1}^{16} y_{i1} = 100 \quad \sum_{i=1}^{16} y_{i2} = 110 \quad \sum_{i=1}^{16} y_{i3} = 116 \quad \sum_{i=1}^{16} y_{i4} = 86$$

$$\sum_{i=1}^{16} y_{i1}^2 = 800 \quad \sum_{i=1}^{16} y_{i2}^2 = 760 \quad \sum_{i=1}^{16} y_{i3}^2 = 810 \quad \sum_{i=1}^{16} y_{i4}^2 = 806$$

Apply the one-way classification ANOVA to determine whether these programs are of equal effectiveness or they differ in their effect, using a significance level of 5%.

SOLUTION: Using expression (7.7) we obtain

$$SST = 3176 - 2652.25 = 523.75$$

$$SSB = 2684.5 - 2652.25 = 32.25$$

$$SSW = 523.75 - 32.25 = 491.5$$

For this example $J = 4$, $N = 64$, so that $MSB = 10.75$, and $MSW = 8.19$. Therefore, the F-ratio equals $10.75/8.19 = 1.31$. The critical value is $F_{1-\alpha,(J-1,N-J)} = 2.76$. Since the F-ratio is smaller than critical value, we cannot reject the null hypothesis of equal population means and conclude that these programs are not of equal effectiveness at a 5% significant level.

We can summarize the calculations in the ANOVA table given below:

Analysis of Variance (ANOVA) for Example, 7.2.1

Source of Variation	Sum of Squares	Degrees of Freedom (d.f.)	Mean Squares	F-Ratio
Between Samples	32.5	3	10.75	1.31
Within Samples	491.5	60	8.19	
Total	523.75	63		

Note that when the number of observations are equal in each sample, that is when $N_1 = N_2 = \ldots = N_j$, we can simplify the calculations and express both MSB and MSW in terms of sample variances. This is demonstrated in Example 7.2.2 below.

EXAMPLE 7.2.2 An expert on sport cars organized a study regarding the mileages obtainable from three different brands of gasoline. Using 15 identical motors set to run at the same speed, the investigator randomly assigned each brand of gasoline to 5 of the motors. Each of the motors was then run on 20 gallons of gasoline, with the total mileages obtained as given below.

	Brand 1	Brand 2	Brand 3
	220	244	252
	251	235	272
	226	232	250
	246	242	238
	260	225	256
$\bar{y}_j$:	240.6	235.6	253.6
s_j^2 :	287.8	59.3	250.8

Test the hypothesis that the average mileage obtained is the same for all three brands. Use a 5% level of significance.

SOLUTION: The average of three sample means is

$$\bar{\bar{y}} = \frac{240.6 + 235.6 + 253.6}{3} = 243.27$$

$$SSB = \Sigma N_j(\bar{y}_j - \bar{\bar{y}})^2 = 5(240.6 - 243.27)^3 + 5(235.6 - 243.27)^2 + 5(253.6 - 243.27)^2 = 863.33$$

$$MSB = SSB/(J-1) = 863.33/2 = 431.667$$

$$SSW = \Sigma(N_j - 1)s_j^2 = 4(287.8 + 59.3 + 150.8) = 1991.6$$

$$MSW = SSW/(N-J) = 1991.6/12 = 165.967.$$

Chapter 7: Analysis of Variance and Experimental Design

The test statistics or F-ratio is then $F = MSB/MSW = 431.667/165.967 = 2.60$. The rejection region is $F > F_{.95,(2,12)} = 3.89$.

Since $F = 2.60 < 3.89$ we cannot reject the null hypothesis that the three brands give equal average mileage.

Analysis of Variance (ANOVA), for Example 7.2.2

Source of Variation	Sum of Squares (SS)	Degrees of Freedom (d.f.)	Mean Squares (MS)	F-Ratio
Between Samples	863.33	2	431.67	2.60
Within Samples	1991.6	12	165.97	
Total	2854.93	14		

EXAMPLE 7.2.3 In this example, we would like to research the annual salaries (in millions) in 2018–2019 season by 20 of baseball's highest paid players from top 4 teams were used for analysis. Of these 4 players from Oklahoma City Thunder, 6 from Golden State Warrior, 5 players from Washington Wizards, and 5 players from Toronto Raptors. The averages for the four different teams were, respectively, $10.33, $10.29, $10.04, and $9.97 million per year. Can we conclude from the differences among these values that the baseball player's annual salaries are significantly different in top 4 teams? Use $\alpha = 0.05$.

SOLUTION: If we let μ_O, μ_G, μ_W, and μ_T denote the true average annual salaries of Oklahoma City Thunder, Golden State Warrior, Washington Wizards, and Toronto Raptors, respectively, the hypotheses to be tested are

$$H_0 : \mu_O = \mu_G = \mu_W = \mu_T \ vs \ H_1 : \text{not all the } \mu_j's \text{ are equal.}$$

Since J = 4, N = 20, and $\alpha = .05$, H_0 will be rejected if $F \geq F_{0.95,(3,16)} = 3.239$. Here $\bar{y} = 10.16$, and

$$SSB = 4(10.33 - 10.16)^2 + 6(10.29 - 10.16)^2 + 5(10.04 - 10.16)^2 + 5(9.97 - 10.16)^2 = 0.47$$

$$SSW = (10.47 - 10.33)^2 + (10.35 - 10.33)^2 + \cdots + (8.93 - 9.97)^2 = 2.54$$

and therefore

$$F = \frac{0.47/3}{2.54/16} = 0.987.$$

Since F does not exceed 3.239, H_0 is not rejected. Therefore, we do not have sufficient evidence to indicate that the true average salaries of top 4 NBA teams were significantly different.

7.3 TWO-WAY CLASSIFICATIONS

In the previous section, we considered a hypothesis involving only one factor with different "levels," for example, blood pressure after administrating different dose of a chemical (treatments). We assumed that all other factors that may effect the blood pressure are controlled or have a insignificant effect.

We now consider a problem involving more than one factor. Suppose that the patients are treated using one of the J different programs (factor A) for reducing the blood pressure and these programs are applied in different, say K, lengths of time (factor B). As a sport example, we may consider an experiment involving three different methods of shooting free throws using two different types of basketballs, or consider an experiment involving four different types of serves in tennis in two different types of courts.

The situation is called a two-way classification or a two-crossed factor model, provided that all possible $J \times K$ combinations (crossings) of treatment A (A levels) and treatment B (B levels) are observed. Thus the population Ω is decomposed into $J \times K$ subpopulations Ω_{jk}, $j = 1, \ldots, J$, $k = 1, \ldots, K$. The model is called nested. The factor B is nested within factor A, if each level of factor B comes at most with one level of factor A. As an example, K different prepared kinds of plants (B) are raised on J plots (A) and the yield Y is measured; say, plants of kind 1 to 5 on plot 1, plant of kind 6 to 10 on plot 2, and so on. Thus, for example, $k = 2$ is always only associated with $j = 1$ and index pairs like $(2,2)$, $(3,2)$, and so on, make no sense.

Consider the crossed model and the $J \times K$ subpopulations Ω_{jk}, from which samples of (for simplicity equal) size N are drawn. The i-th observation in the j,k–sample is denoted by y_{ijk}. Here, we are dealing with several different sample means. They are

$\hat{\mu} = \bar{y}_{...} = \frac{1}{NJK} \sum_{i,j,k} y_{ijk}$ the overall mean (here $\sum_{i,j,k}$ means $\sum_i \sum_j \sum_k$)

$\bar{y}_{.j.} = \frac{1}{NK} \sum_{i,k} y_{ijk}$ the mean over all samples with fixed j-th level of A

$\bar{y}_{..k} = \frac{1}{NJ} \sum_{i,j} y_{ijk}$ the mean over all samples with fixed k-th level of B

$\bar{y}_{.jk} = \frac{1}{N} \sum_i y_{ijk}$ the mean over Ω_{jk}, fixed j-th level of A and k-th level of B

Here each observation may be written (decomposed) as

$$y_{ijk} = \mu + \alpha_j + \beta_k + \gamma_{jk} + \varepsilon_{ijk} \tag{7.8}$$

where μ is the grand mean (in Section 7.1 it was incorporated in $\mu_j = \mu + \alpha_j$). This model assumes that each observation is a linear sum of effects due to factor A, factor B, their interaction AB, and the error (factors that are not considered or are unknown). For example, in this model $\alpha_j = \mu_j - \mu$ presents the effect of the j-th level of factor A, β_k presents the effect of k-th level of factor B, and γ_{jk} the effect of interaction of level j of factor A and level k of factor B. The interaction added to this model plays an important role in many important areas of application, such as medical sciences. For example, certain levels of two chemicals contained in a drug may interact and result in harmful effect. In sports such as tennis interaction of certain type of ball and certain type of racket or court may effect player's performance.

Now, as in Section 7.2, we have to replace the terms involved in model (7.8) by their sample counterparts, such as $\hat{\mu} = \bar{y}_{...}$. Thus

$$\hat{\alpha}_j = \bar{y}_{.j.} - \bar{y}_{...} \quad \text{and} \quad \hat{\beta}_k = \bar{y}_{..k} - \bar{y}_{...} \tag{7.9}$$

Chapter 7: Analysis of Variance and Experimental Design

are estimates for the main effects of the j-th level of factor A and the k-th level of factor B, respectively, subject to the side conditions or constraints

$$\sum_j \alpha_j = \sum_k \beta_k = 0. \tag{7.10}$$

Recall that the sum of the deviations from the mean is always zero. Also,

$$\hat{\gamma}_{jk} = \bar{y}_{.jk} - \bar{y}_{.j.} - \bar{y}_{..k} + \bar{y}_{...} = \bar{y}_{.jk} - \hat{\alpha}_j - \hat{\beta}_k - \hat{\mu} \tag{7.11}$$

is the interaction between the j-th level of factor A and the k-th level of factor B. The errors ε are again assumed to be independent and normally distributed with equal variance. We can summarize the ANOVA by defining different SS. Note that i sums always run from 1 to N, j sums from 1 to J, k sums from 1 to K, respectively. Using the appropriate abbreviations we have

- $SST = SSTotal = \sum_{i,j,k}(\bar{y}_{ijk} - \bar{y}_{...})^2$
- $SSA = SSBetween$ A levels $= NK \sum_j (\bar{y}_{.j.} - \bar{y}_{...})^2$
- $SSB = SSBetween$ B levels $= NJ \sum_k (\bar{y}_{..k} - \bar{y}_{...})^2$
- $SSAB = SSInteraction$ of A and B levels $= N \sum_{j,k}(\bar{y}_{.jk} - \bar{y}_{.j.} - \bar{y}_{..k} + \bar{y}_{...})^2$
- $SSR = SSResidual = SS$ of residuals $= \sum_{i,j,k}(\bar{y}_{ijk} - \bar{y}_{.jk})^2$

Using these, Table 7.3.1 reads as follows:

TABLE 7.3.1 Analysis of Variance (ANOVA) Table For a Two-Way Classification Case

Source of Variation	Sum of Squares (SS)	Degrees of Freedom (d.f.)	Mean Squares (MS)=SS/d.f.	F-Ratio
Factor A	SSA	$J-1$	$MSA = \frac{SSA}{(J-1)}$	$F_A = \frac{MSA}{MSR}$
Factor B	SSB	$K-1$	$MSB = \frac{SSB}{(K-1)}$	$F_B = \frac{MSB}{MSR}$
Interaction	SSAB	$(J-1)(K-1)$	$MSAB = \frac{SSAB}{(J-1)(K-1)}$	$F_{AB} = \frac{MSAB}{MSR}$
Residual	SSR	$JK(N-1)$	$MSR = \frac{SSR}{JK(N-1)}$	
Total	SST	$JKN-1$		

Now, we can test variety of hypotheses using F–ratios as test quantity (statistics).

- There is no effect of A levels. In other words, there is no difference in levels of factor A. This means

$$H_0 : \alpha_1 = \alpha_2 = \ldots = \alpha_J = 0$$

Reject H_0 at significance level α if $F_A > F_{1-\alpha,(J-1,JK(N-1))}$.

- There is no effect of B levels. That is,

$$H_0 : \beta_1 = \beta_2 = \ldots = \beta_K = 0$$

Reject H_0 at significance level α if $F_B > F_{1-\alpha,(K-1,JK(N-1))}$.

- There is no interaction effect between A and B levels, that is,

$$H_0 : \gamma_{jk} = 0 \text{ for all } j,k$$

Reject H_0 at significance level α if $F_{AB} > F_{1-\alpha,((J-1)(K-1),JK(N-1))}$.

Computation Formulas

Again, for calculation of the SS, use

- $SST = \sum_{i,j,k} y_{ijk}^2 - M$
- $SSA = \frac{1}{NK} \sum_j (\sum_{i,k} y_{ijk})^2 - M$
- $SSB = \frac{1}{NJ} \sum_k (\sum_{i,j} y_{ijk})^2 - M$
- $SSR = \sum_{i,j,k} y_{ijk}^2 - \frac{1}{N} \sum_{j,k} (\sum_i y_{ijk})^2$
- $SSAB = SST - SSA - SSB - SSR$

where $M = \frac{1}{NJK} (\sum_{i,j,k} y_{ijk})^2$.

EXAMPLE 7.3.1 Let us demonstrate the application of the two-way classification by considering the data given in Example 7.1.1. The data present the PPG for the top 20 NBA players during the years 2015–2018. The question of main interest was to see if their performances were significantly different in three different seasons or not. To provide an answer, we first decompose the data into three A levels: 2015–2016, 2016–2017, and 2017–2018. So we have $J = 3$. For each year, we further decompose the data into two B levels: top 10 players (1st to 10th) and then 11th to 20th best players. So the value of K is 2. This way we have made six subpopulations with $y_{1,1,1} = 30.1$, $y_{2,1,1} = 29.0$, ..., $y_{10,3,2} = 21.3$. Straightforward calculation yields

$$(\sum_{i,j,k} y_{ijk})^2 = 2,161,194.01, \quad \sum_{i,j,k} y_{ijk}^2 = 36,480.93$$

$$\sum_j (\sum_{i,k} y_{ijk})^2 = 721,414.93, \quad \sum_k (\sum_{i,j} y_{ijk})^2 = 1,088,347.13$$

$$\sum_{j,k} (\sum_i y_{ijk})^2 = 363,313.77, \quad M = 36,019.9$$

Hence

$$SST = 461.03, \quad SSA = 50.85, \quad SSB = 258.34, \quad SSR = 149.55, \quad SSAB = 2.29$$

So we have

$$MSA = 25.42, \quad MSB = 258.34, \quad MSAB = 1.15, \quad MSR = 2.77$$

and

$$F_A = 9.180, \quad F_B = 93.279, \quad F_{AB} = 0.414.$$

Based on the above summary of the data, test at a significance level $\alpha = 0.10$ the following hypotheses:

1. The hypothesis H_0 that there is no effect of A levels
2. The hypothesis H_0 that there is no effect of B levels
3. The hypothesis H_0 that there is no interaction between A and B levels

SOLUTION:

1. The critical value is $F_{0.90,(2,54)} = 2.4$. Since F_A is bigger than the critical value, we reject H_0. That is, there is sufficient evidence to indicate that the players did not perform "equally well" in these 3 years.
2. The critical value is $F_{0.90,(1,54)} = 2.8$. Since F_B is bigger than the critical value, we reject H_0. That is, there is sufficient evidence to indicate that top 10 players did perform significantly better than the next 10 (11th to 20th) best players, which confirms the common sense.
3. The critical value is $F_{0.90,(2,54)} = 2.4$. Since F_{AB} is smaller than the critical value, we do not reject H_0. That is, there is not sufficient evidence to indicate that there is an interaction between A and B levels. In other words, performance and years did not interact. Note that sometimes teams perform better in one season than another for variety of reasons. The following is the ANOVA table for this problem.

Analysis of Variance (ANOVA) Table for Example 7.3.1

Source of Variation	Sum of Squares (SS)	Degrees of Freedom (d.f.)	Mean Squares	F-Ratio
Factor A (Years)	50.85	2	25.42	9.180
Factor B (Ranks)	258.34	1	258.34	93.279
Interaction	2.29	2	1.15	0.414
Residual (Error)	149.55	54	2.77	
Total	461.03	59		

EXAMPLE 7.3.2 Some statisticians suggested the use of two-way classification linear models for analysis of a match schedule for soccer, one model to keep the score of home team, the other to keep the score of the visiting team. For example, for home team μ presents the overall strength, α_j the offensive strength of home team j, and $-\beta_k$ the defensive strength of away team k. They have used two-way classification analysis to obtain a ranking for all national teams.

We finish this section by mentioning that, for ANOVA when normality assumption cannot be justified, it is possible to apply a distribution-free test. The one frequently used is called Kruskal–Wallis H test. We will discuss this in Chapter 10.

7.4 THE EXPERIMENTAL DESIGN

One of the most important issues in statistics is the appropriate way of applying the available theories and methods. People unfamiliar with statistical sciences often collect data and then seek statistical methods for its analysis. This certainly is not recommended by any experienced statistician. In fact, the appropriate way to proceed is to design an experiment before collecting data. By doing this, one makes sure that data collected are relevant and contain information pertaining to the investigation. In short, an experimental design is the set of plans and instructions by which the data in an experiment are collected. An experimental design should be established before any experimental data is collected. If the data are collected without an experimental design, it may not be possible to extract the desired information, and the experimental results will not be reproducible. Without an experimental design, data are usually collected either improperly or in a manner that does not allow the researcher to apply any known method of analysis.

Depending on what the problem objective is, researchers will often design an experiment that will lead to test an appropriate hypothesis or estimate effect of a certain treatment or factor. For example, a completely randomized design, or the independent sampling design, is used when comparing several treatments and independent samples could be made available for each treatment. An example of this is when our objective is to find out if a certain diet helps male and female athletes. Here, independent samples could be made available for each gender. A different design that presents an extension of the paired difference design discussed in Chapter 6 is called the randomized block design, which uses relatively homogeneous blocks of experimental unit in each block to compare the treatment means. For example, when studying the effect of exercise on blood pressure, we cannot use completely randomized design. This is because the factors such as age, amount of exercise, gender, family history, and so on, will effect the value of blood pressure, so we may decide to block on this factor. For instance, we may consider age as a block and divide that into several intervals. This enables an investigator to compare people, in the same age group, who exercise with those who do not.

As discussed in previous sections, the method of analysis for either design involves a comparison of the MSB that measures the variation among the treatment means and MSW that measures the variation among experimental units. Large values of the ratio MSB/MSW is an indication that the means of at least two of the populations differ. The ANOVA may also be used to test existence of an interaction among factors considered. This is usually done through a two-factor factorial experiment.

We now proceed to present some specific examples for each design together with their solutions using the ANOVA. It is important to remember that the three basic principles of experimental design are:

REPLICATION (repetition of the basic experiment)

RANDOMIZATION (allocation and random selection of order in which trials are to be performed)

BLOCKING (a technique to increase the precision by reducing the possible effects of factors other than the one investigated).

Also, when conducting an experiment the following should be included in the set of plans:

Chapter 7: Analysis of Variance and Experimental Design

1. A clear statement of the problem to be addressed
2. A determination of the experimental factors to be studied
3. An identification of the levels of each experimental factor
4. An identification of the response (outcome) of the experiment
5. A statement of how the experiment will be conducted
6. Appropriate methods of analysis

7.4.1 The Completely Randomized Design

The director of a fitness club wants to compare the weekly mean number of satisfaction notes received about the staff offering assistance and training during the three daily shifts: morning, afternoon, and evening. Her plan is to select random samples of weekly records from each shift and count the number of the satisfaction notes received during the week. The first question that arises here is, "what type of experimental design is this?" Since the director is employing an independent sampling design, this is a completely randomized design. Here, the means of $J = 3$ populations (subpopulations) are to be compared by selecting independent random samples from each of the populations.

It should be pointed out that many other factors could effect the number of satisfaction notes (e.g., number of people who come to the club in different shifts, their age, etc.). Although it is possible to include these factors and study their effect, in this section, we will concentrate on only one factor. Using the data presented in Table 7.4.1, we want to set up a test to compare the weekly mean number of satisfaction notes against the three shifts. Here, we wish to test the null hypothesis that the weekly mean number of satisfaction notes is the same for each of the three shifts, against the alternative that at least two of the means are different. That is,

$$H_0 : \mu_1 = \mu_2 = \mu_3 \quad vs \quad H_1 : \text{At least two of the means differ}$$

where $\mu_1, \mu_2,$ and μ_3 are the true mean number of satisfaction notes received weekly against the first, second, and third shifts, respectively. The test is based upon the following assumptions:

TABLE 7.4.1 Data Concerning Weekly Number of Satisfaction Notes

Number of Weekly Complaints		
First Shift	Second Shift	Third Shift
12	9	10
9	11	11
11	12	15
9	8	14
9	8	10
	6	12
Total: 50	54	72

- The distribution of the number of weekly satisfaction notes against each of the three shifts are approximately normal.
- The variance of the number of weekly satisfaction notes is the same for each of the three shifts.

The test statistic is F = MSB/MSW (= MSTr/MSE) and the null hypothesis is rejected at significance level α if $F > F_{1-\alpha,(J-1,N-J)}$.

Using the data and performing the calculations we find

$$SST = 81.88 \quad SSB = 27.88$$

$$\text{MSB} = \frac{SSB}{J-1} = \frac{27.88}{2} = 11.94, \quad \text{MSW} = \frac{SSW}{N-J} = 54/(17-3) = \frac{54}{14} = 3.86,$$

and $F = \frac{\text{MSB}}{\text{MSW}} = \frac{11.94}{3.86} = 3.61.$

The critical value of F is based on $J - 1 = 2$ numerator degrees of freedom and $N - J = 14$ denominator degrees of freedom. Thus, for $\alpha = 0.05$, the rejection region consists of values of F such that $F > F_{0.95,(2,14)} = 3.74$. Since the value of the test statistic does not lie within the rejection region, there is insufficient evidence to conclude that a significant difference exists among the weekly mean number of satisfaction notes against the different shifts.

Let us summarize the results of the ANOVA in an ANOVA table. The ANOVA summary table for this completely randomized design is as follows (Shift = Between, Error = Within)

Analysis of Variance (ANOVA) Table

Source	Sum of Squares (SS)	Degrees of Freedom (d.f.)	Mean Squares (MS)	F
Shift	27.88	2	13.94	3.61
Error	54.00	14	3.86	
Total	81.88	16		

We can also construct a 90% confidence interval for μ_1, the weekly mean number of satisfaction notes about the staff of the first shift. The general form of a confidence interval for μ_1 is

$$\bar{y}_1 \pm t_{1-\alpha/2,(N-J)} \frac{s}{\sqrt{N_1}}$$

where $s = \sqrt{\text{MSW}}$ is the estimate of σ, the common standard deviation of the J populations. For this example, we have

$$\bar{y}_1 = 50/4 = 10, \quad s = \sqrt{\text{MSW}} = \sqrt{3.86} \approx 1.96,$$

and $t_{1-\alpha/2,(N-J)} = t_{0.95,(14)} = 1.761$, assuming that a 90% confidence interval is required. Thus, the required confidence interval is

$$10.0 \pm 1.761 \left(\frac{1.96}{\sqrt{5}} \right) = 10.0 \pm 1.5, \text{ or } (8.5, 11.5).$$

This means that we are 90% confident that the weekly mean number of satisfaction notes about the staff of the first shift lies within this interval.

Chapter 7: Analysis of Variance and Experimental Design

Although there is no significant difference between the true mean number of satisfaction notes, we may for demonstrating purpose, construct a 95% confidence interval for the difference in the weekly mean numbers of satisfaction notes about the third and first shifts. To obtain a 95% confidence interval for $\mu_3 - \mu_1$, we use the following general form:

$$\bar{y}_3 - \bar{y}_1 \pm t_{1-\alpha/2,(N-J)} s \sqrt{\frac{1}{N_3} + \frac{1}{N_1}},$$

where as before $s = \sqrt{MSW}$. Here we have,

$$\bar{y}_3 = \frac{72}{6} = 12.0, \quad \bar{y}_1 = \frac{50}{5} = 10.00, \quad s = \sqrt{SW} = \sqrt{3.86} \approx 1.96.$$

$$\bar{y}_3 - \bar{y}_1 \pm t_{.95,(14)} s \sqrt{\frac{1}{N_3} + \frac{1}{N_1}} = (12.0 - 10.0) \pm 2.145 \, (1.96) \sqrt{\frac{1}{6} + \frac{1}{5}}$$
$$= 2.0 \pm 2.5 \Rightarrow (-0.5, 4.5).$$

Thus, we estimate with 95% confidence that μ_1, the weekly mean number of satisfaction notes about the first shift, could be larger than μ_3, the weekly mean number of satisfaction notes about the third shift, by as much as 0.5, or it could be less than μ_3 by 4.5. Note that, the confidence interval contains zero. This is not surprising since we failed to reject the hypothesis of equality of the weekly mean numbers of satisfaction notes about the three shifts.

A PRACTICE PROBLEM In the 2002 Soccer World Cup, FIFA (Federation Internationale De Football Association) allowed use of new soccer balls. Suppose that prior to this event the FIFA wanted to compare the mean distances associated with four different brands of soccer balls when struck with a driver. A random sample of 10 balls of each brand were chosen and were hit in a random sequence. The data are given in the table below

Distances for Four Brands of Soccer Balls (Meters)

Brand A	Brand B	Brand C	Brand D
63.2	51.2	51.6	69.7
62.9	45.1	48.6	63.2
65.0	48.0	49.4	77.5
54.5	51.1	42.0	67.4
64.3	60.5	46.5	70.5
57.0	50.0	51.3	65.5
62.8	53.9	61.8	70.7
64.4	44.6	49.0	72.9
60.6	54.6	47.1	75.6
55.9	48.8	45.9	66.5

Test an appropriate hypothesis and draw a conclusion using $\alpha = 0.05$.

7.4.2 The Randomized Block Design

Suppose that the following experiment was designed to investigate the effect of information received by a player regarding the quality of instruction to be followed. Five players were asked to perform a set play under each of three experimental conditions, which were presented in a random order: (1) player receives positive feedback and encouragement from coach; (2) player receives no word from coach; and (3) player receives negative feedback and criticism from coach. The experimenter recorded the number of moves correctly performed during the allotted time. (Table 7.4.2)

TABLE 7.4.2 Players Performance Under Different Conditions

Player	Experimental Condition			Totals
	1	2	3	
1	32	30	25	87
2	25	26	21	72
3	19	17	14	50
4	15	12	10	37
5	12	10	8	30
Totals	103	96	78	276

This is an example of a randomized block design, in which the players represent 5 blocks of relatively homogeneous experimental units (Factor B). Note that this design is an extension of paired difference design discussed in Chapter 6. Since the same player is used in three experimental conditions, the collected data sets are not independent. There are $J = 3$ treatments (experimental conditions), each of which is randomly assigned once to each block or player (Factor A). We would like to:

(a) Set up the test to compare the mean number of moves performed correctly for the three experimental conditions.
(b) Set up the test to determine whether blocking is important in this experiment, that is, if the mean numbers of moves performed correctly differ for the five players.

For part (a), it is desired to test the null hypothesis that the mean number of moves performed correctly is the same for the three experimental conditions, against the alternative that at least two of the means are different. That is

$$H_0 : \mu_1 = \mu_2 = \mu_3, \quad vs \quad H_1 : \text{At least two of the means are different,}$$

where μ_1, μ_2, and μ_3 are, respectively, the population mean numbers of moves performed correctly under experimental conditions 1, 2, and 3.

The test is based on the following assumptions:

Chapter 7: Analysis of Variance and Experimental Design

- The probability distributions of the number of moves performed correctly are approximately normal for all player-condition combinations.
- The variances of all the probability distributions are equal.

The test statistic is $F = \frac{MSA}{MSR}$ where MSA is the MS for factor A and MSR is the residual MS. The null hypothesis will be rejected (at significance level α) if $F > F_{1-\alpha,(J-1,N-J-K+1)}$.

For part (b) we perform a test to compare the block means.

H_0: The population mean number of moves performed correctly for the five players are equal (i.e., the five block means are equal).

H_1: The mean number of moves performed correctly are different for at least two of the players (i.e., at least two of the block means differ).

The test requires the same assumptions as the test in part (a), and is based on the test statistic $F = \frac{MSB}{MSR}$.

The null hypothesis is rejected for all values of F such that $F > F_{1-\alpha,(K-1,N-J-K+1)}$.

To set up an ANOVA table the following calculations are required:

$\Sigma_{j,k} y_{jk}^2 = 32^2 + 25^2 + 19^2 + \ldots + 14^2 + 10^2 + 8^2 = 5914$

$M = \frac{(\Sigma_{j,k} y_{jk})^2}{N} = \frac{(276)^2}{15} = 5078.4,$

SSTotal $= \Sigma_{j,k} y_{jk}^2 - M = 5914 - 5078.4 = 835.6$

SS (Treatment) = SSA $= \frac{(\Sigma_k y_{1k})^2}{K} + \frac{(\Sigma_k y_{2k})^2}{K} + \frac{(\Sigma_k y_{3k})^2}{K} - M$
$= \frac{(103)^2}{5} + \frac{(95)^2}{5} + \frac{(78)^2}{5} - 5078.4 = 65.2$

SS (Block) = SSB $= \frac{(\Sigma_j y_{j1})^2}{J} + \frac{\Sigma_j y_{j2})^2}{J} + \frac{\Sigma_j y_{j3})^2}{J} + \frac{\Sigma_j y_{j4})^2}{J} + \frac{\Sigma_j y_{j5})^2}{J} - M$

$= \frac{(87)^2}{3} + \frac{(72)^2}{3} + \frac{(50)^2}{3} + \frac{(37)^2}{3} + \frac{(30)^2}{3} - 5078.4 = 762.27$

SS (Error) = SSR = SSTotal - SSA - SSB $= 835.6 - 65.2 - 767.27 = 8.13$

MSA $= \frac{SSA}{J-1} = \frac{65.2}{2} = 32.6,$ MSB $= \frac{SSB}{K-1} = \frac{762.27}{4} = 190.57$

MSR $= \frac{SSR}{N-J-K+1} = \frac{8.13}{15-3-5+1} = \frac{8.13}{8} = 1.02.$

Then the computed value of the test statistic is

$$F = \frac{MSA}{MSR} = \frac{32.60}{1.02} = 31.96.$$

The critical value of F is based on $J-1 = 2$ numerator degrees of freedom and $N-J-K+1 = 8$ denominator degrees of freedom. Thus, for $\alpha = 0.05$, the rejection region consists of values of F such that $F > F_{0.95,(2,8)} = 4.46$. Since the value of the test statistic lies within the rejection region, we conclude that the population mean number of moves performed correctly differ for at least two of the experimental conditions. Also

$$F = \frac{MSB}{MSR} = \frac{190.56}{1.02} = 186.83.$$

Since the calculated $F = 186.83$ greatly exceeds the critical value of $F_{0.95,(2,8)} = 3.84$, we have strong evidence that the mean number of moves performed correctly differ among the five players. Thus, the decision to use a randomized block design was wise.

The ANOVA summary table for this randomized block design is as follows:

Source	Sum of Squares (SS)	Degrees of Freedom (d.f.)	Mean Squares (MS)	F
Conditions (Treatments)	65.20	2	32.60	31.96
Players(Blocks)	762.27	4	190.57	186.83
Error	8.13	8	1.02	
Totals	835.60	14		

We now construct a 95% confidence interval for $\mu_1 - \mu_2$, the difference in mean numbers of moves performed correctly under conditions 1 and 3. The general form of a 95% confidence interval for $\mu_1 - \mu_3$ is

$$\bar{y}_1 - \bar{y}_3 \pm t_{.975,(8)} s \sqrt{\frac{1}{K} + \frac{1}{K}},$$

where $K = 5$ is the number of blocks, $s = \sqrt{MSR}$, and the distribution of t is based on $N - J - K + 1 = 8$ degrees of freedom. For this example, we have

$$\bar{y}_1 = \frac{103}{5} = 20.6, \ \bar{y}_3 = \frac{78}{5} = 15.6, \ t_{.975,(8)} = 2.306, \text{ and}$$

$$s = \sqrt{MSR} = \sqrt{1.02} \approx 1.01.$$

Substitution yields the desired confidence interval:

$$(20.6 - 15.6) \pm 2.306(1.01)\sqrt{\frac{1}{5} + \frac{1}{5}} = 5.0 \pm 1.47 \Rightarrow (3.53, 6.47).$$

We estimate with 95% confidence that the population mean number of moves performed correctly under condition 1 is between 3.53 and 6.47 higher than the mean number moves performed correctly under condition 3. Since this interval does not include zero, it follows that the true mean number of moves performed correctly under conditions 1 and 3 are different.

EXAMPLE 7.4.1 Three of the currently most popular sport shows (football, basketball, and baseball) produced the ratings presented in Table 7.4.3 (percentage of the television audience tuned into the show) over a period of 4 weeks:

Chapter 7: Analysis of Variance and Experimental Design

TABLE 7.4.3 Ratings for Three Popular Sport Shows

Week	A (Football)	B (Basketball)	C (Baseball)	Totals
1	33.7	27.4	22.8	83.9
2	37.1	31.2	19.7	88.0
3	34.1	31.4	24.8	90.3
4	29.4	27.2	27.9	84.5
Totals	134.3	117.2	95.2	346.7

We would like to answer the following questions:

(a) Is there evidence (at $\alpha = 0.01$) that the mean ratings differ for the three shows?
(b) Is there evidence (at $\alpha = 0.01$) that the use of weeks as blocks is justified in this experiment?
(c) Construct a 95% confidence interval for the difference in mean ratings between shows B and C.
(d) What assumptions are necessary for the validity of the procedures used in parts (a), (b), and (c)?

SOLUTION:

(a) The hypothesis to be tested has the following elements:

$$H_0: \mu_1 = \mu_2 = \mu_3, \quad vs \quad H_1: \text{At least two of the means differ}$$

where μ_1, μ_2, and μ_3 are the population mean ratings for shows A, B and C, respectively.

At $\alpha = 0.01$, the null hypothesis will be rejected for all values of the test statistic F such that $F > F_{0.99,(J-1,N-J-K+1)} = F_{0.99,(2,5)} = 10.92$. Here the following computations are required:

$(\Sigma_{j,k} y_{jk}^2)^2 = (33.7)^2 + (37.1)^2 + \cdots + (24.8)^2 + (27.9)^2 = 10{,}290.65$

$M = \frac{(\Sigma_{j,k} y_{j,k})^2}{N} = \frac{(346.7)^2}{12} = 10{,}016.74$

SSTotal $= \Sigma_{jk} y_{jk}^2 - M = 10{,}290.65 - 10{,}016.74 = 273.91$

SSA $= \frac{(\Sigma_k y_{1k})^2}{K} + \frac{(\Sigma_k y_{2k})^2}{k} + \frac{(\Sigma_k y_{3k})^2}{K} - M = \frac{(134.3)^2}{4} + \frac{(117.2)^2}{4} + \frac{(95.2)^2}{4} - 10{,}016.74 = 192.10$

SSB $= \frac{(\Sigma_j y_{j1})^2}{J} + \frac{(\Sigma_j y_{j2})^2}{J} + \frac{(\Sigma_j y_{j3})^2}{J} + \frac{(\Sigma_j y_{j4})^2}{J} - M$
$= \frac{(83.9)^2}{3} + \frac{(88.0)^2}{3} + \frac{(90.3)^2}{3} + \frac{(84.5)^2}{3} - 10{,}016.74 = 9.11$

SSR = SSTotal - SSA - SSB = 173.91 - 192.10 - 9.11 = 72.70

MSA $= \frac{SSA}{J-1} = \frac{192.10}{3} = 96.05$ 　　MSB $= \frac{SSB}{K-1} = \frac{9.11}{3} = 3.04$

MSR $= \frac{SSR}{N-J-K+1} = \frac{72.70}{12-3-4+1} = \frac{72.70}{6} = 12.12$.

Now the test statistic is

$$F = \frac{MSB}{MSR} = \frac{96.05}{12.12} = 7.92.$$

This value does not lie within the rejection region. There is insufficient evidence (at $\alpha = 0.01$) to conclude that there are differences in the mean ratings for the three shows.

(b) The test statistic is
$$F = \frac{MSA}{MSR} = \frac{3.04}{12.12} = 0.25$$
and the rejection region consists of values of F such that $F > F_{0.99,(K-1,N-J-K+1)} = F_{0.99,(3,6)} = 9.78$. The sample does not provide strong evidence to indicate that blocking is important in this experiment.

(c) The general form of a 95% confidence interval for $\mu_2 - \mu_3$ is
$$\bar{y}_2 - \bar{y}_3 \pm t_{0.975,(6)} s \sqrt{\frac{1}{K} + \frac{1}{K}}$$

where
$$\bar{y}_2 = \frac{117.2}{4} = 29.3, \quad \bar{y}_3 = \frac{95.2}{4} = 23.8, \quad t_{0.975,(6)} = 2.447, \quad \text{and} \quad s = \sqrt{MSR} = \sqrt{12.12} = 3.48.$$

Substitution yields:
$$29.3 - 23.8 \pm 2.447(3.48)\sqrt{\frac{1}{4} + \frac{1}{4}} = 5.5 \pm 6.02, \Rightarrow (-0.52, 11.52).$$

This interval is very wide and includes zero. Thus, we cannot conclude that the population mean ratings for shows B and C are different. This is consistent with the F-test of part (a), in which we failed to reject the null hypothesis of the quality of the population means.

(d) The hypothesis test and confidence interval procedures require the following assumptions:

- The probability distributions of television ratings corresponding to all the show-week combinations are approximately normal.
- The variance of all the probability distributions are equal.

7.4.3 Factorial Experiments

A fitness center employs four trainers, each of whom is trained to work at each of three machines. After a month, the manager noticed that some trainers were more comfortable and effective with certain machines. In an effort to investigate the possibility of interaction between trainer and machine, each trainer was assigned to each machine for two 1-week periods. The average number of mistakes produced during the trials were recorded. What type of experimental design does this represent?

The design is a two-factor factorial experiment where factor A (trainer) is at $J = 4$ levels and factor B (machine) is at $K = 3$ levels. The complete 4×3 factorial experiment is replicated $r = 2$ times. The data for the experiment are shown in Table 7.4.4. We would like

(a) Perform an ANOVA for the data and construct an ANOVA table.
(b) Perform a test to see whether there is sufficient evidence to indicate an interaction between trainers and machines?

Chapter 7: Analysis of Variance and Experimental Design

TABLE 7.4.4 Data for Trainers and Machines

Trainer	Machine 1		Machine 2		Machine 3	
1	5	5	15	20	10	10
2	10	16	6	6	14	18
3	24	16	12	6	12	8
4	8	6	6	9	4	4

SOLUTION:

(a) The following preliminary calculations are required:

$$M = \frac{(\Sigma_{i,j,k} y_{ijk})^2}{N} = \frac{(5+5\ldots+4+4)^2}{24} = \frac{(250)^2}{24} = 2604.16657$$

$$\text{SSTotal} = \Sigma_{i,j,k} y_{ijk}^2 - M = 5^2 + 5^2 + \ldots + 4^2 + 4^2 - M = 3292 - 2604.16667 = 687.83333.$$

$$\text{SSA} = \frac{65^2 + 70^2 + 78^2 + 37^2}{3(2)} - M = \frac{16,578}{6} - 2604.16667 = 158.83333$$

$$\text{SSB} = \frac{90^2 + 80^2 + 80^2}{4(2)} - M = \frac{20,900}{8} - 2604.16667 = 8.33333$$

$$\text{SSAB} = \frac{10^2 + 34^2 + 20^2 + \ldots + 14^2 + 15^2 + 8^2}{2} - \text{SSA} - \text{SSB} - M$$

$$= \frac{6378}{2} - 158.83333 - 8.33333 - 2604.16667 = 417.66667$$

$$\text{SSR} = \text{SSTotal} - \text{SSA} - \text{SSB} - \text{SSAB}$$

$$= 687.83333 - 158.83333 - 8.33333 - 417.66667 = 103.00000$$

$$\text{MSA} = \frac{\text{SSA}}{J-1} = \frac{158.3333}{3} = 52.94444, \quad \text{MSB} = \frac{\text{SSB}}{K-1} = \frac{8.33333}{3} = 4.16667$$

$$\text{MSAB} = \frac{\text{SSAB}}{(J-1)(K-1)} = \frac{417.66667}{(3)(2)} = 69.61111$$

$$\text{MSR} = \frac{\text{SSR}}{JK(r-1)} = \frac{103.00000}{(4)(3)(1)} = 8.58333.$$

The results are summarized in the following ANOVA table.

Source	SS	d.f.	MS	F-Ratio
Trainer (A)	158.83	3	52.94	6.17
Machine (B)	8.33	2	4.17	.49
Trainer–Machine Interaction (AB)	417.67	6	69.61111	8.11
Error	103.00	12	8.58	
Total	687.83	23		

(b) The hypotheses of interest are

H_0: Trainer and machine do not interact in their effect on average number of mistakes per trial

H_1: There is interaction between the factors trainers and machines

The test statistic is the ratio of the MS for interaction and for error:

$$F = \frac{MSAB}{MSR} = \frac{69.61111}{8.5833} = 8.11.$$

From the ANOVA table constructed in part (a), we see that the degrees of freedom for the numerator and denominator of the F statistic are 6 and 12, respectively. Thus, at a significance level 0.05, we will reject H_0 if $F > 3.00$. Since the computed value of the test statistic falls within the rejection region ($8.11 > 3.00$), we reject H_0. There is sufficient evidence to indicate that trainers and machines interact. Suppose now we want to

(a) Construct a 95% confidence interval for the average number of mistakes per trial made by Trainer 2 using Machine 3.

(b) Construct a 90% confidence interval for the difference between the mean numbers of mistakes for Trainers 2 and 3 working using Machine 3.

(a) A 95% confidence interval for the mean associated with Trainer 2 using Machine 3 is

$$\bar{y}_{2,3} \pm t_{0.975,(12)} \left(\frac{s}{\sqrt{r}}\right)$$

where $\bar{y}_{2,3}$ is the mean of the $r = 2$ values given for the factor level combination of trainer 1 and machine 3, $s = \sqrt{MSR} = \sqrt{8.58333} = 2.930$, and $t_{0.975,(12)} = 2.179$ based on 12 degrees of freedom. Substitution of these values into the confidence interval formula yields

$$\frac{32}{2} \pm 2.179 \left(\frac{2.930}{\sqrt{2}}\right) = 16 \pm 4.51 \quad \Rightarrow \quad (11.49, 20.51).$$

We are 95% confident that the mean number of mistakes made by Trainer 2 using Machine 3 is between 11.49 and 20.51.

Chapter 7: Analysis of Variance and Experimental Design

(b) A 90% confidence interval for the difference between the mean numbers of mistakes for Trainers 2 and 3 using Machine 3 is given by

$$\bar{y}_{2,3} - \bar{y}_{3,3} \pm t_{.95,(12)} s \sqrt{2/r}$$

where $\bar{y}_{2,3}$ and $\bar{y}_{3,3}$ means of the values obtained by Trainers 2 and 3, respectively, using Machine 3, $t_{.95,(12)} = 1.782$ and $s = 2.930$ as in Part (I). Substitution yields the interval

$$\frac{32}{2} - \frac{20}{2} \pm 1.782(2.930)\sqrt{2/2} = 6 \pm 5.22 \Rightarrow (.78, 11.22).$$

Since all the values in the interval are positive, we are 90% confident that, when using Machine 3, on average Trainer 2 makes more mistakes than Trainer 3.

A PRACTICE PROBLEM Social issues related to sports are studied by many researchers. In an article published in the *Journal of Sports and Social Issues*, Vol. 21, No. 1, Feb. 1997 Gan, Su-lin and co-researcher compared males and females with respect to their enjoyment of a close college basketball game. The experiment was designed in the following manner: a group of students watched one of the eight National Collegiate Athletic Association (NCAA) games televised live. Four categories were used according to the closeness of scores at the game's conclusion. Having watched the game, each student rated his or her enjoyment from 0 to 10 presenting "not at all" to "extremely." The data are presented in the following table showing the $4 \times 2 = 8$ treatment means

Suspense	Female	Male
Minimal ($\geq$15 points differential)	2.73	1.77
Moderate (10–14 point differential)	4.34	5.38
Substantial (5–9 point differential)	7.52	7.16
Extreme (1–4 differential)	4.92	7.59

The design used here is a 4×2 factorial design with four levels for suspense and two levels for gender.

Suppose that the *F*-test for interaction had the value of 4.42 with degrees of freedom 3 and 68 with *p*-value = 0.007. What conclusion would you draw regarding the possible interaction between suspense and gender?

7.5 FINAL WORDS

Statistical design of experiments help researchers to study the variability in data and discover or understand their possible sources and also identify critical variables. When designing an experiment, researchers pay close attention to the following components:

1. Problem objective (statement of research study)
2. Formulation of hypothesis
3. Design of experiment

4. Determination of statistical tools to be used in conducting research
5. Implementation of experiment
6. Application of statistical techniques
7. Drawing of inference or conclusions
8. Evaluation of whole study and recommendation for future study or replication.

7.6 USING R

There exist build-in functions in R to perform ANOVA analysis. It will be convenience for user to import data from other resource or format, such as comma separated value (CSV) file. Therefore, we prefer to demonstrate how we can read data to R first.

There are different popular formats to store your data. However, the most popular choice will be Excel worksheets. In case, user wish to read Excel worksheet directly to R, one may need to install a third-party plugin/package performing ODBC-based transactions on Windows operating system, or utilize JDBC interfaces on UNIX like system. Yet why not convert the Excel worksheet to a proper format, which does not need any extra package? The choice here will be CSV format. Need to say, plain text format will be also preferred, but CSV format will be better organized than just plain text, thus will be easier to be read.

The R function **read.csv()** will read CSV file to R. As usual, you need to assign a variable to that, so the loaded data will be stored to your variable for future use. You can simply provide the full path, like "C:/data.txt," as argument to the function. You will be curious why we use "/" instead of "\" in a Windows-/DOS- path. It is because R was origin on Linux/UNIX like operating system and then ported to Windows system. Well, you can use "\" to express path if you want, but R will require you to type two "\" to express a single "\" in String. For example, you have to use "C:\\Data.txt."

The following line will read PPG data in example 7.1.1 to R.

ppg = **read.csv**('C:\\Example7.1.1.csv')

Since the following analysis will focus on our loaded ppg data, we can simply use **attach(ppg)** to let R target on ppg data only and thus you can refer its columns using column names directly. Otherwise, we must use **ppgPPG, ppgSeason, ppg$Rank** and **ppg$Top** to refer to those columns in data ppg.

attach(ppg)

Now the data ppg is well prepared for ANOVA, again we will use Season as Factor A, Top as Factor B, and PPG as Y.

model = **glm**(PPG ~ Season + Top + Season:Top)

As we discussed in the chapter, ANOVA is based on a linear model. The function **glm()** will fit a generalized linear model using provided "formula." The "formula" in R is used to define an

Chapter 7: Analysis of Variance and Experimental Design

equation, which tells the "shape" of your model. Instead of "=," R used "~" to separate dependent variable (Y) and independent variables (X), thus you need to write "Y ~ X." When you have more than one independent variables, simple add (+) them to the formula. "Season:Top" indicates the interaction effect in the model. Next, we will extract the ANOVA table from previous model.

summary(aov(model))

The function **aov()** obtains ANOVA results, and **summary()** build the ANOVA table based on extracted results in aov(). You will see

	Df	Sum Sq	Mean Sq	F value	Pr(>F)	
Season	2	50.85	25.42	9.180	0.00037	***
Top	1	258.34	258.34	93.279	2.31e−13	***
Season:Top	2	2.29	1.15	0.414	0.66310	
Residuals	54	149.55	2.77			

Signif. codes: 0 '***' 0.001 '**' 0.01 '*' 0.05 '.' 0.1 ' ' 1

CHAPTER 8

Regression Analysis

Regresssion and correlation analysis is an area of statistics that deals with methods for investigating the existence of associations and, if present, the nature of associations among various observable quantities. It is a commonly used statistical modeling tool and is comparable, in importance, to differential (difference) equations in mathematics, physics, and engineering. For example, as a student you know that there is a relationship between the number of hours of study and the grade you receive on a test. However, you may not know the nature of the relationship and the way it can be expressed. Thus, you may be interested in knowing more about such relationship. A manager may want to know how money spent on advertising and increase on sale of company's product are related. In exercise science, researchers may be interested in the relationship between performance on a distance run and a step test as a measure of cardiovascular fitness. A physical educator may want to know whether there is any association between the height of a basketball player and his or her performance measured, for example, by points scored per game, or PPM played. When existence and the form of such a relationship is uncovered, players performance may be predicted using their age, height, and other variables that have association with the performance. Variables used for prediction are called predictors (explanatory) or independent variables and the variable predicted (in this case performance) is called the response or dependent variable. This chapter is devoted to an introductory presentation of the regression analysis and its applications to sports.

Regression analysis plays a major role in modeling in which data are assumed to be generated as a specific function of the independent variables, some unknown parameters, and error or noise (factors that are not considered or measured). Just as a matter of interest, in recent years a small minority in statistics, and many in areas outside of statistics, taking advantage of computers have begun to use algorithmic modeling, also loosely known as data mining. This approach makes no assumptions about how the data are generated. Some statisticians believe that this area of statistics will become increasingly important in applications, since it deals with large data sets.

8.1 INTRODUCTION

Correlation introduced earlier in Section 1.7 is useful for measuring the association of two variables. Regression goes far beyond this and provides a methodology that uses one or more variables (predictors) to help explain the change in the variable of interest (response).

Developing a regression equation is not an easy task because we need to have some idea about the nature of the relationship between each of the independent variables and the dependent variable. For example, a student needs to know how the number of hours of study affects his or her grade. If we propose a linear relationship, that may imply that as the number of hours of study is increased or decreased, his or her grade will increase or decrease. A quadratic relationship may suggest that his or her grade will increase over a certain range but will decrease over a different range. Perhaps certain combinations of numbers of hours of study and other independent variables influence the grade in one way, while other combinations influence it in other ways. The number of different models that can be proposed is virtually infinite.

You may have encountered various models in other disciplines. For instance, the following represents relationships in natural sciences: $E = mc^2, F = m\gamma,$ and $D = \frac{1}{2}gt^2$, where E, m, c, F, γ, $D, g,$ and t are respectively energy, mass, speed of light, force, acceleration, distance, gravitational acceleration, and time.

In business and economics courses, you might have seen relations such as

$$\text{Profit} = \text{Revenue} - \text{Costs}$$
$$\text{Total cost} = \text{Fixed cost} + (\text{Variable cost} \times \text{Number of units produced})$$

These are examples of deterministic models, so named because—except for small measurement errors—such equations allow us to determine the value of the dependent variable (variable on the left side of the equations) from the values of the independent variables (variables on the right side of the equations). In many practical applications deterministic models are unrealistic. For example, is it reasonable to think that the selling price of a house can be determined solely on the basis of its size? Clearly, the size of a house affects the price, but many other variables (some of which may not be measurable) also influence the price. What must be included in most practical models is a term to represent the randomness (uncertainty) that is a part of real-life process. Such a model is called non deterministic, probabilistic, or stochastic.

This chapter investigates regression relationships between two or more variables based on the available measurements. To start the discussion, consider a runner who tends to run at a certain speed S. Then, the distance D he runs is proportional to the amount of time T he runs ($D = ST$). If for N running sessions $D_1, D_2, \ldots, D_N$ are the distances and $T_1, T_2, \ldots, T_N$ are the times measured, then because of measurement error and the fact that runners may not be able to keep the same speed, the quotient $D_i/T_i(= S_i)$ would be different for each i. The question that arises is the following: which value should we attach to S in the formula $D = ST$ so that D can be "best" predicted for a given value of T? Clearly, such a value needs to be "best" in some sense and for this we need to define which estimate is good. For example, we could use the average value of S_i's or perhaps their median or mode. As was discussed in Chapter 5 in situations like this we need to use the data in some optimum way and provide a point estimate for S and maybe a confidence interval, too.

We note that the same problem arises in any situation where the nature of the law governing the phenomenon under consideration is not known precisely, for example, the relationship between the number of games played by a certain soccer player and the total number of goals she scores, between the amount of time spent to work out and the amount of weight lost, between size of a house and its selling price, between the interest rate and purchasing price of the gold, and so on. In all these cases we need to adapt a special kind of functional dependency between the independent variable (input) X and the dependent variable (output) Y, which seems reasonable. This means consideration of models such as the one given below, known as regression model

$$E(Y) = G(X;a,b,\ldots). \tag{8.1}$$

This model implies that the expected value or average value of Y can be expressed as a function of X that involves some unknown parameters (to be specified). The simplest and perhaps the most frequently used model is when $G(X)$ is a linear function of X, ($E(Y) = a + bX$). Since the true value of Y will have some deviation from its expected value, the regression model (8.1) may be written as

$$\begin{aligned} Y &= G(X;a,b,\ldots) + \text{ Error} \\ &= \text{Deterministic or Systematic Part or Trend } + \text{ Random Part} \\ &= \text{Signal } + \text{ Noise} \\ &= \text{Smooth Part } + \text{ Rough Part}. \end{aligned}$$

This means that to create a probabilistic model, we start with a deterministic model that approximates the relationship we want to model. We then add a random term that measures the error of the deterministic component or deviation from it. For example, the real estate agent knows that the cost of building a new house is about $50 per square foot and that lots of certain size sell for about $30,000. The approximate selling price would then be

$$Y = 30,000 + 50X$$

where Y = selling price and X = size of the house in square feet. A house of 2,000 square feet would therefore be estimated to sell for

$$30,000 + 50(2,000) = \$130,000.$$

We know, however, that the selling price is not likely to be exactly $130,000. Prices may actually range from $90,000 to $200,000. In other words, the deterministic model is not really suitable. To represent this situation properly, we should use the probabilistic model, namely

$$Y = 30,000 + 50X + \varepsilon$$

where ε represents the random term (also called the error variable) – the difference between the actual selling price and the estimated price based on the size of the house. The random term thus accounts for all of the variables, measurable and unmeasurable, that are not part of the model. The value of ε will vary from one sale to the next, even if X remains constant. That is, houses of exactly the same size will sell for different prices because of differences in location, selling season, decorations, and other variables.

In the regression model, the deterministic or the systematic part contains unknown parameters $a, b, \ldots$. In practice, usually all the available information is utilized first for specifying the form of the function $G(X)$. The next step is to use a sample of measurements $(X_1, Y_1), \ldots, (X_n, Y_n)$, and consider questions such as

- What are the "best" point estimates for the unknown parameters $a, b, \ldots$?
- Can we provide confidence intervals for these parameters?
- Is the model in some sense adequate?

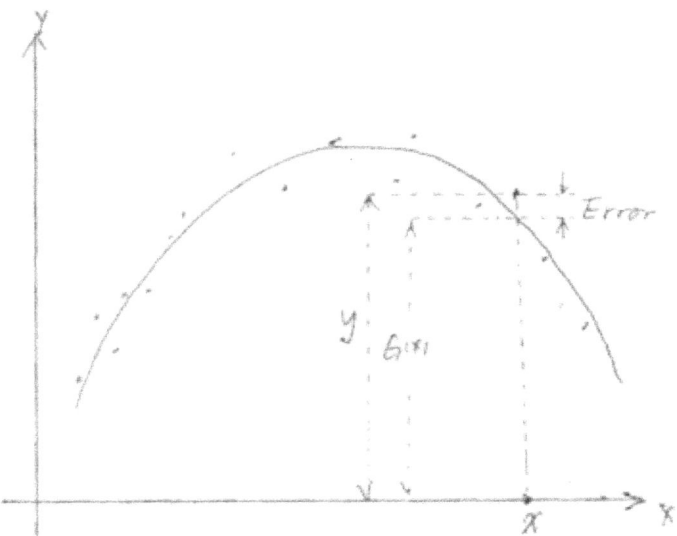

FIGURE 8.1.1 Regression Model

To provide answers to these questions, we first need to specify what we mean by best (good) and an optimization technique. For example, one popular technique minimizes the sum of squares of errors. More importantly we need to make some assumptions regarding the error term so that we can assess the model and estimates of its parameters. It should be pointed out once more that here error refers to the effect of all factors other than X, which are not included in the model either because they were not considered relevant or because of lack of knowledge. Now, in order to make the statistical analysis tractable, we need to make the following assumptions:

- Errors are independent with mean zero
- Errors have a constant variance
- Errors are normally distributed

Note that once such a regression model is established, it may be used to describe the phenomenon under consideration and to predict (within some limits) the values of Y for the values of X, which are not measured.

The function G is called a regression function and if it is linear in the parameters it is called linear regression. An example of a linear regression is

$$Y = a + bX + cX^2. \tag{8.2}$$

It should be pointed out that regression function is different from ordinary functions, which are discussed in calculus. Here, unlike an ordinary function, such as $Y = X^2$ for which Y is exactly equal to 4 when $X = 2$, the value of Y when $X = 2$ may deviate from 4. In fact X^2 may describe only the smooth part of the Y variable. So we need to find out how far the true value of the Y can be from the predicted value of 4, using the information about error term.

EXAMPLE 8.1.1 In exercise sciences many experiments are conducted to examine the relationship between the maximum oxygen consumption (VO_2 max) and time it takes to run a distance such

Chapter 8: Regression Analysis

as a mile or 2 miles. One goal of such a study is to formulate the relationship and use that for a prediction. Regression analysis is used to accomplish this goal.

EXAMPLE 8.1.2 Consider track and field events such as the 100-m dash or the marathon. The times set by athletes in these events will be affected by many different factors such as wind in the case of the 100-m dash and the temperature in the case of the marathon. Using regression analysis, we may be able to express the relationship in a form useful for prediction purposes.

EXAMPLE 8.1.3 The quantity of a product that people are willing to buy during some period of time depends on its price. Generally, the higher the price, the lower the demand; the lower the price, the higher the demand. Similarly, the quantity of a product that a supplier is willing to sell during some period of time also depends on the price. Generally, suppliers are willing to supply more of a product at higher prices and less of a product at lower prices. The simplest supply and demand model is a linear model where the graphs of a demand equation and a supply equation are straight lines. For example, suppose that we are interested in analyzing the daily sales of a T-shirt or hat in a particular city during a baseball season. Using regression analysis, we may arrive at the price-demand and price-supply models and use them to decide how many T-shirts or hats should be produced to maximize the (expected) profit.

EXAMPLE 8.1.4 Consider the application of regression models to the long jump. Long jump is a track and field event in which a jumper tries to jump a maximum distance into a sand pit after a running start. At the edge of the sand pit there is a takeoff board. Jumpers usually try to plant their toes at the front edge of the board to maximize the jumping distance. The absolute distance between the front edge of the takeoff board and the spot where the toe actually lands on the board prior to jumping is called "takeoff error." Here we may ask the following question: how takeoff error is related to the jumping distance? This is an example where a regression analysis may be applied to provide a reasonable answer.

8.2 SIMPLE LINEAR REGRESSION

In this section, we consider the simplest possible linear model, namely

$$Y = a + bX. \tag{8.3}$$

Although very simple, this form of regression occurs frequently in practice. Suppose that a sample of measurement $(x_1, y_1), \ldots, (x_N, y_N)$ is available. This is a realization of (X, Y), where X and Y are random variables on a fixed population Ω. For instance, Ω may consist of the amount of weight lost by using a certain body-fitting equipment. Here, N users may be chosen randomly. The length of time spent working out X and the amount of weight loss Y may be recorded. This gives pairs of measurements (x_i, y_i), $i = 1, \ldots, N$. As the first step of the analysis, the points (x_i, y_i) may be plotted (scatter diagram) to get some hints to whether the simple linear model is adequate and, if so, what values of a and b are reasonable. Analytically, the answer to the question of estimating the unknown parameters is strongly related to the value of the (sample) correlation coefficient defined as (see Section 1.7)

$$r = \frac{\sum_{i=1}^{N}(x_i - \bar{x})(y_i - \bar{y})}{[\sum_{i=1}^{N}(x_i - \bar{x})^2 \sum_{i=1}^{N}(y_i - \bar{y})^2]^{1/2}} = \frac{S_{xy}}{(S_{xx}S_{yy})^{1/2}}. \tag{8.4}$$

Remember that r measures the strength of the linear relationship between the two random variables. Also $|r| \leq 1$, and $|r|$ values close to 1 provide strong evidence for validity of a linear relationship. Small values of r, however, are an indication of no linear relationship. Considering this, we first present a few examples involving the calculation of correlation.

EXAMPLE 8.2.1 The 2016 Rio Men's Olympic Marathon Trials provided marathon qualifying times and the ages. The table below presents the number of minutes and also the ages for a sample of eight runners. A scatter diagram for this data is shown in Figure 8.2.1.

Age	34	29	32	23	27	30	36	24
Time	128.73	129.90	130.08	131.06	131.25	131.50	131.70	131.87

Let X denote the age and Y denote the qualifying time. Use formula (8.4) to compute r. Does the qualifying time relate to the age linearly?

SOLUTION: For this data, $\bar{x} = 29.375$ and $\bar{y} = 130.761$. Using formula (8.4) we get $r = -0.377$. To decide whether correlation is significant, we can apply the test described in Chapter 6.

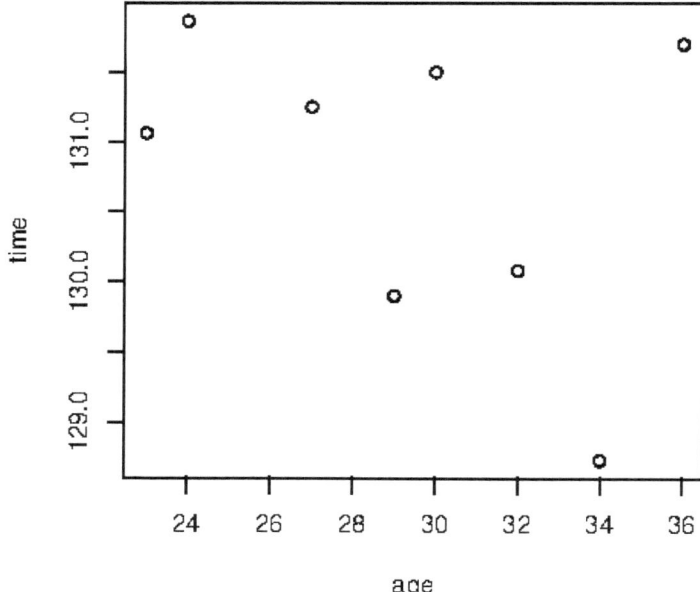

FIGURE 8.2.1 Scatter Diagram for Age and Qualifying Time.

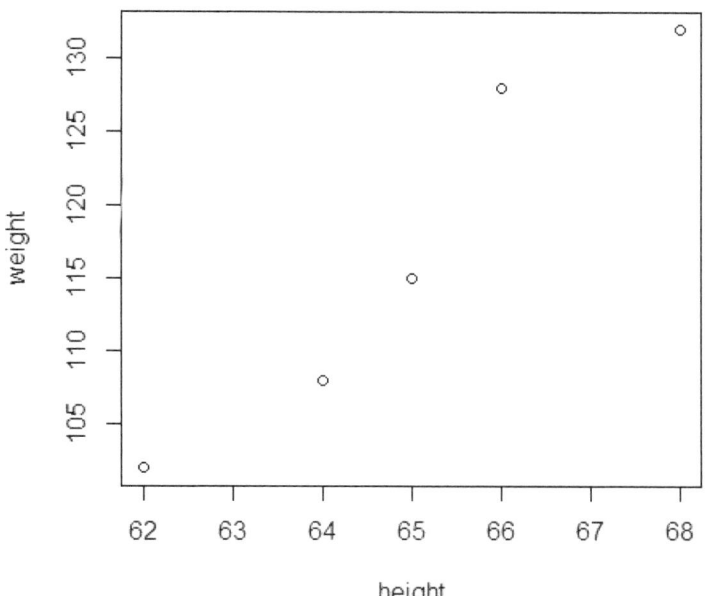

FIGURE 8.2.2 Scatter Diagram for Height and Weight for Women Swimmers

EXAMPLE 8.2.2 The calculation of the correlation coefficient may be carried out using a simple method. Consider the data collected on the height (inches) and weight (pounds) of five women swimmers.

Height	62	64	65	66	68
Weight	102	108	115	128	132

Let X and Y present the height and weight of a randomly selected woman swimmer, respectively. Then the necessary calculations may be carried out as follows:

	x_i	y_i	$(x_i - \bar{x})$	$y_i - \bar{y}$	$(x_i - \bar{x})^2$	$(y_i - \bar{y})^2$	$(x_i - \bar{x})(y_i - \bar{y})$
	62	102	−3	−15	9	225	45
	64	108	−1	−9	1	81	9
	65	115	0	−2	0	4	0
	66	128	1	11	1	121	11
	68	132	3	15	9	225	45
Sum	325	585	0	0	20	655	110
	$\bar{x} = 65$	$\bar{y} = 117$			$= S_{xx}$	$= S_{yy}$	S_{xy}

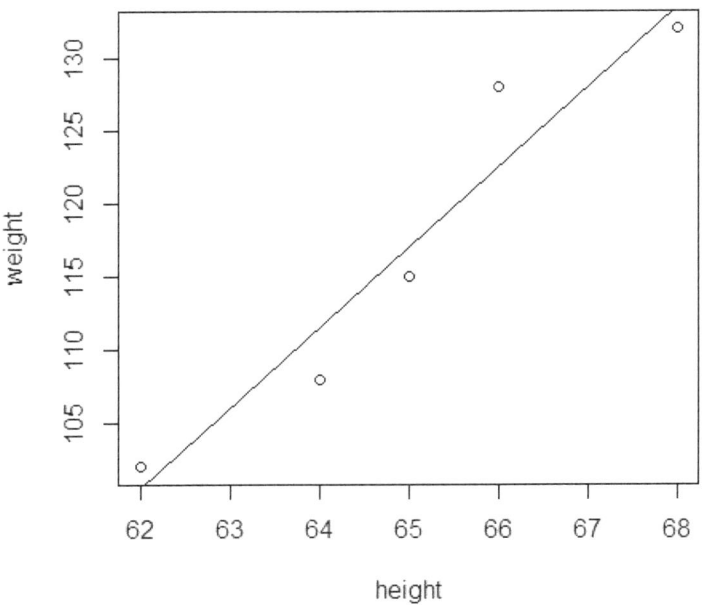

FIGURE 8.2.3 Fitted Line Plot for Example 8.2.2

Then,

$$r = \frac{S_{xy}}{(s_{xx} \cdot s_{yy})^{1/2}} = \frac{110}{114.455} = 0.961.$$

Consider now the realizations of (8.3). When the observed pairs are related to each other linearly we have

$$y_i = a + bx_i + \varepsilon_i, \quad i = 1, \ldots, N. \tag{8.5}$$

Here ε_i's are error terms that are assumed to be independent random variables having normal distribution $N(0, \sigma^2)$. Note that this is a reasonable assumption since, first, we do not expect any deviation from the line so that $E(\varepsilon) = 0$, and second, positive and negative deviations should be equally likely, and finally, small deviations should be more likely than large deviations.

Now as mentioned earlier, given data on X and Y we need to specify values of a and b that are optimum in some sense. There are many optimization techniques that can be employed for this purpose. Clearly, the best estimates for a and b will depend on the technique employed. One commonly used technique is the well-known least squares method. The method of least squares consists of picking the values a and b in such a way that the sum of squares of deviations between the observed or measured (y_i) and the expected or calculated values $(a+bx_i)$ is minimized. Mathematically, this is equivalent to finding the minimum of the function $\Delta(a,b)$ given below by taking derivatives with respect to a and b and setting them equal to zero.

$$\Delta(a,b) = \sum_{i=1}^{N}(y_i - a - bx_i)^2 = \sum_{i=1}^{N}\varepsilon_i^2. \tag{8.6}$$

Chapter 8: Regression Analysis

This leads to the following (least squares) estimates for b and a:

$$\hat{b} = \frac{\sum_{i=1}^{N}(x_i - \bar{x})(y_i - \bar{y})}{\sum_{i=1}^{N}(x_i - \bar{x})^2}, \quad \text{and} \quad \hat{a} = \bar{y} - \hat{b}\bar{x}. \tag{8.7}$$

Note that $\hat{a}$ and $\hat{b}$ are, respectively, the intercept and slope of the regression line. $\hat{b}$ is sometimes called the regression coefficient of Y on X. In this model, b is the important parameter and presents the amount of increase or decrease in Y for a unit increase in X.

EXAMPLE 8.2.3 For the data given in Example 8.2.1, compute the values $\hat{a}$ and $\hat{b}$ using (8.7).

SOLUTION: Straightforward calculation using Formula (8.7) shows that in this case we have

$$\hat{a} = 133.38, \quad \hat{b} = -0.089.$$

The expressions for $\hat{b}$ and $\hat{a}$ may be simplified by using the sample variance for x_i's, namely

$$s_x^2 = \frac{1}{N-1}\sum_{i=1}^{N}(x_i - \bar{x})^2$$

$$= \frac{1}{N-1}[\sum_{i=1}^{N}x_i^2 - (\sum_{i=1}^{N}x_i)^2/N] = \frac{S_{xx}}{N-1} \equiv \frac{SS}{N-1} \tag{8.8}$$

and the sample covariance between x_i's and y_i's, namely

$$S_{xy} = \frac{1}{N-1}\sum_{i=1}^{N}(x_i - \bar{x})(y_i - \bar{y}) = \frac{1}{N-1}[\sum_{i=1}^{N}x_i y_i - (\sum_{i=1}^{N}x_i \sum_{i=1}^{N}y_i)/N]$$

$$= \frac{S_{xy}}{N-1}.$$

This leads to

$$\hat{b} = \frac{S_{xy}}{s_x^2} = \frac{S_{xy}}{S_{xx}} = r(s_{yy}/s_{xx})^{1/2}. \tag{8.9}$$

EXAMPLE 8.2.4 For the data given in Examples 8.2.1 and 8.2.2, compute $\hat{b}$ using (8.8) and (8.9).

SOLUTION: Straightforward calculation shows that for marathon data we have

$$s_x^2 = 21.125, \quad S_{xy} = -1.886.$$

and therefore,

$$\hat{b} = -0.089.$$

Also for the height–weight example for women swimmers, we obtain

$$\hat{b} = \frac{S_{xy}}{S_{xx}} = \frac{110}{20} = 5.5$$

and
$$\hat{a} = 117 - (5.5)(65) = -240.5$$

The equation of line is then
$$\hat{y} = -240.5 + 5.5x.$$

This equation may be used to predict the weight of a women swimmer having a specific height. For example, when $x = 63$, the estimated (predicted) weight is $\hat{y} = 106$. It is also possible to construct an interval estimate (confidence interval). These topics will be discussed in Sections 8.3 and 8.4.

8.3 STATISTICAL INFERENCE FOR LEAST SQUARES ESTIMATORS

In this section, we will briefly discuss statistical inference for estimators of the model parameters and also for the predictions and estimations based on the model. This kind of inference is helpful in assessing the model and the prediction obtained from it. For example, we may want to know whether a linear relationship actually exists between X and Y, that is, if $b = 0$ or if $b \neq 0$. First we will consider confidence intervals.

• **Confidence Intervals**

In order to do statistical inference, we use the ideas discussed in Chapter 5, and consider the pairs (X_i, Y_i), $i = 1, \ldots, N$, as a random sample instead of their realizations (x_i, y_i). This leads to

$$\hat{b} = \frac{S_{XY}}{S_{XX}}, \text{ and } \hat{a} = \bar{Y} - \hat{b}\bar{X}, \tag{8.10}$$

as the least squares estimator of the slope b and of the intercept a. Also, s_{XX}, s_{XY}, and s_{YY} are defined analogously. The estimated regression equation is now

$$\hat{Y} = \hat{a} + \hat{b}X.$$

These replacements stress the point that the values $\hat{a}$ and $\hat{b}$ vary from realization to realization and are therefore random variables. It also reminds us that we need to construct confidence intervals for the true values, a and b.

To construct confidence intervals we use the fact that

$$s^2 = \frac{1}{N-2} \sum (y_i - \hat{a} - \hat{b}x_i)^2 = \left(\sum_{i=1}^{N} y_i^2 - \hat{a} \sum_{i=1}^{N} y_i - \hat{b} \sum_{i=1}^{N} x_i y_i \right) / (N-2) \tag{8.11}$$

is an estimate of σ^2 (variance of the distribution for the errors ε_i). Note that this value is also called the standard error of estimate and measures the variability of the observed values around the regression line. Also

$$s_{\hat{b}}^2 = Var(\hat{b}) = \frac{s^2}{s_{xx}} \tag{8.12}$$

Chapter 8: Regression Analysis

is an estimate of the variance for the regression coefficient. Combining these, we have

$$\hat{b} \pm t_{1-\alpha/2,(N-2)} \cdot s_{\hat{b}} \tag{8.13}$$

as a $100(1-\alpha)\%$ level confidence interval of b. A confidence interval for the intercept a can be constructed in a similar manner using

$$s_{\hat{a}}^2 = Var(\hat{a}) = \frac{s^2 \sum x_i^2}{N s_{xx}}. \tag{8.14}$$

EXAMPLE 8.3.1 For the data given in Example 8.2.1, construct 90% confidence intervals for a and for b.

SOLUTION: In Example 8.2.3, we found that $\hat{a} = 133.38$ and $\hat{b} = -0.089$. Using Formulas (8.11), (8.12) and (8.14) we see that $s^2 = 1.19$, $s_{\hat{b}} = 0.008$, and $s_{\hat{a}} = 7.09$, respectively. The t-critical in this case is 1.94. Hence, a 90% confidence intervals of a and b are, respectively

$$(128.21, 138.55), \quad (-0.263, 0.085).$$

A PRACTICE PROBLEM In recent years many parents and coaches have noticed that boys participating in sports such as basketball wear larger shoe sizes than their older siblings. Consider the following data on six members of the team:

Height (inches):   76   72   65   78   82   67
Shoe size: 12   11   8   13   16   9

Show that the regression equation with height as independent variables has slope equal to 0.0432, intercept equal to -20.18, and standard error of estimate equal to 0.575. Construct a 95% confidence interval for the slope.

• **Tests**

We start with an important test related to the existence of a linear relationship between dependent and independent variables. For a hypotheses such as

$$H_0 : b = b_0$$

we use the test statistic $(\hat{b} - b_0)/s_{\hat{b}}$ which is distributed as t with $N-2$ degrees of freedom. Of special interest is the case $b_0 = 0$, because not rejecting $H_0 : b = 0$ implies that there is no linear relationship between X and Y. This means that the model is inadequate and cannot be used for prediction. If such a hypothesis is rejected, we can use the regression equation to predict Y values for the given X values. As an example, for $X = x_0$, $Y_0 = \hat{a} + \hat{b}x_0$ is the predicted value (point estimate) of Y.

EXAMPLE 8.3.2 For the data given in Example 8.2.1, test whether X and Y are linearly related. That is, test $H_0 : b = 0$ versus $H_1 : b \neq 0$. Use a 10% level of significance.

SOLUTION: The test quantity (statistics) is

$$T = (\hat{b} - 0)/s_{\hat{b}} = -0.089/0.09 = -0.989$$

and the critical value in equals is 1.94. Since $|T| < 1.94$, the null hypotheses H_0 is not rejected. That is, we should not assume that X and Y are linearly related.

• **Prediction Interval**

Suppose that rather than a point estimate (predicted value) for Y (that can be obtained by replacing a value for X_0 in the equation of line), we are interested in a prediction interval. It can easily be shown that confidence limits of a level $100(1-\alpha)\%$, within which a future measurement of Y will lie at $X = x_0$ is

$$y_0 \pm s[1 + \frac{1}{N} + \frac{(x_0 - \bar{x})^2}{s_{xx}}]^{1/2} t_{1-\alpha/2,(N-1)}. \tag{8.15}$$

The length of this interval depends on several factors. For example, the interval becomes wider and wider, as the point x_0 moves out of the range of the sample values (sample mean). Thus, the prediction (extrapolation) becomes more and more inaccurate. One reason for this is that the linear relationship may not hold for values outside the range of the sample values. Note that this is also true for intrapolation. That is, the interval becomes wider as x_0 moves away from $\bar{x}$, even in the range of observed sample values. In fact, the third term in the bracket is exactly zero (minimum) when $x_0 = \bar{x}$.

EXAMPLE 8.3.3 For the data given in Example 8.2.1, construct a 90% prediction interval for future measurement of Y at $x = 34$.

SOLUTION: From previous examples we have $\bar{x} = 29.375$, $s = \sqrt{s^2} = 1.09$, $s_{xx} = (N-1)s_x^2 = 147.875$, and $y_0 = 133.38 - 0.089 \times 34 = 130.35$. In this case the t-critical is 1.90. Using (8.15), the required prediction interval is

$$(131.045, 135.715).$$

A PRACTICE PROBLEM Many authors have used the simple linear regression for modeling and prediction of the athletic records. Consider the following data representing the record times set for the men's mile run during the period 1911 to 1975. Fit a simple regression line using years as the independent variable (rename $1911 \equiv 0$) and times (in seconds) as the dependent variable. Use the model to predict future records (up to now) and compare them to actual records. What reservations do you have about this model and its predictions?

1911	4:15.4	1934	4:06.8	1945	4:01.4	1964	3:54.1
1913	4:14.6	1937	4:06.4	1954	3:59.4	1965	3:53.6
1915	4:12.6	1942	4:06.2	1954	3:58.0	1966	3:51.3
1923	4:10.4	1942	4:04.6	1957	3:57.2	1967	3.51.1
1931	4:09.2	1943	4:02.6	1958	3:54.5	1975	3:51.0
1933	4:07.6	1944	4:01.6	1962	3:54.4	1975	3:49.4

Chapter 8: Regression Analysis

- **Confidence Interval**

The interval obtained above predicts the time for a specific runner who is 34 years old. In practice we may be interested in estimating the time for all the runners whose age is 34. This is called a confidence interval (not prediction interval), and its form is very similar to (8.15). It is

$$y_0 \pm s[\frac{1}{N} + \frac{(x_0 - \bar{x})^2}{S_{xx}}]^{1/2} t_{1-\alpha/2,(N-1)}. \tag{8.16}$$

For Example 8.3.3, the confidence interval when $x = 34$ is

$$(132.304, 134.456).$$

Note that, as expected, this interval is narrower than the corresponding prediction interval. After all, it estimates the average time for all of the runners who are 34 years old, not a particular runner.

PRACTICE PROBLEMS

1. Use the data and model obtained for the men's mile run (see the last subsection) and provide confidence intervals for the future records. Is confidence interval meaningful in this context?
2. A soccer coach gives his players a written test about the strategies he teaches during the off-season training. He has noticed that there is a relationship between the number of absences and player's grade. To study this he has collected the following data:

Number of Absences:	1	2	2	3	3	4	4	5	6
Grade:	98	90	85	88	81	83	76	71	71

Show that the regression equation has a slope equal to -5.233, intercept equal to 110, and standard error of estimate equal to 3.6558. Also show that a 90% confidence interval for Y when $x = 3$ is 84.3 ± 2.37 and the prediction interval for Y when $x = 3$ is 84.3 ± 7.33.

8.4 ANOVA FOR SIMPLE LINEAR REGRESSION

One important question in regression analysis relates to the usefulness of the model developed. For example, we may want to know how much of the variation in the response variable Y can be explained by the variation in predictor variable X. We can provide answers to the questions of this type through analysis of variance (ANOVA), discussed in Chapter 8.

Suppose that (x_i, y_i) is one of the sample points. For this value of X, the corresponding predicted value using the regression line is

$$\hat{y}_i = \hat{a} + \hat{b} x_i. \tag{8.17}$$

The deviation of the actual (observed) value y_i from the regression (predicted) value $\hat{y}_i$ is representable as

$$y_i - \hat{y}_i = (y_i - \bar{y}) - (\hat{y}_i - \bar{y}). \tag{8.18}$$

Squaring both sides and summing over i we obtain

$$\sum(y_i - \hat{y}_i)^2 = \sum(y_i - \bar{y})^2 - \sum(\hat{y}_i - \bar{y})^2 \qquad (8.19)$$

or

$$\sum(y_i - \bar{y})^2 = \sum(\hat{y}_i - \bar{y})^2 + \sum(y_i - \hat{y}_i)^2. \qquad (8.20)$$

This means that the sum of squares about the mean is equal to the sum of squares about the regression line plus the sum of squares due to regression. Thus, we have the following decomposition of the total sum of squares:

$$SSTotal = SSRegression + SSResidual. \qquad (8.21)$$

Note that the better the model fits, the smaller the value of $(y_i - \hat{y}_i)$ and hence the smaller the $SSResidual$.

This formulation leads to Table 8.4.1 for the simple linear regression:

TABLE 8.4.1 ANOVA for Regression

Source of Variation	SS	d.f.	MS = SS/d.f.	F-Ratio
Due to Regression	SSRegression	1	MSRegression	$F = \frac{MSRegression}{MSResidual}$
About Regression	SSResidual	$N-2$	MSResidual	
Total	SSTotal	$N-1$		

Within this context, we can test the hypothesis of no linear relationship, namely

$$H_0: b = 0$$

against the alternative that $b \neq 0$ using the F statistic (the F-Ratio). We reject H_0 and accept that the model is adequate if the F-ratio is greater than the $100(1-\alpha)$ percentile of the $F_{1,(N-2)}$ distribution.

EXAMPLE 8.4.1 The ANOVA table for data given in Example 8.2.1 is shown below. Here, calculations are carried out using Formula (8.8) and Example 8.2.2.

ANOVA for Example 8.2.1

Source of Variation	SS	d.f.	MS	F-Ratio	p-value
Due to Regression	1.1790	1	1.1790	$\frac{1.1790}{1.1896}$	0.3579
About Regression	7.1373	6	1.1896	$= 0.9911$	
Total	9.8750	7			

Since $F_{0.95,(1,6)} = 5.99$, the null hypothesis $H_0: b = 0$ cannot be rejected.

Chapter 8: Regression Analysis

8.5 THE COEFFICIENT OF DETERMINATION: A MEASURE OF THE USEFULNESS OF THE MODEL

Recall the Expression (8.9) namely
$$\hat{b} = r(s_{yy}/s_{xx})^{1/2}.$$

Clearly, since s_{yy} and s_{xx} are not zero, it follows that $\hat{b}$ is zero whenever $r = 0$ and vice versa. This means that the correlation coefficient plays the key roll in assessing the model.

The coefficient of determination, R^2, is the square of the coefficient of correlation, discussed earlier (Section 8.2). It represents the proportion of the total sample variability around $\bar{Y}$ that is explained by the linear relationship between Y and X. Using (8.21), it may be computed as

$$R^2 = \frac{SSRegression}{SSTotal} = \frac{SSTotal - SSResidual}{SSTotal}$$

$$= \frac{Explained\ Sample\ Variability}{Total\ Sample\ Variability} = 1 - \frac{SSResidual}{SSTotal}.$$

Note that $0 < R^2 < 1$. Like correlation, large values of R^2 indicate existence of a linear relationship between X and Y. It can be used to measure the usefulness of the model. All the softwares used for regression analysis report the value of R^2. It is one of the main factors for assessing the model.

EXAMPLE 8.5.1 For the data given in Example 8.2.1, find R^2 and use the ANOVA to test the hypothesis $H_0: b = 0$ at a significance level of 10%.

SOLUTION: As described earlier, based on the results of Example 8.2.2 and the Formula (8.8), a straightforward calculation shows that

$$SSRegression = 1.1790,\ SSResidual = 7.1373,\ SSTotal = 8.3163$$

$$R^2 = 1.1790/8.3163 = 0.1428$$

$$MSRegression = SSRegression/1 = 1.1790$$

$$MSResidual = SSResidual/6 = 1.1896$$

$$F = 0.9911.$$

For this case, the F-critical is 3.78. Since the value of F-ratio is less than 3.78, we fail to reject H_0 and the model is not significant. Note that $R^2 = 0.14$ means that only 14% of total variation is explained by the regression model.

EXAMPLE 8.5.2 Chatterjee et al. (1995) have investigated the effectiveness of NBA guards using data on 105 guards played in 1992–1993 season. Suppose that players performance is measured using points scored per minute (PPM). Then a question of interest is the following: what factors affect the statistical performance of a guard? Clearly, there are many variables (factors) that affect player's performance that may be used as predictors. Let us for now just consider the number of minutes played per game (MPG). These authors have shown that regression analysis leads to the model

$$\text{PPM} = 0.242 + 0.007\text{MPG}$$

and

$$F = 43.53.$$

However, as expected, it was found that $R^2 = 0.32$. Although this a reasonable one-predictor model, it is not satisfactory enough to be used for prediction. In fact there is a need for inclusion of more predictors. The regression analysis involving more than one predictor will be discussed in Section 8.7.

EXAMPLE 8.5.3 A manager of a professional team is interested in trade value of 10 of his top players. He believes that one of the most important determinants is number of times players missed the practice. To understand the situation better and to help make decisions he has collected the following data on these players.

Player	Number of Times Player Missed the Practice (Last 3 Seasons)	Trade Value (100,000 s)
1	59	37
2	92	31
3	61	43
4	72	39
5	52	41
6	67	39
7	88	35
8	62	40
9	95	29
10	83	33

a. Find the regression line for determining how the number of missed practices affects the trade value of the player. Interpret the slope of this line.
b. What does the regression line tell about the relationship between the two variables?
c. Find the standard error of estimate.
d. Can we conclude at a 5% significance level that for all players of the type describe in this experiment, a higher number of misses results in a lower trade value? First specify the null hypothesis and the alternative hypothesis.
e. Measure the strength of the linear relationship by calculating the coefficients of correlation and the coefficient of determination.
f. Predict with 95% confidence the trade value of a player who has missed 60 practices.

Chapter 8: Regression Analysis

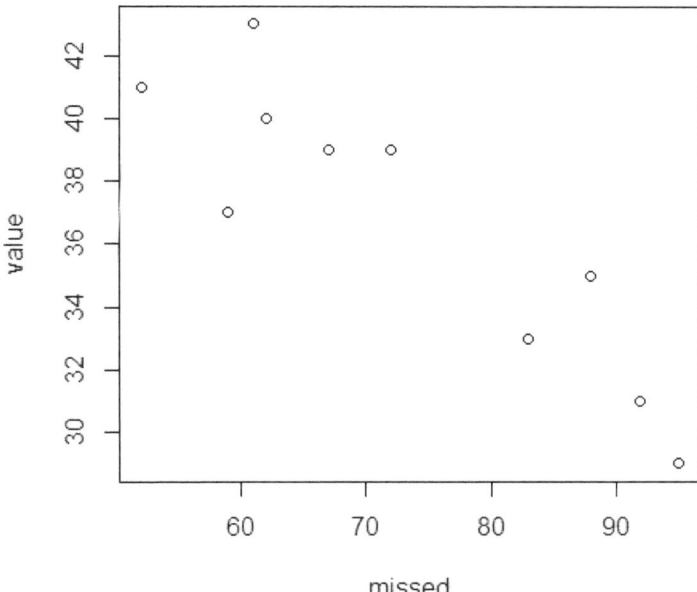

FIGURE 8.5.1 Number of Missed Practice Versus Trade Value

g. Suppose that the organization had a policy of trading its players when they missed 75 practices during the latest three seasons. The organization is about to trade several players for the reason mentioned. The manager would like to know the players' mean trade price. Determine the 95% confidence interval estimate of the expected trade value for all players that had missed 75 practices.

h. Suppose that the sport under consideration is basketball. What other factors do you think will have a positive or negative effect on trade value.

SOLUTION: First we plot the number of practices missed in the last three seasons versus the trade value.

a) The regression equation is

$$\text{Trade Value} = 56.2 - 0.267 \text{ (Number of Practices Missed)}.$$

- For every additional practice missed the trade value for the player decreases by $26,700.
- Although not always applicable, here we may assume that a player who has not missed any practices would have a trade value of $5,620,000.

b) According to the regression line, there is a negative relationship between the trade value and the number of practices missed.

c) The standard error of estimate, 2.178, is fairly small compared to the mean of $\bar{y} = 73.1$. This suggests that the model produces a fairly good fit, with a minimal amount of error.

Summary of the Analysis

Predictor	Coef.	St.Dev.	T	p
Constant	56.205	3.535	15.90	0.000
Practice	−0.26682	0.04743	−5.62	0.000

d) In order to test the hypothesis that a higher number of missed practices results in a lower trade value we must test the slope to determine whether there is a negative relationship. So we test

$$H_0: b = 0 \quad vs \quad H_1: b < 0$$

From the summary p-values = 0.000. Thus, we reject the null hypothesis and conclude that the slope is not equal to zero. Therefore, there is a relationship between the two variables. Note that here the value of test statistics is -5.62. The rejection region for a one-tailed test is $t < -t_{.95,9} = -2.36$. Thus we reject H_0 and accept that $b < 0$. The scatter plot of the number of missed practices versus the trade value confirms that the relationship between these two variables is indeed negative.

e) The coefficient of determination $R^2 = 79.8\%$ suggests that 79.8% of the variation in trade value is a result of the variation in the number of missed practices. The coefficient of correlation is $r = \sqrt{0.798} = -0.893$, with a p-value = 0.000.

The correlation coefficient is a value between -1 and 1, this statistic determines the strength of the relationship between two variables. In this case, $r = -0.893$ suggests that there is a very strong negative relationship between the number of missed practices and the trade value.

Analysis of Variance

Source	SS	d.f.	MS	F	P
Regression	150.14	1	150.14	31.64	0.000
Residential Error	37.96	8	4.74		
Total	188.10	9			

According to the F-ratio = 31.64, P = 0.000, the model is overall significant and may be used for predictions.

Fit	Standard Deviation Fit	95.0% PI
40.195	0.928	(34.736, 45.655)

f) The trade value for one player who has missed 60 practices would be between \$3,473,600 and \$4,565,500 with 95% confidence.

Fit	StDev Fit	95.0% CI
36.193	0.695	(34.591, 37.795)

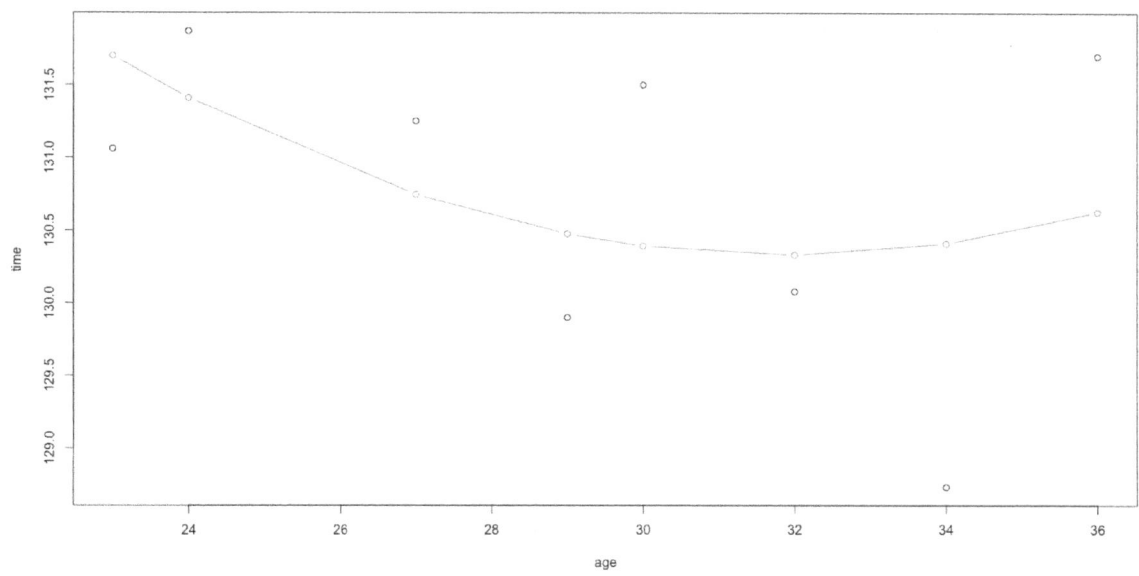

FIGURE 8.5.2

g) The mean trade value for all players missing 75 practices would be between $3,459,100 and $3,779,500 with 95% confidence.

h) Other possible factors that may affect the trade value would be points scored per minutes, age, number of the years played in NBA, position played, number of assists, steals, blocks and leadership ability, and so on.

We end this section by noting that when simple linear regression does not seem appropriate, one can consider other linear models such as a quadratic model taking the form

$$Y = a + bX + cX^2$$

or perhaps a high order polynomial. Figure 8.5.2 presents a quadratic regression model together with 95% confidence and prediction intervals for the data in Example 8.2.1. As can be seen, based on R^2 this model fits the data significantly better than simple linear regression.

Of course, if the form of the regression equation is known based on physical evidence or information coming from the discipline related to the origin of data, then there would be no need to use polynomials. However, in the absence of such information, this class of linear models provides a powerful and flexible alternative.

There are also cases where, rather than additive models, which assume that observations can be expressed as sum of two components (systematic and random), we may consider a multiplicative model. Such models can sometimes be transferred to linear additive models. See Section 8.6 for an example.

8.6 APPLICATION OF SIMPLE LINEAR REGRESSION TO TRACK AND FIELD

Track and field, or athletics as it is known in most countries outside of North America, is one of the world's oldest competitive sports. It has been around since the Ancient Greek Olympics. Every year millions of competitors participate in meets at the interscholastic, collegiate, and international levels throughout the world. Men can compete in 24 events and women have 20 events that can range from running, hurdling, jumping, and throwing. Some new events such as the triathalon were added to the recent Olympics. Track and field is a gold mine of statistical data. Such data can be used to predict future track performances using, for example, trend (regression) analysis. Such data provide wonderful opportunities for statisticians to test their model and the methods. They also provide a chance for physiologists to study the limits of human locomotion. In fact, as pointed out by a physiologist, "one beauty of this field is that the athletic authorities of the world provide one with scientifically-impeccable data, usually accurate to a fraction of one percent, and the guinea pigs select, house and feed themselves. The scientist merely has to sit back and collect the results."

8.6.1 Modeling and Prediction

Track and field being simple to understand has been a wonderful source of data for many scientists. This section includes a brief description of the models developed by statisticians and others for running performances. A valuable source of information used here is Chapter 9 of the book "Statistics in Sports" edited by J. Bennett (1998), Arnold Publication. The chapter written by Schutz and Liu reviews historical models with particular reference to the mile run. Here, we will present part of their research and include some additional models.

1. 1906, Kennelly developed the following model

$$log Y = 1.125 log(D) - 1.236 \quad or \quad Y = 0.588 D^{1.125}$$

   where Y is the time in seconds and D is the distance in meters.

2. 1916, Meade, Hamilton, and Nurmi applied Kennelly's model and presented estimates for the ultimate records for several events. However, all these ultimate records were broken by 1970.

3. Lucy developed an exponential function model

$$Y(t) = a + bc^t$$

   where $Y(t)$ is the time in seconds for the best mile time in a year $1950 + t$ and a and b are parameters to be estimated.

4. Deakin proposed fitting a three-parameter exponential model to the records (21 data points) using the following function:

$$R(n) = \alpha e^{-\beta(n-1)} + \gamma$$

   where $R(n)$ is the nth world record (n = 0 to 20, representing the 21 years in which records were set from 1911 to 1965).

5. Ryder, Carr, and Herget examined the changes from 1910 to 1975 in velocities (m min^{-1}) for distance 100 m to the marathon concluding that the rate of improvement in velocities over the 65-year period was linear, with the 100 m showing the smallest rate of change and the marathon the greatest rate of change.

6. Chatterjee and Chatterjee have fitted a number of models to the winning times in the men's 100, 200, 400, and 800 m run in the Olympic games from 1900 to 1976. They have proposed the following:

$$Y(t) = \alpha + \beta e^{\gamma t}, \quad \beta > 0, \quad \gamma < 0$$

where $Y(t)$ is the time for year t and α, β, and γ are parameters. Note that in this model the parameter α presents the ultimate record since as $t \to \infty$, $Y(t) \to \alpha$.

7. Noubary, instead of models of the type described above, has considered the order statistics corresponding to the winning times and has fitted a model that best described their lower tails (extreme values or values below a threshold). He found that a Pareto (in fact a Generalized Pareto) distribution with a lower bound provided the best fit. This analysis led to a set of estimates for ultimate records.

8. Schutz and McBryde analyzed different models and discussed their shortcomings. They then attempted to overcome some of these by using continuous data. They finally tested the following four models:

(a)
$$Y(t) = \alpha + \beta e^{\gamma t}$$

where $Y(t)$ is the predicted (or velocity) time in year 1899 + t.

(b) The same form as model A, but relating velocity (rather than running time) to year.

(c)
$$Y(t) = \alpha + \beta t$$

where $Y(t)$ is running time for a specific event and $1899 + t$ is the historical year,

(d)
$$Y(t) = \alpha(D(t))^\beta$$

where $D(t)$ is distance. They concluded that all four models provide reasonably good fits and reasonable projection for the shorter races.

9. Smith has considered a regression model of the form

$$Y(t) = Z(t, \theta) + x(t)$$

where $x(t)$'s are independent and identically distributed random variables, and $Z(t, \theta)$ is a nonrandom (trend) term. For $x(t)$, particular distributions considered were normal, Gumbel and generalized extreme value. For $Z(t, \theta)$, linear, quadratic and exponential-decay model were used. He found that a model with linear trend and normal distribution for $x(t)$ worked best for the mile data.

10. Schutz and Liu extended the work of Schutz and McBryde and provided further confirmation that the three-parameter exponential model was the best. They also introduced a new model called the Random Sampling Model. This model assumes that the true mean score of the top performances has now stabilized and is expected to be constant for the next 50 years or more. Under this assumption, new records can still be set and this corresponds to sampling an outlier.

11. Noubary has regarded the fastest times as a realization of a random process, and suggested application of the following models:

    (a): $\log Y(t) = \alpha - \beta t + x(t)$ or $Y(t) = \theta_0 e^{-\theta_1 t} z(t)$
    where $\theta_0 = e^\alpha$ and $z(t) = e^{x(t)}$ with $\beta, \theta_1 > 0$.

    (b): $\log Y(t) = \alpha - \beta t + \gamma \log t + x(t)$ or $Y(t) = \theta_0 t^{\theta_2} e^{-\theta_1 t} z(t)$
    where $\theta_0 = e^\alpha$, $\theta_1 = \beta$, $\theta_2 = \gamma$ and $z(t) = e^{x(t)}$ with $\theta_1 > 0$.

    These models imply that both means (trends) and variances of the times vary with t (in fact, both are decreasing). Note that the latter makes sense since the amount by which records are broken are much less now compared to the past.

12. Blest, like many other authors, has considered the regression model

    $$Y(t) = Z(t, \theta) + x(t)$$

    and some frequently used forms for $Z(t, \theta)$. These forms included a simple linear model and seven nonlinear functions. The three functions (all involving four parameters) that provided the best fits were a logistic, a reparameterized Gompertz, and an antisymmetric exponential model.

13. Noubary and Shi developed a difference equation model for estimating records. They considered an innovative approach to modeling based on difference equations whose complementary solution yield one of the forms for $Z(t, \theta)$ discussed by the author mentioned earlier. They did this by establishing a relationship between the models and certain difference equations. They also showed that each of the models discussed earlier can be converted to the exponential model by applying an appropriate substitution.

14. Berry (Chance Spring 2002) developed a model based on a population of size n, assuming that abilities follow a fixed normal distribution with mean μ and standard deviation σ regardless of when people were born. He then predicted the Olympic records using the $1/n$th quantile. For example, for $n = 10^9$ (1 billion) the 1-billionth quantile is $\mu - 5.9978\sigma$ (in general $\mu - x_{1/n}\sigma$). He then estimates μ and σ for each event using the available data. Although the assumptions made are not very realistic, the model gives a good fit.

We end this part by noting that apart from the above publications, there is another set of literature regarding records and their predictions, which is not as extensive but is equally important and interesting. This includes studies that attempt to combine past data as well as physiological and/or biomechanical factors. The great advantage of this approach over purely statistical approach lies in the fact models developed involve parameters that have some physical interpretation so that validation other than statistical goodness of fit is possible. An example of such models is

$$V(t) = \alpha_1 e^{-\beta_1 t} + \alpha_2 e^{-\beta_2 t} + \alpha_3 e^{-\beta_4 t} + \alpha_5 e^{-\beta_5 t}$$

where $V(t)$ is velocity, t is time in seconds, α_i's are the velocity constant and β_i's are rate constant. The five components of this model represent the energy loss, lactate O_2 debt, locate O_2 debt,

Chapter 8: Regression Analysis

glycogen depletion, and fat metabolization. This model is accepted as a complete and accurate model of the energy requirements for running

8.6.2 More on Ultimate Records

The Olympic and other games provided numerous exciting examples of the ever-increasing desire to "push the envelope." Johnny Weissmuller who played Tarzan in a long-running TV series was the 100-m freestyle champion in the 1924 and 1928 Olympoics. His 1924 gold medal time in the 100-m freestyle was 59.0 seconds. At this pace he would complete 1,500 m in 14:45, which is longer than the Olymmpic record 14:43.48 set in 1992. Other than men, women have sharply improved their performances and surpassed some of the records set by men in the past. So one may wonder whether records will be broken again and again or whether there are limits to human potential and athletic ability. The answer to this would depend on how physiologists can help athletes to run faster, jump higher, or become stronger. It also depends on knowing what factors make an athlete truly exceptional. The debate about the precise mechanism by which athletes improve their performance continues in scientific literature. Numerous physiological variables affect human performance. Some important ones are discussed below using the reference that follows. For details see LeMura et. al., *Journal of Sports Medicine and Physical Fitness*, 2000, 40(1); 1-10.

(A) Maximal Oxygen Consumption (VO_2)

VO_2 is a standard used to judge human capacity to perform prolonged or endurance exercise to maximize oxygen intake. The magnitude of any increase in exercise capacity after training depends on the magnitude of increase in oxygen consumption. The current consensus among exercise physiologists is that the exceptionally high values seen in elite endurance athletes represent a combination of prolonged intense training and genetic factors that results in significant physiological adoptions.

(B) The Lactate Threshold

Researchers have studied intensely another critical metabolic variable—the lactate threshold—to explain the exceptional performances of elite athletes. The chemical lactic acid may be described as a metabolic "waste product," which causes fatigue as its concentration increases in the blood. The lactate threshold represents a physiological phenomenon during activity whereby a significant increase in blood lactate causes a shift from aerobic to anaerobic metabolism. Research has established that most athletes cannot sustain VO_2 max over prolonged exercise. There is an indication that the fraction of VO_2 max sustained during the exercise is significantly related to lactic acid accumulation in active muscles.

(C) Muscles and Genetics

Human muscle fibers are not homogeneous. They are classified according to their physiological, biochemical, and histological properties. In human, and many other mammals, two primary types of muscle fibers each adapted for specialized functions, are classified according to their contraction speed: slow twitch fibers (type I) contract and relax within 100 milliseconds; fast twitch fibers (type II) contract within 50 milliseconds. A sprinter would recruit fast twitch fibers during a race, while a marathoner would use predominately slow twitch fibers. In the average human, skeletal muscles are typically composed of approximately 50% of each type. However, exceptional athletes generally exhibit a disproportionate amount of one fiber type over another as compared to the average individual. Herein lies a distinct genetic advantage. Type I fibers are particularly well suited to activities of low intensity and long duration such as marathon running. Thus, individuals with a preponderance of these fibers are able to generate a significantly high VO_2 max. Type II, on contrast, is well suited to activities requiring maximum force and speed such as weight lifting and sprinting.

(D) Other Factors

Presuming that elite athletes, engaged in intense exercise training, have maximized their training-induced physiological adaptations (i.e., VO_2 max and lactate threshold), why do world records continue to fall? One explanation is that training centers have become technological meccas that seek to determine those factors that could limit elite athletic performances. World class athletes have access to state-of-the-art training facilities equipped with laboratories designed to assess every metabolic, physiologic, and hemodynamic variables that may affect human performance. Computer applications to determine and evaluate perfect form in every sport have augmented significantly the subdiscipline of biomechanics. The technology of sport training also has carried over to the impressive quality and design of athletic equipment and facilities. For example, bumpy cinder tracks have been replaced with faster, rubberized tracks and wavy swimming pools have been replaced with deeper, cooler swimming pools divided by floating lane markers that better absorb the wash from competitors. Technical improvements in track shoes and sport equipment have clearly contributed to record performances in every Olympic sport. A notable example is sprinter Carl Lewis, who runs in four ounce shoes custom made from a mold of his foot. In the Centennial Olympic games in Atlanta, the US cyclists rode the SuperBike 2, a million dollar bicycle designed by computer scientists and tested in the wind tunnels at General Motors. Research indicates that these and other equipment and facilities improvements probably account for as much as 5% of the increase in athletic performance.

Final Words

Knowledge of an athlete's genetic endowment for muscle fiber type distribution obviously would provide a tremendous advantage, particularly in terms of developing precise training recommendations. So how much faster can we get, and, what are the outer limits of athletic performance? Many exercise physiologists maintain that the improvement curve is leveling off and that subsequent mutation performances will be minimal. Still, world records will in all likelihood continue to improve, although slowly. Future athletes will be even bigger and stronger than before due largely to better childhood nutrition and a significant decrease in the impact of various sociological factors that will make a dramatic contribution in world record performances. Specifically, more women, underprivileged children, and people from underdeveloped countries are finding their way into sports. Undeniably, a broader ethnic and genetic pool is entering the athletic arena.

Chapter 8: Regression Analysis

The Limits of Human Performance: Track and Field

Event	Projected Record in 2040 (Men)	Projected Record in 2040 (Women)
100-m	9.44	10.02
200-m	19.19	19.66
400-m	40.76	43.60
800-m	01:37.1	01:48.1
1,500-m	03:23.7	03:40.5
400-m hurdles	42.28	
High jump	8'11"	8'2"
Triple jump	68'4"	
Long jump	33'8"	27'3"
Pole vault	24'10"	

The Limits of Human Performance: Swimming

Event	Projected Record in 2040 (Men)	Projected Record in 2040 (Women)
100-m freestyle	40.15	41.94
400-m freestyle	02:59.0	03:01.7
800-m freestyle		07:21.1
1,500 freestyle	11:01.7	
100-m backstroke	42.36	48.86
200-m backstroke	01:44.4	0:02.01
100-m butterfly	48.77	50.47
200-m butterfly	01:46.2	01.52.8

Tests have been developed to detect the surreptitious drug use that many athletes rely upon to improve their athletic prowess. Steroids, for example, cause faster recovery from training, allowing athletes to train harder and build more muscle. Beta blockers, used primarily by shooters and archers, slow down resting heart rates. Recently, numerous elite athletes in search of record performances have increased significantly the clandestine use of the human growth hormone, convinced it causes growth in bone and muscle density, despite strong messages of potential health hazards. The human growth hormone may create incredibly strong athletes, but at a price.

While the "brave new world" of training promises continuing advances in technology to provide even the smallest performance advantages, it also may bring possibilities that drug and gene technologies will emerge that could suggest a future of genetic tinkering that may take exercise physiology to an entirely new level, far beyond assessing the effects of today's steroids and growth hormones.

Even with all of this in mind, the future improvements will be far less dramatic than those of the past. Based upon the rate of the athletic improvements in events over the past 10 Olympiads, scientists have projected possible times and distances in selected events for the year 2040 (see the above table). It is important to note, of course, that the margin of error is large. With all we know about human biology and genetic potential, the bottom line is whether you want to be an Olympic champion, devote your young life to disciplined, continual training. More importantly, choose your parents wisely.

8.6.3 Some Examples

To predict future performance using mathematical modeling, data from the past are needed. In this section, the best time from each track and field event is used to predict the future times. We note that there are numerous track events from which data can be retrieved. In this and the next Sections 8.6.3 and 8.6.4, we use 10 men's events to predict results for the 1996 and 1997 track seasons. The events chosen are the 100-m dash, 200-m dash, 1,500-m run, 5,000-m run, 10,000-m run, 110-m high hurdles, 3,000-m steeplechase, the triple jump, the hammer throw, and the shot put. Also to see the trends in Olympic games, predictions for year 2002 are obtained for three additional events: high jump, discus throw, and long jump from the Olympic games 1900-1984. The 100-m dash is also analyzed in more detail in Section 8.6.5 using a different set of data. The above events include two sprints, four long distance, one hurdling, one jumping, and two throwing events. The steeplechase is when athletes must run 3,000 meters and jump over five barriers that are set on the track. One of the barriers leads into a water pit. The shot put is a 16-lb metal ball that is thrown for distance. A hammer is a 16-lb sphere that is attached to a 3-foot $11\frac{3}{4}$ inch steel wire which has a grip. A competitor twirls the hammer in a circular motion and then releases it. For the 5,000, and 10,000-m runs we used data from the past 44 years. We considered times since 1980 for the remaining events.

To do trend analysis we used MINITAB. Two simple linear models were produced for each event. One using the performance measures and the other logarithm of these measures. Note that the latter is selected based on the fact that models involving exponential growth or decay can be transferred to simple linear regression through logarithmic transformation. In both cases, the equations were in terms of time presenting years. The first model takes the form $Y(t) = a + bt$ (Linear Model) and the second $\log Y(t) = a + bt$ or $Y(t) = \alpha \beta^t$ (Log-Linear Model). These regression equations were used to predict annual records for 1996 and 1997 to see how close they were to the actual best times of those two years. In most of the cases they were close. The times for 1998 and years thereafter are left as an exercise for students. The models seem to be good. Both regression models produced predictions that are close to the actual records. All the models seem to be adequate except for the hammer throw and shot put. These two equations show a decrease in the record but they should be increasing. The models produced are good predictors for the near future. Thus, they should be used only to predict records within the next few years. They are not good predictors for the distant future.

TABLE 8.6.1 Best Times for Men's 5,000 and 10,000 m

Year	5,000 m (in seconds)	10,000 m (in seconds)
1954	831.21	1,734.20
1955	820.60	1,739.20
1956	816.80	1,710.40
1957	815.00	1,746.40
1958	831.60	1,736.00
1959	822.40	1,743.00
1960	818.10	1,698.00
1961	815.20	1,730.80
1962	818.40	1,698.20
1963	821.20	1,695.20
1964	818.00	1,704.40
1965	804.20	1,659.40
1966	796.60	1,674.00
1967	798.80	1,706.60
1968	807.80	1,669.40
1969	809.00	1,683.60
1970	802.80	1,686.20
1971	802.20	1,667.00
1972	793.00	1,658.40
1973	794.60	1,650.80
1974	794.40	1,663.60
1975	798.60	1,665.40
1976	793.10	1,660.40
1977	792.90	1,650.50
1978	788.40	1,642.47
1979	792.29	1,656.80
1980	796.40	1,649.16
1981	786.20	1,647.70
1982	780.41	1,642.95
1983	788.54	1,643.44
1984	784.78	1,633.81
1985	780.40	1,657.17
1986	780.86	1,640.56

TABLE 8.6.1 Best Times for Men's 5,000 and 10,000 m (*Continued*)

Year	5,000 m (in seconds)	10,000 m (in seconds)
1987	778.39	1,646.95
1988	791.70	1,641.46
1989	784.24	1,628.23
1990	785.59	1,638.22
1991	781.82	1,631.18
1992	780.93	1,634.26
1993	782.75	1,618.38
1994	780.54	1,612.23
1995	764.34	1,603.53
1996	765.09	1,598.08
1997	759.74	1,587.85

TABLE 8.6.2 Best Times for Different Track and Field Events

Year	100 m	200 m	1,500 m	110-m Hurdles	Steeplechase	Triple Jump	Hammer Throw	Shot Put
1980	10.02	19.96	211.36	13.21	489.70	17.35	81.86	21.98
1981	10.00	20.20	211.57	12.93	493.32	17.56	90.56	22.02
1982	10.00	20.15	212.12	13.22	496.16	17.57	83.98	22.02
1983	9.93	19.75	210.77	13.11	492.37	17.55	84.14	22.22
1984	9.96	19.80	211.54	13.15	487.62	17.55	86.34	22.19
1985	9.98	20.07	209.46	13.14	489.18	17.97	84.08	22.62
1986	10.00	20.12	209.77	13.20	490.01	17.80	86.74	22.64
1987	9.93	19.92	210.69	13.17	488.57	17.92	83.48	22.91
1988	9.92	19.75	210.95	12.97	485.51	17.77	85.14	23.06
1989	9.94	19.96	211.00	13.05	485.35	17.62	82.84	22.19
1990	9.96	19.85	208.86	13.04	490.95	17.93	84.48	23.12
1991	9.86	19.88	209.20	12.91	486.46	17.78	84.26	22.03
1992	9.93	19.73	210.61	12.98	482.08	17.72	84.62	21.98
1993	9.87	19.85	210.55	12.92	486.36	17.86	82.79	21.98
1994	9.85	19.87	212.60	13.08	488.80	17.68	83.36	21.09
1995	9.91	19.79	207.37	12.98	479.18	18.29	83.10	22.00
1996	9.84	19.32	209.05	12.92	485.68	18 09	82.52	22.40
1997	9.86	19.77	208.91	12.93	475.72	17.85	83.04	22.03

Chapter 8: Regression Analysis

Equations For Regression Using $Y(t) = a + bt$

100 m:
$$Y(t) = 10 - 0.00921\,t$$

200 m:
$$Y(t) = 20.1 - 0.0237$$

1,500 m:
$$Y(t) = 212 - 0.162\,t$$

5,000 m:
$$Y(t) = 826 - 1.30\,t$$

10,000 m:
$$Y(t) = 1732 - 2.95\,t$$

110 Hurdles:
$$Y(t) = 13.2 - 0.0138\,t$$

Steeplechase:
$$Y(t) = 494 - 0.706\,t$$

Triple Jump:
$$Y(t) = 17.5 + 0.0292\,t$$

Hammer:
$$Y(t) = 83.8 - 0.0109\,t$$

Shot Put:
$$Y(t) = 11.4 - 0.0163\,t$$

Equations For Regression Using

$$\log Y(t) = a + bt \quad (Y(t) = \alpha \beta^t)$$

100 m:
$$Y(t) = 10.0188(0.999073)^t$$

200 m:
$$Y(t) = 20.0999(0.998808)^t$$

1,500 m:
$$Y(t) = 211.896(0.99923)^t$$

5,000 m:
$$Y(t) = 826.049(0.998371)^t$$

10,000 m:
$$Y(t) = 1732.55(0.998237)^t$$

110 Hurdles:
$$Y(t) = 13.1818(0.998943)^t$$

Steeplechase:
$$Y(t) = 494.136(0.998549)^t$$

Triple Jump:
$$Y(t) = 17.4913(1.00165)^t$$

Hammer:
$$Y(t) = 83.8149(0.99989)^t$$

Shot Put:
$$Y(t) = 22.4025(0.999252)^t$$

Predictions For Year 1996

Event	Linear Model	Log-Linear Model	Actual
100 m	9.84	9.86	9.84
200 m	19.7	19.7	19.32
1,500 m	209.25	209.14	209.05
5,000 m	770.1	770.12	765.09
10,000 m	1,605.15	1,605.95	1,598.08
110 hurdles	12.97	12.95	12.92
Steeple	481.99	481.29	485.68
Triple jump	17.99	18.03	18.09
Hammer	83.61	83.72	82.52
Shot put	22.12	22.12	22.4

Predictions For Year 1997

Event	Linear Model	Log-Linear Model	Actual
100 m	9.83	9.85	9.86
200 m	19.67	19.67	19.77
1500 m	209.08	208.98	208.91
5000 m	768.81	769.1	759.74
10,000 m	1,604.9	1,604.11	1,587.85
110 hurdles	12.95	12.95	12.93
Steeple	481.29	481.39	475.72
Triple jump	18.03	83.72	83.04
Hammer	83.6	83.72	83.04
Shot put	22.11	22.1	22.03

Chapter 8: Regression Analysis

8.6.4 Olympic Trends

This dataset contains the gold medal performances in the men's long jump, high jump, and discus throw for the modern Olympic games from 1900 to 1984. Plots of this data show that the simple linear regression provides a good model for all three events. One general observation applicable to almost all Olympic events is that records after 1950 improved at a faster pace than those prior to 1950. This means that the lines fitted to the data before and after 1950 would have different slopes. This point is important for prediction based on modeling of trend. In fact, whenever possible, prediction of future Olympics should be based on data after 1950.

The Data (in Inches)

Renamed Year	High Jump	Discuss Throw	Long Jump
0	74.8	1,418.9	282.875
4	71	1,546.5	289
8	75	1,610	294.5
12	76	1,780	299.25
20	76.25	1,759.25	281.5
24	78	1,817.125	293.125
28	76.375	1,863	304.75
32	77.625	1,948.875	300.75
36	79.9375	1,987.375	317.3125
48	78	2,078	308
52	80.32	2,166.85	298
56	83.25	2,218.5	308.25
60	85	2,330	319.75
64	85.75	2,401.5	317.75
68	88.25	2,550.5	350.5
72	87.75	2,535	324.5
76	88.5	2,657.4	328.5
80	92.75	2,624	336.25
84	92.5	2,622	336.25

Although data for many recent Olympic games were available, here we used data for 19 games to see how trend fits the future games and how the predictions are.

Equations For Regression Using $Y(t) = a + bt$

High Jump	$Y(t) = 71.7 + 0.22379t$	$R^2 = 90.9\%$
Discus Throw	$Y(t) = 1480 + 14.314t$	$R^2 = 97.9\%$
Long Jump	$Y(t) = 283 + 0.61755t$	$R^2 = 75.7\%$

Predictions for Year 2000

	Fitted Values	95% Prediction Interval	Actual Values
High Jump	94.079	(89.713, 99.341)	95.6
Discus Throw	2,911.14	(2,798.9, 3081.1)	2,963.2
Long Jump	344.755	(322.44, 369.54)	356
Predictions for Year 2040:	103.03	3483.96	369.46
Projections by Physiologists 2040:	107	3699.10	404

Of course, we should not expect a better agreement since these are long-term predictions based on only 19 Olympic games.

A PRACTICE PROBLEM For the above three events use data sets from all the Olympic games and predict the gold medal records for 2040. Compare your findings with the above projections.

In the next section, we focus on one particular event and present a more detailed analysis of the data.

8.6.5 An Alternative Regression Model

In an interesting article that appeared in *Chance*, Spring 2002, Vol. 15, No. 2, S.M. Berry presented an alternative analysis of Olympic records for several different events including the men's 100-m run. Rather than using years as independent variables, he considered population of the world as a predictor. In fact, he assumes that records are improved not because athletes are getting better or they are better trained but because there are just more of them. He has based his analysis on the argument that records set by the best of 1 billion people should be better than the one set by the best of 1 million people.

Statistically we can look at this in the following way: Suppose that it is now year 1924 and the best record for men's 100-m run is b so that the chance of breaking record for the population of males in that year is $P(X > b) = p$. Now suppose further that in year 2000 the population of males has tripled. If we divide this population to three groups of males then, each group could break the record with probability p. Thus, the probability that at least one group breaks the record is equal to one minus the probability that non of them break the record; $1 - (1-p)^3$. Now from the following table, we see that for small values of p the chance of breaking record almost triples if population of males triple.

Chapter 8: Regression Analysis

p	$1-(1-p)^3$
0.01	0.030
0.02	0.059
0.03	0.087
0.04	0.115
0.05	0.143
0.10	0.271
0.15	0.386

Returning to the regression modeling, by regressing Olympic records (1900–2000) on $\mu + z_{1/n}\sigma$ Berry arrived at a reasonable model with $R^2 = 81.3\%$. Here μ and σ are, respectively, the mean and the standard deviation of fixed normal distribution presenting the athletic ability and $z_{1/n}$ is the $1/n$th quantile of a standard normal distribution. So the model used is of the form

$$\text{Record} = \mu + z_{1/n}\sigma$$

where n is the size of the population.

8.6.6 Estimation of Ultimate Record, An Example

In Sections 8.6.1 and 8.6.2, we presented a historical overview of the regression models used for prediction of future records including the ultimate records. Here an example from swimming is used to illustrate yet another application of regression.

"Records are made to be broken." This common saying in sports implies that there is no limit to athletic performances. But when we observe data on world records, we see that new records are not being set as often as they once were. This implies that, perhaps, one day there would be no hope for new records, unless some unexpected breakthroughs take place.

For example, following is a study on the men's 100-m freestyle competition. They noted that in the 54-year time period between 1956 and 2009, 19 new records were set in this competition.

Year, x	1956	1957	1957	1961	1961	1964	1967	1968	1970	1972
Time, y	55.4	55.2	54.6	54.4	53.6	52.9	52.6	52.2	51.9	51.47

Year, x	1972	1975	1975	1975	1976	1976	1976	1981	1985	1985
Time, y	51.22	51.12	51.11	50.59	50.39	49.99	49.44	49.36	49.24	48.95

Year, x	1986	1988	1994	2000	2000	2008	2008	2008	2008	2008	2009	2009
Time, y	48.74	48.42	48.21	48.18	47.84	47.6	47.5	47.24	47.20	47.05	46.94	46.91

In Larson, Hostetler and Edwards in the TI-92 Lab Manual, they used every record except the one for the year 2000 to produce the model: $y = \frac{38{,}504.5 + 44.375x^2}{1+x+x^2}$, where y is time (in seconds) and x is the year with $x = 0$ representing 1,900. Note that in this model as $x \to \infty$, y approaches 44.375. Thus, according to this model, the ultimate record for the 100-meter freestyle is 44.375 seconds.

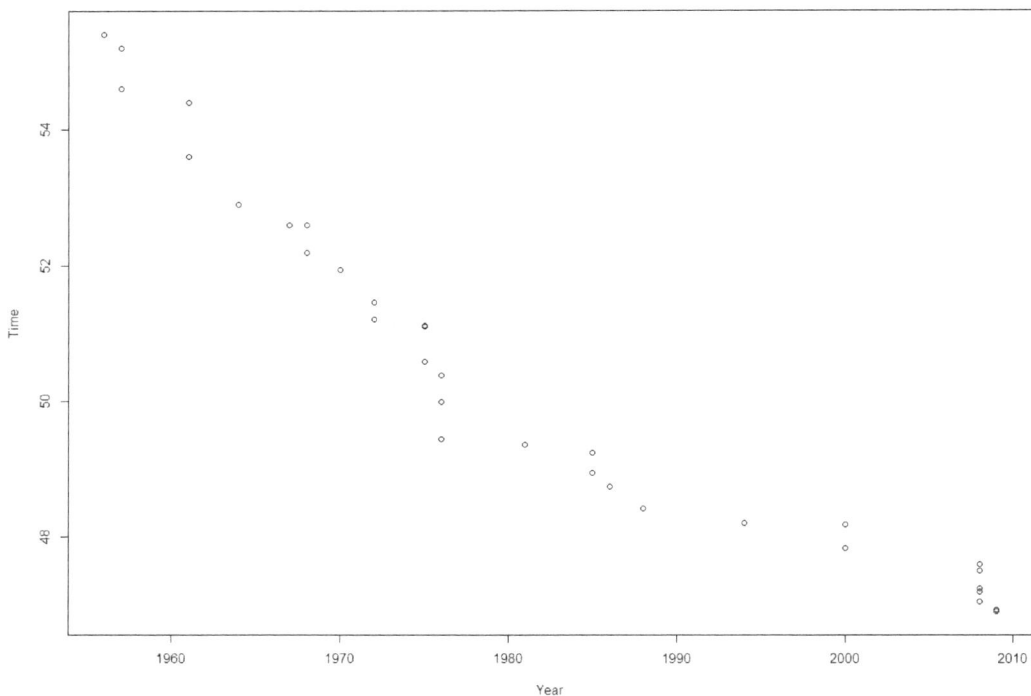

FIGURE 8.6.1 Time Versus Year for the Men's 100-Meter Freestyle

Now using this data and a record set at the 2000 Olympics, a regression analysis can be performed to find a model for the times in seconds. Here we tried a model of the form

$$y = \frac{a + bx + cx^2}{1 + x + x^2} = a\frac{1}{1 + x + x^2} + b\frac{x}{1 + x + x^2} + c\frac{x^2}{1 + x + x^2}$$

so that as $x \to \infty$, $y \to c$. Thus, estimate of c is an estimate for the ultimate record based on the records set so far. Note that as new records are set and added to the data, updated estimates may be obtained.

The regression equation we found is $y = \frac{-89.2 + 148x + 45.7x^2}{1 + x + x^2}$, where y presents the time (in seconds) and x is the year with $x = 0$ representing 1956. Here as $x \to \infty$, y approaches 45.7. Thus, according to this model, an estimate for the ultimate record for the 100-m freestyle based on the past data is 45.7 seconds.

8.6.7 Least Squares Using Matrices

This section describes (through an example) the use of matrix algebra in regression analysis. The material presented here is not new. It is only different in that matrices are used in its presentation.

A researcher who is interested in health issues related to professional athletes and their training location has collected data on atmospheric pollutants. The following table shows the data (atmospheric pollutant y_i) at half-hour intervals t_i:

t_i	1	1.5	2	2.5	3	3.5	4	4.5	5
y_i	−0.15	0.24	0.68	1.04	1.21	1.15	0.86	0.41	−0.08

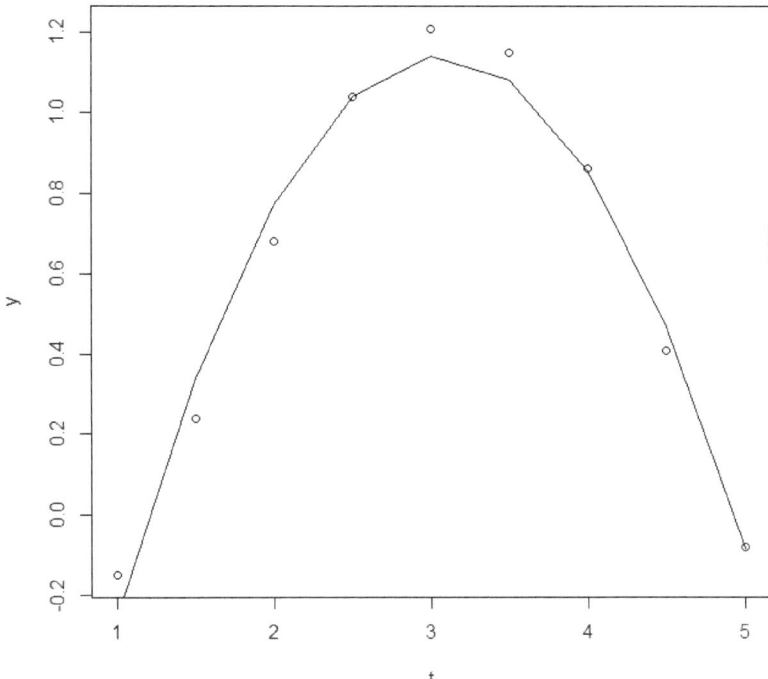

FIGURE 8.6.2 Pollution Data and Least Squares Quadratic Fit

The plot of this data, shown below suggests that a quadratic polynomial namely
$$y(t) = a_2 t^2 + a_1 t + a_0$$
may produce a good model. From the expression $y(t_i) = a_2 t_i^2 + a_1 t_i + a_0 = y_i$, we get the system of equations

$$a_2(1)^2 + a_1(1) + a_0 = 0.15$$
$$a_2(1.5)^2 + a_1(1.5) + a_0 = 0.24$$
$$\vdots \quad \vdots \quad \vdots \quad \vdots$$
$$a_2(5)^2 + a_1(5) + a_0 = 0.08.$$

These equations can be written in matrix form as $A\mathbf{a} = \mathbf{y}$, where

$$A = \begin{bmatrix} 1 & 1 & 1 \\ 2.25 & 1.5 & 1 \\ 4 & 2 & 1 \\ 6.25 & 2.5 & 1 \\ 9 & 3 & 1 \\ 12.25 & 3.5 & 1 \\ 16 & 4 & 1 \\ 20.25 & 4.5 & 1 \\ 25 & 5 & 1 \end{bmatrix}, \quad \mathbf{a} = \begin{bmatrix} a_2 \\ a_1 \\ a_0 \end{bmatrix}, \quad \mathbf{y} = \begin{bmatrix} -0.15 \\ 0.24 \\ 0.68 \\ 1.04 \\ 1.21 \\ 1.15 \\ 0.86 \\ 0.41 \\ -0.08 \end{bmatrix}.$$

Multiplying both sides by A^T we get a system of equations known as the normal equation $A^T A \mathbf{a} = A^T \mathbf{y}$. For the above data, normal equation leads to

$$\begin{bmatrix} 1{,}583.25 & 378 & 96 \\ 378 & 96 & 27 \\ 96 & 27 & 9 \end{bmatrix} \begin{bmatrix} a_2 \\ a_1 \\ a_0 \end{bmatrix} = \begin{bmatrix} 55.4 \\ 16.86 \\ 5.39 \end{bmatrix}$$

Solving this system, we obtain $\hat{\mathbf{a}} = (A^T A)^{-1} A^T \mathbf{y}$, that is

$$\hat{\mathbf{a}} = \begin{bmatrix} a_2 \\ a_1 \\ a_0 \end{bmatrix} \approx \begin{bmatrix} -0.3238 \\ 1.9889 \\ -1.9137 \end{bmatrix}$$

so the best fit least squares quadratic polynomial model is

$$y(t) = -0.3238 t^2 + 1.9889 t - 1.9137.$$

The figure above shows the data set indicated with ◇, and the graph of $y(t)$. We see that $y(t)$ is close to each data point but is not required to go through any of the data.

8.7 MULTIPLE LINEAR REGRESSION

In Sections 8.2 through 8.6, regression analysis of dependent variable (response) Y on one independent variable (predictor) X was considered. In many applications the response variable has association with several independent variables (predictors). For example, the price of a house depends not only on its size but also location, number of bedrooms, and so on. Your grade depends not only on the number of hours of study, but your attendance, interest, background, and so on. As a more specific example, consider the problem of predicting the body composition. Equations to predict body composition are usually population-specific. An equation developed for one subset of the population may not be accurate for another subset. Variation in age, gender, ethnicity, fitness levels, nutrition, and other factors influence the equation. If an equation to predict percent body fat in sedentary mid-life females is needed, and one is not available in the literature, we would have to develop our own specific equation. To do this, we may use hydrostatically determined percent body fat as the dependent variable, and measurements for the five skinfold sits (triceps, subscapular, abdominal, suprailiac, and calf) as independent variables. Of course, some of the independent variables have more predictive power than others and their inclusion in the regression model result in a greater improvement. So instead of one independent variable X we may consider m such variables X_1, ..., X_m and have measurements of the form $(x_{1i}, x_{2i}, \ldots, x_{mi}, y_i)$, $i = 1, \ldots, N$, where x_{mi} is the ith measurement of X_m. Our task is now to see whether we can establish a relationship of the following form known as multiple linear regression

$$y_i = b_0 + b_1 x_{1i} + b_2 x_{2i} + \ldots + b_m x_{mi} + \varepsilon_i, \quad i = 1, \ldots, N. \tag{8.22}$$

The following is an example of multiple linear regression

$$LBW = 10.38 + 0.9259 \,(\text{wt}) - 0.1881 \,(\text{thigh skinfold}) + 0.637 \,(\text{bi-iliac diameter})$$
$$+ 0.4888 \,(\text{neck circumference}) - 0.5951 \,(\text{abdominal circumference}).$$

In this equation, a man's lean body weight (*LBW*) is being predicted from several anthropometric measures, including skinfold thicknesses, circumferences, and diameters. The regression model, developed by Behnke and Wilmore (1974), have an R^2 equal to 0.918 and a standard error of estimate of 2.358, which is interpreted just the same as in the regression equation with only one predictor variable.

As with the simple linear regression, here too we make an important assumption that ε_i's are independent and identically distributed random variables (errors) having $N(0, \sigma^2)$ distribution. $\hat{b}_1$, $\hat{b}_2, \ldots, \hat{b}_m$ are called least squares regression coefficients, if the choice $b_i = \hat{b}_i (i = 1, \ldots, m)$ minimizes the sum of squares of the deviations (errors), namely

$$\Delta = \sum_{i=1}^{N}(y_i - b_0 - b_1 x_{1i} - \ldots - b_m x_{mi})^2 = \sum_{i=1}^{N} \varepsilon_i^2. \quad (8.23)$$

The minimal value of Δ is equal to the *SSResidual*. The residual (or error) sum of squares has $N - (m+1) = N - m - 1$ degrees of freedom. The *SSRegression*, the sum of squares due to regression, is as before $\sum_{i=1}^{N}(\hat{y}_i - \bar{y})^2$ where

$$\hat{y}_i = \hat{b}_0 + \hat{b}_1 x_{1i} + \ldots + \hat{b}_m x_{mi} \quad (8.24)$$

is the corresponding regression value. *SSRegression* has now m degrees of freedom since there are now $m+1$ parameters in the model. Using these entities we can set up an ANOVA table. The hypothesis of no linear relationship between the dependent variable and the independent variables, namely

$$H_0: b_1 = b_2 = \ldots = b_m = 0,$$

is equivalent to saying that the independent variables are not useful for the prediction of the dependent variable y. The hypothesis is tested using the *F*-ratio,

$$\frac{MSRegression}{MSResidual},$$

as before. Also, here the multiple correlation coefficient (square root of R^2) indicates the relationship between the response variable and a weighted sum of predictor variables. It follows then that R^2 represents the amount of variation of the response that is explained or accounted for by the combined predictors. This is the same concept as the coefficient of determination, which was discussed earlier with regard to the common association between two variables. Now, however, we have the amount of association between one variable (the response) and weighted combination predictors.

Note that for multiple regression, the analysis and calculation can easily be done by many excellent software programs available. MINITAB is used in the examples presented below.

EXAMPLE 8.7.1 In this example, we are interested in predicting the racing performance of trained female distance runners. The measured variables include the running performance in a 10-km road race, body composition (such as height, weight, skinfold sum, and relative body fat), and maximal aerobic power. The data including the top 14 runners are shown in Table 8.7.1, where X_1 stands for the height, X_2 for the weight, X_3 for skinfold sum, X_4 for relative body fat, X_5 for the maximal

TABLE 8.7.1 Data for Example 8.7.1

x_1	x_2	x_3	x_4	x_5	y
163	53.6	76.4	17.9	61.32	39.37
167	56.4	62.1	15.2	55.29	39.80
166	58.1	65.0	17.0	52.83	40.03
157	43.1	44.9	12.6	57.94	41.32
150	44.8	59.7	13.9	53.31	42.03
151	39.5	59.3	19.2	51.32	42.37
162	52.1	98.7	19.6	52.18	43.93
168	58.8	73.1	19.6	52.37	44.90
152	44.3	59.2	17.4	57.91	44.90
161	47.4	51.5	14.4	53.93	45.12
161	47.8	61.4	7.9	47.88	45.60
165	49.1	62.5	10.5	47.41	46.03
157	50.4	60.3	12.6	47.17	47.83
154	46.4	76.7	19.6	51.05	48.55

Source: Conley et al. (1981)

aerobic power, and Y for the running time. Use the multiple linear regression model and test the hypothesis of no linear relationship at a significance level 10%. In other words, test

$$H_0 : b_1 = b_2 = \ldots = b_5 = 0.$$

against the alternative that at least one b is non-zero.

SOLUTION: For this example $N = 14$. By minimizing the sum of squares given in (8.23), we get

$$\hat{b}_0 = 80.9212, \quad \hat{b}_1 = -0.0635, \quad \hat{b}_2 = -0.0851,$$
$$\hat{b}_3 = 0.0274, \quad \hat{b}_4 = 0.0465, \quad \hat{b}_5 = -0.4790.$$

Using formula (8.24), a straightforward calculation provides

$$SSRegression = 57.9867$$
$$SSResidual = 54.2308$$
$$MSRegression = SSRegression/5 = 11.597$$
$$MSResidual = SSResidual/8 = 6.779$$
$$F = 11.597/6.779 = 1.711.$$

ANOVA Table for Example 8.7.1

Source of Variation	SS	d.f.	MS	F-Ratio
Due to Regression	57.987	5	11.597	
				1.711
About Regression	54.231	8	6.779	
Total	112.218	13		

The F-critical $F_{0.90,(5,8)}$, is equal to is 2.73. Since the value of the F-ratio is smaller than 2.73, we cannot reject H_0. This means that the hypothesis of no linear relationship between Y and the five predictors cannot be rejected at 10% significance level.

One interesting problem is to test whether a special variable, say X_1, actually improves the prediction of the dependent variable. A test quantity for the hypothesis $H_0:b_1 = 0$ against the alternative $H_1:b_1 \neq 0$ is $\hat{b}_1^2/Var(\hat{b}_1)$, which is distributed as $F_{1-\alpha,(1,N-m-1)}$ (partial F test). It is important to note that we cn equivalently use $\hat{b}_1/\sqrt{Var(\hat{b}_1)} = \hat{b}_1/s_{\hat{b}_1}$, which is distributed as t with $N - m - 1$ degrees of freedom. Clearly the considerations of confidence intervals for $\hat{b}_i$'s carry over to the case of multiple linear regression. In fact, a $100(1 - \alpha)\%$ confidence interval for b_i is

$$\hat{b}_i \pm t_{\alpha/2,(N-m-1)} s_{\hat{b}_i}.$$

EXAMPLE 8.7.2 For the data in Example 8.7.1, carry out a partial F test for the hypothesis $H_0:b_1 = 0$ against the alternative $H_1:b_1 \neq 0$ at a significance level 10%.

SOLUTION: Similar to formula (8.12), we have

$$Var(\hat{b}_1) = \frac{s^2}{S_{x_1 x_1}}$$

where $s^2 = SSResidual/(N - m - 1) = 54.2308/8 = 6.7789$ and

$$S_{x_1 x_1} = \sum_{i=1}^{N}(x_{1i} - \bar{x}_1)^2 = 485.4286.$$

Hence $Var(\hat{b}_1) = 0.0139$ and $\hat{b}_1^2/Var(\hat{b}_1) = 0.2891$. The critical value equals $F_{1-\alpha,(1,N-m-1)} = F_{.90,(1,8)} = 3.46$. Clearly in this case we cannot reject H_0. In other words, we do not have sufficient evidence to indicate that X_1 (the height) would contribute much to the prediction of Y (running time). Note that, as mentioned, we can compare the absolute value of $-0.0635/\sqrt{0.0139} = -0.538$ to $t_{0.95,(8)}$. The conclusion will, of course, be the same.

Following the same line, we have

$$\hat{b}_2^2/Var(\hat{b}_2) = 0.4696$$
$$\hat{b}_3^2/Var(\hat{b}_3) = 0.2438$$
$$\hat{b}_4^2/Var(\hat{b}_4) = 0.0572$$
$$\hat{b}_5^2/Var(\hat{b}_5) = 7.5514$$

Therefore, only X_5 (the maximal aerobic power) will contribute to the prediction of running time. In other words, it has a predictive power. Thus, a linear regression using only X_5 as an independent variable should be applied in this case. Note that although the model developed based on all variables is not a useful one (based on the F-test), the variable X_5 is useful for prediction of Y. This seemingly contradictory result could happen when applying multiple regression and will be discussed later in this chapter.

EXAMPLE 8.7.3 Recall Example 8.5.2 regarding the guards played in NBA. To predict PPM we can consider several independent variables. Chatterjee et al. (1995) have considered minutes per game (MPG), height, and free throw percentage (FTP). This led to the model

$$\text{PPM} = -0.781 + 0.006\ \text{MPG} + 0.004\ \text{Height} + 0.003\ \text{FTP}$$

Here we have

$$SSResidual = 0.523$$

$$MSRegression = 0.142$$

$$MSResidual = 0.006$$

$$F = 24.43$$

Like for a single predictor case (MPG) the value of F is large. As a result of including height and FTP the value of R^2 is increased from 0.32 to 0.45, which means that we now have a better model. Note that adding new predictors should, in general, improve the model, but as will be discussed in Section 8.10, it presents many difficulties, and some of them may be serious.

8.8 BEST SUBSET SELECTION AND STEPWISE REGRESSION

When applying multiple regression, we may find that some of the predictors do not have any significant predictive power. Thus from a set of independent variables, we may need to select a subset that is useful for our problem objective.

Suppose that we have a response variable Y and a set of k predictor variables $X_1, X_2, \ldots, X_k$, and we want to determine the best subset of the k predictors and the corresponding best-fitting regression model describing the relationship between Y and X's. Our goal is to find a model that provides the best prediction of Y, given $X_1, X_2, \ldots, X_k$, for some new observation or for a batch the value of of new observations. To illustrate this, consider the data in the following table that includes statistics for 10 children. Suppose that we want to predict the weight of a child using his or her height and age.

Child	1	2	3	4	5	6	7	8	9	10
AGE	8	10	6	11	8	10	9	10	6	12
HGT	57	59	49	62	51	55	48	42	42	61
WGT	64	71	53	67	55	77	57	56	51	76

Chapter 8: Regression Analysis

Possible models we could consider are

1. $\text{WGT} = b_{01} + b_{11}\text{HGT} + \varepsilon_1$.
2. $\text{WGT} = b_{02} + b_{12}\text{AGE} + \varepsilon_2$.
3. $\text{WGT} = b_{03} + b_{13}\text{HGT} + b_{23}\text{AGE} + \varepsilon_3$.

Also the nature of the growth may suggest consideration of some nonlinear term such as inclusion of at least one quadratic term like $(\text{AGE})^2$. Note that inclusion of this term will not make the model nonlinear since we call a model linear or nonlinear with respect to its parameters. By including such a term we now have several new possible models. This, of course, makes the decision regarding the "best" model harder.

In order to decide which of the possible models is more suitable for predicting the weight, we have to decide on a selection criterion. For example, we can use R^2 to make such a selection. In general, the goal should be to select a model that has a minimum number of predictors and at the same time make sure that they are not carrying or producing the same information. For example, we know that height and age are correlated (collinearity); that is, they both may contribute the same information as far as prediction of weight is concerned. Several excellent software programs are available both for selection of best subset and for performing regression analysis.

Summarizing these, often a dependent variable Y may be related to many independent variables $X_1, \ldots, X_m$ some of which have a higher predictive power. The problem is then to select a subset of these (using an appropriate procedure) to produce a most "effective" linear regression model. One possibility is to use R^2 for subset selection. There are also methods based on ideas from information theory that determines which predictors carry more information for prediction of response variable. Here, one measures gain and loss of adding a new predictor. The other possibility is the stepwise procedure that checks the variables one-by-one to decide whether they should enter or exit the regression using F-to-enter and F-to-remove tests at a given level α. This procedure is lengthy but can be done using one of the known softwares such as MINITAB, SAS, and SPSS.

Regression and Contract Negotiation

When negotiating their contracts many players or their agents argue that, in addition to contributing to the success of their teams, they had also been an attraction at the box office. They sometimes claim that people come to the games specifically to see them play, and that they should be compensated accordingly.

Although many managers may believe there could be some truth to such claims, they look carefully at figures so that they can make an informed decision. Smart managers usually put together the information, recording everything they think could possibly influence the attendance. They then carefully analyze the data and get ready to meet the players and their agents. Some, of course, hire a statistician to analyze the data and, through applying statistical techniques such as regression and correlation analysis, specify factors that are most important for predicting ticket sales.

It is not hard to see that a relatively large number of the factors effect the ticket sales. Some of these include the date of the game, the opposing team and their positions, home or away, day of the week, average temperature, precipitation, time of the game, if it is televised, if it is promoted, if the player negotiating played, and so on. But clearly not all of these factors are equally important as far as ticket sales are concerned. In this respect, regression analysis can prove particularly useful in decision-making.

Some Problems Associated With Multiple Regression

Like simple linear regression, in multiple regression the basic determiner is the size of the correlation. For a higher correlation, a more accurate prediction will be obtained. However, some other factors affecting the final analysis should be mentioned. One limitation of prediction relates to generalizability. Regression equations developed with a particular sample often lose considerable accuracy when applied to others. This loss of accuracy in prediction is called shrinkage. The term population specificity is also related to this phenomenon. Thus, we need to recognize that the more accuracy one seeks through selection procedures that capitalize on specific characteristics of the sample, the more difficult it is to generalize to other populations. For example, a formula for predicting percent body fat from skinfold measurements that were developed with adult males would lose a great deal of accuracy if it is used for adolescents. Thus, the researcher should select the sample carefully with regard to the population for which the results are to be generalized.

In prediction studies, the number of subjects in the sample should be sufficiently large. Usually, the larger the sample, the more likely that the sample will represent the population from which it is drawn. However, another problem with small samples in multiple regression studies is that the correlation may be spuriously high. A direct relationship exists between the correlation and the ratio between the number of subjects (sample size) and the number of variables. In fact, the degree to which the expected value of R^2 will exceed zero when it is zero in the population is dependent on two things: the size of the sample (N) and the number of variables (m). More precisely, it is equal to the ratio of $(m-1)/(N-1)$. To illustrate, suppose you read a study in which $R^2 = 0.90$. You are, of course, impressed. However, the results would be meaningless if the study involved 40 subjects and 30 variables, because we could expect an R^2 of 0.74 just on the basis of chance alone $(R^2 = (m-1)/(N-1) = (30-1)/(40-1) = 0.74)$. One should be aware of the relationship between the number of subjects and number of variables when reading research that uses multiple regression. In the most extreme case with the same number of variables as subjects, $(m-1)/(N-1)$ would yield an R^2 of 1.0! A subject-to-variable ratio of 10:1 or higher is often recommended. For this problem, application of information theory or entropy is usually recommended.

8.9 APPLICATION

8.9.1 Why NBA Teams Win

The regression analysis presented in this section is an attempt to determine whether certain statistical factors have a noticeable effect on the winning percentage of NBA teams. The NBA is enjoying a period of record revenues and growth. There is a long established correlation between a team's winning percentage and revenues generated, "fans come out to watch a winning

Chapter 8: Regression Analysis

team." There are many other suppositions pertaining to why teams win. These predictors will be examined to see which have the greatest effect on a team's success.

There are many possible uses for a model of this type. A coach could use this information to learn which are the best areas of improvement he should concentrate on. A general manager could draft players, which are most likely to provide more of certain capabilities and services. Sportswriter can learn which cliches are worthless.

The general model we use here is

$$Y = a + b_1 X_1 + b_2 X_2 + b_3 X_3 + b_4 X_4$$

where

$X_1 = $ Offensive points scored per game
$X_2 = $ Defensive point given up per game
$X_3 = $ Turnovers permitted per game

Here a turnover, being defined as the other team stealing/otherwise receiving the ball without a shot being taken and

$X_4 = $ Rebound percentage

being defined as the percent of all rebounds a team gets in each game. For example, if there were 100 rebounds by both teams in a game, and Team A out rebounded Team B 55-45, Team A's rebound percentage would be 0.55.

Why are these variables reasonable predictors? The use of offensive points as a predictor is obvious—the team who scores the most points wins ("they've got to add some offensive scoring punch if they hope to win"). However, some teams try hard to score a lot of points, but at the expense of the defense. For example, consider the team Denver in 1990–1991. They led the league in scoring, but their opponents scored even more, and they had one of the league's worst records. Defense is thus a good variable to consider ("offense may be flashy, but defense wins championships").

Some teams have both good offenses and defenses, but play sloppily and have a frustrating tendency to turn the ball over too often. A good example of this was Michigan's "Fab Five" in the early stages of 1991, "they just can't expect to win with so many turnovers."

Finally, rebounding should be considered. Both defensive and offensive points can be improved with an increase in rebounds. Rebounds allow teams to get a second chance and take more shots.

It would be expected, a priori, that offensive points and rebounds would have a positive effect on winning percentage, and that defensive points and turnovers would have a negative effect.

Tables 8.9.1 and 8.9.2 are the source data for the dependent variable (winning percentage) and independent variable. The abbreviation of team names are used in Table 8.9.1 to fit the size of page. The full name of teams are listed in the table on page 399. We used a set of relatively old data so that students carry (if they like) the analysis for more recent data. We do not foresee any discrepancies or biases in the data. The first model to be run is the one described above. This model involves four predictors, $X_1, X_2, X_3,$ and X_4.

TABLE 8.9.1 NBA Team Comparison. *USA TODAY*, April 21, 1992

Team	Points Per game Own	Points Per game Opp.	Field Goal Percent Own	Field Goal Percent Opp.	Turnovers Per game Own	Turnovers Per game Opp.	Rebounds Percent Off.	Rebounds Percent Def.	Tot.	Below 100 pts Own	Below 100 pts Opp.	OT Gms W-L	Decided by 3 pts W-L	Decided by 10 pts W-L
Atl	106.2	107.7	0.467	0.480	15.3	15.3	0.336	0.663	0.509	28	24	2-6	7-10	19-22
Bos	106.6	103.0	0.492	0.456	14.2	13.6	0.319	0.687	0.503	28	33	2-1	7-23	0-15
Cha	109.5	113.4	0.477	0.496	15.5	17.0	0.311	0.654	0.482	16	9	2-3	6-9	13-27
Chi	109.9	99.5	0.508	0.460	13.3	15.7	0.351	0.693	0.522	7	39	2-1	7-8	44-4
Cle	108.9	103.4	0.488	0.470	13.1	14.0	0.305	0.665	0.485	16	28	4-1	11-9	25-7
Dal	97.6	105.3	0.439	0.470	14.7	12.9	0.312	0.678	0.495	50	27	2-1	8-8	6-38
Den	99.7	107.6	0.442	0.480	17.6	17.0	0.348	0.680	0.514	40	20	2-2	4-8	11-37
Det	98.9	96.9	0.465	0.453	14.8	13.6	0.349	0.685	0.517	43	47	3-1	5-5	25-13
Gol	118.7	114.8	0.507	0.482	16.5	*18.4	0.318	0.642	0.480	6	9	4-3	7-6	25-11
Hou	102.0	103.7	0.475	0.463	16.8	14.3	0.309	0.664	0.486	36	30	3-3	10-5	12-23
Ind	112.2	110.3	0.494	0.468	17.1	15.2	0.318	0.674	0.496	7	13	5-5	4-10	22-14
LAC	102.9	101.9	0.473	0.459	15.5	17.0	0.316	0.675	0.496	34	32	5-1	11-4	21-20
LAL	100.4	101.5	0.456	0.480	13.3	15.3	0.327	0.640	0.480	40	39	2-3	12-5	20-18
Mia	105.0	109.2	0.461	0.493	16.8	16.0	0.330	0.676	0.503	23	17	2-4	6-6	16-26
Mil	105.0	106.7	0.460	0.498	16.5	17.3	0.359	0.650	0.504	25	25	2-6	6-6	18-21
Min	100.5	107.5	0.458	0.485	14.1	14.8	0.315	0.623	0.469	36	20	2-3	4-11	4-37
NJ	105.4	107.1	0.458	0.477	17.0	16.4	0.391	0.663	0.527	30	26	1-3	7-4	13-19
NY	101.6	97.7	0.477	0.458	15.1	15.2	0.346	0.711	0.528	37	52	4-2	6-8	25-13
Orl	101.6	108.5	0.453	0.486	16.9	15.5	0.324	0.678	0.501	35	19	1-0	8-3	7-38
Phi	101.9	103.2	0.471	0.483	15.1	14.4	0.309	0.664	0.486	36	31	1-2	8-8	17-17
Pho	112.1	106.2	0.492	0.459	15.1	15.3	0.317	0.677	0.497	11	20	3-2	4-7	29-10
Por	111.4	104.1	0.473	0.454	16.2	16.7	0.356	0.699	0.528	15	31	2-4	3-5	31-9
Sac	104.3	110.3	0.466	0.479	16.6	16.5	0.285	0.642	0.464	26	21	2-3	8-9	8-30
San	104.0	100.6	0.476	0.452	16.0	15.3	0.342	0.698	0.520	29	38	1-2	5-8	28-13
Sea	106.5	104.7	0.474	0.475	16.1	16.1	0.365	0.668	0.517	24	30	4-3	8-6	19-11
Uta	108.3	101.9	0.492	0.459	15.4	14.7	0.327	0.689	0.508	20	34	2-4	9-6	31-9
Was	102.4	106.8	0.461	0.478	15.3	16.2	0.289	0.662	0.475	39	23	5-1	3-8	12-29

TABLE 8.9.2 NBA Team Comparison, *Philadelphia Daily News*, April 21, 1992

Eastern Conference Atlantic Division								
	W	L	Pct.	GB	Last 10	Streak	Home	Away	Conf.
y-Boston	51	31	0.622	—	9-1	Won 8	34-7	17-24	35-21
x-New York	51	31	0.622	—	5-5	Won 1	30-11	21-20	34-22
x-New Jersey	40	42	0.488	11	7-3	Won 1	25-16	16-26	29-27
x-Miami	38	44	0.463	13	4-6	Lost 1	28-13	10-31	27-29
Sixers	35	47	0.427	16	4-6	Won 2	23-18	12-29	23-33
Washington	25	57	0.305	26	2-8	Lost 1	14-27	11-30	15-41
Orlando	21	61	0.256	30	4-6	Lost 2	13-28	8-33	16-40

Central Division								
	W	L	Pct.	GB	Last 10	Streak	Home	Away	Conf.
z-Chicago	67	15	0.817	—	8-2	Won 2	36-5	31-10	47-9
x-Cleveland	57	25	0.695	10	7-3	Won 2	35-6	22-19	42-14
x-Detroit	48	34	0.585	19	6-4	Lost 2	25-16	23-18	31-25
x-Indiana	40	42	0.488	27	5-5	Lost 1	26-15	14-27	27-29
Atlanta	38	44	0.463	29	3-7	Lost 2	23-18	15-26	23-33
Charlotte	31	51	0.378	36	2-8	Lost 2	22-19	9-32	21-35
Milwaukee	31	51	0.378	36	1-9	Lost 3	25-16	6-35	22-34

Western Conference Midwest Division								
	W	L	Pct.	GB	Last 10	Streak	Home	Away	Conf.
y-Utah	55	27	0.671	—	8-2	Won 7	37-4	18-23	37-17
x-S. Antonio	47	35	0.573	8	5-5	Lost 1	31-10	16-25	34-20
Houston	42	40	0.512	13	5-5	Lost 3	28-13	14-27	26-28
Denver	24	58	0.293	31	1-9	Lost 2	18-23	6-35	14-40
Dallas	22	60	0.268	33	4-6	Won 2	15-26	7-34	15-39
Minnesota	15	67	0.183	40	3-7	Won 1	9-32	6-35	11-43

Pacific Division								
	W	L	Pct.	GB	Last 10	Streak	Home	Away	Conf.
z-Portland	57	25	0.695	—	6-4	Lost 1	33-8	24-17	37-17
x-Golden State	55	27	0.671	2	6-4	Won 3	31-10	24-17	35-18
x-Phoenix	53	29	0.646	4	6-4	Won 3	36-5	17-24	34-20
x-Seattle	47	35	0.573	10	7-3	Lost 1	28-13	10-22	34-20
x-LA Clippers	45	37	0.549	12	6-4	Lost 1	29-12	16-24	29-24
x-LA Lakers	43	39	0.524	14	3-7	Won 1	23-17	19-22	26-27
Sacramento	29	53	0.347	28	5-5	Lost 1	21-20	8-33	17-37

x—Clinched playoff birth; y—Clinches division title; z—Clinched conference title.

TABLE 8.9.3 Variable Used in Regression Analysis

No.	Winning Percent (Y)	Points Per Game Own (X_1)	Points Per Game Opposition (X_2)	Turnover Percent Opposition (X_3)	Rebound Percent (X_4)
1	0.463	106.2	107.7	15.3	0.509
2	0.622	106.6	103.0	14.2	0.503
3	0.378	109.5	113.4	15.5	0.482
4	0.817	109.9	99.5	13.3	0.522
5	0.695	108.9	103.4	13.1	0.485
6	0.268	97.6	105.3	14.7	0.495
7	0.293	99.7	107.6	17.6	0.514
8	0.585	98.9	96.9	14.8	0.517
9	0.671	118.7	114.8	16.5	0.480
10	0.512	102.0	103.7	16.8	0.486
11	0.488	112.2	110.3	17.1	0.496
12	0.549	102.9	101.9	15.5	0.496
13	0.524	100.4	101.5	13.3	0.483
14	0.463	105.0	109.2	16.8	0.503
15	0.378	105.0	106.7	16.5	0.504
16	0.183	100.5	107.5	14.1	0.469
17	0.488	105.4	107.1	17.0	0.527
18	0.622	101.6	97.7	15.1	0.528
19	0.256	101.6	108.5	16.9	0.501
20	0.427	101.9	103.2	15.1	0.486
21	0.646	112.1	106.2	15.1	0.497
22	0.695	111.4	104.1	16.2	0.528
23	0.347	104.3	110.3	16.6	0.464
24	0.573	104.0	100.6	16.0	0.520
25	0.573	106.5	104.7	16.1	0.517
26	0.671	108.3	101.9	15.4	0.508
27	0.305	102.4	106.8	15.3	0.475

Chapter 8: Regression Analysis

Table Including Team Names

Abbreviation	Team	Abbreviation	Team	Abbreviation	Team	Abbreviation	Team
Atl	Atlanta	Bos	Boston	Cha	Charlotte	Chi	Chicago
Cle	Cleveland	Dal	Dallas	Den	Denver	Det	Detroit
Gol	Golden State	Hou	Houston	Ind	Indiana	LAC	L.A.Clippers
LAL	L.A.Lakers	Mia	Miami	Mil	Milwaukee	Min	Minnesota
NJ	New Jersey	NY	New York	Orl	Orlando	Phi	Philadelphia
Pho	Phoenix	Por	Portland	Sac	Sacramento	San	San Antonio
Sea	Seattle	Uta	Utah	Was	Washington		

We begin our analysis with statistical evaluation of regression results. The analysis is carried out using MINITAB. Here we are not planning to describe everything at length. An example including details of such an analysis is presented elsewhere.

As a first step we like to know whether independent variables are producing the same information, at least partially (collinearity). For this we need to examine the correlation matrix, namely,

Correlation Matrix

	Y	X_1	X_2	X_3
X_1	0.581			
X_2	−0.452	0.416		
X_3	−0.313	0.088	0.503	
X_4	0.453	0.020	−0.490	0.153

We will refer to this matrix as we proceed.

The regression equation is

$$Y = 0.4265 + 0.0304 X_1 - 0.0309 X_2 + 0.0035 X_3 + 0.1486 X_4.$$

$s = 0.04927 \qquad R\text{-sq} = 92.0\% \qquad R\text{-sq(adj)} = 90.5\%$

Analysis of Variance

Source	SS	d.f.	MS	F-Ratio	P
Regression	0.60996	4	0.15249	62.82	0.000
Error	0.5340	22	0.00243		
Total	0.66336	26			

Source	d.f.	SEQ SS
X_1	1	0.22362
X_2	1	0.38560
X_3	1	0.00066
X_4	1	0.00009

The last column (SEQ SS) presents the contribution of different predictors to the sum of squares.

TABLE 8.9.4 Minitab Output Including ANOVA for All Four Predictors

Predictors	Coef.	St.Dev.	T	p
Constant	0.04265	0.5696	0.75	0.462
X_1	0.030353	0.002442	12.43	0.462
X_2	−0.030875	0.004216	−7.32	0.000
X_3	0.00346	0.01151	0.30	0.766
X_4	0.1486	0.7926	0.19	0.853

TABLE 8.9.5 Minitab Output Including ANOVA for Points Scored and Points Allowed

	Correlation Matrix	
	Y	X_1
X_1	0.581	
X_2	−0.452	0.416

Unusual Observations

Obs	C3	C2	Fit	Stdev.Fit	Residual	St.Resid
16	101	0.18300	0.27648	0.02468	−0.09348	−2.19R

From Table 8.9.4, it is clear that we cannot reject the null hypotheses of zero coefficients for X_3 and X_4. This is because the T-values are small, P-values are large, and SEQ SS values are small. Thus, next we drop these two variables and try a model that includes only X_1 and X_2. The results are summarized in Table 8.9.5.

The regression equation is

$$Y = 0.534 + 0.0304\, X_1 - 0.0307\, X_2$$

Predictor	Coef	St.Dev	T	P
Constant	0.5338	0.25451	2.10	0.046
X_1	0.030375	0.002096	14.49	0.000
X_2	−0.030699	0.002348	−13.07	0.000

$s = 0.04750$ R-sq = 91.8%   R-sq(adj) = 91.2%

Analysis of Variance

Source	SS	d.f.	MS	F-Ratio	P
Regression	0.60922	2	0.30461	135.03	0.000
Error	0.05414	24	0.00226		
Total	0.66336	26			

Source	d.f.	SEQ SS
X_1	1	0.22362
X_2	1	0.38560

Unusual Observations

Obs.	Off. Pts.	Win. Pct.	Fit	St.Dev. Fit	Residual	St. Resid.
9	119	0.67100	0.61510	0.02910	0.05590	1.49 x
14	105	0.46300	0.37087	0.01312	0.09213	2.02R
16	101	0.18300	0.28637	0.01596	−0.10337	−2.31R

R denotes an observation with a large standard residual
x denotes an observation whose X value gives it large influence.
Next we examine (assess) the models developed.

(A) Residual Analysis
After the first run, there was only one unusual observation noted. This was observation 16 in Table 8.9.1, for the Minnesota Timber wolves. The first residual was −2.19. This could be possibly explained by the fact that the Timber wolves had the NBA's lowest winning percentage as shown in Table 8.9.2 and by the fact that apparently, they lost several close games.

(B) Model Analysis
The first run displayed appears to confirm that the model is good. With an R^2 value of 92.0%, most of the variation is explained by the independent variables identified. The regression equation including these four predictors has coefficients that appear similar to what was expected. Interestingly, though, the regression coefficient for turnovers, which are something a team tries to avoid, was positive. This could possibly be explained by the effect that "run and gun" offenses have on both points and turnovers. However, it must be noted that turnovers do not appear to have a strong predictive contribution based on the high P-value of 0.766. The same can be said, interestingly enough, for rebounding (P-value of 0.853). The T-ratios and F-statistic, however, appear to confirm the validity of the overall model.

(C) Correlation Analysis
The fact that none of the correlation coefficients is above 0.8 (in fact none is above 0.6) is a further indication that this is a good model. The negative correlations are as they would be expected.

(D) Model Revision
Overall, the initial run of the model is quite satisfactory. It appears the best step to improve the model is to run it once using only offensive points scored and defensive points allowed as predictors. While the initial R^2 value was high 92.0%, it appeared that turnovers and rebounds were not strong indicators of winning percentage. Therefore, we decided to run the model again using just points scored and points allowed as predictors. This provided an R^2 value of 91.8%, which nearly accounted for all the variation that original model did. The P-values of the two predictors were still extremely low.

The model followed our prior expectations in that, as expected, offensive production and strong defense both contribute greatly to a winning percentage. The surprise in the results, though, is how little impact turnovers and rebounding appear to have. This appears to back the conventional wisdom.

As a conclusion, we see that the model gives a good indication of why NBA teams win. It also provided some insight into some factors (turnovers and rebounding) that might not be as important as some think they are.

One additional question might be the age old debate regarding whether offense or defense is most important. In the first run, the regression coefficient for offense was -0.30375, while the coefficient for defense was 0.30875. These are probably too close to make a call, but if one were made it would probably go to the defense. The second run produced similar results, with regression coefficients of 0.30375 and -0.30699, respectively.

The value of prior thinking and data analysis appeared to be good based on the results of the first run. However, the results were improved by the process of analyzing the initial output and determining that it could be rerun without turnovers and rebounding as predictors.

Finally, when using software we can apply best subset selection or stepwise regression. These methods will provide a model that include the independent variables with the most predictive power. This is left for students to try.

8.9.2 Prediction of Medal Totals for Olympic Games

In an interesting article, *Bernard and Busse (2000)* considered the role of population and economic development in determining medal totals won by different countries using data of Olympic games from 1960 to 1996. They also provided predictions for the 2000 Olympics in Sydney. Clearly population should play a role in determining country medal totals. Larger countries have a deeper pool of talented athletes and thus a greater chance at finding medal winners. However, pure population levels were not sufficient enough to explain national totals. The author then turned to available resources in enabling gifted athletes to train for, attend, and succeed in the games. The population-based model was extended to include this measure for resources in the form of gross domestic product (GDP) per capita. GDP is most of the story, but not all. To improve their model, the authors included the factors of hosting advantages; the medal premium enjoyed by the former Soviet Union and its satellites, a vote of large-scale boycotts.

To find a regression equation to predict Olympic medals, they turned to economic resources first (GDP). The production function for generating Olympic caliber athletes for a country i in year t requires people (N), money (Y), and some organizational ability (A). This is represented as

$$T_i = f(N_i, Y_i, A_i).$$

The share of Olympic medals, M_i, won by a country is a function of the talent(T) in a given country. This is expressed as

$$\frac{medals_i}{\Sigma_i medals_i} = M_i = g(T_i), \quad T_i \geq T^*$$
$$= 0, \quad T_i < T^*.$$

Chapter 8: Regression Analysis

They also have chosen the so-called Cobb-Douglas production function in population and national income for the production of Olympic talent and a log function for the translation of relative talent to medal shares. This had led to the following choices:

$$T_i = A_i N_i^{\tau} Y_i^{\theta}$$

$$M_i = ln(\frac{T_i}{\Sigma_i T_i}), \quad T_i \geq T^*.$$

This choice yields the following regression equation

$$M_{it} = lnA_{it} + \tau lnN_{it} + \theta lnY_{it} - ln(\sum_j T_{jt}), \quad T_i \geq T^*$$

$$= 0, \quad T_i < T^*.$$

Since national income can be expressed as the product of population and per capita income, the authors actually estimated a specification of the form given below which leads to a regression equation

$$M_{it} = C + \alpha lnN_{it} + \beta ln(\frac{Y}{N})_{it} + \delta_t + v_i + \varepsilon_{it}, \quad T_i \geq T^*$$

$$= 0, \quad T_i < T^*.$$

where δ_t is year dummy included in the total pool of talent in the number of countries participating, as well as the changing number of sports, v_i, a country random effect, and ε_{it} it is a normally distributed error term.

To even better the regression model, the authors included the factors of hosting advantages, the medal premium enjoyed by the former Soviet Union and its satellites, and the role of large scale boycots. The resulting regression equation was

$$M_{it} = C + \alpha lnN_{it} + Bln(\frac{Y}{N})_{it} + Host_{it} + Soviet_{it} + Planned_{it} + \varepsilon_{it}.$$

The authors used the model to predict medals for the 2000 Sydney Olympics. They made predictions using recent numbers on population and GDP growth for a subset of countries. First, they used the model to predict 1996 medal winners and compared their predictions to actual medals won. Of course, there were deviations from the actual medals won and the predicted number of medals. The results of predictions are meant to answer the question of how many medals a county "ought" to earn on the basis of its population and income resources. They then used the last regression equation to predict the numbers of medals each country would win in Sydney. These results are listed in Table 8.9.6 following this summary. Also included in the table are 1996 medals for comparison.

To have an idea of how good the proposed models are, we have also included two figures, one presenting the frequency distribution of medals in the 2000 Olympics, and one presenting a comparison of the predicted and actual medals. As can be seen, the regression models provide a reasonable description and prediction.

TABLE 8.9.6 Total Medal Predictions for Sydney and the Actual Numbers

	Actual 1996 Medals	Predicted 2000 Medals(1)	Predicted 2000 Medals(2)	Actual 2000 Medals
Australia	41	52*	47	58
Belarus	15	12	0	17
Belgium	6	7	8	5
Brazil	15	17	19	12
Bulgaria	15	10	0	13
Canada	22	23	21	14
China	50	49	51	59
Cuba	25	20	59	29
Czech Republic	11	9	2	8
Denmark	6	7	6	6
France	37	38	33	38
Germany	65	63	50	57
Greece	8	8	5	13
Hungary	21	18	7	17
Italy	35	35	31	34
Jamaica	6	1	0	7
Japan	14	19	28	18
Kazakhstan	11	8	0	7
Kenya	8	5	0	7
Korea	32	30	29	28
Netherlands	19	19	17	25
New Zealand	6	5	0	4
Nigeria	6	5	0	3
Norway	7	7	5	10
Poland	17	16	11	14
Romania	20	17	4	26
Russia	63	59	40	88
South Africa	5	6	5	5
Spain	17	18	18	11
Sweden	8	9	8	12
Switzerland	7	8	9	9
Turkey	6	7	7	4
Ukraine	23	21	11	23
United Kingdom		18		28
United States	101	97	75	97

8.10 DIFFICULTIES OF USING MULTIPLE REGRESSION

Regression analysis, in general, and linear regression in particular, is part of statistics used frequently by practioners, including those interested in sport-related questions. In fact, with availability of computer and numerous software programs, regression analysis has become accessible to nearly everyone. The use of such software is normally very easy, however, their purely mechanical application is not appropriate. In fact, this has been one reason for having doubts about the conclusions drawn by statistical analysis. Although description of the methods presented earlier may seem to suggest that regression analysis is a straightforward exercise and is without pitfalls, unfortunately this is not always the case. In short, regression analysis is perhaps the most widely used and also widely misused statistical technique applied to data.

As pointed out earlier, regression analysis, especially the hypothesis testing part of it, is based on several important assumptions. Among them are (1) that the correct equation is being used—that is, the proper variables were included as independent variables and the proper functional form was used; (2) that the variables are measured accurately; (3) that the predictor variables are independent of each other; (4) that the data constitute a random sample; and (5) that the residuals are normally distributed and have a constant variance. Difficulties arise in regression analysis when any of these assumptions are violated. Software such as SAS, SPSS, or MINITAB do not automatically take care of these difficulties, although they point out their existence. It is, therefore, up to the user to handle the problems they cause and find remedies for them. Note that even very qualified and experienced statisticians who are familiar with data analysis techniques but are not familiar with the discipline related to the origin of data may end up drawing incorrect conclusions.

Thus, many analysts recognize the shortcomings of linear regression and often attempt to overcome the resulting problems. The following section addresses some of the more common problems associated with linear regression, the implications each problem has on the outcome, and some of the methods that analyst use to circumvent the difficulties. We begin with the issue that faces any analyst namely–specification of the model. After examining the issues associated with data, we discuss various problems related to the form of the error term in the regression equation. Discussion of issues are mostly due to Schroeder et al. (1986) "Understanding Regression Analysis" Sage University Paper series on Quantitative Applications in the Social Sciences 07-057.

8.10.1 Exclusion of a Relevant Variable

We start with a difficulty related to the so-called parsimony, which is the principle of explaining the most with the least. This means that if we can explain a response variables with a few explanatory variables nearly as "well" as a larger number of explanatory variables, we should do so. In many cases achieving this requires omitting some of the variables involved.

When a variable is omitted from a regression, coefficients on the included variables will, in general, be unreliable or invalid, since they will be "biased" estimates of the true population regression coefficients. While this conclusion stems from statistical theory and is not proven here, the idea underlying this result is intuitively plausible. Suppose that two variables, height and age, are the sole determinants of performance of a high school basketball player and that all other variability in performance is purely a random occurrence. If the analyst uses only height to explain the variability in performance and if height and age are correlated, the estimated coefficient on height will reflect the effects of both height and age on the performance.

Since the task of regression analysis is to estimate the response in the dependent variable to changes in an independent variable, an incorrect estimate of the response may be serious. Unfortunately, there is little an analyst can do to detect whether an important variable has been left out of the equation. Because of the uncertainty regarding omitted variables, researchers often include results from two or more different specifications of the same phenomenon. If, under alternative specifications, there is little change in the size of the estimated coefficients, the estimates are said to be robust. Such experimentation strengthens the analyst's belief in the model used, even then one can never be absolutely certain that a relevant variable has not been omitted.

8.10.2 Inclusion of an Irrelevant Variable

Since exclusion of a relevant variable is "bad," one might think that the solution to the problem is to throw every available variable into the equation. Of course, this solution also has pitfalls. If a variable is included in the equation but is not in fact relevant, the estimates of the coefficients will be unbiased. However, if the irrelevant variable is corrected with the included relevant variables, the size of the estimated standard errors of the coefficients of the relevant variables will increase. This in turn means that the ratios will be smaller than if the correct specifications were used. Hence, the analyst is more likely to conclude that the coefficient of a relevant variable is not significantly different from zero (i.e., the researcher will not be able to reject the null hypothesis that there is no association with the dependent variable). Thus, adding unnecessary variables causes a loss in precision of the estimated coefficients on the relevant variables.

8.10.3 Incorrect Functional Form

The least squares linear regression is not restricted to simple linear relationships among variables. There are, in fact, many possible functional forms that are amenable to estimation using least squares techniques. The issue is which form to use. If the underlying relationship between variables is actually nonlinear but a linear function is estimated, the resulting coefficient will be biased.

One way in which nonlinearities may be detected is to plot the residual error. In other words, plot the difference between the actual value of the dependent variable and its value as estimated from the equation. If there are large negative (positive) residuals at low and high values of an independent variable and large positive (negative) residuals at intermediate levels of the independent variable, a nonlinear relationship is suggested.

8.10.4 Stepwise Regression

Since decisions regarding which of numerous possible variables to include in a regression equation are difficult, stepwise regression techniques are sometimes used. These techniques allow the computer to experiment with different combinations of independent variables. The stepwise regression method is a variation of the forward technique except that each time a new predictor variable is stepped in, the new relationship between the response and the predictor variables is reevaluated to see whether the predictor variables already selected still significantly contribute when variables are added later. It is possible, then that a predictor entered earlier may be dropped out later when new predictors are brought into the equation. In most cases, however, the stepwise method is identical to the forward selection method. In one method of stepwise regression, the computer first estimates simple linear regressions using each of all the possible independent variables specified by the analyst. For example, if there were 20 possible independent variables, the computer program would estimate 20 different simple linear regressions. From the set of 20 results, the program would choose which one is "best." This selection, which is a part of the computer program, usually relies on the coefficient of determination, R^2.

In step 2, the program would try each of the 19 remaining independent variables together with the variable chosen in step 1 and produce 19 different regression results, each with two independent variables. Again, the rule regarding which of these 19 is "best" would be invoked and results from this second step would be printed. This process continues until either all 20 variables are included in the equation or not remaining variable increases the R^2 statistic sufficiently to permit the inclusion of additional variables.

Although R^2 statistics can be tested using an F distribution, it should be recognized that changes in R^2 attributable to any particular variable usually depend on what variables are already in the equation. Without careful thought, stepwise regression analysis can turn into a fishing expedition that is void of theory.

In summary, specification is one of the most perplexing programs faced by most analysts. Misspecification can produce misleading or imprecise results. Furthermore, computational techniques relying heavily on computers and devoid of theory do not provide the solution. It is still innovative thought and theory that must be relied on most to surmount problems.

8.10.5 Proxy Variables and Measurement Error

While theorizing about appropriate variables is not always easy, actually observing some variables and measuring them accurately can be equally difficult. Appropriate data are often not available. In such cases analysts often turn to alternative, second-best measures of the phenomenon at hand. The variables chosen are termed proxy variables since they are being used to approximate the real thing. The degree of approximation will influence the estimated impact of the variable of actual interest.

There are many examples of uses of proxy variables in the literature. Whenever dummy variables are substitute for what is really a continuous variable, a proxy is being used. For example, some analysts of player behavior may think that the "liberalism" of the coach affects particular types of behavior, but, in the absence of a direct measure of liberal tendencies, they use a dummy variable set equal to 1 if the coach is soft and 0 if he is strict. Note that these attitudes are seldom easy to measure directly. For that reason, numerous scaling variables have been developed, which are constructed from responses to attitudinal surveys.

Variables that are available are often substituted for unobserved variables. For example, even though theory may suggest that players experience influences their wages, experiences may not be available in a data set. In such instances, researchers often substitute age under the assumption that the older the player, the greater his or her playing history.

Use of imperfect proxy variables can introduce errors of measurement into the analysis. Another form of measurement error is simply mismeasurement of the variables that are available. For example, respondents to a survey may deliberately understate their age or not report accurately the candidate for whom they voted. Measurement error can also occur if survey questions are asked in an ambiguous way.

Measurement errors can result in biased estimates of regression coefficients. Sometimes these errors can be avoided through more accurate data collection procedures, however, when analysts use data collected by others, it is unlikely that much can be done to improve the quality of the data. Instead, cognizance should be taken of the probable measurement errors and how systematic over or underreporting of either the independent or dependent variables might influence the estimated coefficients.

8.10.6 Selection Bias

There are instances in which, even though every variable is measured accurately, the nature of the sample is such that the observations are a nonrepresentative sample of the population. All results based on questionnaires that can be completed by anyone who is willing to put forth the effort are potentially nonrepresentative, since the participants have been self-selected. Similarly, when studying player's wages, players not in the market are systematically excluded from the analysis. In such a case, the results of the regression analysis cannot readily be used to predict the wage that a player currently not playing could get if he were to get a contract. This is because there is some systematic difference between players who are playing on a team and those who are not playing for wages. Any regression based only on the former group will not capture this influence. If the regression results from the censured sample (players on a team) are to be used to make inferences about all players, it is necessary to adjust for the selection bias that exists.

8.10.7 Multicollinearity and Singularity

A final problem associated with data used in regression is multicollinearity. Its presence can make estimation of regression coefficients difficult. It arises whenever two or more independent variables used in a regression are not independent but are correlated. Unfortunately, when using sport data this problem arises often since, for example, in basketball many performance variables such as points per minute, assists per minute, rebounds per minute, age, and height for a basketball player are likely to be interrelated. Data varying with time such as score of and individual player or a team in successive games are likely to exhibit multicollinearity.

When two or more independent variables are correlated, the statistical estimation techniques discussed earlier are incapable of sorting out the independent effects of each on the dependent variable. While regression coefficients estimated using correlated independent variables are unbiased, they tend to have larger standard errors than they would have in the absence of multicollinearity. This in turn means that the t-ratios will be smaller. Thus, it is more likely that one will find that the regression coefficients are less significant than in the case where no multicollinearity plagues the data. In essence, there is less precision associated with estimated coefficients.

Multicollinearity is probably present in all regression analysis, since the independent variables are unlikely to be totally uncorrelated. Thus, whether or not multicollinearity is a problem depends on the degree of collinearity. The difficulty is that there is no statistical test that can determine whether or not it really is a problem. One method for searching for the problem is to look for "high" correlation coefficients between the variables included in a regression equation. This can easily be done using software like MINITAB. Even then, however, this approach is not foolproof, since multicollinearity also exists if linear combinations of variables are used in a regression equation. There is no single preferable technique for overcoming multicollinearity, since the problem is due to the form of the data. If two variables are measuring the same thing, however, one of the variables is often dropped, since little information is lost by doing so.

Singularity means two or more independent variables are perfectly related to each other ($r = 1.00$). This may occur if one variable is created from another by a mathematical manipulation such as squaring, taking the square root, or adding, subtracting, multiplying, or dividing by a constant. Most advanced computer programs screen for multicollinearity and singularity and warn the user in the printout if these relationships are detected by producing the squared multiple correlation values for all variables.

8.10.8 Autocorrelation

Measurement errors, selection bias, and multicollinearity are all attributable to the data available to a researcher. The next set of issues pertains to assumptions regarding the residual error term. Recall that the residual error term is the difference in the observed value of the dependent variable for the ith observation, and the value of the dependent variable predicted from the estimated regression for the ith observation. The discussion of regression analysis is based on the ordinary least squares regression model. This model assumes that (1) even though some errors are small and others are large, some are positive and others are negative, they have a mean of zero; (2) the error term associated with one observation is uncorrelated with the error term associated with all other observations; (3) while some of the error terms may be small and others large, the variability of the error terms is in no way related to the independent variables used; and (4) the error term is not correlated with the independent variables. Violations of any of these assumptions produce undesirable properties in the results obtained when regression coefficients are estimated without regard for these assumptions. While a full discussion of all these topics is beyond our purpose here, it is useful to review the most common problems that arise in the course of regression analysis and to indicate the steps that analysis take in response to these problems.

The first of these issues is termed autocorrection or serial correlation. Autocorrelation refers to the case in which the residual errors terms from different observations are correlated. If the terms are positively correlated, positive autocorrelation is said to exist, while if they are negatively correlated, negative autocorrelation is present.

Autocorrelation and the problem it presents is more likely to appear with time series data (data ordered in time), and most commonly the problem is restricted to error terms associated with successive time periods.

Autocorrelation can be caused by several factors, including omission of an important explanatory variable or the use of an incorrect functional form. It may also simply be due to the tendency of effects to persist over time or for dependent variables to behave cyclically. Whatever the cause, autocorrelation influences the outcome of the hypothesis-testing procedure. The effect of positive autocorrelation is underestimation of the standard error of the estimated coefficient. This in turn yields an inflated t-ratio, which means that it is possible that coefficients will be found to be significantly different from zero when in fact they are not.

While simply looking at the residual terms may provide some clue to the existence of autocorrelations, application of a formal test statistic called the Durbin-Watson (an option in

MINITAB) is recommended, especially when time series data are being analyzed. This statistic can be used to test the null hypothesis that the successive error terms are not autocorrelated.

When serially correlated error terms are detected, there are special techniques available to circumvent the problem. Many analysts use a technique called generalized least squares (GLS) regression to overcome the problem. This method is based on ordinarily least squares regression techniques but uses variables that have been transformed. Sometimes existence of a trend in data causes autocorrelation. When this the case, differencing the neighboring measurements could solve the problem.

8.10.9 Heteroskedasticity

Heteroskedasticity refers to another nonrandom pattern in the residual error term. Assumption (3) in the discussion of the regression model is that the variability in the error term does not depend on any factor included in the analysis. This assumption is known as the assumption of homoscedastic errors; when it is violated, heterokedasticity is said to exist. The problem arises most frequently in the analysis of cross-sectional data.

Consider the relationship between the number of players in an organization and the number of trainers. One might specify that the number of trainers is a function of the number of players. While a general positive relationship will probably be found (i.e., organizations with larger number of players have a greater number of trainers). It is possible to find some large organizations with few trainers. This is a situation where the variability in the residual error terms is not constant for all values of the independent variables. The residuals are said to be heteroskedastic.

As with autocorrelation, heteroskedasticity affects the size of the standard error of the regression coefficient, thereby biasing hypothesis-test results. The effect on standard error of slope will depend on the exact manner in which the heteroskedasticity was formed. Several different tests are available for detecting the problem of heteroskedasticity. All tests depend on an examination of the residuals. Again, when the problem is detected, generalized least squares can be used to give differential weights to the observations and thereby circumvent its effects on tests of hypothesis.

8.10.10 Outliers

Outliers are observations that are not well fitted by the assumed model. Such observations will have large residuals. A crude "rule of thumb" is that an observation with a standardized residual greater than 2.5 in absolute value indicates a possible outlier, and the source of the data should be investigated. If there is doubt about the accuracy or veracity of the observation, then it should be omitted and the model should be refitted. If a large difference in the model parameter estimates is thereby obtained, then the observation is said to be influential. Influential outliers are more important than ones which do not cause much perturbation in model parameters on omission, because they raise the possibility of model instability. However, it is not only outliers that can be influential, so this topic deserves further discussion.

8.10.11 Influential Observations

An influential observation is defined to be one which, for whatever reason, causes a large change in some or all of the estimated regression parameters when it is omitted from the data set. We have seen above that sometimes a large outlier is an influential observation. However, outliers reflect disparities between observed and fitted response variable values, and these are not the sole causes of model instability. If an observation is well separated from the others in terms of its values on the regressor variables, then this observation is also likely to influence the fitted regression model. Finally, a combination of regressor variable separation and response variable disparity also has the potential for influencing regression fit.

8.10.12 Misconception

In any regression analysis, it is important to interpret the estimate of the parameters correctly. A typical misconception is that b_i, for example, measures the effect of x_i on $E(Y)$, independent of the other independent variables in the model. This, of course, could be the case in some models, but it is not true in general.

Another misconception is that a statistically significant b_i value establishes a cause-and-effect relationship between $E(Y)$ and x_i. That is, if b_i is found to be significantly greater than 0, one might infer that an increase in x_i causes an increase in the mean response $E(Y)$.

8.10.13 Ethical Issues

When applying regression it is easy to manipulate the process of developing the regression model. The key here is intent. Examples include the following:

1. Prediction with the intent of excluding certain independent variables from consideration in the model so that a desired conclusion may be reached.
2. Deleting observations to obtain a better fit without giving reasons.
3. Applying the model for any purpose without providing an evaluation of the assumptions made regarding the validity of the model.
4. Keeping independent variables in the model that exhibit a high degree of multicollinearity.

8.10.14 Cross Validation

In any regression analysis and prediction based on models, sound research design requires that the results be tested for accuracy. In the development of either bivariate or multivariate regression equations, the equation should be developed on one sample of the population and then tested on another equivalent sample. This may require dividing the sample in half, which would reduce the size of data used for modeling. But without cross validation on an equivalent sample to test the accuracy of the prediction, the results will be suspect.

8.10.15 Some Remedies

As pointed out earlier, in many practical situations one or more of conditions required for validity of the regression analysis may be violated. The most commonly used method to remedy non-normality and heteroscedasticity is to transform the dependent variable. There are several points to note about this procedure. First, the actual form of the transformation depends on which condition is not satisfied and on the specific nature of the violation. Because there are many different ways to violate the required conditions of the statistical techniques, the list of transformations given here is unavoidably incomplete. Second, these transformations can be useful in improving the model. That is, if the linear model appears to be quite poor, we often can improve the model's fit by transforming y. Third, many computer software programs allow us to make transformations quite easily. Students might want to experiment to see the effect these transformations have on the statistical results.

A brief list of the most commonly used transformations is as follows:

1. **Log Transformation:** $y' = \log y$ (provided $y > 0$). The logarithmic transformation is used when
   (a) the variance of the error variable increases as y increases
   (b) the distribution of the error variable is positively skewed
   (c) the variance is proportional to square of the expected value of y
2. **Square Transformation:** $y' = y^2$. Use this transformation when
   (a) the variance is proportional to the expected value of y
   (b) the distribution of the error variable is negatively skewed
3. **Square-Root Transformation:** $y' = \sqrt{y}$ (provided that $y > 0$). The square-root transformation is helpful when the variance is proportional to the expected value of y.
4. **Reciprocal Transformation:** $y' = 1/y$. When the variance appears to significantly increase when y increases beyond some critical value, the reciprocal transformation is recommended.
5. **Arcsine Transformation:** $y' = \sin^{-1}(\sqrt{y})$. This is useful when
   (a) responses are proportions with constant denominators
   (b) the variance is proportional to $\mu(1-\mu)$ where $\mu = E(Y)$

The following example illustrates an application of these remedies.

EXAMPLE A coach wanted to know whether time limits on evaluation of players affected their marks. Accordingly, he took a random sample of players and split them into five groups of 20 players each. All players took a test that involved questions regarding the history and rules of the sport and also some set moves. Each group was given a different time limit. Group 1 was limited to 40 min; Group 2, to 45 min; Group 3, to 50 min; Group 4, to 55 min; and Group 5, to 60 min. The tests were marked (out of 40).

To analyze the data, he considered the simple regression model and used MINITAB. The following presents the output:

Regression Analysis

The regression equation is Mark $= -2.20 + 0.550$ Time

Predictor	Coef.	St.Dev.	T-Ratio	p
Constant	-2.200	1.646	-1.34	0.184
Time	0.55000	0.03259	16.88	0.000

$s = 2.305 \quad$ R-sq $= 74.4\% \quad$ R-sq(adj) $= 74.1\%$

Analysis of Variance

Source	d.f.	SS	MS	F	p
Regression	1	1512.5	1512.5	284.77	0.000
Error	98	520.5	5.3		
Total	99	2033.0			

The standard error of estimate, the coefficient of determination, and the *t*-test of β_1 (and the *F*-test) all indicate a relatively good model. The residuals and the predicted values were calculated. The histogram of the residuals and the plot of the residuals versus the predicted values of *y* were produced by MINITAB.

The error variable appeared to be normal. However, the variance was clearly not constant; it increased as the predicted marks increased. The remedy he applied was to transform the dependent variable. He attempted the following two transformations:

1. $y' = log(y)$ (new variable LOGMARK), 2. $y' = 1/y$ (new variable 1/MARK.)

The following presents the outputs:

Regression Analysis

The regression equation is

LogMark = 2.13 + 0.0217 Time

Predictor	Coef.	St.Dev.	T-Ratio	p
Constant	2.12958	0.06030	35.32	0.000
Time	0.021716	0.001194	18.19	0.000

$s = 0.084444$ $R-sq = 77.1\%$ $R-sq(adj) = 76.9\%$

Analysis of Variance

Source	d.f.	SS	MS	F	p
Regression	1	2.3579	2.3579	330.72	0.000
Error	98	0.6987	0.0071		
Total	99	3.0566			

The histogram of the residuals indicated that the error variable may be normal. The plot of the residuals versus the predicted values of the dependent variable showed some change in the variance. However, the change was smaller than in the original model. Thus, the transformation has decreased the degree of heteroscedasticity.

For the second transformation he obtained the following:

Regression Analysis

The regression equation is

1/Mark = 0.0846 − 0.000876 Time

Predictor	Coef.	St.Dev.	T-Ratio	p
Constant	0.084574	0.002389	35.40	0.000
Time	−0.00087647	0.00004731	−18.53	0.000

$s = 0.003345$ $R-sq = 77.8\%$ $R-sq(adj) = 77.6\%$

Analysis of Variance

Source	d.f.	SS	MS	F	P
Regression	1	0.0038410	0.0038410	343.21	0.000
Error	98	0.0010968	0.0000112		
Total	99	0.0049377			

The problem of heteroscedasticity has been resolved. However, the error variable did not appear to be normal. Thus, the logarithmic transformation is judged to be superior.

In practice, statisticians often experiment with different transformations to determine which one works best. Ideally, they look for transformations where the required conditions are well satisfied and whose fit is best.

8.10.16 Final Word

When applying regression, the modeling and analysis of the data may be improved by considering items introduced below which were not discussed so far. Here we restrict the list to inclusion of dummy variable, interaction term, and multiplicative model. The following is a brief description of each:

- **Dummy Variables** (also known as indicator or 0-1 variables) are used to include categorical (qualitative) variables that are related to the response variable in a regression equation. An example of a categorical variable is a "gender" variable that has two categories "male" and "female." If a categorical variable has n categories, then it can be included in a regression equation by using $(n-1)$ dummy variables. The "gender" variable can be included in a regression equation by using a single dummy variable called, for example, gender. This variable may be coded as 1 for females and 0 for males, or 0 for females and 1 for males.

  If a single dummy variable is included in a regression equation, then two lines with different intercepts and the same slopes (forcing lines to be parallel) are allowed. Consider, for example, simple linear regression

  $$Y = b_0 + b_1 t.$$

  Let us assume that this model is selected to predict record for high jump. The advantage of using a 0-1 coding scheme is that the b coefficients are easily interpreted. The above model allows us to compare the mean $E(Y)$ for males with the corresponding mean for females:

  $$\text{Males}(t=1): \quad E(Y) = b_0 + b_1$$
  $$\text{Females}(t=0): \quad E(Y) = b_0.$$

  Note that this makes sense since in the above model, $b_1 > 0$. Also note that when 0-1 coding is used b_0 will always represent the mean response associated with the level of the qualitative variable assigned to the value 0 often referred to as base level. Here, b_1 denotes the difference between the mean records.

Chapter 8: Regression Analysis

- **Interaction Variable** is the product of two variables and it allows the effect on Y (response variable) of one of the explanatory variables to depend on the value of the other explanatory variables. Consider, for example, interaction variable XD, whether X is a quantitative variable and D is a dummy variable, and then consider the equation.

$$y = b_0 + b_1 x + b_2 d + b_3 x d.$$

When $d = 1$, the above equation becomes $y = (b_0 + b_2) + (b_1 + b_3)x$. Also $y = b_0 + b_1 x$ when $d = 0$. This interaction variable XD has allowed the slope of regression line to differ between the two categories. For example, rate of change in records with age or height will be driven upward by a large number of participants or by increase in reward. Note that in this model $(b_1 + b_3 d)$ represents the change in $E(Y)$ for every one unit increase in x, holding d fixed. $(b_2 + b_3 x)$ represents the change in $E(Y)$ for every one unit increase in d, holding x fixed.

- **Multiplicative Model (Constant Elasticity)** This relationship is useful in economic theory, and has the form $y = b_0 x_1^{b_1} \cdots x_k^{b_k}$. Unlike the additive relationship for a multiplicative relationship, the effect of change in any explanatory variable on Y depends on the levels of other variables. A multiplicative relationship can be modeled using linear regression by taking natural logarithms of all variables of this relationship. The resulting equation $log(y) = log(a) + b_1 log(x_1) + \cdots + b_k log(x_k)$ is linear in the coefficient $log b_0, b_1, b_2, \cdots, b_k$ and can be estimated with multiple regression. The coefficients of the explanatory variable (log variables) represent elasticities.

8.11 THE LOGISTIC MODEL

Logistic regression is a mathematical model approach that can be used to describe the relationship of several predictor variables $x_1, x_2, \cdots, x_k$ (covariates) to a dichotomous dependent variable Y, where Y is typically coded as 1 or 0 (e.g., presence or absence) for its two possible categories. The logistic model expresses or describes the expected value of Y in terms of the following "logistic" formula:

$$E(Y) = \frac{1}{1 + \exp\left[-\left(b_0 + \sum_{j=1}^{k} b_j x_j\right)\right]}. \tag{8.25}$$

Since $E(Y) = 0 \cdot P(Y = 0) + 1 \cdot P(Y = 1) = P(Y = 1) = \pi$, the logistic model can be written in a form that describes the probability of occurrence of one of the two possible outcomes of Y. In other words,

$$P(Y = 1) = \frac{1}{1 + \exp\left[-\left(b_0 + \sum_{j=1}^{k} b_j x_j\right)\right]}. \tag{8.26}$$

The logistic model is useful in many important practical situations where the response variable takes only one of two possible values. For example, a study of the development of a particular chronic type of injury in sports could employ a logistic model. Here the model describes, in probabilistic terms whether a given player in the study group will ($Y = 1$) or will not ($Y = 0$) develop the injury in question during a follow-up period of interest. As a different example, students trying out for a college basketball team will either make it or not and this depends on many measurable factors.

The first step of a logistic regression analysis is to postulate (based on knowledge about and experience with the process under study) a mathematical model describing the mean of Y as a function of x_j and the b_j values. The model is then fitted to the data by maximum likelihood, and eventually appropriate statistical inference are made (after the model's adequacy of fit is verified, including consideration of the relevant regression diagnostic indices).

The mathematical expression given on the right side of the logistic model (8.26) is of the general mathematical form

$$f(z) = \frac{1}{1+e^{-z}}$$

where $z = b_0 + \sum_{j=1}^{k} b_j x_j$. The function $f(z)$ is called the logistic function. This function is well suited for modeling probability, since the values of $f(z)$ range from 0 to 1 as z varies from $-\infty$ to $+\infty$. In epidemiologic studies, such a probability can be used to state an individual's risk of developing a disease. The logistic model, therefore, is set up to ensure that, whatever estimate of risk we get, it always falls between 0 and 1. This is not true for other possible models, which is why the logistic model is often used when a probability must be estimated.

Another reason why the logistic model is popular relates to the general sigmoid shape of the logistic function (see Figure 8.11.1). A sigmoid shape is particularly appealing, for example, to epidemiologist if the variable z is viewed as representing an index that combines the contributions of several risk factors, so that $f(z)$ represents the risk for a given value of z.

Where does logistic regression fit in here? To answer this, we must consider an equivalent way to write the logistic regression model, called the logit form of the model. The "logit" is a

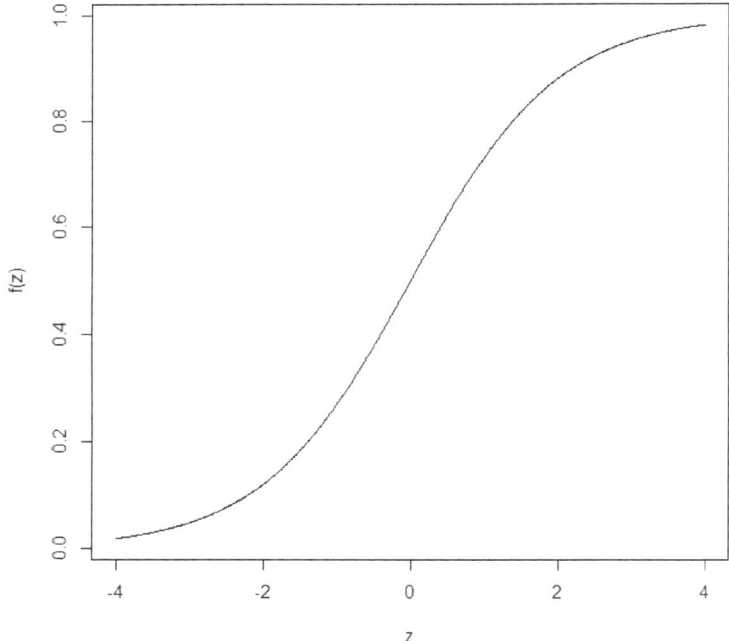

FIGURE 8.11.1 The Logistic Function $f(z) = \frac{1}{1+e^{-z}}$.

transformation of the probability $P(Y=1)$, defined as the natural log odds of the event $\{Y=1\}$. In other words

$$\text{logit}[P(Y=1)] = \log_e[\text{odds}(Y=1)] = \log_e\left[\frac{P(Y=1)}{1-P(Y=1)}\right] = \log_e\left(\frac{\pi}{1-\pi}\right) = \pi'. \quad (8.27)$$

If we then substitute the logistic model (8.26) for $P(Y=1)$ into the equation (8.27), we find

$$\pi' = \text{logit}[P(Y=1)] = b_0 + \sum_{j=1}^{k} b_j x_j. \quad (8.28)$$

(8.28) is called the logit form of the model. The logit form is given by the linear function $b_0 + \sum_{j=1}^{k} b_j x_j$. For convenience, many books describe the logistic model in its logit from given by (8.28) rather than in its original form (8.26). Note that the logistic transformation in fact linearize the response function.

EXAMPLE Suppose that Y denotes lung cancer status (1 = Yes, 0 = No) and there is only one predictor say, smoking status (1 = Smoker, 0 = Nonsmoker), then the logistic model (8.26) can be written equivalently in logit from as

$$\text{logit}[P(Y=1)] = b_0 + b_1 x_1 = b_0 + b_1(\text{smoking status}).$$

To obtain an expression for the odds ratio from a logistic model, we must compare the odds for two groups of individuals. For the preceding example involving one predictor, the two groups are smokers ($x_1 = 1$) and nonsmokers ($x_1 = 0$). Thus, for this example, the log odds for smokers and nonsmokers can be written as

$$\log_e \text{odds}(\text{smokers}) = b_0 + (b_1 \cdot 1) = b_0 + b_1$$

and

$$\log_e \text{odds}(\text{nonsmokers}) = b_0 + (b_1 \cdot 0) = b_0$$

respectively. It follows that the odds ratio comparing smokers to nonsmokers is given by

$$OR_{Svs.NS} = \frac{\text{odds}(\text{smokers})}{\text{odds}(\text{nonsmokers})} = \frac{e^{(b_0+b_1)}}{e^{b_0}} = e^{b_1}.$$

In other words, for the simple example involving one (0-1) predictor, the odds ratio comparing the two categories of the predictor is obtained by exponentiating the coefficient of the predictor in the logistic model.

EXAMPLE 8.11.1 Consider the population of soccer players with certain injury that requires at least 42 days of rest for possible recovery. Suppose that in a sample of 223 such players drawn from the population, 147 were recovered and 76 developed a chronic type of the injury that led to their retirement. Using this data we would estimate the probability that a player with such injury in this population develops a chronic form of the injury by the sample proportion $\hat{p} = 76/223 = 0.341$.

Now the trainers might suspect that there are certain factors that affect the chance that a particular player will develop the chronic type of injury. If they can classify a player according to these factors, it is possible that they could estimate his or her chance of developing chronic type of injury with a greater precision than just using $\hat{p} = 0.341$. Note that in this sense the problem is similar to that of unconditional versus conditional probability. Here, 0.341 is unconditional probability and applies to all the players with this injury regardless of their other characteristics. Once we find out some information regarding a particular player, the probability becomes conditional and different from 0.341 unless we pick a characteristic, which is independent from the injury under consideration. To clarify, suppose that the factor of interest is the player's weight, which in this case means body structure or body muscle since soccer players usually have a very low body fat. If the response Y were continuous, we would begin an analysis by constructing a two-way scatter plot of the response versus the continuous explanatory variable. Note that in this case, all points lie on one of two parallel lines, depending on whether Y takes the value 0 or 1. Thus, we might instead begin to explore whether an association exists subdividing the population of players into say , three categories. We could then estimate the probability that a player will develop the chronic type of injury in each of these subgroups individually.

Weight (Pounds)	150–165	166–175	>175	
Sample Size	68	80	75	223
Numbers With Chronic Injury	49	18	9	76
$\hat{p}$	0.721	0.225	0.120	0.341

Here the estimated probability of developing the chronic type of injury decreases with weight. Thus, this factor helps us to predict the likelihood that a player will develop the chronic type of injury. Now to improve our prediction further we may fit the model (8.28) with $k = 1$ using $P(Y = 1) = p$. Although this model is very similar to ordinary linear regression, here we cannot apply the ordinary least squares to estimate b_0 and b_1 (see Example 8.11.2). Assuming that the response is continuous and normally distributed, to fit a logistic model; instead, we use maximum likelihood estimation (see Hosmer and Lemeshow (1989), "Applied Logistic Regression" Wiley for detail). Recall that this technique uses the information in a sample to find the parameter estimates that are most likely to have produced the observed data.

For the sample of 223 players, suppose that the estimated logistic regression equation is

$$log_e \left[\frac{\hat{p}}{1 - \hat{p}} \right] = 4.0343 - 0.193x.$$

The coefficient of weight implies that for each one-pound increase in weight, the log odds that the player develops the chronic type of injury decrease by 0.0193, on average. When the log odds decrease, the probability p decreases as well.

In order to estimate the probability that a player with a particular weight develops the chronic type of injury, we simply substitute the appropriate value of x into the preceding equation. To estimate the probability that a player weighing 163 pounds develops the chronic type of injury, for

Chapter 8: Regression Analysis

example, we substitute the value $x = 163$ pounds to find

$$\log_e\left[\frac{\hat{p}}{1-\hat{p}}\right] = 4.0343 - .0193(163) = 0.8884.$$

Taking the antilogarithm of each side of the equation, we get

$$\frac{\hat{p}}{1-\hat{p}} = e^{0.8884} = 2.4312.$$

Finally, solving for $\hat{p}$ we obtain

$$\hat{p} = \frac{2.4312}{1+2.4312} = \frac{2.4312}{3.4312} = 0.7086.$$

If we repeat the calculation for other weights and plot the weight versus $\hat{p}$, we find a curve similar to the one presented in Figure 8.11.1.

EXAMPLE 8.11.2 (USE OF WEIGHTED LEAST SQUARES) In order to increase home game attendance, a sport organization has a promotion policy of issuing coupons. To study the effectiveness of coupons offering 1,000 fans were selected and a coupon and advertising material for the games were mailed to each. The coupons offered different price reductions (5, 10, 15, 20, and 30 dollars), and 200 fans were assigned at random to each of the price reduction categories. The independent variable X in this study is the coupon value and the dependent variable Y is a binary variable indicating whether or not the coupon was redeemed in that season.

It was expected that the logistic response function would be an appropriate description of the relation between price reduction and probability that the coupon is utilized. Since there were repeat observations at each x_j, and since the number of repeat observations at each x_j was large ($n_j = 200$, for all j), the procedure described earlier could be used for fitting the logistic response function.

The table below contains the basic data for this experimental study in columns 1 through 4. The transformed proportions p'_j are shown in column 5. For instance, for $x_1 = 5$, we have

$$p'_1 = \log_e\left(\frac{0.160}{1-0.150}\right) = -1.65823.$$

Recall that $\pi = E(Y)$, $\pi' = \log_e\left(\frac{\pi}{1-\pi}\right)$, and $\pi' = b_0 + b_1 x$.

Now the logistic transformation, while linearizing the response function, does not eliminate the unequal variances of the error term. Hence, weighted least squares should be applied here. So, unlike ordinary least squares where all the terms had an equal weight, here we need to weigh them according to their value for variance. It can be shown that when n_j is reasonably large, the approximate variance of p'_j is

$$\hat{\sigma}(p'_j) = \frac{1}{n_j \pi_j (1 - \pi_j)}$$

which is estimated by

$$s^2(p'_j) = \frac{1}{n_j p_j (1 - p_j)}.$$

Hence, the estimated weight to be used in the weighted least squares computations are

$$\hat{\omega}_j = n_j p_j (1 - p_j).$$

As an example for $x_1 = 5$, we have

$$\hat{\omega}_1 = n_1 p_1 (1 - p_1) = 200(0.16)(1 - 0.16) = 26.880$$

These values are shown in column 6 in the table.

Reduction x_j	Number of Fans n_j	Number of Coupons Redeemed R_j	Proportion of Coupons Redeemed p_j	Transformed Proportion p'_j	Weight $\hat{\omega}_j$
5	200	32	0.160	−1.6582	26.880
10	200	51	0.255	−1.0721	37.995
15	200	70	0.350	−.6190	45.500
20	200	103	0.515	0.0600	49.955
30	200	148	0.740	1.0460	38.480

Prior to the fitting of the logistic model, the transformed proportions p'_j were plotted against x_j. It appeared that a linear response function would fit the transformed proportions well. Hence, it was decided to proceed with fitting the transformed logistic model (8.28).

The fitting of the transformed logistic model by weighted least squares is straightforward. A computer run using weighted least squares, with the weights being those in column 6, led to the following fitted response function:

$$\hat{\pi}' = 2.18506 + 0.108700x.$$

The fitted response function is used in the ordinary manner. To estimate the probability of a coupon redemption if the price reduction is, say, 25 dollars, we first substitute in the model

$$\hat{\pi}' = 2.18506 + 0.108700(25) = 0.53244$$

and then transform to the original variable.

$$antilog_e(.53244) = 1.70308 = \frac{\hat{\pi}}{1 - \hat{\pi}}$$

so that

$$\hat{\pi} = 0.630$$

when $x = 25$ dollars. Hence, we estimate that about 63% of 25 dollar coupons will be redeemed with the season.

Unlike the straightforward interpretation of the slope in a linear regression model the interpretation of the slope $b_1 = 0.1087$ in the fitted logistic response function is not simple. The reason is that the effect of increasing x by a unit varies for the logistic model according to the location of

Chapter 8: Regression Analysis

the starting point on the x scale. One interpretation of b_1 is found in the property of the logistic function that the "odds" $\hat{\pi}/(1-\hat{\pi})$ are multiplied by $\exp(b_1)$ for any unit increase in x.

PRACTICE PROBLEMS

(1) Repeat Example 8.11.2 by changing column 3(R_j) to 42, 61, 60, 113, 138.
(2) Consider the following data, which shows number of the individuals who played sports and the number with permanent injuries.

Age Group (Years)	Number of Players	Number with Permanent Injury
25–34	522	41
35–44	330	51
45–54	344	81
55–64	219	81
65–74	114	50

Fit a logistic regression model. Show that slope and intercept are respectively 0.0573 and −4.0388 and an estimate of p for an individual 29.5 years of age is 0.08719.

A RESEARCH PROBLEM Pick a sport and a tournament of your choice. Use Logistic regression to study the home-court advantage. Do you think that home-court advantage for different sports has the same effect on the outcome? In fact some fans think that other than teams and their performance, calls made by officials are affected by the crowd, level of the noise, and more importantly closeness of the fans to the officials and the players.

8.11.1 Effect of the Star Player

The Philadelphia 76ers finished the 2001-2002 NBA season with a record of 43 wins and 39 losses. Larry Brown, the team's head coach, wanted to find out what were best predictors for the outcome of his team's games. In particular, he wanted to investigate the roles that three main factors played: the location of the game (home or away), the presence (or absence) of Allen Inverson (team's star player), and the opponent's winning percentage.

SOLUTION: This is a situation in which logistic regression can be used. In this case, the binary response is win/loss. Coach Brown's three predictors serve as covariates. Note that a home game was coded as a 1, an away game as a 0. Similarly, a game in which Allen Iverson played was coded as a 1, while one in which he did not play was coded as a 0. Opponent's winning percentage was obtained from the final standings of the 2001-2002 season. Using the data from the 82 games of the 2001-2002 season MINITAB produced the following results:

Binary Logistic Regression: Win/Loss Versus Home/Away, Iverson Present, Opponent's Winning Percentage

Link Function: Logit

Response Information

Variable	Value	Count	
Win/Loss	1	43	(Event)
	0	39	
	Total	82	

Logistic Regression Table

Predictor	Coef.	SE Coef.	z	P	Odds Ratio	95% CI Lower	95% CI Upper
Constant	0.620	1.024	0.61	0.545			
Home/Away	0.0647	0.4646	0.14	0.889	1.07	0.43	2.65
Iverson	1.2245	0.5405	2.27	0.023	3.40	1.18	9.82
Opponent	−2.934	1.890	−1.55	0.120	0.05	0.00	2.16

Log-Likelihood = −52.887
Test that all slopes are zero: G = 7.707, d.f. = 3, p-value = 0.052
(G stands for global null hypothesis)

From the results, we can see that the only significant predictor of a game's outcome for the 76ers is the availability of Allen Iverson (P-value = 0.023). In fact, in this model, the odds of the 76ers winning with Iverson are 3.4 times those of them winning without him.

It is interesting to note that at 5% (even 10%) significance level neither the location of the game nor the opponent's winning percentage turned out to be significant predictors. The former is of interest because home-court advantage is so highly valued in sports. The latter is of intuitive interest: one would expect the odds of winning a game to decrease significantly as the opponent's winning percentage is increased. However, relative to each other opponent's winning percentage has a much greater effect of $\hat{p}$ (check the P-values).

Note that here the regression model is

$$log_e \left[\frac{\hat{p}}{1-\hat{p}} \right] = 0.620 + 0.0647 x_1 + 1.2245 x_2 - 2.934 x_3$$

which may be used for predictions. For example, for a home game with Iverson present and the opponent's winning percentage of 50, we have

$$log_e \left[\frac{\hat{p}}{1-\hat{p}} \right] = 0.44175.$$

This, taking the antilogarithm of each side of the equation, results in

$$\frac{\hat{p}}{1-\hat{p}} = 1.55543 \rightarrow \hat{p} = 0.608675.$$

A PRACTICE PROBLEM Find $\hat{p}$ for a home game with Iverson present and for an opponent with a winning percentage of 70.

We end this section by demonstrating the role of the logistic function introduced earlier. For this consider all home games with Iverson present, then replacing $x_1 = x_2 = 1$ in above model we get

$$log_e\left[\frac{\hat{p}}{1-\hat{p}}\right] = 1.9092 - 2.934x_3 \approx 1.91 - 2.934x_3$$

or

$$\hat{p} = \frac{6.753}{6.753 + e^{2.934x_3}} \quad 0 \leq x_3 \leq 0$$

which is a logistic function.

8.12 USING R

We will still utilize R functions, **lm()** and **glm()**, mentioned in Chapter 7 to fit a linear model using data. As we discussed, user must specify the desired model's "formula" as the first argument in the function call. In addition, we will demonstrate **plot()** function; therefore, user will draw scatter plots to research relationships of variables in dataset.

First of all, let's see how to create a plot using data. For example, one wish to draw logistic function plot. We will need to make a vector of x using **seq()** function. The first parameter is *Start*, the second parameter is *End*, and the third one will be *Step length*. By default, R will perform a point-wise calculation with given algorithm. **plot()** function is used to create a plot, the syntax is **plot(x, y, ...)**

```
# Make a vector of x
x = seq(-4, 4, 0.01)
# Calculate function values
f = 1/(1 + exp(-x))
# Plot, here 'l' means a smooth curve instead of points.
plot(x, f, 'l')
```

Let's represent the analysis of Example 8.2.1 using R here. We would like to build a linear model to predict Time using Age. Here **abline()** function will add a reference line on your plot using fitted model.

```
# Import data, you need to replace following
# file name and path using your own
d1 = read.csv('Suppmental\textbackslash\textbackslash Example8.2.1.csv')
attach(d1)
plot(age, time)
model = lm(time ~ age)
anova(model)
abline(model)
```

You should be able to see following outputs:

Analysis of Variance Table

Response: **time**

	Df	Sum Sq	Mean Sq	F value	Pr(>F)
age	1	1.1790	1.1790	0.9911	0.3579
Residuals	6	7.1373	1.1896		

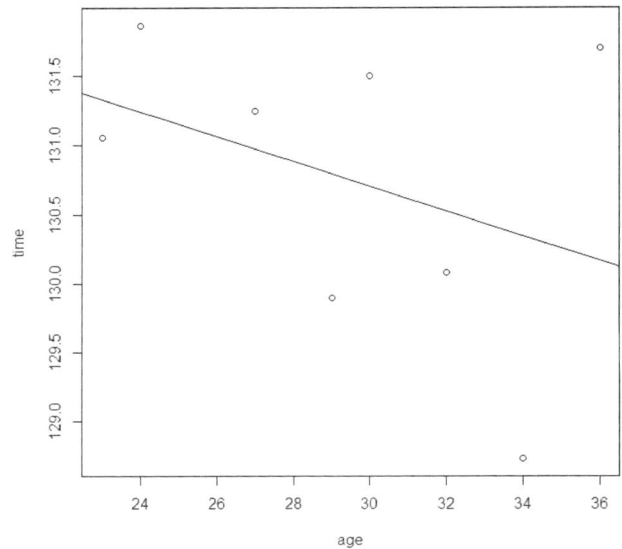

If you would like to see more details about your model, apply **summary()** function to your fitted model, you will get:

Call:
lm(formula = time ~ age)

Residuals:

Min	1Q	Median	3Q	Max
−1.61828	−0.55883	0.00311	0.67025	1.53030

Coefficients:

	Estimate	Std. Error	t value	Pr(>|t|)	
(Intercept)	133.38414	2.66272	50.093	4.25e−09	***
age	−0.08929	0.08969	−0.996	0.358	

Signif. **codes**: 0 *** 0.001 ** 0.01 * 0.05 . 0.1   1

Residual standard error: 1.091 **on** 6 degrees of freedom
Multiple **R**−squared: 0.1418, Adjusted **R**−squared: −0.001274
F−statistic: 0.9911 **on** 1 and 6 DF, p−value: 0.3579

Chapter 8: Regression Analysis

We always need to diagnose the model whenever it was fitted. For example, we want to validate the normality of residuals. **residuals()** will extract the residuals from your model, and **qqnorm()** will create a Q-Q plot of the variable.

```
r = residuals(model)
qqnorm(r)
```

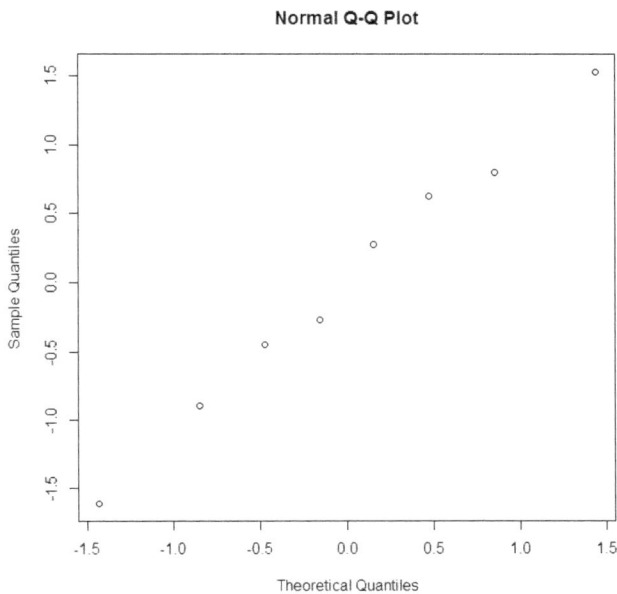

Alternatively, a histogram of residuals can be used.

```
hist(r)
```

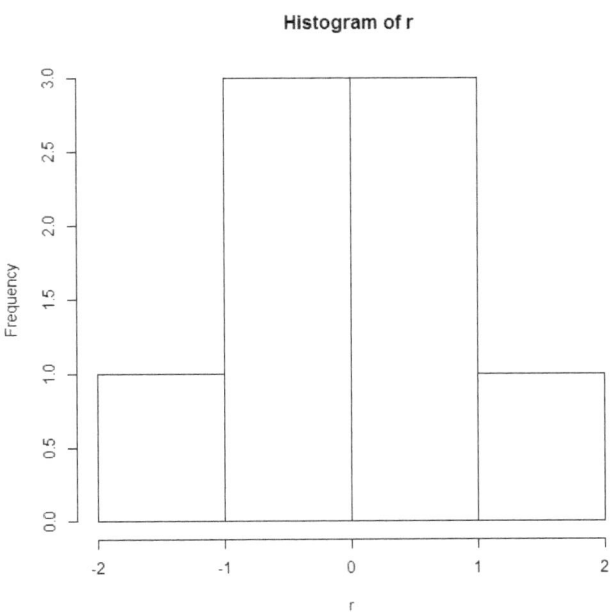

Let's assume there is another variable age^2 in the model, we can fit a new model as below.

```
# Calculate age square as a new predictor
age2 = age^2
model2 = lm(time ~ age + age2)
anova(model2)
summary(model2)
```

Analysis of Variance Table

Response: **time**
Df Sum Sq Mean Sq F value Pr(>F)
age   1 1.1790 1.17896 0.9109 0.3837
age2   1 0.6658 0.66576 0.5144 0.5053
Residuals 5 6.4716 1.29431

Call:
lm(formula = time ~ age + age2)

Residuals:
1   2   3   4   5   6   7   8
−1.6794 −0.5757 −0.2519 −0.6399 0.5049 1.1069 1.0743 0.4608

Coefficients:
Estimate Std. Error t value Pr(>|t|)
(Intercept) 147.95967 20.51173   7.213 0.000798 ***
age   −1.10581   1.42043 −0.779 0.471489
age2   0.01734   0.02418   0.717 0.505349
———
Signif. **codes**: 0 *** 0.001 ** 0.01 * 0.05 . 0.1 1

Residual standard error: 1.138 **on** 5 degrees of freedom
Multiple **R**–squared: 0.2218,   Adjusted **R**–squared: −0.08945
F–statistic: 0.7126 **on** 2 and 5 DF, p–value: 0.5342

REFERENCE

Chatterjee, S., Handcock, M. S., and Simonoff, J. S. 1995. *A Casebook for a First Course in Statistics and Data Analysis*. Wiley and Sons.

CHAPTER 9

Time Series Analysis

Most of the measurable quantities in sports have a time index. In other words, they form a numerical sequence in which individual values are generated at regular (often equally spaced) intervals of time. Thus, they form a time series. A time series is a succession of observations indexed by time. Although not essential, it is desirable for the times to be equidistant. For sports, this distance is usually 1 year or one season. This chapter describes probability models for time series together with some statistical methods for their analysis. Time series methods are applied to uncover the time correlation structures and are useful tools for analysis of real-life problems as well as prediction of future values. They are commonly used when several years of data exist. Because time series analysis is totally dependent upon historical data, its implicit assumption is that, the past is a good guide to the future. Consequently, time series analysis performs better in the short-term than in long-term forecast. In sports, records are naturally ordered in time and any future prediction requires specification of a particular time in the future. So when applying time series methods to sport records or any other quantities of interest, predictions of the next year or the year after (in fact, immediate future) are much more reliable than prediction of, for example, the ultimate record.

Some theoretical results about time series are presented in Sections 9.1 through 9.5. Applied techniques useful in separating components of time series and their applications to forecasting future values are discussed in Sections 9.6 through 9.10. When a response variable varies with time, it often has a rather complex structure and the data require a more elaborate analysis than those discussed in previous chapters. The standard time series model expresses the observed y_t as a sum of several components. Some of these components represent the macroscopic properties or behavior and others the microscopic properties or parts. As was pointed out, time series methods assume that what has occurred in the past will continue to occur in the future (stationarity). Nonstationary time series pose many problems when it comes to forecasting.

9.1 STOCHASTIC PROCESSES

This section describes the probability models for time series, which are called stochastic processes. Most processes evolving with time in the real world have a random or stochastic element in their structure. In most cases, we can think of them as made up of a deterministic component (trend or signal) and a random (noise) component. Here, deterministic refers to a process whose future values can be determined with certainty. The modeling and analysis of a random element is critical for any individual or organization involved in planning for the future. It is also critical for assessing predictions made from such processes.

We begin with a simple definition or description of a stochastic process. The term stochastic process can be described as a phenomenon or process that evolves with time according to some probabilistic laws (future values cannot be predicted with certainty). Many authors use the term stochastic process to describe both the real physical process and the mathematical model of it. The word stochastic, which is of Greek origin, means "pertaining to chance" (as opposed to deterministic). Many writers use the phrase random process as a synonym for stochastic process. Mathematically, a stochastic process may be defined as a collection of random variables $\{X(t), t \in T\}$, where T denotes the set of time—points at which the process is defined. We will denote the random variable at time t by $X(t)$ if T is continuous (usually $-\infty < t < \infty$) and by X_t if T is discrete (usually $t = 0, \pm 1, \pm 2, \ldots$). Thus, a stochastic process is a collection of random variables, which are ordered in time. For a single outcome of the process, we have only one observation on each random variable and these values evolve with time according to some probabilistic laws.

EXAMPLE 9.1.1 Let X_t denote the number of people who play a certain sport in a given city during day t. Then $\{X_t\}$ forms a stochastic process. This means that the number of people who play that sport in a given day is a random variable and varies from day to day.

EXAMPLE 9.1.2 Let $X(t)$ denote the average (maximum or minimum) temperature in day t in a certain area. Then $\{X(t)\}$ forms a stochastic process. Note that, unlike the previous example, here $X(t)$ is a continuous random variable.

EXAMPLE 9.1.3 Let X_t denote the number of records broken in year t in a certain sport or in all sports for which records are kept. Then X_t is a random variable and $\{X_t\}$ forms a stochastic process.

EXAMPLE 9.1.4 Let X_t denote the stock price of Apple Inc. at minute t today. Then X_t is a random variable and $\{X_t\}$ forms a stochastic process.

In most statistical problems, we are concerned with estimating the properties of a population from a sample. In time series analysis, however, it is often impossible to make more than one observation at a given time so that we only have one observation on the random variable at time t. Nevertheless, we may regard the observed time series as just one sample of the infinite set of time series, which might have been observed. This infinite set of time series is sometimes called the ensemble. Every member of the ensemble is a possible realization of the stochastic process. The observed time series can be thought of as one particular realization, and will be denoted by $x(t)$ for $(0 \leq t \leq T)$ if the observations are continuous, and by x_t for $t = 1, \ldots, N$, if the observations are discrete. In the case described in Example 9.1.3, the observed value of X_t in the t-th year will be denoted by x_t.

Time series analysis is essentially concerned with evaluating the properties of the probability model, which generated the observed time series. One way to describe a stochastic process is to specify the joint probability distribution of $X_{t_1}, \ldots, X_{t_n}$ for any set of time $t_1, \ldots, t_n$ and any value of n. But this is rather complicated and is not usually attempted in practice. A simpler way of describing a stochastic process is to give the moments of the process, particularly the first and second moments namely, the mean, variance, and a measure for association for X_t's, (covariance or autocovariance functions).

If the joint distribution for any set of times are multivariate normal, then the process is called a Gaussian (normal) process and the latter is completely determined by its first two moments. This property, which is possessed only by a Gaussian process, has a very important practical implication.

Higher moments of a stochastic process may be defined in an obvious way. However, they are rarely used in practice since, as mentioned, knowledge of two functions, namely the mean function and the variance–covariance (autocovariance) function, is sufficient for characterizing the most useful process, namely the Gaussian process. Even for non-Gaussian processes the estimated higher moments are usually unreliable and not very useful. The next section discusses an important class of stochastic processes that have proved extremely useful in the analysis of real-life time series.

9.2 STATIONARY PROCESS

An important class of stochastic processes for describing time series is the so-called stationary process. A mathematical definition of a stationary time series will be given later. Here, we introduce the idea of stationarity from an intuitive point of view. Roughly speaking, a time series is said to be stationary if there is no systematic change in mean (no trend), no systematic change in variance, and no periodic variations. The case described in Example 9.1.2, in general, may meet such conditions (assuming that the climate is not changed significantly) and therefore can be considered as a stationary process.

Most of the theories developed for time series analysis are based on stationarity, and for this reason time series analysis often requires one to transform a non stationary series into a stationary one (whenever possible) so that these theories become applicable either exactly or approximately. One simple departure from stationarity occurs when a time series has trend or perhaps a seasonable component. For such cases, one may remove the trend and seasonal variation from a set of data and then try to model the residuals by means of a stationary stochastic process. This approach will be discussed in detail later in this section.

A time series is said to be **strictly stationary** if the joint distribution of $X(t_1), \ldots, X(t_n)$ is the same as the joint distribution of $X(t_1+\tau), \ldots, X(t_n+\tau)$ for all $t_1, \ldots, t_n$ and τ. In other words, shifting the time origin by an amount τ has no effect on the joint distributions that must therefore depend only on the intervals between $t_1, \ldots, t_n$. The above definition holds for any value of n. In particular, if $n = 1$, it implies that the distribution of $X(t)$ must be same for all t so that

$$\mu(t) = \mu, \quad \sigma^2(t) = \sigma^2$$

are both constants independent of the value of t. Here $\mu(t)$ and $\sigma^2(t)$ are defined by

$$\mu(t) = E(X(t))$$

and
$$\sigma^2(t) = Var(X(t)).$$

If the correlation between $X(t_i)$ and $X(t_j)$ is zero, then the resulting stochastic process is called a **purely random process** or a **white noise process**. Usually this process is denoted by $Z(t)$. A process that is made up of a linear combination of a white noise process in different times is called a **moving average (MA) process**. For example,

$$X(t) = Z(t) + \beta_1 Z(t-1) + \ldots + \beta_q Z(t-q)$$

is a MA process of order q (MA(q)).

As an example, suppose that you win or lose \$1 if a coin you flip shows heads or tails, respectively. Let X_t denote the amount of money you win in game number t. Then X_t is either 1 or -1 independent from its value in the previous or future games. This is an example of a discrete white noise process.

EXAMPLE 9.2.1 Consider the number of goals scored in a soccer tournament in year t. If we assume that, each year's number has the same distribution and is independent of any other year then $\{X_t\}$ forms a white noise process. In this case, if we let,

$$Y_t = X_t + X_{t-1} = Z_t + Z_{t-1}$$

present the sum of this year's and the last year's scores, then $\{Y_t\}$ is an MA process of order one, (MA(1)).

It should be pointed out that the white noise process (or purely random process) is mainly of theoretical interest. Mathematically, it can be looked at as the infinite dimensional orthogonal basis for the space of the stationary process. This means that any stationary process may be expressed as a linear combination (finite or infinite) of white noise process.

Since strict stationarity is a very strong requirement, we may weaken the requirement for practical purposes as follows: a discrete stochastic process is said to be stationary in the wide sense or simply stationary if its mean, variance, and covariances are invariant under a common translation of the time arguments. That is, if

$$E(X_t) = \mu, \quad Var(X_t) = \sigma^2 \text{ both independent of t}$$
$$Cov(X_t, X_{t+r}) = E\{(X_t - \mu)(X_{t+r} - \mu)\}$$
$$= E(X_t X_{t+r}) - E(X_t)E(X_{t+r}) = \gamma_X(r), \qquad (9.1)$$

In other words, γ_X is a function of r only for all integer values of t and r, with $\gamma_X(0) = \sigma^2 < \infty$. Obviously, any process that is strictly stationary and has finite variance is also stationary in the wide sense. For the normal process defined above, strictly stationary and stationary in the wide sense are equivalent. For example, the process discussed in Example 9.2.1 is a stationary process. To see this, we need to check the required conditions. For simplicity assume that $E(Z_t) = 0$ for all t. Then

$$E(Y_t) = E(Z_t + Z_{t-1}) = E(Z_t) + E(Z_{t-1}) = 0$$
$$Var(Y_t) = Var(Z_t + Z_{t-1}) = Var(Z_t) + Var(Z_{t-1}) = 2\sigma_z^2$$

Chapter 9: Time Series Analysis

$$Cov(Y_t, Y_{t+r}) = E(Y_t Y_{t+r}) - E(Y_t)E(Y_{t+r}) = E(Y_t Y_{t+r})$$

$$= E[(Z_t + Z_{t-1})(Z_{t+r} + Z_{t+r-1})] = \begin{cases} 2\sigma_z^2 & r = 0 \\ \sigma_z^2 & r = 1 \\ 0 & r > 1. \end{cases}$$

For processes in discrete time, $\gamma_X(r)$ in (9.1) defines the autocovariance function. Similar definition applies to a continuous time series. Let us, with no loss of generality, assume that $\mu = 0$. Then

$$Cov(X_t, X_{t+r}) = E(X_t X_{t+r}) = \gamma_X(r) \tag{9.2}$$

which is a real-valued function of r is called the autocovariance function of the time series X_t. It is sometimes convenient to work with the autocorrelation function (ACF) defined as

$$\rho_X(r) = \frac{\gamma_X(r)}{\gamma_X(0)}. \tag{9.3}$$

Clearly,

$$\rho_X(r) = \rho_X(-r) \tag{9.4}$$

and

$$|\rho_X(r)| \leq \rho_X(0) = 1 \tag{9.5}$$

For white noise process

$$\rho_X(r) = \begin{cases} 1 & r = 0 \\ 0 & r \neq 0 \end{cases}$$

For a first-order MA process

$$X_t = Z_t + \beta Z_{t-1}$$

$E(X_t) = 0$ since $E(Z_t) = 0$. Also noting that $E(Z_t Z_{t+r}) = E(Z_t)E(Z_{t+r}) = 0$ for $r \neq 0$

$$\gamma_X(r) = E(X_t X_{t+r}) = E[(Z_t + \beta Z_{t-1})(Z_{t+r} + \beta Z_{t+r-1})]$$

$$= \begin{cases} (1+\beta^2)\sigma_z^2 & r = 0 \\ \beta \sigma_z^2 & r = 1 \\ 0 & r \geq 2 \end{cases}$$

Hence

$$\rho_X(r) = \begin{cases} \frac{\beta}{1+\beta^2} & r = 1 \\ 0 & r \geq 2 \end{cases}$$

A PRACTICE PROBLEM Consider a second-order MA process given below

$$X_t = Z_t + \beta_1 Z_{t-1} + \beta_2 Z_{t-2}.$$

Show that $\gamma_X(0) = (1+\beta_1^2+\beta_2^2)\sigma_z^2$ and

$$\rho_X(r) = \begin{cases} \frac{\beta_1(1+\beta_2)}{1+\beta_1^2+\beta_2^2} & r=1 \\ \frac{\beta_2}{1+\beta_1^2+\beta_2^2} & r=2 \\ 0 & r \geq 3. \end{cases}$$

Also show that for a general MA(q) we have (assuming $\beta_0 = 1$)

$$\rho_X(r) = \begin{cases} \sum_{i=0}^{q-|k|} \beta_i \beta_{i+|k|} / \sum_{i=0}^{q} \beta_i^2 & |k| \leq q \\ 1 & k=0 \\ 0 & \text{otherwise.} \end{cases}$$

The analysis of ACF has an important practical implication. For example, suppose we have developed a method for estimating the ACF for a time series and actually calculated its values using a realization. Then if the estimated values beyond $r = 1$ are small (not significant at a chosen level), we could conclude that the time series is generated by a MA process of order one. This type of inference is called identification, since it identifies the type of the model and also its order.

Now the ACF characterizes a process in the so-called time domain. In some application, areas such as engineering a "frequency domain" presentation may prove more useful. This is particularly true when a time series has a periodic or cyclical component. Such a presentation is possible via the so-called Fourier transform, which is applied when certain conditions are satisfied. In fact, in this case, (9.4) and (9.5) are among the requirements that a function be a cosine (Fourier) transform of a probability distribution. If these conditions are satisfied, it can be shown that

$$\rho_X(r) = \int_{-\pi}^{\pi} \cos r\omega \, dF_X(\omega) = \int_{-\pi}^{\pi} e^{ir\omega} dF_X(\omega) \qquad (9.6)$$

where $F_X(\omega)$ is the called (normalized) spectral distribution function. If $F_X(\omega)$ is absolutely continuous, the derivatives of $F_X(\omega)$, that is, $F_X'(\omega) = f_X(\omega)$ is called the (normalized) spectral density function. The information contained in the ACF and the spectral density function are equivalent. In fact, corresponding to (9.6) we have the inversion formula

$$F_X(\omega) = \frac{1}{2\pi} \sum_{r=-\infty}^{\infty} \rho_X(r) e^{-ir\omega}. \qquad (9.7)$$

The spectral density accomplishes a mapping of the properties of the stochastic process from the time domain to corresponding properties in the frequency domain. This enables us to study a given time series in either domain. It should be pointed out that the spectral density function is often defined as the Fourier transform of the autocovariance function rather than the ACF, that is

$$\begin{aligned} f_X(\omega) &= \frac{1}{2\pi} \sum_{r=-\infty}^{\infty} \gamma_X(r) e^{-ir\omega} \\ &= \frac{1}{2\pi} \sum_{r=-\infty}^{\infty} \gamma_X(r) \cos(r\omega), \quad -\pi \leq \omega \leq \pi \end{aligned} \qquad (9.8)$$

Chapter 9: Time Series Analysis

$f_X(\omega)$ is usually referred to as the power spectrum, or simply spectrum. We shall use this latter definition and refer to it as spectrum. We note that (9.8) is of special interest when the series $\gamma_X(0)$, $\gamma_X(1), \ldots$ converges. If

$$\gamma_X(0) + 2\sum_{r=1}^{\infty} |\gamma_X(r)| = \sum_{r=-\infty}^{\infty} |\gamma_X(r)| < \infty \qquad (9.9)$$

then $f_X(\omega)$ is bounded and uniformly continuous. Some studies place condition (9.9) upon the stationary time series and refer to it as a mixing condition. In fact, this is the requirement that a time series should have a "weak memory." That is, the present value of the time series has a high correlation with the "recent values" and almost no correlation with "past values." In other words, the process has a short history (or memory) and recent values have much more predictive power than the past value.

9.3 AUTOREGRESSIVE PROCESSES

Autoregressive (AR) processes are one of the most useful and frequently used stationary processes. They are, in fact, the stochastic versions of differential and difference equations. Consider a discrete time random process. Suppose that $\{Z_t\}$ is a white noise process with mean zero and variance σ_z^2. A process $\{X_t\}$ is an **autoregressive process** of order p if it satisfies the following relation (difference equation):

$$X_t = \alpha_1 X_{t-1} + \ldots + \alpha_p X_{t-p} + Z_t. \qquad (9.10)$$

Note that (9.10) is like a multiple regression model where X_t is regressed not on a set of related variables but on the past values of X_t itself. This explains use of the term AR. An AR process of order p is abbreviated as $AR(p)$. Because of the similarities between AR processes and difference equations, they have proved extremely useful for situations where difference equations allowing for randomness may be used as a model.

EXAMPLE 9.3.1 Let X_t denote the best record of t-th year in a sport such as the men's 400-m run. Suppose, we believe that the records will eventually tend to a limit (a) and fit the following exponential model:

$$X_t = a + be^{-ct} + Z_t$$

where a, b, and c are constant parameters and Z_t is a random process with mean zero. Note that in this case

$$E(X_t) = a + be^{-ct}$$

and $E(X_t)$ tends to limit (in this case a) as t tends to infinity. So here a presents the ultimate record. For this model we have

$$\begin{aligned} X_t - X_{t-1} &= e^{-c}(be^{-c(t-1)} - be^{-c(t-2)}) + (Z_t - Z_{t-1}) \\ &= e^{-c}(X_{t-1} - X_{t-2} - Z_{t-1} + Z_{t-2}) + (Z_t - Z_{t-1}) \end{aligned}$$

or equivalently,

$$X_t = (1 + e^{-c})X_{t-1} + (-e^{-c})X_{t-2} + Z'_t$$

where $Z'_t = Z_t - Z_{t-1} - e^{-c}(Z_{t-1} - Z_{t-2})$, which is still a random process with mean zero. By definition, X_t is an AR process of order 2.

9.4 FIRST-ORDER AR PROCESS

For simplicity, we begin by examining the $AR(1)$, where $p = 1$, that is

$$X_t = \alpha X_{t-1} + Z_t. \tag{9.11}$$

To find the autocorrelation we multiply both sides of (9.11) by X_{t-r} and take expectations. We get

$$E(X_{t-r}X_t) = \alpha(X_{t-r}X_{t-1}) + E(X_{t-r}Z_t).$$

Since X_{t-r} depends on values of $Z_{t-r}, Z_{t-r-1}, \cdots$ (past values) and these are independent of Z_t we obtain

$$\gamma_X(r) = \alpha \gamma_X(r-1), \quad \rho_X(r) = \alpha \rho_X(r-1).$$

Finally the successive substitution for r (replacing $r-1, r-2, \cdots$) leads to

$$\rho_X(r) = \alpha^r \quad |\alpha| < 1$$

EXAMPLE 9.4.1 Consider the following model

$$X_t = be^{-ct} + Z_t.$$

It is easy to see that

$$X_t = e^{-c}(X_{t-1} - Z_{t-1}) + Z_t = e^{-c}X_{t-1} + Z_t^*$$

where $Z_t^* = Z_t - e^{-c}Z_{t-1}$ and Z_t^* is a random process with mean zero. Hence, in this case X_t can be expressed as an $AR(1)$ process.

The $AR(1)$ process is sometimes called a Markov process. A process is Markov if it's present value can be specified from it's immediate past value and does not require the knowledge of its history. To see the relation, suppose that a basketball player tries a free throw and wins or loses $1 according to whether he makes or misses the shot. Starting from zero Z_1 could be 1 or -1. The same is true for $Z_2, Z_3, \cdots$. At time t (after t attempts), the player has X_t dollars that equals to $Z_1 + Z_2 + \cdots + Z_t$ where each Z is either 1 or -1. Since $X_{t-1} = Z_1 + Z_2 + \cdots Z_{t-1}$ we can write this as

$$X_t = X_{t-1} + Z_t.$$

Thus the value of X_t is specified by X_{t-1} plus a random component. This explains the similarity to the Markov chain. Now returning to (9.11) we see that by successive substitution in the equation, we can write X_t, for example, as

$$X_t = \alpha[\alpha X_{t-2} + Z_{t-1}] + Z_t = \alpha^2[\alpha X_{t-3} + Z_{t-2}] + \alpha Z_{t-1} + Z_t. \tag{9.12}$$

Chapter 9: Time Series Analysis

This means that eventually X_t can be expressed as an infinite-order MA process (infinite linear combination of Z_t's) having the form (provided that $-1 < \alpha < 1$)

$$X_t = Z_t + \alpha Z_{t-1} + \alpha^2 Z_{t-2} + \ldots \tag{9.13}$$

It is interesting to note that in a similar way a first-order MA process can be expressed as an AR model of infinite order. This duality between AR and MA processes is useful for a variety of purposes. To see this, let us rather than using successive substitution, use the backward shift operator B defined as $BX_t = X_{t-1}$. Then equation (9.11) may be written as

$$(1 - \alpha B)X_t = Z_t \tag{9.14}$$

or

$$\begin{aligned} X_t &= Z_t/(1 - \alpha B) \\ &= (1 + \alpha B + \alpha^2 B^2 + \ldots)Z_t \\ &= Z_t + \alpha Z_{t-1} + \alpha^2 Z_{t-2} + \ldots \end{aligned} \tag{9.15}$$

provided that $|\alpha| < 1$. It follows from (9.15) that

$$E(X_t) = 0 \tag{9.16}$$

and

$$Var(X_t) = \sigma_z^2(1 + \alpha^2 + \alpha^4 + \ldots). \tag{9.17}$$

Here the variance is finite provided that $|\alpha| < 1$. When this is the we have

$$Var(X_t) = \sigma_z^2/(1 - \alpha^2). \tag{9.18}$$

A PRACTICE PROBLEM Consider an AR(2) namely

$$X_t = \alpha_1 X_{t-1} + \alpha_2 X_{t-2} + Z_t.$$

Show that

$$\rho_X(1) = \frac{\alpha_1}{1 - \alpha_2} \quad \rho_X(2) = \alpha_2 + \frac{\alpha_1^2}{1 - \alpha^2}$$

and

$$Var(X_t) = \sigma_z^2/(1 - \alpha_1 \rho_X(1) - \alpha_2 \rho_X(2)).$$

9.5 GENERAL-ORDER AR PROCESS

As demonstrated above an AR process of finite order can be expressed as an MA process of infinite order. This may be done either by successive substitution, or by using backward shift operator. Using the latter, an AR of order p may be written as

$$(1 - \alpha_1 B - \ldots - \alpha_p B^p)X_t = Z_t \tag{9.19}$$

or
$$X_t = Z_t/(1 - \alpha_1 B - \ldots - \alpha_p B^p) = \phi(B) Z_t \qquad (9.20)$$

where
$$\phi(B) = (1 - \alpha_1 B - \ldots - \alpha_p B^p)^{-1} = (1 + \beta_1 B + \beta_2 B^2 + \ldots). \qquad (9.21)$$

Note that βs can be found for given values of αs. Since $E(Z_t) = 0$, it follows that $E(X_t) = 0$. The variance of X_t is finite provided that $\sum \beta_i^2$ converges, and this is a necessary condition for stationarity. An equivalent condition for stationarity condition is to say that the roots of the equation

$$\phi(B) = 1 - \alpha_1 B - \ldots - \alpha_p B^p = 0 \qquad (9.22)$$

must lie outside the unit circle. If this condition is satisfied, then the variance of the X_t can be expressed in terms of β_i's.

AR models have been applied to many situations in which it was reasonable to assume that the present value of a time series can be approximated by a weighted average of its past values. For example, in regression-based forecasting, AR models are used whenever the independent variables are all previous values of the same time series. This occurs in situations where recent values of time series carry the most information for prediction of its value in the next period. A good example for this occurs in sport where one wants to predict the best time for an event such as men's 100-m dash for the next Olympic competition. Clearly, here times of recent Olympics provide more information for prediction than those of early Olympics. Also, it is reasonable to express the future times in terms of past times or records. Forecasting using time series is discussed further in the next section.

So far we concentrated on stationary processes. In practice, most time series are non stationary. In order to apply analysis of a stationary model, it is necessary to remove the nonstationarity. There are several powerful techniques for removing nonstationarities. Box and Jenkins (1981) have developed a popular method based on differencing the neighboring values. This transformation removes the slowly varying part of the time series. In other words, it acts like a high-pass filter.

If instead of a single-valued time series, we consider an s-vector-valued series $X_t^{(j)}$, $t = 0, \pm 1, \ldots$, (a time series with s components), then stationarity requirements are

$$E(X_t^{(j)}) = \mu_j$$
$$Cov(X_t^{(j)}, X_{t+r}^{(k)}) = \gamma_{jk}(r) < \infty \qquad (9.23)$$

for all $t, r = 0, \pm 1, \ldots$ and $j, k = 1, 2, \ldots, s$. Let

$$f_{jk}(\omega) = \frac{1}{2\pi} \sum_{r=-\infty}^{\infty} \gamma_{jk}(r) e^{-ir\omega}, \quad -\pi \leq \omega \leq \pi \qquad (9.24)$$

then $\gamma_{jk}(r)$ and $f_{jk}(\omega)$ are called cross-covariance and cross-spectrum of the series $X_t^{(j)}$ with the series $X_t^{(k)}$, respectively. Corresponding to (9.9) we require that

$$\sum_{r=-\infty}^{\infty} |\gamma_{jk}(r)| < \infty, \quad j, k = 1, 2, \ldots, s \qquad (9.25)$$

Chapter 9: Time Series Analysis

It is sometimes useful to consider the parameters

$$R_{jk}(\omega) = \frac{f_{jk}(\omega)}{[f_{jj}(\omega)f_{kk}(\omega)]^{1/2}} \tag{9.26}$$

for $1 \leq j, k \leq s$. $R_{jk}(\omega)$ is called the coherency of the series $X_t^{(j)}$ with the series $X_t^{(k)}$ at frequency ω. It is similar to correlation in time domain. Finally, $f_{jk}(\omega)$ is called the phase spectrum, while the absolute value $|f_{jk}(\omega)|$ is called the amplitude spectrum.

We stop the discussion of theoretical results here and turn to statistical aspects of the time series analysis.

Estimation

Turning to the time series data analysis, suppose that we are given a time series of length N. Then the estimate for mean, variance, and autocovariances may be obtained as

$$\bar{X} = \frac{1}{N}\sum_{t=1}^{N} X_t, \qquad s_X^2 = \frac{1}{N}\sum_{t=1}^{N}(X_t - \bar{X})^2$$

$$K(r) = \frac{1}{N}\sum_{t=1}^{N-r}(X_t - \bar{X})(X_{t+r} - \bar{X}), \quad r = 0, 1, 2 \cdots$$

Also estimation for autocorrelations may be obtained as

$$c(r) = K(r)/K(0).$$

For example, suppose that in a certain school district the number of students who participated in sport activities in a given school district are recorded for the past 10 years (in 100s) as

$$X_t: \quad 49 \quad 66 \quad 25 \quad 73 \quad 40 \quad 66 \quad 57 \quad 43 \quad 61 \quad 50$$

To find $C(1)$, first we find $\bar{X}$. This is 53. Next we calculate the deviations

$$X_t - \bar{X}: \quad -4 \quad 13 \quad -28 \quad 20 \quad -13 \quad 13 \quad 4 \quad -10 \quad 8 \quad -3$$

Thus

$$K(0) = [(-4)^2 + (13)^2 + \cdots + (-3)^2]/10 = 189.6$$
$$K(1) = [(-4)(13) + (13)(-28) + (-28)(20) + \cdots + (8)(-3)]/10 = -149.7$$

and

$$c(1) = K(1)/K(0) = -0.79.$$

A PRACTICE PROBLEM For above data calculate $K(2)$ and $c(2)$.

EXAMPLE 9.5.1 The following data present the number of points scored by a professional basketball team in their last 70 games. Estimate the first 15 autocorrelations and plot that versus r.

1-15	16-30	31-45	46-60	61-70
97	94	100	112	118
114	130	121	94	98
73	105	106	114	100
121	87	124	93	110
88	127	100	102	89
114	101	108	88	109
105	07	95	109	90
91	100	104	105	107
109	110	86	91	104
98	95	104	103	73
121	107	98	99	
85	100	105	84	
107	95	95	85	
90	75	107	104	
108	109	100	95	

SOLUTION:

Estimated ACF

r	$c(r)$	r	$c(r)$	r	$c(r)$
1	-0.39	6	-0.05	11	0.11
2	0.30	7	0.04	12	-0.07
3	-0.17	8	-0.04	13	0.15
4	0.07	9	-0.01	14	0.04
5	-0.10	10	0.01	15	-0.01

Estimated ACF

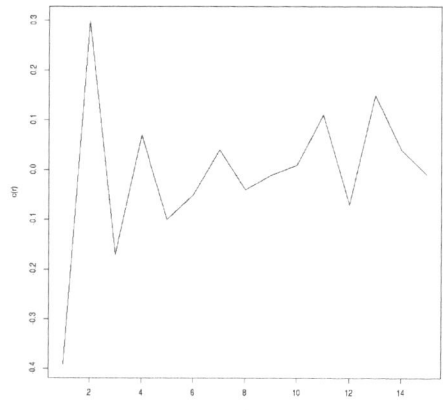

Autoregressive-Moving Average Process

Although both the MA and AR have their own appeal, in some cases we may have to use a relatively long AR or MA model to capture the complex structure of a time series. This may become undesirable since we usually want to fit a parsimonious model; a model with relatively few unknown parameters. To achieve this goal, we can combine the AR and MA parts to form an autoregressive-moving average (ARMA) model of the form

$$\phi(B)Y_t = \theta(B)Z_t.$$

where

$$\phi(B) = 1 - \alpha_1 B - \cdots - \alpha_p \beta^p$$
$$\theta(B) = 1 + \beta_1 B + \cdots + \beta_q \beta^q.$$

The usefulness of ARMA models lies in their parsimonious representation. As in the AR and MA cases, properties of ARMA models can usually be characterized by their ACFs. To this end, some lucid discussions of the various properties of the ACF of simple ARMA models are given in most books on time series.

We now turn to some simple methods for analysis and forecasting time series, which are useful in practice. In general, it is possible to fit both AR and MA and even mixed ARMA to (at least stationary) time series data and estimate their order and the unknown parameters. We will not discuss this here, but just mention that most statistical software will do this.

9.6 FORECASTING USING TIME SERIES

As mentioned earlier, any variable that is measured over time in sequential order is called a time series. One objective of analyzing time series is to detect existing patterns and use them to forecast the future values of the time series. The following lists some examples of such applications:

1. Financial institutions that invest in the stock market may want to predict the future values of interest rates and unemployment rates.
2. Physiologists and physical educators may like to forecast future records of athletic events.
3. Companies may attempt to predict the demand for their products and their shares of the market.
4. Sport organizations may like to forecast how many fans will come to watch a particular home game, or all home games during the next season.

Consider example 4 in the above list. Clearly, forecasting ticket sales helps the organizations to come up with a better plan, for example, determine the number of ticket windows, food stands, and parking spaces. This aspect of managing a sport organization is critical in planning for the future since the long-run success of such an organization is closely related to how well the management is able to foresee the future and develop appropriate strategies.

There are many different forecasting methods. These methods can generally be classified into three groups

(1) Judgmental methods
(2) Extrapolation (or time series) methods
(3) Econometric (or causal) methods

Judgmental methods are nonquantitative and are used by managers, coaches, and experts. Extrapolation methods are quantitative methods and use past data of a time series (e.g., best annual records for men's 100-m run). Basically, these methods search for patterns in historical data and extrapolate these patterns into the future. Examples of extrapolation methods are trend-based regression, exponential smoothing, MAs, and autoregression models. Finally, econometric (causal) models use regression to forecast a time series variable by means of other explanatory time series variables.

Let y_t denote the observed value of a time series at time t, for $t = 1, 2, \cdots, T$, where T is the number of historical observations on Y. Let $y_{t-k,t}$ be a "k-period-ahead" forecast (forecast of y_t based on information until time period $t - k$ that was just observed). Moreover let e_{t-k} be the forecast error associated with $y_{t-k,t}$ where $e_{t-k,t} = y_t - y_{t-k,t}$.

Now, in order to decide how good a forecast is, one of the several available summary measures may be used. Summary measures are based on the expectation or idea that a model that forecasts the historical data well will also forecast future observations well. Let N be the number of terms used in each sum (typically slightly less than T). Recall that T is the number of historical observations. Here, we consider the summary measures known as, mean absolute error (MAE), root mean square error (RMSE), and mean absolute percentage error (MAPE or mean absolute deviation [MAD]). The formulas for these measures are, respectively,

$$MAE = \left(\sum_{i=1}^{N} |e_t| \right) / N$$

Here, the average of the absolute value of errors is computed.

$$RMSE = \sqrt{\sum_{t=1}^{N} e_t^2 / N}$$

This is somehow similar to MAE except that squared errors are used instead of absolute values of errors. In the latter formula, the square root of average of squared error is obtain. The square root is taken here to make the units of RMSE same as that of forecast variable. Finally

$$MAPE = 100\% \times \sum_{t=1}^{N} |e_t/y_t| / N.$$

Here, the average of the absolute value of the proportion of errors is computed, and then the result is multiplied by 100%. Note that MAPE does not depend on the units of measurement of y.

It is generally true that the forecasting models that make one of the above measures small tend to make the others small, too. Therefore, any of the measures introduced above can be minimized.

The forecasting methods discussed here (Sections 9.7 and 9.8) are all based on time series analysis. Here, the first step is to analyze the components of a time series. This includes the methods for detecting and measuring whatever components exist. Once this is accomplished, we can develop forecasting tools and answer the questions regarding the future values and their accuracies. As we shall see, there are many different ways of extracting information from a time series. Here, we will discuss four different methods, namely, simple MA, weighted MA, exponential smoothing, and regression analysis (Section 9.9). The MA and exponential smoothing techniques are appropriate for stationary time series—time series with, for example, no significant upward or downward long-term trend. The regression, on the other hand, is more appropriate for nonstationary time series, although it is also used for trend analysis too.

In general, forecasts themselves may be classified according to the time frame of interest to (1) short range, (2) medium range, and (3) long range. This type of classifications should, of course, be viewed only as a generalization. The line of demarcation between these ranges is often arbitrary and varies from case to case. It may also vary from application area to application area, for example, sport to sport.

It should be pointed out that discussion of this and the next sections does not involve the theory described in previous sections, although some relationship is apparent. Our objective here is to expose students to the concepts of forecasting and to introduce some of the popular techniques used in practice. The level of this book precludes our investigating the more complicated methods.

9.7 COMPONENTS OF TIME SERIES

When trying to study behavior of a time series, it is often helpful to think of it as consisting of four different components: long-term trend (T), cyclical component (C), seasonal component (S), and irregular component or random variation (R).

The trend is a steady increase or decrease in the time series. It is a long-term, relatively smooth pattern or direction that a series exhibits. It reflects the effect of forces making for gradual growth or decline. These forces operate over a long period of time and are not subject to sudden reversals in direction. For example, during the past 50 years, the number of sport fans in the United States has exhibited a trend of relatively steady growth. The same is true about participation in sport and sales of sporting goods. There has also been a trend in records, downwards, or upwards, depending on the sport. Note that the trends exhibited in time series are not always linear.

Many time series exhibit cyclical behavior with regular runs of observations below and above the trend line. Formally, a cycle is a wavelike pattern in a long-term trend, Examples of cycles include the well-known business cycles that record periods of economic recession and inflation, long-term product–demand cycles, and cycles in the monetary and financial sectors. For example, periods of moderate inflation followed by periods of rapid inflation can lead to many time series that alternate below and above a generally increasing trend line (e.g., housing costs). If cyclical patterns are within a 1-year time frame, they are viewed as seasonal variations.

Seasonal variations are like cycles, but they occur over short repetitive calendar periods and, by definition, have durations of less than 1 year. The term seasonal variation may refer to the four traditional seasons or to systematic patterns that occur during a shorter period. For example, a manufacturer of swimming pools expects low sales activity in the fall and winter months, with peak sales occurring in the spring and summer months. Similarly, the number of swimmers are always higher in the summer than in other seasons, and present a seasonal component. Another simple example is the sale of Christmas trees.

The irregular component or random variation of the time series is the residual, or "catchall" factor that exists in almost all time series. It accounts for the deviation of the actual time series value from what we would expect if the trend, cyclical, and seasonal components completely explains the time series. It is caused by the short-term, unanticipated, and nonrecurring factors. Such variation tends to hide the existence of the other, more predicable components. We will benefit from learning how to remove the random variation, thereby making it possible to describe and measure the other components and ultimately to make accurate forecasts. Most time series only include trend and random variation (signal and noise). Because of this, we will only consider techniques for analysis of these components. To this end, we first mention two frequently used models.

The time series model is generally expressed either as an additive model, in which case the value of the time series at time t (denoted y_t) is specified as

$$y_t = T_t + C_t + S_t + R_t \tag{9.27}$$

or as a multiplicative model, in which case the value of the time series at time t is specified as

$$y_t = T_t \cdot C_t \cdot S_t \cdot R_t \tag{9.28}$$

Both models are acceptable and useful. However, it is frequently easier to understand the techniques associated with separating different components of time series by referring to the multiplicative model. We note that multiplicative models can be transferred to additive models by transformations such as taking logarithms.

EXAMPLE 9.7.1 Consider the performance of a basketball player. Let P_t and S_t be the number of points scored and number of assists made in the t-th game, respectively. If we use

$$y_t = P_t + S_t$$

as an index to measure player's performance, then we are considering an additive model.

EXAMPLE 9.7.2 Let P_t and G_t denote, respectively, the total number of points scored and number of game played by an National Basketball Association (NBA) superstar in the t-th year. Also let $D_t = 1/G_t$, then the player's average PPG (point per game) in the t-th year can be expressed as

$$y_t = P_t \cdot D_t$$

This is an example of a multiplicative model.

We now discuss methods useful for removing random variation (noise). We note that once we have identified the data and labeled them properly, it is helpful to display them using a scatter plot with the x-axis representing time and the y-axis representing the variable of interest. In the analysis of time series data, five common types of forecasting techniques are used. They are, MA, Exponential Smoothing, Trend Projection, Classical Decomposition, and Box–Jenkins method. In future sections, some of these techniques will be discussed.

Chapter 9: Time Series Analysis

9.8 SMOOTHING TECHNIQUES

Determination of the components that actually exist in a time series helps to develop a better forecast. Unfortunately, the existence of the random component often hides the other components and makes it hard to identify them. One of the simplest ways of removing the random fluctuation is to smooth the time series. Since in many real-life applications, random components make up the rough part (high-frequency) of the observed time series, application of these methods seems reasonable. In this section, we will describe two methods for doing this: MAs and exponential smoothing. These techniques are appropriate for a fairly stable time series; that is, one that exhibits no rapidly changing trend, cyclical, or seasonal effects.

MAs: The simple MA forecasting method eliminates the effects of irregular or random fluctuations by averaging the historical data points. Suppose that one is given the job of forecasting the attendance of the next game for a particular sport organization. If one suggests finding the average attendance of all of the recent games and using that as the forecast for the next game, then one would be thinking along the right line. A MA for a time period is the simple arithmetic average of the values in that time period and its neighboring periods. It may be expressed as

$$\text{MA} = \frac{\text{Sum of the Most Recent } k \text{ Data Values}}{k}$$

This may be called a k-period MA. For example, to compute the three-period MA for any time period, we would sum the value of the time series in that time period, the value in the previous time period, and the value in the following time period, and divide that by 3. We can calculate the three-period MA for all time periods except the first and the last ones. To compute the five-period MA, we average the value in that time period, the values in the two previous time periods, and the values in the two following time periods. We can choose any number of periods for which we wish to calculate MAs.

Many authors consider the MA curves as the most important technique for summarizing and interpreting time series. Simple to construct, MA curves serve two quite different proposes. They can be defined so as to suppress the seasonal effect, thereby highlighting the trend effect, or they can be used to quantify the seasonal effect, that is, to estimate the $S'_t s$.

EXAMPLE 9.8.1 As part of an effort to forecast future attendance, the manager of a fitness club recorded the quarterly attendance (in hundreds) for the past 4 years. These are shown in Table 9.8.1. Calculate the 3-quarter and 5-quarter MAs. Then graph the quarterly attendance and the MAs.

SOLUTION: To compute the 3-quarter MAs, we group the attendance into periods 1, 2, and 3, then find their average. Thus, the first MA is

$$\frac{39+37+61}{3} = (1/3)39 + (1/3)37 + (1/3)(61) = \frac{137}{3} = 45.7$$

TABLE 9.8.1 Quarterly Attendance of a Fitness Club

Time Period	Year	Quarter	Attendance (in hundreds)
1	1	1	39
2		2	37
3		3	61
4		4	58
5	2	1	18
6		2	56
7		3	82
8		4	27
9	3	1	41
10		2	69
11		3	49
12		4	66
13	4	1	54
14		2	42
15		3	90
16		4	66

The second MA is calculated by dropping the first period's attendance (39), adding the fourth period's attendance (58), and then computing the new average. Thus, the second MA is

$$\frac{37 + 61 + 58}{3} = \frac{156}{3} = 52.0$$

The process continues as shown in Table 9.8.2. Similar calculations are used to produce the 5-quarter MAs shown also in that table. With this example, it should be now clear that why this method is called MA.

Notice that we place each MA in the center of the group of values that are averaged. It is for this reason that the use of an odd number of periods in calculating MAs is preferred. The quarterly attendance as well as MAs are plotted in Figure 9.8.1. Later in this section, we will discuss how to deal with an even number of periods.

To see how the MAs remove some of the random variation, examine Figure 9.8.1 that depicts quarterly attendances. It is difficult to discern any of the time series components because of the large amount of random variation. Now consider the 3-quarter MA in Figure 9.8.1. You should be able to detect a seasonal pattern that exhibits peaks in the third quarter of each year (periods 3, 7, 11, and 15) and valleys in the first quarter of each year (periods 5, 9, and 13). There is also a small but discernible long-term trend of increasing attendance.

Chapter 9: Time Series Analysis

TABLE 9.8.2 Moving Averages (3-Quarter and 5-Quarter)

Time Series	Attendance	3-Quarter Moving Average Forecast	Forecast Error	5-Quarter Moving Average Forecast
1	39	–	–	–
2	37	45.7	$(37 - 45.7 =) - 8.7$	–
3	61	52.0	11	42.6
4	58	45.7	12.3	46.0
5	18	44.0	−26	55.0
6	56	52.0	4	48.2
7	82	55.0	27	44.8
8	27	50.0	−23	55.0
9	41	45.7	−4.7	53.6
10	69	53.0	16	50.4
11	49	61.3	−12.3	55.8
12	66	56.3	9.7	56.0
13	54	54.0	0	60.2
14	42	62.0	−20	63.6
15	90	66.0	24	–
16	66	–	–	–
Total			9.3	

Note, also that in Figure 9.8.1, the 5-quarter MA produces more smoothing than the 3-quarter MA. In general, the longer the time period over which we average, the smoother the series becomes. Unfortunately, in this case, we have smoothed too much, since the seasonal pattern is no longer apparent in the 5-quarter MA. All that we can see is the long-term trend. It is important to realize that our objective is to smooth the time series sufficiently to remove the random variation and to retain and reveal the other components present (trend, cycle, and/or seasonality). With too little smoothing, the random variation disguises the real pattern. With too much smoothing, however, some or all of the other effects may be eliminated along with the random variation. In fact, a logical issue here is to decide how to select the value of k. In other words, should we use a 3-period MA, a 4-period MA, or some other number of periods. The answer to this question involves the concepts that were discussed in selection of the best regression model in Chapter 8.

Centered MAs: if we use an even number of periods in calculating the MAs, we have to figure out where to place the MAs in table or graph. For example, suppose that we want to calculate the four-period MA of the data in Table 9.8.3. The data present the points scored by a football team in six successive games.

The first MA is

$$\frac{15 + 27 + 20 + 14}{4} = \frac{76}{4} = 19.0$$

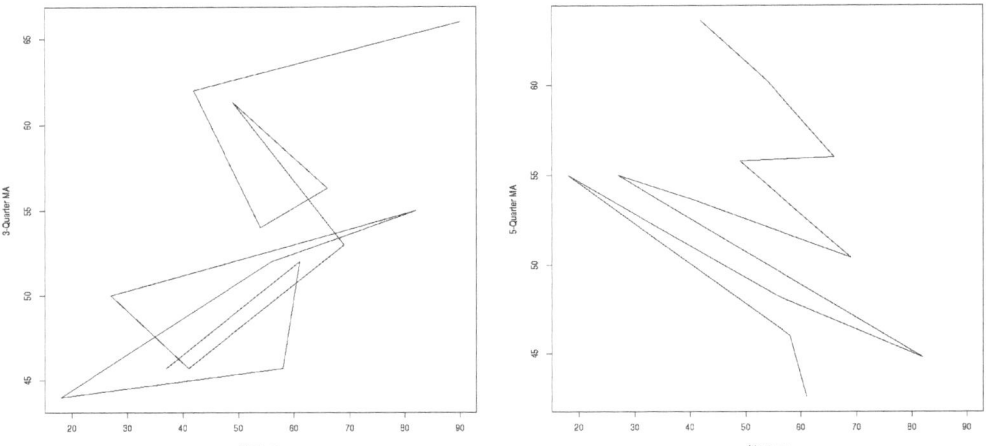

FIGURE 9.8.1 Quarterly Attendance and the 3-Quarter and 5-Quarter Moving Averages

TABLE 9.8.3 Points Scored in Six Games

Period	1	2	3	4	5	6
Time Series	15	27	26	14	25	11

However, since this value represents time periods 1, 2, 3, and 4 we must place this value between periods 2 and 3. The next MA is

$$\frac{27+20+14+25}{4} = \frac{86}{4} = 21.5$$

and it must be placed between periods 3 and 4. The MA that falls between periods 4 and 5 is

$$\frac{20+14+25+11}{4} = \frac{70}{4} = 17.5$$

Having the MAs fall between the time periods causes various problems, including difficulty in graphing. Centering the MAs corrects the problem. This is performed by computing the two-period MA of the MAs. Thus, the centered MA for period 3 is

$$\frac{19.0+21.5}{2} = 20.25$$

Similarly, the centered MA for period 4 is

$$\frac{21.5+17.5}{2} = 19.50$$

Table 9.8.4 summarizes these results. In short, centered MA is the MA of size 2 of the MAs.

Weighted MA: Often, it is desirable to vary the weights given to historical data to forecast future values such as sport records or game attendances. Past history may show that a significantly better

TABLE 9.8.4 Four-Period-Centered Moving Average

Period	Time Series	Four-Period Moving Average	Four-Period-Centered Moving Average
1	15		
2	27		
3	20	19.0	20.25
4	14	21.5	19.50
5	25	17.5	–
6	11	–	–

forecast is obtained when the more recent data are given heavier weight. The weighted MA is computed (constructed) as follows:

$$\hat{y}_t = \frac{\sum_{i=1}^{N} \omega_{t-i} y_{t-i}}{\sum_{i=1}^{N} \omega_{t-i}}$$

where

$\hat{y}_t$ = Forecast for time period t

y_{t-i} = Actual value for time period $t-i$

N = Number of time periods used in the averaging process

ω_{t-i} = Weight given to the $(t-i)th$ period in the averaging process

For instance, assume that in Example 9.8.1 a weight of 0.5 is assigned to the most recent attendance, a weight of 0.35 to the next most recent attendance, and weights of 0.15 to the next most recent attendance. Then the fourth period MA for the fourth quarter is computed as follows:

$$\hat{y}_4 = (0.5)(39) + (0.35)(37) + (0.15)(61) = 41.6$$

The main advantage of the MA technique is its simplicity, low cost, and the little time necessary for its implementation. In general, however, the MA technique does a poor job for long-range and medium-range forecasts. For short-term forecasts, that do not require a great deal of accuracy, a MA forecast may be a workable alternative to other forecasting methods.

Forecast Error: As pointed out earlier an important consideration in using any forecasting method is the accuracy of the forecast. By their inherent nature, all forecasts are incorrect and building a forecast model devoid of error is impossible regardless of the sophistication of the technique being used. Clearly, we would like the forecast errors to be small, but in real life this may be very costly. For time series, columns such as column 4 of Table 9.8.2, which contains the forecast errors, are usually used to develop measures of accuracy.

Although we have already discussed measures for accuracy in Section 9.6, here we discuss them in more detail and present some examples. Also, rather than error, here we use deviation and replace D for E used in Section 9.6 mainly because most softwares use this notation.

One measure of forecast accuracy that seems reasonable is simply sum of the forecast errors over time. However, the problem with this measure is that, if the errors are random (as they should be if the forecasting method is appropriate), some errors will be positive and some negative, resulting in a sum near zero, regardless of the size of the individual error. Indeed, we see from Table 9.8.2 that the sum of forecast errors for the attendance time series is only 9.3. Thus, in practice, the direct sum of forecast errors may not provide a good measure for assessing the method used. Another problem is that two sets of errors 1, -1, 2, -1 and 3, -3, 4, -3 result in the same measure of accuracy. These difficulties can be avoided by squaring each of the individual forecast errors.

For the attendance time series, we can compute the average of the sum of the squared errors. By doing so we obtain

$$\text{Average of the Sum of Squared Errors} = \frac{3,797.45}{14} = 271.25$$

The average of the sum of squared errors, like in regression analysis, is commonly referred to as the mean squared error (MSE). The MSE is a frequently used measure that is used frequently for accuracy assessment of a forecasting method.

As was indicated previously, to use the MAs method, one must first select the number of data values to be included in the MA. Clearly, for a particular time series, different length MAs will differ in their ability to accurately forecast the time series. That is, they will differ in their accuracy measure.

Another common method for measuring the forecast error is the MAD. The MAD is simply the average of the absolute differences between the forecast and the actual values, that is

$$MAD = \frac{\sum_{t=1}^{N} |y_t - \hat{y}_t|}{N}$$

where, y_t, $\hat{y}_t$, and N, respectively, present the actual value for time period t, the forecast value for time period t, and the number of time periods.

To illustrate how MAD is calculated consider the following table that presents demand for tennis balls in a college during a 6-month period.

Demand for Tennis Balls and Mean Absolute Deviation (MAD)

Month	Actual Demand	Forecast Demand	MAD = $\|\hat{y}_t - y_t\|$
1	2,150	2,500	350
2	3,210	2,465	745
3	2,622	2,540	82
4	2,475	2,548	73
5	2,910	2,541	369
6	3,412	2,579	833
			2,452

For this data we have
$$\text{MAD} = 2{,}452/6 = 408.7$$

If we assume that MAD of a forecast is normally distributed, we can estimate the probability that it (MAD) will fall within a specific confidence interval. It can be shown that 1 MAD = 0.8 standard deviations. Using the foregoing relationship, we can compute the relative frequency of MADs falling within specified ranges.

For this tennis ball data the MSE is

$$\text{MSE} = \frac{\sum_{t=1}^{N}(y_t - \hat{y}_t)^2}{N} = \frac{1{,}519{,}628}{6} = 253{,}271.$$

When analyzing the comparative merits of several different forecasting models, both measures (MAD and MSE) are often used. The choice of which measure to use is up to the user of the forecast. It is possible that model A could outperform model B as measured by MAD and under-perform model B if measured by MSE. Large forecast errors are magnified using MSE, and therefore, if the decision maker wants to avoid large forecast errors, he or she may choose MSE as the measure of forecast error when comparing models.

Confidence Interval for Mean Absolute Deviations (MADs)

Range of MAD	Relative Frequency of MADS Falling Within the Range
±1	0.5705
±2	0.8895
±3	0.9833
±4	0.9986

Demand for Tennis Balls and Mean Squared Error (MSE)

Month	Actual Demand	Forecast Demand	$(y_t - \hat{y}_t)$	$(y_t - \hat{y}_t)^2$
1	2,150	2,500	-350	122,500
2	3,250	2,465	745	555,025
3	2,622	2,540	82	6,724
4	2,475	2,548	-73	5,329
5	2,910	2,541	369	136,161
6	3,412	2,579	833	693,889
				1,519,628

A variation of MAD is the mean absolute percent deviation (MAPD). It measures the absolute error as a percentage of demand rather than per period. It is computed according to the following formula:

$$\text{MAPD} = \frac{\sum |y_t - \hat{y}_t|}{\sum y_t}$$

We end the discussion of simple MAs by restating that this method uses the simple average of the most recent k observations to predict the next time period. In fact, when applying this method, each observation in the MA calculation need not receive the same weight. The variation, known as weighted MAs (discussed earlier), involves selecting different weights for each data value and then computing a weighted average (mean) as the forecast. In the majority of cases, the most recent observation receives the largest weight and the weight decreases for older data values. As an example, recall once more the attendance data and consider a 3-quarter weighted MA with weights 0.50, 0.25, and 0.25. Rather than

$$\frac{39+37+61}{3} = (1/3)39 + (1/3)37 + (1/3)61 = 45.7$$

we now have

$$(1/2)39 + (1/4)37 + (1/4)61 = 44.$$

Exponential Smoothing (Single Parameter) Exponential smoothing is a forecasting technique that uses a weighted average of past values of a time series to smooth the data. The exponential smoothing is different from the weighted MA in that here all of the historical data in the time series are used to generate the forecast for the next period. Recall that, two drawbacks are associated with the ordinary MA method of smoothing a time series. First, there are no MAs for the first and the last time periods. This is particularly important for short time series that includes only few observations. Second, the MA "forgets" most of the previous time series values. In fact, only information contained in the most recent values contribute to forecasting. This problem is addressed by exponential smoothing.

The exponentially smoothed time series is defined as

$$S_t = \omega y_t + (1 - \omega) S_{t-1} \text{ for } t \geq 2$$

where

$S_t = $ The forecast of the time series (predicted value) at time (period) t

$y_t = $ Actual value of the time series at time (period) t

$S_{t-1} = $ The forecast of the time series (predicted value) at time (period) $t - 1$

$\omega = $ Smoothing constant $(0 \leq \omega \leq 1)$ assigned to the most recent observation

Note that, since the sum of the weights equals one, this is also an averaging technique. Here, the weights are assigned in such a way that the most recent observation y_t carries the largest weight; the second most recent observation y_{t-1} carries the second largest weight; and the weight assigned to the other data points decreases according to some pattern (exponentially).

Let us describe how exponential smoothing is carried out. We begin by setting

$$S_1 = y_1$$

Then we find S_2 as

$$S_2 = \omega y_2 + (1 - \omega) S_1 = \omega y_2 + (1 - \omega) y_1$$

and S_3 as

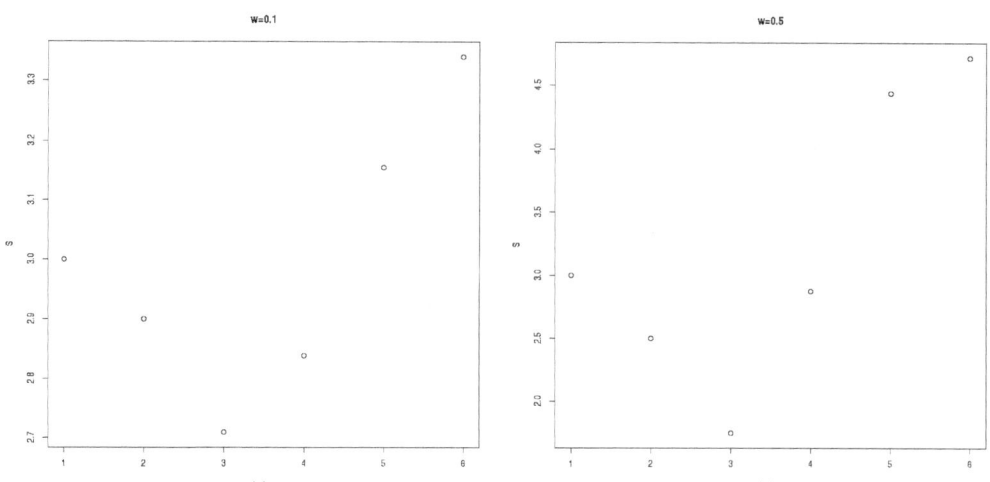

FIGURE 9.8.2 A Time Series {3,2,1,4,6,5} and Two Exponentially Smoothed Series With $\omega = 0.1$ and $\omega = 0.5$

$$S_3 = \omega y_3 + (1 - \omega)S_2$$

and so on. The smoothing constant ω is chosen on the basis of how much smoothing is required. A small value of ω produces a great deal of smoothing. A large value of ω results in very little smoothing. Figure 9.8.2 depicts a time series and two exponentially smoothed series with $\omega = 0.1$ and $\omega = 0.5$. The setting of ω is typically not a scientific process and is usually done by trial and error. One popular approach is to try to forecast the already known most recent value or values from the past data and adjust the value of ω to produce minimal error.

To assess the smoothing effect one can calculate

$$\frac{\Sigma(y_t - S_t)^2}{N - p}$$

where N is the number of observations and p is the number of parameters.

EXAMPLE 9.8.2 Apply the exponential smoothing technique with $\omega = 0.2$ and $\omega = 0.7$ to the data in Example 9.8.1 and graph the results.

SOLUTION: The exponential smoothed values are calculated from the formula

$$S_t = \omega y_t + (1 - \omega)S_{t-1}$$

The results with $\omega = 0.2$ and $\omega = 0.7$ are shown in Table 9.8.5.

Figure 9.8.3 depicts the graph of the original time series and the exponentially smoothed series. We see that $\omega = 0.7$ results in very little smoothing, while $\omega = 0.2$ results in perhaps too much smoothing. In both smoothed time series, it is difficult to discern the seasonal pattern that we detected by using MAs. A different value of ω (perhaps $\omega = 0.5$) is likely to produce more satisfactory results.

TABLE 9.8.5 Exponentially Smoothed Time Series

Time Period	Attendance in Hundreds	Exponentially Smoothed Attendance (with $\omega = 0.2$)	Exponentially Smoothed Attendance (with $\omega = 0.7$)
1	39	$= 39.0$	$= 39.0$
2	37	$0.2(37) + 0.8(39) = 38.6$	$0.7(37) + 0.3(39) = 37.6$
3	61	$0.2(61) + 0.8(38.6) = 43.1$	$0.7(61) + 0.3(37.6) = 54.0$
4	58	$0.2(58) + 0.8(43.1) = 46.1$	$0.7(58) + 0.3(54.0) = 56.8$
5	18	$0.2(18) + 0.8(46.0) = 40.5$	$0.7(18) + 0.3(56.8) = 29.6$
6	56	$0.2(56) + 0.8(40.5) = 43.6$	$0.7(56) + 0.3(29.6) = 48.1$
7	82	$0.2(82) + 0.8(43.6) = 51.3$	$0.7(82) + 0.3(48.1) = 71.8$
8	27	$0.2(27) + 0.8(51.3) = 46.4$	$0.7(27) + 0.3(71.8) = 40.4$
9	41	$0.2(41) + 0.8(46.4) = 45.3$	$0.7(41) + 0.3(40.4) = 40.8$
10	69	$0.2(69) + 0.8(45.3) = 50.0$	$0.7(69) + 0.3(40.8) = 60.5$
11	49	$0.2(49) + 0.8(50.0) = 49.8$	$0.7(49) + 0.3(60.5) = 52.5$
12	66	$0.2(66) + 0.8(49.8) = 53.0$	$0.7(66) + 0.3(52.5) = 61.9$
13	54	$0.2(54) + 0.8(53.0) = 53.2$	$0.7(54) + 0.3(61.9) = 56.4$
14	42	$0.2(42) + 0.8(53.2) = 51.0$	$0.7(42) + 0.3(56.4) = 46.3$
15	90	$0.2(90) + 0.8(51.0) = 58.8$	$0.7(90) + 0.3(46.3) = 76.9$
16	66	$0.2(66) + 0.8(58.8) = 60.2$	$0.7(66) + 0.3(76.9) = 69.3$

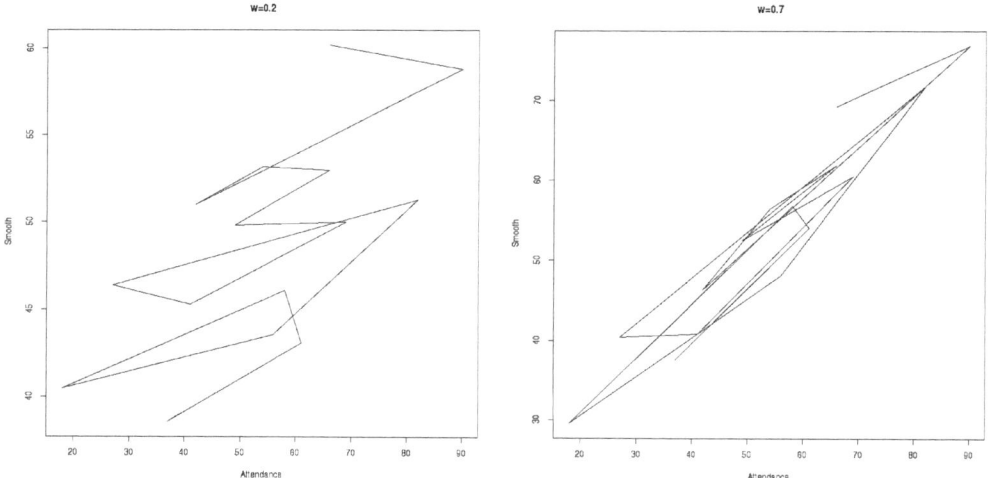

FIGURE 9.8.3 Quarterly Attendance and Exponentially Smoothed With $\omega = 0.2$ and $\omega = 0.7$

Chapter 9: Time Series Analysis

To see how exponential smoothing works, consider the case $\omega = 0.7$. The weight of y_{16}, that is the most recent observation, is 0.7. Since S_t can also be written as

$$S_t = \omega y_t + (1-\omega)[\omega y_{t-1} + (1-\omega)S_{t-2}]$$
$$= \omega y_t + \omega(1-\omega)y_{t-1} + (1-\omega)^2 S_{t-2}$$

we see that the weight of y_{15} is $\omega(1-\omega) = (0.7)(0.3) = 0.21$. Similarly, the weight of y_{14} is $\omega(1-\omega)^2 = (0.7)(0.3)^2 = 0.063$. So the forecast for quarter 17 is calculated as a weighted average of all of the historical data, that is

$$\hat{y}_{17} = (0.7)(66) + (0.7)(1-0.7)(90) + (0.7)(1-0.7)^2(42) + \ldots + (0.7)(1-0.7)^{15}(39)$$
$$= 66.99$$

The general formula for the forecast of the next period, $t+1$, is

$$\hat{y}_{t+1} = \sum_{i=1}^{t} \alpha_i y_i$$
$$= \omega y_t + \omega(1-\omega)y_{t-1} + \ldots \omega(1-\omega)^{\ell} y_{t-\ell} + \ldots$$
$$+ \omega(1-\omega)^{t-1} y_1 \tag{9.29}$$

Replacing t for $t+1$, we see that the formula for $\hat{y}_t$ is

$$\hat{y}_t = \omega y_{t-1} + \omega(1-\omega)y_{t-2} + \ldots + \omega(1-\omega)^{t-2} y_1$$

From this it follows that

$$\hat{y}_{t+1} - (1-\omega)\hat{y}_t = \omega y_t$$

This means that the formula presented earlier is algebraically equivalent to

$$\hat{y}_{t+1} = \hat{y}_t + \omega(y_t - \hat{y}_t)$$

If we use backward shift operator defined as

$$By_t = \begin{cases} y_{t-1} & \text{if } t > 1 \\ 0 & \text{if } t = 1 \end{cases}$$

then Formula (9.29) can also be written as

$$\hat{y}_{t+1} = \omega[1 + (1-\omega)B + \ldots + (1-\omega)^{t-1} B^{t-1}] y_t$$

or equivalently as

$$\hat{y}_{t+1} = \omega \frac{1 - (1-\omega)^t B^t}{1 - (1-\omega)B} y_t$$

By definition of the operator B we have $B^t y_t = 0$. Thus, the above formula is further simplified to

$$\hat{y}_{t+1} = \omega \frac{1}{1-(1-\omega)B} y_t$$

or equivalently to

$$\hat{y}_t = \omega \frac{1}{1-(1-\omega)B} y_{t-1}$$
$$= \omega \frac{1}{1-(1-\omega)B} By_t.$$

From this, we can further deduce that

$$\hat{y}_{t+1} - (1-\omega)\hat{y}_t = \omega y_t$$

or

$$\hat{y}_{t+1} = \hat{y}_t + \omega(y_t - \hat{y}_t).$$

This means that forecast for period $t+1$ equals forecast for period t plus ω times the forecast error for period t.

One of the limitations of the simple exponential smoothing model is that it only provides forecast for one period ahead of the time series data (unless one is willing to use the same forecast for periods $t+1$, $t+2$, etc,). Furthermore, if the data exhibit any trend, the simple exponentially smoothed forecasts will lag behind the actual time series data and incur significant forecast errors.

To account for trend and gain capability to forecast beyond one period, exponential smoothing models adjusted for trend can be used. Moreover, double and triple exponential smoothing models can be used to address linear trend and nonlinear trend, respectively. Software packages contain several of the more complex exponential smoothing models. The packages also search for the optimum smoothing constants for exponential smoothing applications that minimizes the forecast error.

In short, exponential smoothing is considered superior to the MA methods previously discussed. Its cost is equivalent and its accuracy, especially in the short term, is usually better.

Double Exponential Smoothing (Two Parameters) Double exponential smoothing, also known as the adjusted exponential smoothing forecast consists of the exponential smoothing forecast with a trend adjustment factor added to it. This method uses the formula

$$V_t = \omega y_t + (1-\omega)(V_{t-1} + F_{t-1})$$
$$F_t = \gamma(V_t - V_{t-1}) + (1-\gamma)F_{t-1}$$
$$S_{t+1} = V_t + F_t$$

where

$V_t =$ The value of the observation after smoothing
$\omega =$ Smoothing constant
$F_t =$ Trend after smoothing,
$\gamma =$ Trend smoothing constant, $0 < \gamma < 1$

Here high γ reflects trend changes more than a low γ.

EXAMPLE 9.8.3 Smooth the time series given in the following table (column 2) using $\omega = 0.2$ and $\gamma = 0.3$.

SOLUTION:

Period	y_t	V_t	F_t	S_t
1	4,890			
2	4,910	4,890	20	
3	4,970	4,922	23.6	4,910.0
4	5,010	4,958	27.3	4,945.6
5	5,060	5,000	31.7	4,985.3
6	5,100	4,045	35.7	5,031.7
7	5,050	5,075	34.0	5,080.7
8	5,170	5,121	37.6	5,109.0
9	5,180	5,163	38.9	5,158.6
10	5,240	5,210	41.3	5,201.9

For example, $F_3 = 0.3(49.22 - 4,890) + 0.7(20) = 23.6$ and $S_4 = 4,922 + 23.6 = 4,945.6$.

EXAMPLE 9.8.4 This example demonstrates the computation of exponentially smoothed and adjusted exponentially smoothed forecasts. A small university runs a tennis camp every summer. The number of participants in the last 8 years are given below:

Year:	1	2	3	4	5	6	7	8
Participants:	56	61	55	70	66	65	72	75

Develop an exponential smoothing forecast using $\omega = 0.40$, and an adjusted exponential smoothing forecast using $\omega = 0.40$, and $\gamma = 0.20$. Compare the accuracy of the two forecasts using MAD and cumulative error.

SOLUTION: Using exponential smoothing we have the following:
For year 2 the forecast (assuming $V_1 = 56$) is

$$V_2 = (0.40)(56 + (0.60)(56) = 56$$

For year 3 the forecast is

$$V_3 = (0.40)(61) + (0.60)(56) = 58$$

The remaining forecasts are computed similarly and are shown in the table below (column 3). Now for double exponential smoothing we have for year 3

$$S_3 = 58 + 0.40 = 58.40$$

The remaining adjusted forecasts are computed similarly and are shown in the following table (column 4).

Year	y_t	V_1	S_t	$y_t - V_t$	$y_t - S_t$
1	56	–	–	–	–
2	61	56.00	56.00	5.00	5.00
3	55	58.00	58.40	-3.00	-3.40
4	70	56.80	56.88	13.20	13.12
5	66	62.08	63.20	3.92	2.80
6	65	63.65	64.86	1.35	0.14
7	72	64.18	65.26	7.81	6.73
8	75	67.31	68.80	7.68	6.20
9	–	70.39	72.19	–	–
				37.97	30.60

$$\text{MAD}(V_t) = \frac{\Sigma |y_t - V_t|}{n} = \frac{41.97}{7} = 5.99$$

$$\text{MAD}(S_t) = \frac{\Sigma |y_t - S_t|}{n} = \frac{37.39}{7} = 5.34$$

Also the cumulative error is $E(V_t) = 35.97, E(S_t) = 30.60$. Because both MAD and the cumulative error are less for the adjusted forecast, it would appear to be the most accurate.

We end this part by noting that MAs and exponential smoothing are relatively crude methods of removing random variations. In the next section, we will present a method that attempts to separate the components of a time series such as trend, in a more precise fashion. Of course, as expected, methods that provide more precise results require more information too.

9.9 TREND ANALYSIS (TREND PROJECTION)

The smoothing techniques presented in the last section are helpful in providing a clearer picture of which components are present in a time series. In order to forecast, however, we often need more precise measurements of trend, cyclical effects, and seasonal effects. In this section, we will discuss methods that allow us to describe trend more precisely. We note that there are some other techniques for modeling trend in the data. Two of these are second-order MA, and double exponential smoothing, which are extensions of MA and exponential smoothing techniques covered in the previous section.

As noted earlier, a trend (growth) can be linear or nonlinear and, indeed, can take on a host of many functional forms. The easiest way of isolating the long-term trend is the use of regression analysis, where the independent variable is time. If we believe that the long-term trend is essentially linear, we use the following model

$$y_t = \beta_0 + \beta_1 t + \varepsilon \tag{9.30}$$

where y_t is the dependent or response variable, t is the independent variable (time period), β_0 is the theoretical, unknown y intercept of the line, β_1 is the theoretical, unknown slope of the line, and ε is the independent error term, assumed to have a normal distribution with a mean of zero and a constant standard deviation.

Using observations on y_t, estimates of β_0 and β_1 can be obtained using the least squares method. The prediction equation is given as

$$\hat{y}_t = \hat{\beta}_0 + \hat{\beta}_1 t$$

where $\hat{\beta}_0$ is the estimate of the y intercept and $\hat{\beta}_1$ is the estimate of the slope. Note that in this model the rate of change in y_t from one time period to the next is relatively constant.

When the time series appears to be slowing down or accelerating as time increases, a nonlinear trend may be present. For such cases, we may consider polynomial models, which are often used to describe nonlinear time series with certain slowly varying character. Note that in most cases, we do not need to consider polynomials of order higher than 3. In fact, in many cases a quadratic trend given below suffix.

$$y_t = \beta_0 + \beta_1 t + \beta_2 t^2 + \varepsilon \tag{9.31}$$

There are also occasions that a decaying trend such as

$$y_t = \beta_0 + \beta_1 (1/t)$$

or an exponentially decaying trend given below may be appropriate

$$y_t = \beta_0 + \beta_1 e^{-t}$$

Note that

$$y_t = \beta_0 \beta_1^t$$

also presents an exponential trend or growth. Finally the S-shaped trend

$$y_t = \alpha / (\beta_0 + \beta_1 \beta_2^t)$$

may be used, which is a form of the logistic model discussed in Chapter 8.

These are just examples of functional forms that could be considered in trend analysis. Figure 9.9.1 depicts one form of the polynomial model. This may apply, for example, to present participation in a sport that has experienced a slow start in popularity followed by a more rapid growth. Such models are generally useful for sport data since most sport records experience a rapid early growth followed by an inevitable leveling off. Examples 9.9.1 and 9.9.2 illustrate how and when these models are used.

EXAMPLE 9.9.1 FOOTBALL ATTENDANCE Annual football game attendances (in thousands) at a small university have been recorded for selected 10 years. These data are shown in Table 9.9.1. The director of the sport programs believes that the trend over this period is basically linear. Use regression analysis to measure the trend.

SOLUTION: It is easier (though not necessary) to change (code) the times from the years 1983 through 1992 to time periods 1 through 10. Here, we can use software to estimate the model. We

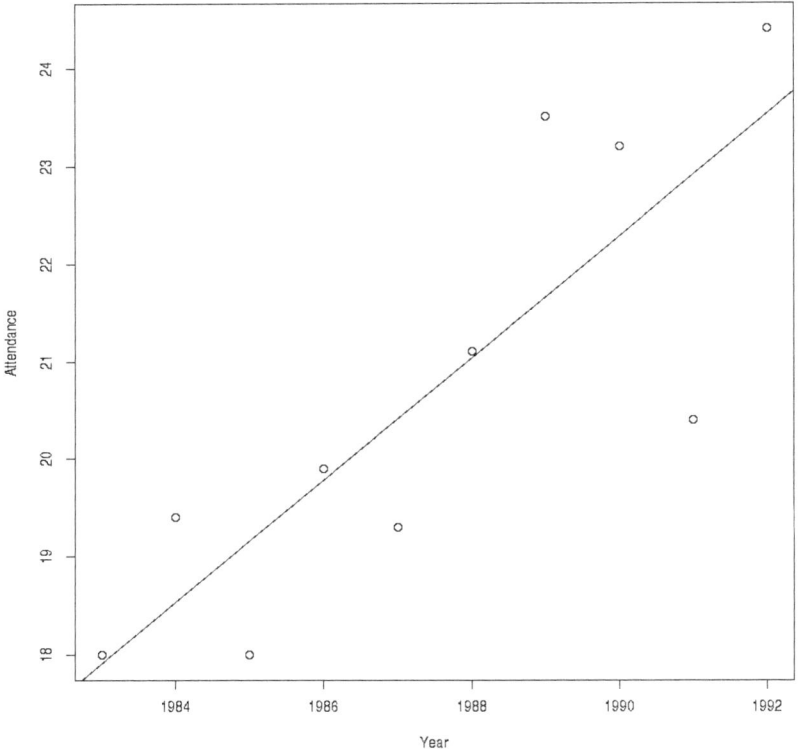

FIGURE 9.9.1 Polynomial Model (9.31) With Coefficients 1, 3, 1 And $t = 1, 2, 3, 4, 5$

TABLE 9.9.1 Football Attendance

Year	Attendance	Year	Attendance
1983	18.0	1988	21.1
1984	19.4	1989	23.5
1985	18.0	1990	23.2
1986	19.9	1991	20.4
1987	19.3	1992	24.4

found that the fit of the line is relatively poor, with $R^2 = 68.8\%$. Because of the possible presence of cyclical and seasonal effects and because of random variation, we do not usually expect to have a very good fit with such a small data set. Remember that we are only measuring the trend in this analysis, and not any other components. The time series is shown graphically in Figure 9.9.2. The estimated trend line is superimposed on the graph, showing a clear upward trend to the right. As was mentioned earlier, one of the purposes of isolating the trend, is to use it for forecasting. For example, we may want to forecast the attendance of the year 1993 ($t = 11$). From the trend equation, we get

$$\hat{y} = 17.28 + 0.6255t = 17.28 + 0.6255(11) = 24.16.$$

Chapter 9: Time Series Analysis

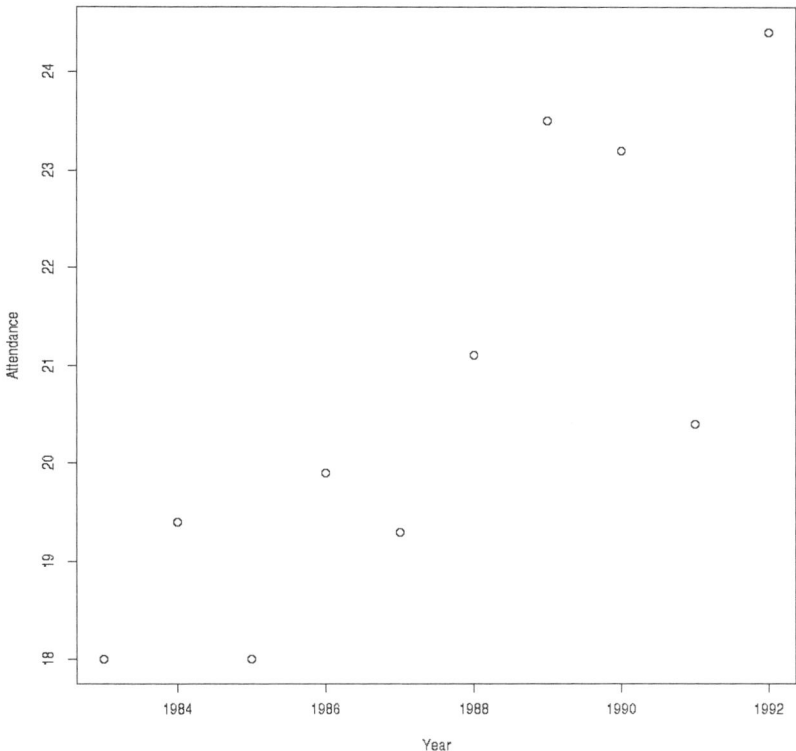

FIGURE 9.9.2 Time Series and Trend Line in Example 9.9.1

This value, however, represents the forecast based only on trend. If we believe that a cyclical pattern also exists, we should incorporate that into the forecast.

EXAMPLE 9.9.2 In the last few decades, sport magazines have enjoyed an increased number of readers. In the last decade, however, because of the Internet, many fans have either decreased their magazine reading or eliminated it. The effect has been serious for publishers. To help analyze the problem, the number of sport magazines (in millions) sold has been recorded for the period 1970–1999. These data are shown in Table 9.9.2. To help forecast future readership, apply regression analysis to determine the trend.

SOLUTION: Graph of y_t versus t suggests that a second-order polynomial may be appropriate to describe the trend. Considering this the model

$$y_t = \beta_0 + \beta_1 t + \beta_2 t^2 + \varepsilon$$

was fitted. All statistics (R^2, F, t-values) indicate that the second-order polynomial given below fits the data quite well.

$$y_t = 79.6832 + 4.41895t - 0.1348t^2 \text{ and } (R^2 = 0.813)$$

TABLE 9.9.2 Annual Sale of a Sport Magazine (in Millions)

Year	t	y_t	Year	t	y_t
1970	1	85.0	1985	16	118.8
1971	2	87.8	1986	17	127.6
1972	3	88.8	1987	18	123.9
1973	4	94.3	1988	19	117.9
1974	5	99.8	1989	20	105.5
1975	6	99.5	1990	21	103.4
1976	7	104.2	1991	22	104.3
1977	8	106.2	1992	23	104.4
1978	9	109.7	1993	24	105.8
1979	10	110.8	1994	25	105.5
1980	11	113.5	1995	26	107.0
1981	12	113.0	1996	27	106.7
1982	13	116.1	1997	28	99.6
1983	14	109.6	1998	29	93.7
1984	15	116.8	1999	30	91.8

To forecast the sales for 2000, replace $t = 31$ in regression equation. We have

$$\hat{y}_{31} = 79.6832 + 4.4189(31) - 0.1348(31)^2 = 87.151$$

with a 95% prediction interval of (76.019, 98.181). The actual 2000 sale was 91.8. Figure 9.9.3 shows the time series and the trend line for this example. Note that we can also find the prediction and confidence intervals for the years after 2000.

Box–Jenkins Method: Thus, far in this chapter we have examined basic time series analysis techniques that can be useful in a variety of time series forecasting situations. There is a more complex time series technique called the Box–Jenkins method described in their book and almost all books in applied time series analysis. It involves a three-step model development procedure namely; identification, estimation, and diagnostic checking. At the heart of the Box–Jenkins method is the autoregressive integrated moving average (ARIMA) model.

The Box–Jenkins method is often considered the most accurate time series forecasting technique available. Its more complex time series modeling techniques can generally fit the time series movement patterns more closely and thereby achieve the lowest forecast error through the existing data. However, while the Box–Jenkins technique is more powerful in modeling past time series movements, there is no guarantee that the forecasts generated will be more accurate than the projections of simpler time series techniques.

Chapter 9: Time Series Analysis

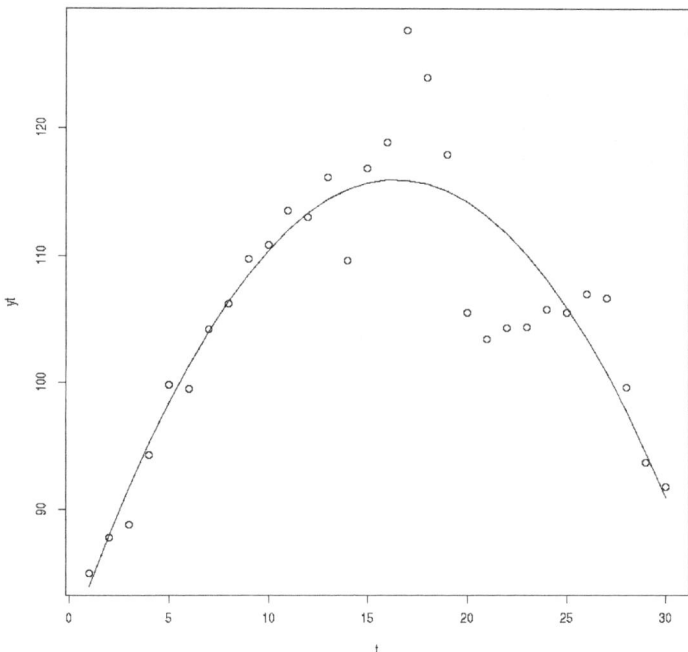

FIGURE 9.9.3 Time Series and Polynomial Trend for Example 9.9.2

A Complete Example As part of an effort to forecast future production, a consulting firm recorded the quarterly demand for golf balls for the years 1999–2002 shown in the table below:

Quarterly Demand for the Golf Balls

Time Period	Year	Quarter	Demand
1	1	1	1,000
2		2	2,500
3		3	3,000
4		4	4,000
5	2	1	10,000
6		2	12,000
7		3	15,000
8		4	14,000
9	3	1	20,000
10		2	25,000
11		3	40,000
12		4	50,000
13	4	1	65,000
14		2	75,000
15		3	80,000
16		4	100,000

(1) Calculate the four-quarter MA for the first quarter of 2003.
Forecast for the first quarter of 2003 = $\frac{65,000+75,000+80,000+100,000}{4} = 80,000$ balls

(2) Using a weighted five-quarter MA with the weights 0.45, 0.25, 0.15, 0.10 and 0.05 from most recent to old, respectively, calculate the forecast for the first quarter of 2003.
The forecast for the first quarter of 2003 is computed as

$$(0.05)(50,000)+(0.1)(65,000)+(0.15)(75,000)+(0.25)(80,000)+(0.45)(100,000) = 85,250 \text{ balls}$$

(3) Using a simple exponential smoothing model, calculate the forecast for the first quarter of 2003, given that the forecast for the last quarter of 2002 was 80,250 units. Use $\omega = 0.15$.
The exponentially smoothed forecast for the first quarter of 2003 is computed as

$$S_t = \omega y_t + (1 - \omega)S_{t-1} = 0.15(100,000) + 0.85(80,250) = 83,212.5$$

(4) Find the regression line and calculate the forecast for the first quarter of 2003.
By fitting a line using simple linear regression we get

$$y_t = 6,308.0882t - 21,337.5.$$

To calculate the first quarter 2003 demand, we merely substitute the t-value corresponding to the first quarter of 2003, which is 17.
The demand for first quarter of 2003 = $\hat{y}_{17} = (6,308.0882)(17) - 21,337.5 = 85,900$ balls

Question If the consulting firm feels that in addition to the time period, other factors such as price, season, location, and so on, effect the demand, what do you think they should do?

Answer: This question arises in most real-world problems. In fact, the inherent complexity of most real-world problems necessitates the consideration of more than one variable to predict the variate of interest. Here, we should use the multiple regression analysis discussed in Chapter 9.

A PRACTICE PROBLEM Table 9.9.3, presents Stephen Curry's free throw percentages for 10 seasons, 2008–2018. Use a 3-period MA, exponential smoothing, and regression analysis to forecast his percentage for the year of 2019, and compare that with his performance for that year.

9.10 ANALYSIS OF DATA FOR 100, 400, AND 800-M RUNS

This section considers a specific problem. The analysis presented may require knowledge beyond elementary concepts discussed so far. It provides useful material for students who may want to use software and also for those interested in research and a higher level of data analysis. Section 9.10.1 presents an advanced analysis and may be omitted for undergraduate teaching. Section 9.10.2 includes a more elementary presentation and discussion of the results. Section 9.10.3 presents similar results using data for the men's 100-m run and includes a comparison between the regression and a more advanced time series analysis.

TABLE 9.9.3 Stephen Curry's Free Throw Percentage 2008–2018

Year	Free Throw Percentage
2017–2018	0.926
2016–2017	0.926
2015–2016	0.899
2014–2015	0.909
2013–2014	0.894
2012–2013	0.885
2011–2012	0.903
2010–2011	0.809
2009–2010	0.934
2008–2009	0.885

9.10.1 Advanced Analysis (400 m and 800 m)

Athletic records and related issues have received considerable attention by physiologists and physical educators. One aspect is concerned with the improvement of the records over time in order to address the question of forecasting future records. Consider an event (e.g., the 400-m run) and a time series showing the records (the fastest time) for that event over a period of time. The series provides some information about the likely records for that event in the future.

Predicting future records is of great interest and there have been many attempts to forecast them via statistical modeling. Tryfos and Blackmore (1985) derived a method of generalized least square analysis, under the assumption that the records are derived from an underlying independent and identically distributed sequence of random variables. Smith (1988) has considered a more general approach based on a regression model of the form

$$y_t = Z(t,\theta) + x_t \tag{9.32}$$

where the x_t's are independent and identically distributed, $Z(t,\theta)$ is a non random term, and θ is the vector of parameters to be determined. For x_t, particular distributions considered were normal, Gumbel, and the generalized extreme value. For $Z(t,\theta)$ the linear, quadratic, and exponential-decay models are often examined. The form of linear model is

$$Z(t,\theta) = \theta_0 - \theta_1 t, \quad \theta_1 > 0 \tag{9.33}$$

The form of quadratic model is

$$Z(t,\theta) = \theta_0 - \theta_1 t + \frac{\theta_2}{2} t^2, \quad \theta_1 > 0 \tag{9.34}$$

and the form of exponential–decay model is

$$Z(t,\theta) = \theta_0 - \theta_1[1-(1-\theta_2)^t]/\theta_2, \quad \theta_1 > 0, \ 0 < \theta_2 < 1 \tag{9.35}$$

Equation (9.35) is equivalent to the model studied by Chatterjee and Chatterjee (1982) and Morton (1983). In fact, it is easy to see that (9.35) can be converted into the form

$$Z(t,\theta) = a - be^{-ct} \tag{9.36}$$

with $\theta = (a,b,c)$ in this case. Note that all these models are special cases of regression model discussed in Section 8.1.

Using the method of maximum likelihood (see Section 5.4) and numerical approximation procedures, these models were applied to the records for mile and marathon races (see Smith 1988, for details). The normal distribution for x_t was the most appropriate distributions for x_t.

In this section, we will also consider an innovative model composed of an envelope function and a stationary stochastic process. We will discuss the model and its advantages over the models mentioned earlier. We will also consider the problem of fitting and forecasting based on data consisting of a series of best yearly performance records for a particular event. Then statistical inference and examples will be considered. Data from the 400- and the 800-m races are used. Predictions based on models considered will also be presented.

We first remark that fitting a regression model of type (9.32) with independent and identically distributed errors to the time series of fastest times may neither be appropriate nor satisfying. First, these models imply no relation between variation in x_t and changes in $Z(t,\theta)$. In fact, it is reasonable to expect less variation in the latter portion of the data. This is because recent fastest times are closer to the ultimate record and significant improvements are less likely now than, say, 50 years ago. Second, these models do not consider the possibility of a correlation among the record times. In fact, since most world-class runners remain competitive for a number of years (usually between 3 and 6 years) it is reasonable to assume a correlation between the fastest times.

We regard the fastest times as a realization of a random process, and suggest application of the following models:

$$\log y_t = \theta_0 - \theta_1 t + x_t, \quad \theta_1 > 0 \tag{9.37}$$

or allowing for an extra parameter

$$\log y_t = \theta_0 - \theta_1 t + \theta_2 \log t + x_t, \quad \theta_1 > 0 \tag{9.38}$$

where $\{x_t, t=1,2,\ldots\}$ is a zero-mean stationary process. Note that the models (9.37) and (9.38) can alternatively be written in multiplicative form as

$$y_t = \bar{\theta}_0 e^{-\theta_1 t} \bar{x}_t, \quad \theta_1 > 0 \tag{9.39}$$

and

$$y_t = \bar{\theta}_0 t^{\theta_2} e^{-\theta_1 t} \bar{x}_t, \quad \theta_1 > 0 \tag{9.40}$$

where $\bar{\theta}_0 = e^{\theta_0}$ and $\bar{x}_t = e^{x_t}$. This implies that for these models both means and variances vary with time. In fact, unlike additive models where variance remains unchanged, here it decreases as t increases. As a result, the standard errors of the predicted records are smaller than those from additive models and therefore the likelihood of obtaining a meaningful prediction is higher.

Before discussing the statistical inference for a general linear model with stationary dependent residual x_t, let us first note that rather than a single universal model it is also possible to consider a piecewise linear model for running times. However, since, (1) such models contain an unnecessarily large number of parameters, (2) it is not clear how the points separating the lines can be determined, (3) only the line fitted to the last segment of data can be used for prediction, and (4) not a long series is available to guarantee a reasonable estimate for each section, we decided not to include them in our discussion.

Now let us suppose that an observed series $\{y_t; t = 1, 2, \ldots, N\}$ is generated by the regression model

$$y_t = \sum_{k=1}^{p} \theta_k Z_k(t) + x_t \tag{9.41}$$

where the θ_k's are unknown parameters and x_t is a zero-mean stationary process possessing a continuous spectrum (see Section 9.2). The models (9.33), (9.34), (9.37), and (9.38) are all special cases of (9.41). Let $y = (y_1, y_2, \ldots, y_N)^T$, $\theta = (\theta_1, \theta_2, \ldots, \theta_p)^T$, and Z^T be a matrix with $Z_k(t)$ as the entry in the k-th row and the t-th column. Also, let $\Sigma = \{\gamma(s-t); s, t = 1, 2, \ldots, N\}$ denote the autocovariance matrix of x_t. That is,

$$\Sigma = \begin{bmatrix} \gamma(0) & \gamma(1) & \cdots & \gamma(N-1) \\ \gamma(1) & \gamma(0) & \cdots & \gamma(N-2) \\ \cdots & & & \\ \gamma(N-1) & \gamma(N-2) & \cdots & \gamma(0) \end{bmatrix}$$

Then applying the least squares method it can be shown that the best linear unbiased estimator, also known as the generalized least square estimate, of θ is given by

$$\hat{\theta} = (Z^T \Sigma^{-1} Z)^{-1} Z^T \Sigma^{-1} y. \tag{9.42}$$

Note that for simple least squares, that is when x_t is independent and identically distributed, Σ is an identity matrix. In practice, Σ is often unknown and $\hat{\theta}$ is unavailable. Even if Σ is known its inversion introduces computational problems, especially for long series. To avert these difficulties a common approach is to replace $\hat{\theta}$ by the ordinary least squares estimator (OLS),

$$\tilde{\theta} = (Z^T Z)^{-1} Z^T y \tag{9.43}$$

which does not involve Σ. Since $\tilde{\theta}$ is easy to calculate the question is whether precision is lost by using it. It has been shown that loss of precision depends on $Z_k(t)$'s. In fact, it has been known for some time (Grenander, 1954) that there are certain cases where $\tilde{\theta}$ is efficient in the sense that

$$[var(\tilde{\theta})][var(\hat{\theta})]^{-1} \longrightarrow I_p \text{ as } N \longrightarrow \infty \tag{9.44}$$

where $var(\tilde{\theta})$, $var(\hat{\theta})$ are the covariance matrices of $\tilde{\theta}$ and $\hat{\theta}$, respectively, and I_p denotes the unit matrix of order p (a matrix whose ij-th entry represents the covariance between the i-th and j-th random variables). An important specific case where the required conditions are satisfied occurs when $\sum_{k=1}^{p} \theta_k Z_k(t)$ is a polynomial in t (Hannan, 1960, p.122). It can also be shown that the required conditions are still satisfied if $\log t$ is added to a polynomial. For both cases, the limiting form of $var(\tilde{\theta})$ and $var(\hat{\theta})$ is given by

$$V = 2\pi f(0)(Z^T Z)^{-1} \tag{9.45}$$

where $f(\omega)$ denotes the spectral density function of the x_t process.

Having obtained $\tilde{\theta}$, the inference concerning x_t can be carried out using classical methods based on fitting stationary time series models to the least squares residuals. Hannan (1970, Ch.VII) has shown that for linear models, with some regularity conditions on the design matrix (which are satisfied for models considered here), the estimated autocorrelations from the OLS residuals converge to the true autocorrelations for long time series. Further asymptotic efficiency of the two-stage estimation procedures described above is demonstrated in several publications, see for example, Kobayashi (1985) and Seber and Wild (1989, Ch.6).

In this section, data from the 400-m and 800-m races are used to illustrate an application of the proposed models. The fastest times for each race were recorded every year from 1860 to 1985, yielding a sample of size $n = 126$. These data are presented in Tables 9.10.1 and 9.10.2. See also Figure 9.10.1. In order to judge the models, we also considered the fastest times for the years 1986–1988. These additional times were not used for modeling, but for comparison with forecasts from alternative models. Each observation used for modeling is recorded in seconds.

The models (9.33), (9.34), (9.37), (9.38) were fitted to the data using ordinary least squares. The coefficient of determination, R^2, was calculated for each case. They are, respectively, $R^2 = 0.897$, 0.906, 0.903, and 0.917 for the 400-m race, and $R^2 = 0.903$, 0.909, 0.913, and 0.933 for the 800-m race. We note that the same data sets were used in Ballerini and Resnick (1987) to predict the rate of the future records. Next, the estimated residuals from each model were tested for randomness using the Portmanteau, turning points, difference-sign and rank test (for a description of these tests, see Brockwell and Davis, 1991) and for normality using Shapiro–Wilkes test. These tests are now part of software such as R, MINITAB, SAS, and SPSS. For the 400-m race the residuals from models (9.33), (9.34), and (9.37) showed significant departure from randomness and normality. The significant peaks in the ACF and partial autocorrelation function (PACF) motivated the trial fitting of AR or MA models to the estimated residuals. In each case, $AR(2)$ and $MA(4)$ processes were the best fit. The $AR(2)$ models had one significant coefficient and the $MA(4)$ models had two significant coefficients. We were guided in the selection of these models by significance of the coefficients and AICC statistics (see e.g. Brockwell and Davis 1991). The AICC statistics indicated that the $MA(4)$ process was better.

The residuals of the model (9.38) passed all the tests of randomness and normality easily. The ACF and PACF showed no significant peaks and the p-value of the Shapiro–Wilkes test was 0.4573 indicating a good fit. This, together with the highest coefficient of determination for this model and justification given earlier, confirms its appropriateness as a model for 400-m times.

For the 800-m race, the residuals were significantly non random in each case. Also, except for model (9.38), the residuals were significantly non normal. Significant peaks in the ACF and PACF plots guided the selection of AR and MA models for the residuals. For the models (9.33), (9.34), and (9.37), AR and MA models were fitted to the residuals that removed the non randomness but barely improved non normality. Note that the validity of inference and prediction from these models is dependent upon the assumption that the resultant residuals are normally distributed. The residuals of the model (9.38) failed the Portmanteau test of randomness at the 5% significance level, but the residuals were well fitted to a normal distribution. Significant peaks in the ACF and PACF plots led to the selection of an $AR(3)$ and $MA(4)$ model. These models removed the non randomness from the residuals and further conformed the residuals to a normal distribution. Again $MA(4)$ produced a better fit.

Chapter 9: Time Series Analysis

TABLE 9.10.1 Data For 400-m Run (Time is in Seconds): 1860–1988

Year	Time	Year	Time	Year	Time	Year	Time	Year	Time
1860	53.7	1861	50.2	1862	53.2	1863	51.7	1864	51.7
1865	50.2	1866	52.5	1867	51.4	1868	50.0	1869	51.9
1870	50.7	1871	50.2	1872	49.5	1873	50.3	1874	50.2
1875	50.5	1876	50.5	1877	50.1	1878	51.3	1879	48.9
1880	49.3	1881	48.3	1882	49.9	1883	49.0	1884	48.9
1885	48.5	1886	49.5	1887	49.9	1888	49.7	1889	48.2
1890	48.7	1891	49.1	1892	49.2	1893	48.9	1894	48.7
1895	48.2	1896	48.5	1897	48.7	1898	48.5	1899	49.1
1900	47.5	1901	49.3	1902	49.3	1903	48.7	1904	48.9
1905	48.2	1906	48.5	1907	48.5	1908	47.9	1909	48.3
1910	48.5	1911	48.5	1912	47.7	1913	46.9	1914	48.1
1915	47.7	1916	47.1	1917	48.7	1918	47.3	1919	48.9
1920	48.1	1921	47.7	1922	47.7	1923	47.9	1924	47.4
1925	47.6	1926	48.3	1927	47.5	1928	47.0	1929	47.4
1930	47.6	1931	47.1	1932	46.1	1933	46.6	1934	46.5
1935	46.8	1936	46.1	1937	46.6	1938	46.3	1939	46.0
1940	46.4	1941	46.0	1942	46.6	1943	47.5	1944	47.5
1945	46.7	1946	45.9	1947	45.9	1948	45.7	1949	46.2
1950	45.8	1951	46.0	1952	45.9	1953	45.9	1954	46.1
1955	45.4	1956	45.2	1957	46.0	1958	45.4	1959	45.8
1960	44.9	1961	45.7	1962	45.5	1963	44.6	1964	44.9
1965	45.5	1966	44.7	1967	44.5	1968	43.8	1969	44.4
1970	44.9	1971	44.2	1972	45.0	1973	45.2	1974	45.2
1975	44.93	1976	44.26	1977	45.36	1978	45.47	1979	44.00
1980	44.60	1981	45.12	1982	45.00	1983	45.44	1984	44.27
1985	44.96	1986	44.45	1987	44.32	1988	43.29		

Listed below are the fastest times of the 400-m run from 1986 to 1988 with the 95% prediction interval based on the model (9.38):

Year	Time	95% Prediction Interval
1986	44.45	(43.22, 45.56)
1987	44.32	(43.18, 45.52)
1988	43.29	(43.13, 45.47)

TABLE 9.10.2 Data For 800-m Run (Time is in Seconds): 1860–1988

Year	Time	Year	Time	Year	Time	Year	Time	Year	Time
1860	131.3	1861	122.3	1862	127.3	1863	122.3	1864	121.3
1865	121.8	1866	124.3	1867	120.3	1868	121.3	1869	121.1
1870	121.3	1871	122.0	1872	119.1	1873	119.0	1874	119.7
1875	119.5	1876	116.8	1877	118.9	1878	119.5	1879	118.5
1880	115.7	1881	114.9	1882	114.9	1883	115.3	1884	114.7
1885	114.7	1886	117.3	1887	116.1	1888	113.7	1889	114.6
1890	116.3	1891	113.8	1892	115.4	1893	114.6	1894	115.1
1895	112.7	1896	115.7	1897	113.7	1898	111.3	1899	116.7
1900	117.1	1901	114.1	1902	116.5	1903	114.9	1904	114.1
1905	115.3	1906	112.9	1907	114.5	1908	112.8	1909	112.1
1910	113.9	1911	113.5	1912	111.9	1913	113.7	1914	112.7
1915	112.9	1916	111.5	1917	115.2	1918	114.9	1919	112.5
1920	113.4	1921	114.1	1922	114.3	1923	113.6	1924	112.4
1925	112.7	1926	110.9	1927	112.8	1928	110.6	1929	111.5
1930	111.7	1931	111.9	1932	109.7	1933	110.2	1934	109.1
1935	111.3	1936	109.7	1937	108.9	1938	108.4	1939	106.6
1940	107.8	1941	109.0	1942	110.1	1943	108.9	1944	110.1
1945	109.3	1946	110.0	1947	108.7	1948	108.3	1949	109.6
1950	108.5	1951	108.9	1952	108.0	1953	107.9	1954	107.1
1955	105.7	1956	106.4	1957	105.8	1958	106.6	1959	106.2
1960	106.3	1961	106.4	1962	104.3	1963	106.1	1964	105.1
1965	106.3	1966	104.2	1967	105.0	1968	104.3	1969	105.2
1970	104.8	1971	104.7	1972	107.3	1973	105.6	1974	103.9
1975	104.9	1976	103.5	1977	104.0	1978	103.41	1979	102.4
1980	105.4	1981	101.73	1982	102.7	1983	103.65	1984	103.0
1985	104.01	1986	103.73	1987	102.64	1988	103.45		

The fastest times for these years all fall within the 95% prediction intervals. The new world record time, set in 1988, was more than 1 second less than the fastest time of the previous year, yet it still fell comfortably within the 95% prediction interval model for that year. Note that the breadth of the prediction intervals is not surprising as a sample size of 126 is just too small to expect narrow intervals. Moreover, the data include times set during the First and the Second world war, which are larger than expected times. Further comparison predicted values for each fitted model, for years 1986–1988 with the 95% prediction intervals, are presented in Table 9.10.3.

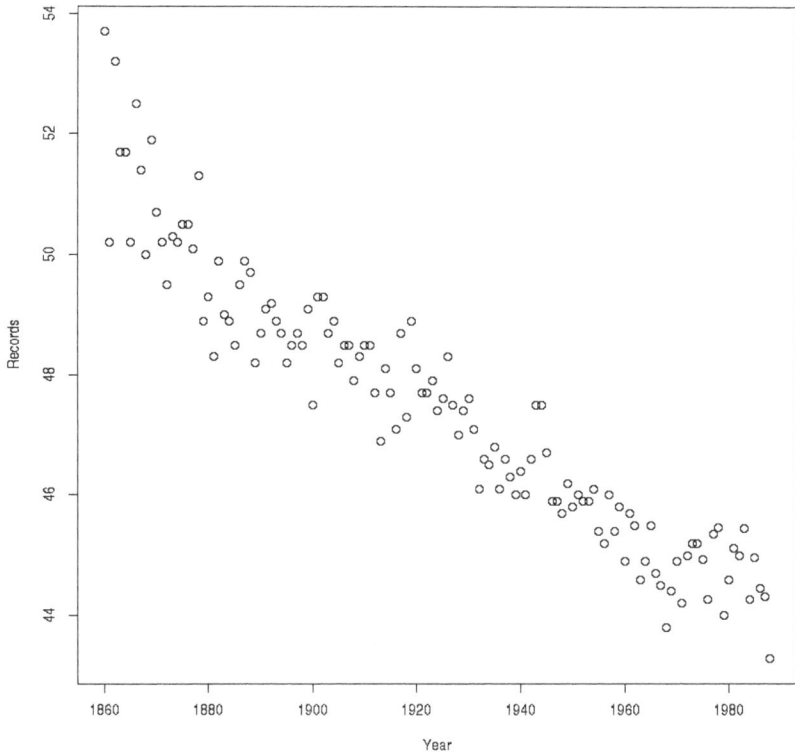

FIGURE 9.10.1 Mens' 400-m Run (1860–1988)

We note that the MSE (estimate for error variance) for model (9.38) is the smallest. Moreover, the prediction intervals based on this model are narrower than those for the other models.

Below are listed the fastest times for 1986 to 1988 and the 95% prediction intervals from the model for those years:

The world record of 101.73, set in 1981, also falls in these prediction intervals. Table 9.10.4 is similar to Table 9.10.3 and presents information for the 800-m data, using different models.

TABLE 9.10.3 Prediction For 400-m Run

Year	Model (9.33)		Model (9.34)	
1986	44.024	(42.631, 45.412)	44.492	(43.131, 45.854)
1987	43.968	(42.575, 45.361)	44.459	(43.094, 45.823)
1988	43.913	(42.519, 45.306)	44.426	(43.058, 45.793)
Year	Model (9.37)		Model (9.38)	
1986	44.120	(42.896, 45.380)	44.376	(43.220, 45.562)
1987	44.068	(42.845, 45.327)	44.332	(43.177, 45.518)
1988	44.018	(42.795, 45.275)	44.288	(43.134, 45.474)

TABLE 9.10.4 Prediction For 800-m Run

Year	Model (9.33)		Model (9.34)	
1986	102.387	(98.752,106.021)	103.389	(99.779,106.998)
1987	102.236	(98.600,105.871)	103.286	(99.668,106.903)
1988	102.085	(98.448,105.722)	103.183	(99.557,106.809)
Year	Model (9.37)		Model (9.38)	
1986	102.682	(99.592,105.869)	103.516	(100.750,106.359)
1987	102.545	(99.456,105.728)	103.405	(100.639,106.245)
1988	102.406	(99.322,105.588)	103.292	(100.528,106.133)

TABLE 9.10.5 Predictions For Years 1989–95 Based on Model (9.10.7)

Year	Fit (400 m)	95% Prediction Interval	Fit (800 m)	95% Prediction Interval
1989	44.245	(43.090,45.431)	103.273	(100.418,106.020)
1990	44.202	(43.047,45.387)	103.069	(100.306,105.907)
1991	44.158	(43.004,45.343)	102.958	(100.196,105.795)
1992	44.115	(42.961,45.300)	102.847	(100.087,105.684)
1993	44.072	(42.918,45.257)	102.736	(99.977,105.572)
1994	44.029	(42.875,42.213)	102.626	(99.867,105.460)
1995	43.986	(42.833,45.170)	102.515	(99.757,105.349)

Year	Time	95% Prediction Interval
1986	103.73	(100.75, 106.36)
1987	102.64	(100.64, 106.25)
1988	103.45	(100.53, 106.13)

Finally Table 9.10.5 lists the 95% prediction intervals from model (9.38) for the years 1989 to 1995 both for the 400- and 800-m runs. Predictions for more recent years, and their comparison with actual times, are left as excessive for students.

As a conclusion, we see that of the models considered, (9.38) gave the best fit to the annual fastest running times for the 400- and 800-m events. The 400-m data was well fitted using only the non-random part of the model (9.38). After fitting the model, the residuals were random and normal. The prediction intervals for the model embraces the recent fastest times that include a new world record. The 800-m data also fit well by the same model with a MA(4) process used for modeling the residuals. Final residuals were random and normal. The prediction intervals of the model captures the fastest times of the recent year. Compared with the other models, model (9.38) provided a smaller MSE and a narrower interval for predicted times of the years 1986 to 1988, both for the 400- and 800-m data.

Although model (9.38) fits both data, the residuals have a different structure. This could be due to a physiological difference between these two events. In fact, it is well-known that various athletic events require specific demand for energy production and that many events require a combination of both a high- and low-power output. The different energy demands of skeletal muscle tissue in response to various competitive events are met by the interaction of three "energy producing systems" (see, e.g., Fox 1979 and references therein). The first system is referred to as the ATP-PC Phosphagen System. This system's function is to supply short-term energy to working muscle tissue. If one was to sprint 100- or 400-m much of the phosphagen stores would be depleted. The second energy system, the Lactic Acid of Anaerobic Glycolysis System, has the ability to supply energy to skeletal muscle in the absence of oxygen. This system has the capacity to provide working muscle with energy for between 1 and 3 min during maximal exercise. Events such as the 400 and 800 meters rely heavily upon this system for energy production. The third system, the Oxygen, or Aerobic System, complements the Phosphagen and Lactate Systems. Athletic events requiring maximal effort lasting longer than 3 to 4 min will require the utilization of this third system in addition to the first two to supply energy to working muscle.

Given the time course of the 400- and 800-m events and the energy systems necessary for working muscle tissue to function at these maximal levels, it becomes evident that a combination of these energy systems provide the energy for competition.

From a physiological perspective, the difference between the 400 and 800 m centers around the energy systems required to complete the event. As the 400 m has become a sprint event, it calls upon the ATP-PC Phosphagen System and the Anaerobic/Lactic Acid System for energy production. The 800 m is longer in duration, therefore it is requiring the interaction of all three systems including the Oxygen or Aerobic System.

9.10.2 Elementary Analysis Using Minitab (400 m)

Regression Analysis

We start by applying regression using four classical trend models. MINITAB is used in each case.

- **Linear Trend**

    First we fit a simple linear trend equation. Using MINITAB we have

    $$y_t = 51.0947 - 0.0556t$$

    If chosen as the best model, this equation will be used for forecasting the best record for a particular year, in the future. MINITAB calculates several different accuracy measures. Some of these measures are given below.

    Accuracy Measures: MAPE: 1.07745, MAD: 0.516052, MSD: 0.466114

    The values for the MAPE, MAD, and mean squared deviation (MSD) are all used in the process of determining the best model. Because the MSD involves the sample size N, it can be used for comparison across models and is the best measure when determining which of the four trend models best fits the data.

- **Quadratic Trend**

    The fitted quadratic trend equation is

    $$y_t = 51.5346 - 0.0758t = 0.000155t^2$$

    If chosen as the best model, this equation will be used to forecast future records or best times for a particular year. For this model we have

    Accuracy Measures: MAPE: 1.05051,   MAD: 0.503297,   MSD: 0.429178

- **Exponential Growth Trend**

    The fitted trend equation is

    $$y_t = 51.1694(.998833)^t$$

    For this model we have
    Accuracy Measures: MAPE: 1.06464,   MAD: 0.510121,   MSD: 0.452269

- **S-Curve Trend analysis:**

    The fitted trend equation is

    $$y_t = 1000/(10.8013 + 8.7976(1.00238)^{t-1})$$

    For this model we have
    Accuracy Measures: MAPE: 1.07346,   MAD: 0.514479,   MSD: 0.468175

Conclusions

According to the MSD, among the models considered, the model that produces the best fit is the quadratic trend. This model, with the lowest MSD = 0.42918, also has relatively low MAPE and MAD values. However, because it is unlikely that future records will continue to be broken with a significant amount, the S-curve trend should also be examined further. This model is similar to the well-known logistic equation used frequently for modeling population growth. It has a great advantage over other models since it assumes that records have a non zero lower bound and as we approach this bound the amount by which records may be broken decreases.

In order to examine the trend with the best fit, in this case the quadratic, a regression analysis using this model could be performed. This would allow one to fully predict future records through the use of confidence and prediction intervals.

Chapter 9: Time Series Analysis

Time Series Analysis

We now turn to analysis of the data as a time series. Here, we only present results for two of the smoothing techniques discussed. Recall that when the form of the trend is known, it is better to apply regression. However, when this is not the case, smoothing techniques may be applied.

- **Double Exponential Smoothing**

Smoothing Constants: Omega (level): 0.827470,   Gamma (trend): 0.023628

Accuracy Measures: MAPE: 1.29774,   MAD: 0.62437,   MSD: 0.72096

Values for the MAPE, MAD, and MSD remain low which suggests that this model is accurate and produces a good fit.

Row	Period	forecast	Lower	Upper
1	130	43.3890	41.8593	44.9187

According to this method, the best time for the 400-m run in 1989 would be roughly 43.3890 ≈ 43.39. With 95% confidence, the upper and lower limits of time for year 1989 are, respectively, 41.86 and 44.92. The interval includes the actual time 44.27.

- **Moving Average**

MA Length: 1

Accuracy Measures: MAPE: 1.34974,   MAD: 0.64633,   MSD: 0.7448

Although the values of the MAPE, MAD, and MSD have all greater than those for the double exponential smoothing, they are still rather small, which indicates that this model is fairly accurate and could be used for predictions.

Row	Period	Forecast	Lower	Upper
1	130	43.29	41.5985	44.9815

According to this method, the best time for the 400-m run in 1989 would be roughly 43.29. With 95% confidence, the upper and lower limits of this interval are those given below.

It should be pointed out that the MA of length 1 basically is the same as the original data. Here, we also considered 3- and 5-year MAs and used them for prediction. Table 9.10.6 presents predictions for the years 1989–2038 from the different models and methods discussed above. The actual times for the years 1989–2000 are also included for comparison. We can use these, for example, to find the sums of squares of the differences between the predicted and actual times for each model. The one with the smallest sum of squares may be considered the best. We leave this for students to try.

TABLE 9.10.6 Prediction for 1989-2039 Using Different Models (400m)

Year	Moving Average		Exponential		Linear	Quadratic	Exponential Growth	S-Curve	Actual Results
	Length 3	Length 5	Single	Double					
1989	44.02	44.258	44.172	43.385	43.861	44.301	43.961	43.902	44.27
1990	44.02	44.258	44.172	43.311	43.806	44.266	43.910	43.847	44.06
1991	44.02	44.258	44.172	43.237	43.750	44.231	43.858	43.792	44.17
1992	44.02	44.258	44.172	43.163	43.695	44.196	43.807	43.737	43.50
1993	44.02	44.258	44.172	43.090	43.639	44.162	43.757	43.683	43.65
1994	44 02	44.258	44.172	43.016	43.583	44.128	43.705	43.626	43.90
1995	44.02	44.258	44.172	42.942	43.528	44.094	43.654	43.573	43.39
1996	44.02	44.258	44.172	42.868	43.472	44.060	43.603	43.518	43.44
1997	44.02	44.258	44.172	42.795	43.416	44.027	43.552	43.464	43.75
1998	44.02	44.258	44.172	42.729	43.361	44.994	43.501	43.409	43.68
1999	44.02	44.258	44.172	42.647	43.305	43.962	43.451	43.354	43.18
2000	44.02	44.258	44.172	42.573	43.249	43.930	43.400	43.300	43.68
2001	44.02	44.258	44.172	42.500	43.194	43.898	43.349	43.245	
2002	44.02	44.258	44.172	42.426	43.138	43.866	43.299	43.190	
2003	44.02	44.258	44.172	42.352	43.082	43.835	43.248	43.136	
2004	44.02	44.258	44.172	42.279	43.027	43.804	43.198	43.081	
2005	44.02	44.258	44.172	42.204	42.971	43.773	43.147	43.026	
2006	44.02	44.258	44.172	42.131	42.916	43.743	43.097	42.972	
2007	44.02	44.258	44.172	42.057	42.860	43.713	43.046	42.917	
2008	44.02	44.258	44.172	41.983	42.804	43.683	42.996	42.862	
2009	44.02	44.258	44.172	41.909	42.749	43.653	42.946	42.808	
2010	44.02	44.258	44.172	41.836	42.693	43.624	42.896	42.753	
2011	44.02	44.258	44.172	41.762	42.637	43.595	42.846	42.699	
2012	44.02	44.258	44.172	41.688	42.582	43.567	42.796	42.644	
2013	44.02	44.258	44.172	41.614	42.526	43.539	42.746	42.589	
2014	44.02	44.258	44.172	41.540	42.470	43.511	42.696	42.535	
2015	44.02	44.258	44.172	41.467	42.415	43.483	42.646	42.480	
2016	44.02	44.258	44.172	41.393	42.359	43.456	42.596	42.426	
2017	44.02	44.258	44.172	41.319	42.303	43.429	42.547	42.371	
2018	44.02	44.258	44.172	41.245	42.248	43.402	42.497	42.317	
2019	44.02	44.258	44.172	41.172	42.192	43.376	42.447	42.262	
2020	44.02	44.258	44.172	41.098	42.137	43.350	42.398	42.208	
2021	44.02	44.258	44.172	41.024	42.081	43.324	42.348	42.153	
2022	44.02	44.258	44.172	40.950	42.025	43.299	42.299	42.099	
2023	44.02	44.258	44.172	40.876	41.970	43.274	42.249	42.044	
2024	44.02	44.258	44.172	40.803	41.914	43.249	42.200	41.990	
2025	44.02	44.258	44.172	40.729	41.858	43.224	42.151	41.935	
2026	44.02	44.258	44.172	40.655	41.803	43.200	42.102	41.881	
2027	44.02	44.258	44.172	40.581	41.747	43.176	42.053	41.826	
2028	44.02	44.258	44.172	40.508	41.691	43.153	42.003	41.772	
2029	44.02	44.258	44.172	40.434	41.638	43.320	41.954	41.718	
2030	44.02	44.258	44.172	40.360	41.580	43.106	41.905	41.663	
2031	44.02	44.258	44.172	40.286	41.524	43.084	41.857	41.609	
2032	44.02	44.258	44.172	40.212	41.469	43.062	41.808	41.554	
2033	44.02	44.258	44.172	40.139	41.413	43.0405	41.75	41.500	
2034	44.02	44.258	44.172	40.065	41.358	43.018	41.710	41.446	
2035	44.02	44.258	44.172	39.991	41.302	42.996	41.661	41.391	
2036	44.02	44.258	44.172	39.917	71.246	42.975	41.613	41.337	
2037	44.02	44.258	44.172	39.844	41.191	42.955	41.564	41.283	
2038	44.02	44.258	44.172	39.770	41.135	42.934	41.516	41.229	

Chapter 9: Time Series Analysis

9.10.3 Time Series Analysis of the Men's 100-m Run

In this section, methods for analyzing time series are applied to the men's 100-m annual records, which was discussed in Chapter 8. MINITAB is used and results are summarized below. For data and detailed regression analysis of it see Section 8.6.5.

Choosing a Method for Analysis

The annual records are the same as the one used for regression analysis in Section 8.6.1. First data were plotted to help determine the type of trend and seasonal pattern. The plot appears to have a linear trend with no seasonal component. The goal is to predict the winning time for the next year, a short-term forecast, and the ultimate record (a long-term forecast), if possible.

Trend Analysis

Trend analysis is a useful tool for both short- and long-term forecast. Here, we want to forecast the next 50 years. MINITAB is used to fit linear, quadratic, exponential growth, and S-curve trend models for this purpose. To decide which of these four models provide a better fit, we looked at the MAPE, the MAD, and the MSD. The MSD (0.000864) is smallest for the quadratic model. The MAPE (0.249445) and MAD (0.024789) are also relatively small. Clearly, we cannot predict the ultimate record using a quadratic trend model. We assume that there will be an ultimate record because it is impossible to run 100 m in 0 sec. The quadratic model, predicts the record time for the year 2049 as 9.04175 sec.

Double Exponential Smoothing

This method is usually used for short-term projections. We chose to forecast the next 5 years. Use of double exponential smoothing provides a 5-year forecast and their respective confidence intervals. The annual record for the year 2000 is predicted as 9.81810 seconds with a 95% prediction interval of $(9.75954, 9.91379) \approx (9.76, 9.91)$.

MA

Moving averages are also good for short-term forecasts. We chose a MA of length 3 to smooth the data and to forecast the next year and perhaps the years after. The MAPE, MAD, and MSD are still low that indicates that the moving model is pretty accurate. The annual record for the year 2000 is predicted as 9.83667 sec.

Autocorrelation and Partial Autocorrelation

The autocorrelation and partial autocorrelation plots are used to choose AR and MA components for the ARIMA and to determine whether or not the data are stationary. They are also used to determine whether differencing is needed remove the non stationarity and to make the data stationary.

ARIMA Modeling

Stationary Data

To determine whether the data were stationary, we checked the mean and variances to see if they are constant. Their plot suggests that the data have a negative slope. The autocorrelation plot has two of the six spikes beyond the 95% confidence interval lines. This suggests that the data are not stationary and need to be transformed using differencing. We therefore applied one-step differencing. To see if the difference data were stationary, we did a time series plot and an autocorrelation plot for the difference data. The time series plot appears to be predominately horizontal. The difference autocorrelation plot has only 1 spike beyond the 95% confidence interval. These factors allowed us to conclude that the difference data were stationary.

AR and MA Parameters

The autocorrelation of the difference data (ACF) was plotted. The partial autocorrelation of the difference data (pact) was plotted too. A moving average component is indicated in these two plots. A moving average parameter of 1 is suggested in the acf since 1 spike is beyond the 95% confidence interval limits. A moving average parameter of 2 is suggested in the pact plot because 2 spikes are beyond this level.

Performing an ARIMA Analysis

Based on the analysis, we chose the models (0,1,1), (0,1,2), and (2,1,0). We rejected the ARIMA model (0,1,2) because the chi-square statistic (5.7) of lag 12 was significant, showing the residuals are correlated. So, we looked at the mean square (MS) values to find the best fit. The MS for ARIMA (0,1,1) is 0.001062, for ARIMA (1,1,0) is 0.0016823, and for ARIMA (2,1,0) is 0.0012731. Comparing these, we found that the MS for ARIMA (0,1,1) was the smallest. So, we chose this one as the best-fit model.

Forecasting Using the ARIMA Model

Forecasts, intervals, residuals, and fits were obtained together with 15 annual forecasts for the ARIMA (0,1,1) model. The forecast for the year 2000 is 9.82491 sec.

Regression Versus Time Series Forecasts for the Year 2000

We now compare results from regression and times series analysis. There are four forecasts. One from linear regression, one from double exponential smoothing, one from the moving average, and one from the best ARIMA model, namely (0,1,1). Listed below are the 95% prediction intervals for the year 2000's annual record, in seconds

    Regression: (9.7953, 9.8481)
    Double Exponential: (9.74888, 9.88731)
    MA: (9.75954, 9.91379)
    ARIMA (0,1,1) Model: (9.76101, 9.88881)

Chapter 9: Time Series Analysis

The 95% prediction interval for the regression model is narrower than the rest. The range is only 0.0528 (i.e., 9.7953 + 0.0528 = 9.8481). The ranges for the other models are 0.13843, 0.15425, and 0.1278, respectively.

9.11 USING R

We will demonstrate time series analysis using R in this section. R defined an object to handle time series data, one can create the object by applying function **ts()** to a vector. For example,

```
dat = c(49, 66, 25, 73, 40, 66, 57, 43, 61, 50)
dat.ts = ts(dat)
```

Here the object dat.ts will be a time series object containing data declared in vector dat.

There are functions defined in R to calculate fundamental results about the time series, such as autocovariance and autocorrelation.

```
# Autocorrelation
c = acf(dat.ts)
# Autocovariance
K = acf(dat.ts, type = c(''covariance''))
```

You should be able to see following outputs

```
> c
```

Autocorrelations of series dat.ts, by lag

0	1	2	3	4	5	6	7	8	9
1.000	−0.790	0.462	−0.164	−0.123	0.253	−0.227	0.120	−0.037	0.006

```
> K
```

Autocovariances of series dat.ts, by lag

0	1	2	3	4	5	6	7	8	9
189.6	−149.7	87.6	−31.1	−23.4	47.9	−43.0	22.8	−7.1	1.2

Recall that in the Example 9.8.1, we applied exponential smooth to time series 3, 2, 1, 4, 6, 5. Besides built-in functions, it is not hard to code the algorithm by yourself.

```
# Time series data
y = c(3, 2, 1, 4, 6, 5)
# Smooth
S = c()
# S1 = y1
```

```
S[1] = y[1]
# Omega value in the model, here we use w=0.5
w = 0.5
for(i in 2:length(y)){
        S[i] = w*y[i] + (1-w)*S[i-1]
}
```

The new created vector S is exponential smooth of time series y.

One can draw the plot of a given time series simply using **plot.ts()**, which is a variant of plot() function for time series. Therefore you may imagine the parameter in plot.ts() should be a time series object rather than regular vector(s). For example,

```
# Please select Table.9.10.1.csv for 400M run data
run400 = read.csv(file.choose())
run400ts = ts(run400$Record)
plot.ts(run400ts)
```

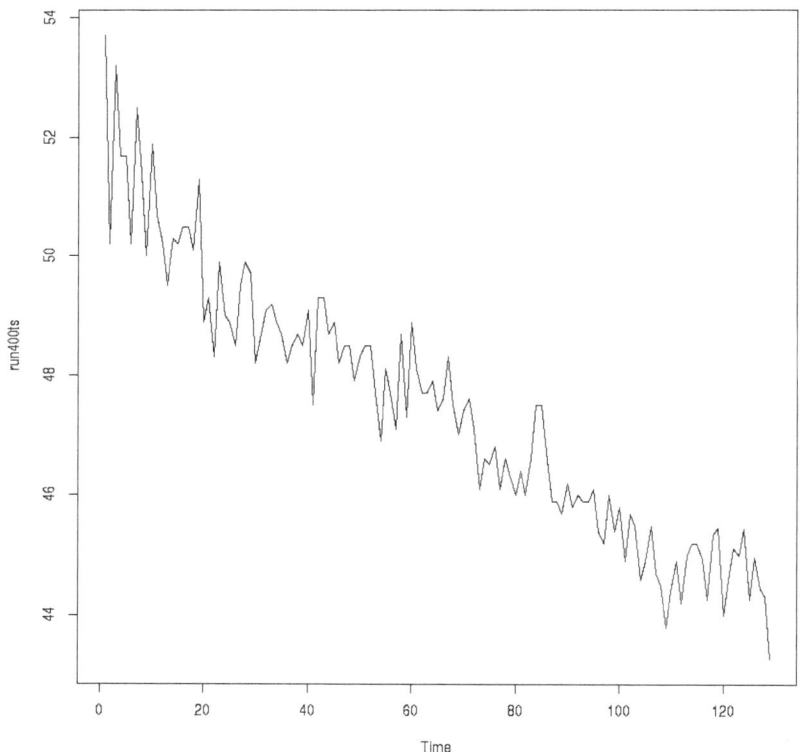

Next, let us focus on another data set: the number of births per month in New York city, from January 1946 to December 1959 (originally collected by Newton). As usual, we need to read it to R first.

```
# Please select nybirths.dat file
births = scan(file.choose())
```

Chapter 9: Time Series Analysis

```
# Convert to ts object, start with year 1946 quarter 1
births.ts <- ts(births, frequency=12, start=c(1946,1))
plot.ts(births.ts)
```

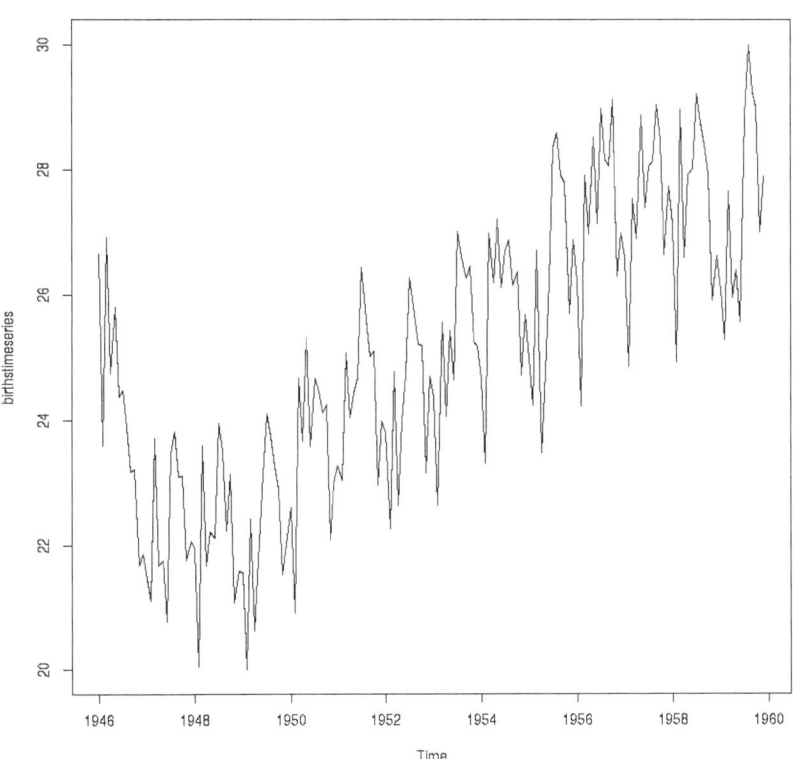

Well we found seasonal component in this time series. So we will apply decompose() function to data.

```
births.ts.comp <- decompose(births.ts)
births.ts.comp$seasonal
```

We obtain seasonal component as below.

```
> births.ts.comp$seasonal
      Jan        Feb        Mar        Apr        May        Jun        Jul        Aug        Sep        Oct        Nov        Dec
1946 -0.6771947 -2.0829607  0.8625232 -0.8016787  0.2516514 -0.1532556
      1.4560457  1.1645938  0.6916162  0.7752444 -1.1097652 -0.3768197
1947 -0.6771947 -2.0829607  0.8625232 -0.8016787  0.2516514 -0.1532556
      1.4560457  1.1645938  0.6916162  0.7752444 -1.1097652 -0.3768197
1948 -0.6771947 -2.0829607  0.8625232 -0.8016787  0.2516514 -0.1532556
      1.4560457  1.1645938  0.6916162  0.7752444 -1.1097652 -0.3768197
1949 -0.6771947 -2.0829607  0.8625232 -0.8016787  0.2516514 -0.1532556
      1.4560457  1.1645938  0.6916162  0.7752444 -1.1097652 -0.3768197
```

```
1950  -0.6771947  -2.0829607   0.8625232  -0.8016787   0.2516514  -0.1532556
       1.4560457   1.1645938   0.6916162   0.7752444  -1.1097652  -0.3768197
1951  -0.6771947  -2.0829607   0.8625232  -0.8016787   0.2516514  -0.1532556
       1.4560457   1.1645938   0.6916162   0.7752444  -1.1097652  -0.3768197
1952  -0.6771947  -2.0829607   0.8625232  -0.8016787   0.2516514  -0.1532556
       1.4560457   1.1645938   0.6916162   0.7752444  -1.1097652  -0.3768197
1953  -0.6771947  -2.0829607   0.8625232  -0.8016787   0.2516514  -0.1532556
       1.4560457   1.1645938   0.6916162   0.7752444  -1.1097652  -0.3768197
1954  -0.6771947  -2.0829607   0.8625232  -0.8016787   0.2516514  -0.1532556
       1.4560457   1.1645938   0.6916162   0.7752444  -1.1097652  -0.3768197
1955  -0.6771947  -2.0829607   0.8625232  -0.8016787   0.2516514  -0.1532556
       1.4560457   1.1645938   0.6916162   0.7752444  -1.1097652  -0.3768197
1956  -0.6771947  -2.0829607   0.8625232  -0.8016787   0.2516514  -0.1532556
       1.4560457   1.1645938   0.6916162   0.7752444  -1.1097652  -0.3768197
1957  -0.6771947  -2.0829607   0.8625232  -0.8016787   0.2516514  -0.1532556
       1.4560457   1.1645938   0.6916162   0.7752444  -1.1097652  -0.3768197
1958  -0.6771947  -2.0829607   0.8625232  -0.8016787   0.2516514  -0.1532556
       1.4560457   1.1645938   0.6916162   0.7752444  -1.1097652  -0.3768197
1959  -0.6771947  -2.0829607   0.8625232  -0.8016787   0.2516514  -0.1532556
       1.4560457   1.1645938   0.6916162   0.7752444  -1.1097652  -0.3768197
```

Oops, it looks messy, let's make a graph.

plot(births.**ts**.comp)

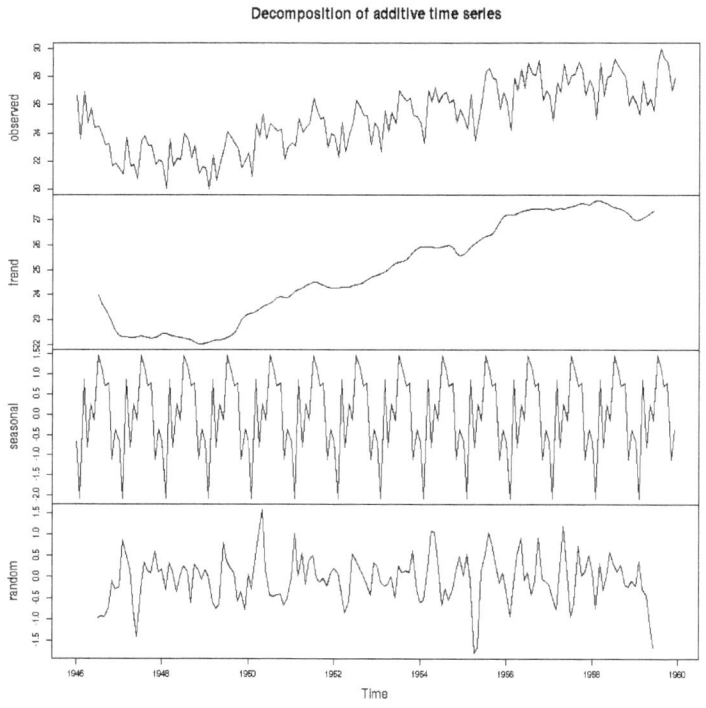

REFERENCES

Ballerini, R., and S. I. Resnick. 1987. "Embedding Sequences of Successive Maxima in Extremal Processes, with Applications." *Journal of Applied Probability* 24:827–37.

Box, G. E. P., and G. M. Jenkins. 1981. *Time Series Analysis: Forecasting and Control*. San Francisco: Holden-Day.

Brockwell, P. J., and R. A. Davis. 1991. *Time Series: Theory and Methods*. Springer.

Chatterjee, S. and S. Chatterjee .1982. "New Lamps for Old: An Exploratory Analysis of Running Times in Olympic Games." *Applied Statistics* 31:14–22.

Fox, E. L. 1979. *Sports Physiology*. Saunders College Publishing.

Grenander, U., and M. Rosenblatt. 1954. "Regression Analysis of Time Series with Stationary Residuals." *Proceedings of the National Academy of Sciences of the United States of America* 40(9):812–6.

Hannan, E. J. 1960. *Time Series Analysis*. London/NY.

Hannan, E. J. 1970. *Multiple Time Series*. John Wiley and Sons, Inc.

Kobayashi, K. 1985. "Cohort Life Tables based on Annual Life Tables for the Japanese Nationals Covering the Years 189101982." NUPRI Research Paper Series no. 23.

Morton F. I. 1983. "Operational Estimates of Areal Evapotranspiration and Their Significance to the Science and Practice of Hydrology." *Journal of Hydrology* 66:1–76.

Seber, G. A. F., and C. J. Wild. 1989. *Nonlinear Regression*. New York: Wiley.

Smith, P. J. 1988. "Asymptotic Properties of Linear Regression Estimators Under a Fixed Censorship Model." *Australian Journal of Statistics* 30(1):52–66.

Tryfos, P., and R. Blackmore. 1985. "Forecasting Records." *Journal of the American Statistical Association* 80:46–50.

CHAPTER 10

Nonparametric Statistics

In Chapters 5 through 9, we presented statistical inference for data that were classified as numerical (quantitative) or categorical (qualitative). There is still an another type of data known as ranked data. The statistical methods developed for this type of data are called nonparametric or distribution-free. Although, we do not intend to discuss these methods in a great detail, it is useful to know something about them and the way they work.

Chapters 5 through 9 noticed that most of the statistical methods discussed in earlier chapters are valid for normally distributed random variables or situations where the central limit theorem is applicable. In practice, however, there are many cases where the use of normal or any other known distribution cannot be justified. In such a situation, we can apply nonparametric technique that are valid regardless of the form of the underlying distribution. For this reason, nonparametric procedures are also called distribution-free. Note that theoretically these techniques are weaker than parametric methods, but then they require less assumptions for their valid application.

Nonparametric methods rely on the properties of the order statistics, that is data arranged in ascending (descending) order. In Chapter 1, we talked about finding the median by ordering the data and selecting the observation that falls in the middle. In fact, in nonparametric statistics the median is used as a measure for central tendency or location and for tests regarding the population location.

In general, nonparametric methods can be applied to both nominal (e.g., classification of students according to their major) and ordinal (e.g., classification of students according to their letter grade) data. They are usually easier to understand and compute than their parametric counterparts and are used to test hypotheses that do not involve population parameters. However, they are usually less efficient and less sensitive than their parametric counterparts.

Sections 10.1 to 10.4 present some theoretical background regarding order statistics. These sections may be ignored by students who are interested only in nonparametric methodologies. Section 10.5 includes distribution-free methods for estimation and testing. Finally, Section 10.6 presents some applications of the methods discussed.

10.1 ORDER STATISTICS, RANKING THE BEST

Before presenting the formal definition of order statistics, we present a problem related to order statistics that arises in sports. It seems that human beings have a basic urge to rank individuals, groups, and objects. People like to know who is the best, who is the second best, and so on, in any area of human endeavor. Clearly, nowhere ranking is more important to the general public than in sports. It is a controversial subject involving a never-ending dispute. Is there any reasonable way to rank the players and the teams? More importantly, is there any need for this? One problem with ranking is that, attempts to utilize complicated mathematical formulae arouse suspicion, because sports fans cannot follow the reasoning behind their developments. So far two simple ways of producing sports ranking has been (1) by adding up points for every tournament and (2) by applying the law of averages. To see how difficulties may arise, and what type of the problem we face consider the ranking summarized in the following table:

Team Name	Games Played	Won	Tie	Loss	Points
Jim	20	12	2	6	38
John	20	11	3	6	36
Mike	20	9	3	8	30
Mark	20	8	1	11	25
Fred	20	7	2	11	23
Bob	20	4	5	11	17

The table looks like results of a tournament, but in fact, it is simply the result of drawing balls from a box with 5 balls numbered 1, 2, 3, 4, and 5. Here, each player (team) draws a ball from the box with replacement 20 times. Drawing 4 or 5 is a win. Drawing 1 or 2 is a loss and drawing 3 is tie. Win, tie, and loss are, respectively, 3, 1, and 0 points. Thus, the object of ranking is to see if a table like this is produced purely by chance or in reality some players or teams are better than the others. In this section, we study order statistics related to a random sample from a distribution and use the result to decide if ranking presents something more than randomness.

As mentioned, the order statistics plays an important role in statistics, especially in non parametric or distribution-free part of that. Let $X_1, X_2, \ldots, X_n$ be a random sample from a probability density function $f_X(x, \theta)$ where θ denotes a vector of unknown parameters (note that here we are using n not N for the sample size). Suppose now that the n observations are arranged in ascending order so that $X_{(1)} \leq X_{(2)} \leq \ldots \leq X_{(n)}$, where $X_{(1)}$ is the smallest observation and $X_{(n)}$ is the largest. Note that this notation is different from the one used in Chapter 1. $X_{(1)}$ is called the first-order statistics, and $X_{(n)}$ is called the last or the n-th order statistics. In general, $X_{(i)}$ is called the i-th order statistics. In sport, such as men's 100-m run, data constitute the best times or records set by different athletes and observations $X_{(1)} \leq X_{(2)} \leq \ldots \leq X_{(n)}$, are already ordered. Thus, the first observation, $X_{(1)}$, is the smallest time or the present record. For men's 100 m in the period 2000 to 2018 the best times were, respectively, 9.58, 9.63, 9.69, 9.74, 9.76, 9.77, 9.78, 9.79, 9.80, 9.82, and 9.86. This is the order statistics and we have $X_{(1)} = 9.58$, $X_{(2)} = 9.63$, ..., and $X_{(11)} = 9.86$.

10.2 DISTRIBUTION OF THE *i*-TH ORDER STATISTICS

Consider an order statistic $X_{(i)}$ that has arisen from a random variable with probability density $f_X(x, \theta)$ and distribution function $F_X(x, \theta)$. Let us assume that n observations have been recorded,

Chapter 10: Nonparametric Statistics

and we want to find the probability density function of $X_{(i)}$ say $f_{X_{(i)}}(x,\theta)$. Consider the event E that the i-th ordered observation $X_{(i)}$ lies between x and $x+dx$, and from the rest of observations, $i-1$ are less than x and $n-i$ are greater than $x+dx$. Noting that dividing n observations into three groups including $i-1$, 1, and $n-i$ observations can be done in $n!/(i-1)!1!(n-i)!$ many different ways and that $P(X < x) = F_X(x)$ and $P(X > x) = 1 - F_X(x)$, we have

$$P(E) = P(x \leq X_{(i)} \leq x+dx) = \frac{n!}{(i-1)!1!(n-i)!}[F_X(x)]^{i-1}f_X(x)dx[1-F_X(x)]^{n-i}. \tag{10.1}$$

In the limit, as $dx \longrightarrow 0$ we obtain

$$f_{X_{(i)}}(x) = i\binom{n}{i}[F_X(x)]^{i-1}[1-F_X(x)]^{n-i}f_X(x). \tag{10.2}$$

For $i=1$, $f_{X_{(1)}}(x)$ gives the probability density function of the first (smallest) order statistic, namely

$$f_{X_{(1)}}(x) = n[1-F_X(x)]^{n-1}f_X(x). \tag{10.3}$$

For $i=n$, $f_{X_{(n)}}(x)$ gives the probability density function of the last (largest) order statistics as

$$f_{X_{(n)}}(x) = n[F_X(x)]^{n-1}f_X(x). \tag{10.4}$$

Clearly in most sport-related problems one is interested in either $X_{(1)}$ or $X_{(n)}$ and the probability questions regarding them. To derive distribution functions of $X_{(1)}$ and $X_{(n)}$ without using calculus we can consider the following approach. By definition

$$F_X(x) = P(X \leq x) = 1 - P(X > x).$$

Now for the smallest observation to be larger than x, all the observations should by larger than x. Thus we have

$$P(X_{(1)} > x) = P(X_1 > x, X_2 > x, \ldots, X_n > x)$$
$$= P(X_1 > x)P(X_2 > x)\ldots P(X_n > x)$$
$$= [1-F_X(x)]^n$$

or

$$F_{X_{(1)}}(x) = 1 - [1-F_X(x)]^n. \tag{10.5}$$

Similarly, for the largest observation to be smaller than x, all the observations should be smaller than x, that is,

$$F_{X_{(n)}}(x) = P(X_{(n)} \leq x)$$
$$= P(X_1 \leq x, X_2 \leq x, \ldots, X_n \leq x)$$
$$= P(X_1 \leq x)P(X_2 \leq x)\ldots P(X_n \leq x)$$
$$= [F_X(x)]^n. \tag{10.6}$$

As an example of application of the above results, let X denote the time set by a randomly selected participant in men's 100-m run. Then the best annual record of that year is $X_{(1)}$ with a distribution function (10.5). If n and $F_X(x)$ are known, then it is possible to answer probability questions regarding $X_{(1)}$.

EXAMPLE 10.2.1 Consider a training process for men's 100-m run in a given year. A runner is to try n times and the results (in seconds) will be recorded as $X_1, X_2, \ldots, X_n$. This is regarded as a random sample from the distribution function $F_X(x)$. Here X stands for the time it takes for the runner to complete the 100 m, and $F_X(x)$ is the distribution function of X. Let us assume that X has approximately a normal distribution $N(\mu, \sigma^2)$. This is reasonable since the probability of small deviations from his expected time (average performance) μ is greater than large deviations. Moreover deviations in either direction may be considered equally likely. Thus we have

$$F_X(x) \approx \int_{-\infty}^{x} \frac{1}{\sigma\sqrt{2\pi}} e^{-(1/2\sigma^2)(s-\mu)^2} ds.$$

Let $X_{(1)}, \ldots X_{(n)}$ be defined as above, then $X_{(1)}$ stands for the best time or record of the n tries. Using (10.5) we have

$$F_{X_{(1)}}(x) \approx 1 - [1 - \int_{-\infty}^{x} \frac{1}{\sigma\sqrt{2\pi}} e^{-(1/2\sigma^2)(s-\mu)^2} ds]^n.$$

Also the density function of $X_{(1)}$ is

$$f_{X_{(1)}}(x) \approx n[1 - \int_{-\infty}^{x} \frac{1}{\sigma\sqrt{2\pi}} e^{-(1/2\sigma^2)(s-\mu)^2} ds]^{n-1} \frac{1}{\sigma\sqrt{2\pi}} e^{-(1/2\sigma^2)(x-\mu)^2}.$$

This is not the density function of a normal variable. It is interesting to note that if distribution of X is exponential then the distribution of $X_{(1)}$ is also exponential. We leave this for students to verify.

10.3 JOINT DISTRIBUTION OF THE FIRST r-ORDER STATISTICS

Consider now a more general case, namely

$$P(E) = P(x_1 \leq X_{(1)} < x_1 + dx_1, x_2 \leq X_{(2)} < x_2 + dx_2, \ldots, x_r \leq X_{(r)} < x_r + dx_r)$$

where the event E denotes occurrence of no observations prior to x_1, occurrence of the first observation between x_1 and $x_1 + dx_1$, no observations between $x_1 + dx_1$ and x_2 and so on, and finally occurrence of $n - r$ observations after $x_r + dx_r$. Extending the idea used for the first order statistics, we have

$$P(E) = \frac{n!}{0!1!0!\ldots(n-r)!} f_X(x_1)dx_1 f_X(x_2)dx_2 \ldots f_X(x_r)dx_r [1 - F_X(x_r)]^{n-r}. \tag{10.7}$$

or

$$f_{X_{(1)},X_{(2)},\ldots X_{(r)}}(x_1, x_2, \ldots x_r) = \frac{n!}{(n-r)!} [1 - F_X(x_r)]^{n-r} \prod_{i=1}^{r} f_X(x_i) \tag{10.8}$$

where $-\infty < x_1 < x_2 < \ldots < x_r < \infty$. For $r = n$, we have from (10.8)

$$f_{X_{(1)},X_{(2)},\ldots X_{(n)}}(x_1, x_2, \ldots x_n) = n! \prod_{i=1}^{n} f_X(x_i) \quad -\infty < x_1 < x_2 < \ldots < x_n < \infty. \tag{10.9}$$

Chapter 10: Nonparametric Statistics

This result implies that ordering destroys the independent property of a random sample. Recall that the joint density function of a random sample of size n is the same as (10.9) without the term $n!$.

A special case of interest and importance in nonparametric statistics are results for an underlying uniform distribution on the interval $(0,1)$. For this case $f_X(x) = 1$ and we have

$$f_{X_{(1)},X_{(2)},\ldots X_{(n)}}(x_1, x_2, \ldots x_n) = n! \quad 0 < x_1 < x_2 < \ldots < x_n < 1$$
$$= 0 \quad \text{otherwise.}$$

EXAMPLE 10.3.1 Continue with Example 10.2.1. Let us consider the joint distribution of the two best records, that is $f_{X_{(1)},X_{(2)}}(x_1, x_2)$. From (10.8) we have

$$f_{X_{(1)},X_{(2)}}(x_1, x_2) = n(n-1)[1 - \int_{-\infty}^{x_2} \frac{1}{\sigma\sqrt{2\pi}} e^{-(1/2\sigma^2)(s-\mu)^2} ds]^{n-2}$$

$$\prod_{i=1}^{2} \frac{1}{\sigma\sqrt{2\pi}} e^{-(1/2\sigma^2)(x_i-\mu)^2} \quad x_1 < x_2.$$

A PRACTICE PROBLEM Please find the joint distribution of the two best records, $f_{X_{(1)},X_{(2)}}(x_1, x_2)$, using exponential distribution.

10.4 THE PROBABILITY INTEGRAL TRANSFORMATION AND UNIFORM ORDER STATISTICS

As pointed out, the uniform distribution on $(0,1)$ plays a special role in nonparametric statistics. This is primarily due to a result referred to as the probability integral transformation presented below. For a random variable X with distribution function $F_X(x)$ define the inverse distribution function $F_X^{-1}(y)$ as

$$F_X^{-1}(y) = \inf\{x | F_X(x) \geq y\} \quad 0 < y < 1. \tag{10.10}$$

Note that if $F_X(x)$ is strictly increasing between 0 and 1, then there is only one x such that $F_X(x) = y$. In this case, the infimum is unnecessary and $F_X^{-1}(y) = x$ without ambiguity.

Suppose that there is some x such as $F_X(x) = y$. Since $F_X(x)$ is continuous from the right $F_X(F_X^{-1}(y)) = y$. In particular, this shows that if $F_X(x)$ is continuous then $F_X(F_X^{-1}(y)) = y$ for every y satisfying $0 < y < 1$. However, if $F_X(x)$ is the distribution function of a discrete random variable, then for a given y there may be no x for which $F_X(x) = y$. In such cases, $F_X^{-1}(y)$ is the smallest x yielding an $F_X(x)$ value larger than y, and hence, in general, we have the relationship

$$y \leq F_X(F_X^{-1}(y)) \quad \text{for } 0 < y < 1. \tag{10.11}$$

We now present a theorem known as "probability integral transformation," that relates any arbitrary distribution to the uniform distribution.

THEOREM 10.4.1 Let X be a continuous random variable with distribution function $F_X(x)$. Then the random variable $Y = F_X(x)$ has a uniform distribution on $(0,1)$.

PROOF: Since $F_X(x)$ is continuous, $F_X(F_X^{-1}(y)) = y$ for $0 < y < 1$. Using the monotonicity of $F_X(x)$ we see that $x \leq F_X^{-1}(y)$ implies $F_X(x) \leq F_X(F_X^{-1}(y)) = y$. Also

$$\{x; F_X(x) \leq y\} = \{x; x \leq F_X^{-1}(y)\} \cup \{x; x > F_X^{-1}(y) \text{ and } F_X(x) = y\}$$

The continuous distribution of x implies that $P(F_X(x) = y)) = 0$. Thus

$$P(F_X(x) \leq y) = P(x \leq F_X^{-1}(y)).$$

Let $H(y)$ be the distribution function for Y. Since Y assumes values only in $[0,1]$, we know that

$$H(y) = 0, \quad y < 0$$
$$= 1, \quad y \geq 1.$$

Also

$$H(y) = P(Y \leq y) = P(F_X(x) \leq y)$$
$$= P(x \leq F_X^{-1}(y)) = F_X(F_X^{-1}(y))$$
$$= y, \quad 0 < y < 1.$$

Since $H(y)$ is nondecreasing we see that it is the distribution function of a uniform distribution.

One application of this theorem is that, if $X_{(1)}, X_{(2)}, \ldots, X_{(n)}$ are order statistics for a random sample from a continuous distribution with distribution function $F_X(x)$, then $F_X(X_{(1)}) \leq \ldots \leq F_X(X_{(n)})$ are distributed as the order statistics from a uniform distribution on $(0,1)$. Hence the joint density function of $v_i = F_X(X_{(i)})$, $i = 1, 2, \ldots, n$, is

$$f_{v_1, v_2, \ldots, v_n}(v_1, v_2, \ldots, v_n) = n!, \quad 0 < v_1 < v_2 < \ldots < v_n < 1$$
$$= 0, \quad \text{otherwise} \tag{10.12}$$

and the marginal density for each v_j, $1 \leq j \leq n$ is

$$f_{v_j}(v) = \frac{n!}{(j-1)!(n-j)!} v^{j-1}(1-v)^{n-j}, \quad 0 < v < 1$$
$$= 0, \quad \text{otherwise.} \tag{10.13}$$

This is the probability density function of the Beta distribution (see Section 3.8). Here for a positive number r we have

$$E(v_j^r) = \frac{n!}{(j-1)!(n-j)!} \int_0^1 t^{r+j-1}(1-t)^{n-j} dt$$
$$= \frac{n!\, \Gamma(r+j)}{(j-1)!\, \Gamma(n+r+1)} \int_0^1 \frac{\Gamma(n+r+1)}{\Gamma(r+j)(n-j)!} t^{r+j-1}(1-t)^{n-j} dt$$
$$= \frac{n!\, \Gamma(r+j)}{(j-1)!\, \Gamma(n+r+1)} \tag{10.14}$$

Chapter 10: Nonparametric Statistics

where $\Gamma(k) = (k-1)!$. Thus when v_j is the j-th order statistic from a uniform distribution we have

$$E(v_j) = \frac{j}{n+1} \quad \text{and} \quad Var(v_j) = \frac{j(n-j+1)}{(n+1)^2(n+2)}. \qquad (10.15)$$

In particular, for v_1 these are, respectively,

$$E(v_1) = \frac{1}{n+1} \quad \text{and} \quad Var(v_1) = \frac{n}{(n+1)^2(n+2)}. \qquad (10.16)$$

EXAMPLE 10.4.1 Consider a situation similar to the Example 10.2.1. Assume that the runner will make 101 attempts during the summer. What is the probability that he will set a record in the last attempt(number 101)?

SOLUTION: The probability of this event is v_1, the area to the left of his best time in the first 100 attempts. This is a random variable. Using (10.13) its density function is

$$f_{v_1}(v) = 100(1-v)^{99}, \quad 0 < v < 1.$$

Also using (10.16) we have

$$E(v_1) = \frac{1}{100+1} = 0.0099$$

$$Var(v_1) = \frac{100}{(101)^2(102)} = 0.000096$$

Now since v_1 is an unknown random quantity we may estimate that by its expected value 0.0099. Another option is to use its median $1 - 2^{1/100} = 0.007$. We can also construct a confidence interval. For example, a 90% confidence interval for the required probability is

$$(0.0005, 0.0295)$$

10.5 DISTRIBUTION-FREE CONFIDENCE INTERVALS AND TESTS

In this section, we introduce some frequently used nonparametric methods for estimation and testing hypothesis. Parametric methods were discussed in details in Chapters 5 and 6.

Confidence Intervals

In Chapter 5, we discussed parametric approach to both point estimation and the confidence interval for population mean based on the sample mean. Here in place of mean we talk about the median, and also percentage points. Using the results of previous section it is possible to construct interval estimates or the percentage points. For example, a confidence interval for median v can be constructed using binomial distribution. From the definition of median, the probability that an observation falls to the left or the right of v is $1/2$. Thus, in a random sample of size n the probability that exactly i observations fall to the left of v is

$$\binom{n}{i}(1/2)^i(1/2)^{n-i} = \binom{n}{i}(1/2)^n. \tag{10.17}$$

Using this the probability that $X_{(r)}$, the r-th order statistic, exceeds v is therefore

$$P(X_{(r)} > v) = \sum_{i=0}^{r-1}\binom{n}{i}(1/2)^n. \tag{10.18}$$

Similarly the probability that $X_{(s)}$ does not exceed v is

$$P(X_{(s)} < v) = \sum_{i=s}^{n}\binom{n}{i}(1/2)^n. \tag{10.19}$$

Now, assuming that $s > r$, adding (10.18) and (10.19), and subtracting both sides from one, we obtain

$$P(X_{(r)} < v < X_{(s)}) = \sum_{i=r}^{s-1}\binom{n}{i}(1/2)^n \tag{10.20}$$

which provides a confidence interval for v. Ordinarily s is taken to be $n - r + 1$ so that r-th observations in order of magnitude from the top and from the bottom are used. The following theorem summarizes the results for v:

THEOREM 10.5.1 Let $X_{(1)}, X_{(2)}, \ldots, X_{(n)}$ be the order statistic corresponding to a sample of size n, then the probability that the random interval $(X_{(r)}, X_{(n-r+1)})$ contains the median v is

$$\sum_{i=r}^{n-r}\binom{n}{i}(1/2)^n. \tag{10.21}$$

To construct a confidence interval for the median v with confidence coefficient $1 - \alpha$ the integer r is chosen so that (10.21) is nearest $1 - \alpha$ or so that probability in (10.21) is greater than $1 - \alpha$. If r' is the largest integer so that (10.21) is greater than or equal to $1 - \alpha$, then the required confidence interval is

$$P(X_{(r')} < v < X_{(n-r')}) \geq 1 - \alpha.$$

As an example, for a sample of size 6, we could compute

$$P(X_{(1)} < v < X_{(6)}) = 1 - (1/2)^5 = 0.97 \tag{10.22}$$

and

$$P(X_{(2)} < v < X_{(5)}) = 1 - 14/2^6 = 0.78.$$

If we are interested in a 95% confidence interval, (10.22) is the one we should use. If, for instance, a 90% confidence interval is desired, we may use some randomized device between the two intervals.

Chapter 10: Nonparametric Statistics

If the sample size is small, one only has a few confidence levels available. In particular, when $n=2$, the 50% confidence interval given below is the only possibility

$$P(X_{(1)} < v < X_{(2)}) = 0.50. \tag{10.23}$$

For moderate sample sizes the binomial sum in (10.20) may be computed directly. For a large n, one would use the normal approximation to the binomial. Since the index i in (10.18) is approximately normal with mean $n/2$ and standard deviation $\sqrt{n}/2$, a 95% confidence interval is obtained by counting $1.96\sqrt{n}/2$ observations to the left and to the right of the sample median.

A similar technique can be employed to obtain confidence intervals for other percentage points. If ξ_p is the $100p$ percentage point of a distribution, then using the same argument, corresponding to (10.20) we find

$$P(X_{(r)} < \xi_p < X_{(s)}) = \sum_{i=r}^{s-1} \binom{n}{i} (p)^i (1-p)^{n-i}. \tag{10.24}$$

Thus for a sample of size 6, a possible confidence interval for the 25th percentage point is given by

$$P(X_{(1)} < \xi_{0.25} < X_{(4)}) = \sum_{1}^{3} \binom{6}{i} (1/4)^i (3/4)^{6-i} = 0.78. \tag{10.25}$$

A 96% upper bound for $\xi_{0.25}$ is

$$P(\xi_{0.25} < X_{(4)}) = \sum_{0}^{3} \binom{6}{i} (1/4)^i (3/4)^{6-i} \approx 0.96.$$

Tests of Hypotheses

To test the null hypothesis $v = v_0$ against the alternatives $v > v_0$, we use the relation (10.17), choosing in advance an integer r so that the probability of a Type I error is as near to the desired value as possible. Thus for a sample of size 6, we can make the probability of a Type I error $7/64 \approx 0.11$ by choosing $r = 2$. If after drawing the sample we find $x_r < v_0$, the null hypothesis is not rejected; if $x_r > v_0$, it is rejected. In the same fashion two-sided tests of $v = v_0$ may be constructed; the two-sided test is obviously equivalent to constructing a confidence interval for v and rejecting or not rejecting $v = v_0$ according to whether confidence interval does not or does cover v_0. Tests on a percentage point ξ_p would be carried out similarly using probabilities p and $(1-p)$ instead of $1/2$ and $1/2$.

Chapters 5 through 9 a more formal test for median that is simple to apply and is in line with methods discussed in previous chapters.

10.5.1 Single Sample Sign Test

Small Samples

The sign test for single samples, is considered the simplest nonparametric test and is used to test the value of a median (as opposed to mean in parametric case) for a specific sample or a nonnormal probability distribution. When using the sign test, the researcher hypothesizes a specific value for the median of a population; selects a sample and compares each value with the conjectured median. If the data value is above the conjectured median, it is assigned a + sign. If it is below the conjectured median, it is assigned a − sign. Finally if it is exactly the same as the conjectured median, it is assigned a 0. Next the number of + and − signs are compared. If the null hypothesis is true, the number of + signs should be approximately equal to the number of − signs. If the null hypothesis is not true, there will be a disproportionate number of + and − signs.

The test quantity (statistics) is the smaller number of + or − signs. For example, if there are 7 positive signs and 4 negative signs, the test quantity is 4. When the sample size is 25 or less, Table 10.5.1 below is used to determine the critical value. For a specific α, if the test quantity is less than or equal to the critical value obtained from the table, the null hypothesis is rejected. The values in Table 10.5.1 are obtained from binomial distribution.

To demonstrate the applications of the sign test, let us consider an example.

TABLE 10.5.1 Critical Values for the Sign Test

	One-tailed			
	$\alpha=0.005$	$\alpha=0.01$	$\alpha=0.025$	$\alpha=0.05$
	Two-tailed			
n	$\alpha=0.01$	$\alpha=0.02$	$\alpha=0.05$	$\alpha=0.10$
8	0	0	0	1
9	0	0	1	1
10	0	0	1	1
11	0	1	1	2
12	1	1	2	2
13	1	1	2	3
14	1	2	3	3
15	2	2	3	3
16	2	2	3	4
17	2	3	4	4
18	3	3	4	5
19	3	4	4	5
20	3	4	5	5
21	4	4	5	6
22	4	5	5	6
23	4	5	6	7
24	5	5	6	6
25	5	6	6	7

Chapter 10: Nonparametric Statistics

EXAMPLE The owner of store selling sporting goods hypothesize that the median number of hats sold per day is 40. A random sample of 20 days yields the following data for the number of hats sold each day:

22	16	40	43	18
30	29	32	37	36
39	34	39	45	28
36	40	34	39	52

Test the owner's hypothesis using $\alpha = 0.05$.

SOLUTION: The hypotheses to be tested are

$$H_0: \text{median} = 40 \quad vs \quad H_1: \text{median} \neq 40$$

The test statistics (quantity) is T = Number of days that less than or more than 40 hats were sold (whichever is smaller). Comparing each data value with the median, we obtain

−	−	0	+	−
−	−	−	−	−
−	−	−	+	−
−	0	−	−	+

Here 0 is replaced for tie. The test quantity is the number of + signs, which in this case is 3 or simply the number of measurements greater than the median (40). Here $n = 18$ (the total number of + and − signs omitting the zeros) and $\alpha = 0.05$. Thus using the Table 10.5.1 the critical value for a two-tailed test is 4.

Comparing the value of test quantity 3 with the critical value 4, leads to rejection of the null hypothesis since $3 < 4$. Thus, there is sufficient evidence at 5% significant level to reject the claim that the median number of hats sold per day is 40.

Note that if we want to conduct the test at the $\alpha = 0.05$ level of significance, the rejection region can be expressed in terms of the observed significance level, or *p*-value of the test, that is

$$\text{Rejection Region}: p\text{-value} \leq 0.05$$

In this example in three $(T = 3)$ of the 20 days, the number of hats sold were greater than 40. To determine the observed significance level (*p*-value) associated with this outcome, we use the fact that the number of days with sales greater than 40 is a binomial random variable with $p = 0.5$ since if H_0 is true, the probability that a measurement lies below (or above) the median is equal to 0.5. We want to calculate the probability that a result is as contrary or more contrary to H_0 than the one observed? That is, we want to calculate the probability that 3 or more of 20 binomial measurements will result in success (be greater than 40) if the probability of success is 0.5? Binomial Table B1 (using $n = 20$ and $p = 0.5$) indicates that

$$p(T \leq 3) = 0.001$$

Taking into the account that this is a two-tailed test we have *p*-value = 0.002. This leads to the rejection of H_0.

A PRACTICE PROBLEM A trainer has developed an index to measure the athletic ability of the students. The index is normalized to have a median equal to 10. Suppose that the following is a sample of measurements on 8 students. Test to determine whether the population median is less than 10. Find the *p*-value using binomial distribution and make decision based on that too.

$$5.1, 2.3, 7.7, 9.6, 8.9, 9.8, 19.5, 7.8.$$

Large Samples

In the example discussed above, we had a "small" sample. For sample sizes greater than 26, the normal approximation may be used to find the test statistics. This leads to the following test statistics:

$$T = \frac{(S+0.5) - (n/2)}{\sqrt{n}/2}$$

where S = smaller number of + or − signs and n represents the sample size. Note that here mean and the standard deviation are obtained as $np = n/2$ and $\sqrt{npq} = \sqrt{n(1/2)(1/2)} = \sqrt{n}/2$ under the null hypothesis. Since under the null hypothesis $p = 1/2$, the normal distribution provides a good approximation to the binomial distribution even for $n \geq 10$.

EXAMPLE A manufacturer of sport shoes claims that the median lifetime of their sneakers is at least 8 years. A sample of 50 sneakers showed that 21 lasted more than 8 years. At $\alpha = 0.05$, is there sufficient evidence to reject the manufacturer's claim?

SOLUTION: The hypotheses to be tested are;

$$H_0 : \text{median} \geq 8 \quad vs \quad H_1 : \text{median} < 8$$

or just

$$H_0 : \text{median} = 8 \quad vs \quad H_1 : \text{median} < 8$$

Here the test statistics is

$$T = \frac{(21+0.5) - (50/2)}{\sqrt{50}/2} = \frac{-3.5}{3.5355} = -0.99.$$

Since this is a one-tailed test with $\alpha = 0.05$, and $n = 50$, the critical value is -1.645 which is less than -0.99. Thus, we do not reject the null hypothesis. There is not sufficient evidence to reject the claim that the median lifetime of the sneakers is at least 8 years at 5% significance level.

In this example, the sample size was 50 and 21 of the shoes lasted more than 8 years, so 29 of the sneakers did not last 8 years. The value of S corresponds to the smaller of the two numbers 21 and 29. In this case, $S = 21$ is used in the formula. This is because there would be 21 positive signs, since subtracting 8 years from the value in years of a sneaker that lasted longer than 8 years would result in a positive answer. When 8 is subtracted from the value in years of a sneaker that did not last 8 years, the answer would be negative. Assuming that no sneaker lasted exactly 8 years would result in 21 positive answers and 29 negative answers. Since 21 is the smaller of the two numbers, the value of S is 21.

Chapter 10: Nonparametric Statistics

A PRACTICE PROBLEM

(1) A manufacturer of treadmill has established that the median time to failure for their treadmills is 1,000 h of utilization. A sample of 30 treadmills made by this manufacturer are tested to their failure. The failure times range from 10 to 1,200, and 21 of the 30 exceeded 1,000 h. Is there sufficient evidence to indicate that the median failure time differs from 1,000 h at 5% significant level?

(2) A professional fan of the college basketball competitions NCAA claims that more than half of the players in the top 30 teams play less than 10 min per game. That is, the median playing time for such players is less than 10 min. You randomly sample 30 players and find that 18 of them actually played less than 10 min. Set up and conduct an appropriate test using $\alpha = 0.10$. State all the necessary assumptions.

Based on what we discussed so far, it is apparent that nonparametric methods, besides being extremely general in that they require few assumptions about the form of the distribution function, are also extraordinarily simple. No complex analysis or distribution theory is needed; the simple binomial provides much of the necessary equipment for estimation when dealing with a single population. The only inconvenience is in the paucity of confidence levels when the sample size is quite small.

We now present few other well-known nonparametric tests including those used for comparing two or more populations. Examples describing further applications of the tests discussed here are presented in Section 10.6.

10.5.2 Run Test

We already noted that randomness plays a key role in statistical analysis of any data set. For example, in regression analysis tests such as t or F are valid when residuals are independent. How a sequence of observations or residuals can be tested for presence or absence of randomness? One way to do this is to apply the run test.

The run test examines a sequence containing an arrangement of two symbols like hit (H) and miss (M) to determine whether such a sequence was generated randomly. Here a run consists of a string of identical symbols or values. For example, consider the following data that represents the free-throw attempts by three basketball players:

Player 1	H	H	H	H	H	M	M	M	M	M
Player 2	H	M	H	M	H	M	H	M	H	M
Player 3	H	H	M	H	M	M	H	M	M	H

These sequences have, respectively, 2, 10, 7 runs. Looking at the first sequence, one might think that the player 1 had a hot hand (at least after the first five attempts). The sequence 2, also, presents a pattern (although not as interesting as the first sequence). In fact, only sequence 3 may appear to be generated randomly.

To see how actual testing is done, consider a sequence of n observations, containing n_1 hits (symbol of the first type) and $n_2 = n - n_1$ misses (symbol of the second type). Let R denote the number of runs within these n observations. Clearly, the small values of R provide evidence against randomness. If R is used as a test quantity, we need to know its distribution so that its critical values can be determined. For small samples (in this case $n_1 \leq 10$ and $n_2 \leq 10$), the critical values are tabulated (see standard statistical table books). For samples exceeding 10, the test statistics R is approximately normally distributed with mean $\mu_R = 1 + 2n_1 n_2/(n_1 + n_2)$ and standard deviation $\sigma_R = \sqrt{\frac{2n_1 n_2 (2n_1 n_2 - n_1 - n_2)}{(n_1 + n_2)^2 (n_1 + n_2 - 1)}}$. Thus for large samples testing can be carried out in usual manner.

To see how this test is applied we present an example.

EXAMPLE 10.5.1 A basketball player has attempted 45 three-pointers in a row. The result is given below

$$MMHHHHHHMHHMMMMMMHHHHMMH$$

$$MMHHMHHHHHMMMMMHHHHMMM$$

Based upon this sequence, would you conclude that this sample consists of 45 randomly occurring hits and misses? Use $\alpha = 0.05$.

SOLUTION: The preceding sequence contains $R = 15$ runs. Also, there are

$$n_1 = \text{Number of hits} = 22$$
$$n_2 = \text{Number of misses} = 23$$

For these values of n_1 and n_2, the mean (expected) number of runs if H_0 is true is

$$1 + \frac{(2)(22)(23)}{45} = 23.49.$$

This implies that, on the average, whenever $n_1 = 22$ and $n_2 = 23$, we expect to have 23.49 runs.

The above sample, however, contains only 15 runs, so perhaps this sequence exhibits a nonrandom pattern. But this depends heavily on the standard deviation of R, therefore, to complete the analysis, we need to find the standard deviation. We have

$$\sqrt{\frac{(2)(22)(23)[(2)(22)(23) - 45]}{(45)^2 (44)}} = \sqrt{10.9832} = 3.314.$$

To determine whether $R = 15$ is sufficiently small, that is, it provides evidence against random sequence hypothesis, we calculate the test statistic (quantity).

$$T = \frac{15 - 23.49}{3.314} = -2.56.$$

The test procedure here (using $\alpha = 0.05$) is to reject H_0 if $|T| > 1.96$. The computed test statistics T does have an absolute value larger than 1.96, and so we reject H_0. There is sufficient evidence to indicate that the sequence is nonrandom, indicating a lack of randomness in making or missing the shots. In other words, player's talent had something to do with the result.

A PRACTICE PROBLEM On January 7, 2003, the Los Angeles Lakers' Kobe Bryant hits an National Basketball Association (NBA) single-game record nine consecutive three-pointers against Seattle Super Sonics. Bryant's total of 12 three-pointers was also a single-game record. Laker coach Phil

Jackson was quoted as saying "That was perhaps the greatest streak shooting I think I have ever seen in my life."

Here is his sequence

$$MHHHHHHHHHHMHHMMHMM$$

Apply the run test and show that the hypothesis of randomness cannot be rejected at 5% significance level.

10.5.3 Wilcoxon (Mann–Whitney) Rank Sum Test

This test is analogous to the parametric independent t-test for equality of population means and is one of the more powerful nonparametric tests available. It can be used with very small or fairly large groups and requires only rank (ordinal) measurements. It is used when there are two independent groups and one dependent variable. For this test, there are more than one way for computing the test statistics.

This test is applicable to the problems with the following characteristics:

1. The problem objective is to compare two populations.
2. The data are either ranked or numerical and conditions required for application of parametric t-test are not satisfied.
3. The samples are independent.

Suppose that two independent samples are available. Label the sample with the smaller number of observations 1 and the other sample 2. If the two sample sizes are equal, arbitrarily assign the labels. Let n_1 and n_2 represent the sample sizes ($n = n_1 + n_2$). Rank all the observations, with 1 being the smallest observation and n being the largest observation. In case of ties, average the ranks of the tied observations. Then calculate the sum of the ranks of sample with a smaller number of measurements, which is the test statistics T. As usual we judge the value of T relative to its sampling distribution. For small samples (both sample sizes are less than or equal to 10), special tables developed by Mann and Whitney are used. These tables can be found in standard books of statistical tables. Table B8 is a version of this table for few values of α. For samples exceeding 10, the test statistics is approximately normally distributed with mean $\mu_T = n_1(n_1 + n_2 + 1)/2$ and standard deviation $\sigma_T = \sqrt{\frac{n_1 n_2 (n_1 + n_2 + 1)}{12}}$. Thus to apply this test, one needs to calculate the z-score of T and proceed as usual. Note that here the goal is to determine whether two populations differ in location (mean)using two independent samples. So the null hypothesis states that sampled populations have identical probability distributions and the alternative states that one distribution is shifted to the left or to the right.

EXAMPLE A physical educator claims that the order in which warm up moves are practiced affects a player's performance. To investigate this assertion, he randomly divides a team of 13 players into two groups—7 in one group and 6 in the other. He prepares one set of warm up moves but arranges them in two different orders. In one, moves are arranged in order of increasing difficulty (that is, from easiest to most difficult), while on the other the order is reverse. On a scale devised to measure the performance the scores are as follows:

| Arrangement 1: | 90 | 71 | 83 | 82 | 75 | 91 | 65 |
| Arrangement 2: | 66 | 78 | 50 | 68 | 80 | 60 | |

Do data provide sufficient evidence to indicate a difference (a shift in location) in the probability distributions of scores on the two arrangements?

SOLUTION:

Arrangement 1	Rank	Arrangement 2	Rank
90	12	66	4
71	6	78	8
83	11	50	1
82	10	68	5
75	7	80	9
91	13	60	2
65	3		29
	62		

H_0: The probability distributions of scores are identical

H_1: There is a shift in the locations of the probability distributions of scores

The test statistic is $T = 29$.

The null hypothesis will be reject if $T \leq T_L$ or $T \geq T_U$ where $\alpha = 0.5$ (two-tailed), $n_1 = 6$ and $n_2 = 7$. From Table B8 $T_L = 28$ and $T_v = 56$. Thus reject H_0 if $T \leq 28$ or $T \geq 56$.

Since T does not fall in rejection region, do not reject H_0. There is insufficient evidence to indicate a shift in location at $\alpha = 0.05$.

Instead of T, Mann and Whitney use the criterion (test quantity) U, which is defined as the total number of times first sample's observations preceed second sample's observations. For example, if the arrangement is

$$x_2 x_1 x_2 x_1 x_2 x_2$$

then $U = 5$ because the first x_1 precedes three $x_2's$ and the second x_1 precedes two $x_2's$. For this example, $T = 2 + 4 = 6$. Note that U and T have a simple linear relation, namely

$$U = n_1 n_2 + \frac{n_1(n_1 + 1)}{2} - T.$$

For sample sizes exceeding $10, U$ is approximately normal with mean $n_1 n_2/2$ and standard deviation given above (σ_T).

EXAMPLE 10.5.2 Due to the popularity of sport and its growth as a scientific subject and also a business matter, many specific research questions has been raised by different investigators. Suppose a researcher wishes to test the hypothesis that experienced instructors or coaches require less time (duration of eye fixation) than novice instructors while observing skill performance. To study this, a group of golf instructors with more than 10 years of experience are compared with a group of novice golf instructors. Both groups observe the same individuals performing a golf

Chapter 10: Nonparametric Statistics

drive. An eye movement recorder is used to measure the duration of eye fixation in milliseconds. The scores for the 11 experienced and 12 novice instructors are shown the table below. The median time were, respectively, 117 ms and 128 ms for experienced and novice golf instructors. Use this information to test researcher's hypothesis.

Table for Example 10.5.2. Group 1; Experienced Instructors, Group 2; Inexperienced Instructors

Group 1	Rank	Group 2	Rank
111	4	130	19
114	5.5	123	12
111	4	130	19
114	5.5	123	12
120	10	124	13
101	1	138	21
118	8	142	23
128	17	120	10
125	14	127	15.5
117	7	140	22
106	2	136	20
120	10	129	18
110	3	127	15.5
		114	5.5
	81.5		194.5

SOLUTION: First, we rank the shortest time as 1 and the longest as 23. Of course, we must maintain group identities in the ranking process. When we have ties in ranks, we give each score the average of the ranks they occupy. For example, two instructors have score of 114, which occupy the fifth and the sixth ranks. So, we give each instructor the average of those two ranks (5.5), and the next rank is now 7. Three instructors have a score of 120, which occupy the 9th, 10th, and 11th ranks. Therefore, all three are given the middle rank (10), and the next rank is 12. This calculation is summarized in the above table.

We can check the accuracy of ranking by using the formula $n(n+1)/2$, where n is the total number of scores in all groups combined. For this example, $23(23+1)/2 = 276$, which agrees with the sum of the ranks of Group 1 (81.5) plus the sum the ranks of Group 2 (194.5), or 276.

Since both sample sizes are greater than 10 the distribution of test statistics is approximately normal with mean 132 and standard deviation 16.248. The test statistics is then

$$z = \frac{81.5 - 132}{16.248} = -3.11$$

Noting that the p-value $= p(z < -3.11) < 0.01$, we conclude that the difference cannot happen by chance.

Next, we calculate U statistics defined above. The difference between the groups will be maximum when there is little overlap between samples. In other words, if there is no overlap, then all measurements for one sample would be greater than the other sample. On the other hand, if the null hypothesis is true, scores from the two groups will be mixed, in which case the ranks of one sample will be greater than the other sample about half the time and the same applies to the other sample. Thus, U would be near $n_1 n_2/2$ if there is no difference between groups. Conversely, a real difference between groups will show a high U for one group and a low U for the other group. The formula for calculating U for Group 1 is

$$U = n_1 n_2 + [n_1(n_1+1)/2] - T$$

where n_1 = the number of subjects in Group 1, n_2 = the number of subjects in Group 2, and T = the sum of ranks for Group 1. In our example, U for Group 1 is

$$U = (11)(12) + [11(11+1)/2] - 81.5 = 116.5.$$

The formula for U for Group 2 is by symmetry

$$U = n_1 n_2 + [n_2(n_2+1)/2] - T^*$$

where T^* is the sum of the ranks of Group 2. In this example, U for Group 2 is

$$U = (11)(12) + [12(12+1)/2] - 194.5 = 15.5.$$

Next we determine the probability. This can be accomplished either by consulting a table of U values (for the smaller of the two U's) or by calculating z and consulting the table for the standard normal distribution (Table B3). The U tables are used with small samples. If samples are of moderate size (say 10 or more in each sample), the sampling distribution of U will be approximately normal, and the calculation of z is appropriate. Most statistic textbooks contain tables for Mann–Whitney U test for small samples. The calculation of z is as follows:

$$z = \frac{U - (n_1 n_2)/2}{\sqrt{n_1 n_2 (n_1 + n_2 + 1)/12}}.$$

Substituting the values from the example, and using U for Group 1 we find

$$z = \frac{116.5 - (11)(12)/2}{\sqrt{(11)(12)(11+12+1)/12}} = \frac{116.5 - 66}{\sqrt{(132)(24)/12}}$$

$$= \frac{50.5}{\sqrt{264}} = 50.5/16.25 = 3.11.$$

Note that, it does not matter which U is used. If we had used U for Group 2, the resulting z would be -3.11. So the absolute value of z is the same (only the sign is different).

Now to find the probability, we consult Table B3 and find the area to the right of $z = 3.11$ (p-value). We see that the probability that a difference this large would occur by chance is less than 1%. Thus, there appears to be a real difference between the two groups of instructors on duration of eye fixation, with novice golf instructors watching significantly longer than experienced instructors.

Chapter 10: Nonparametric Statistics

A PRACTICE PROBLEM A physiologist is planning to introduce a new training technique. In a preliminary experiment to determine its effectiveness, 30 players were randomly selected of which 15 followed the new technique and 15 followed the old technique. All 30 players were told to indicate which of the following statements most accurately represented the effectiveness of the technique they followed.

> 5 = The technique was extremely effective. 4 = The technique was very effective.
> 3 = The technique was somewhat effective. 2 = The technique was slightly effective.
> 1 = The technique was not at all effective.

The responses are listed below using the codes. Can we conclude at the 5% significance level that the new technique is perceived to be more effective?

$$\text{New Technique}: 3, 5, 4, 3, 2, 5, 1, 4, 5, 3, 3, 5, 5, 5, 4$$
$$\text{Old Technique}: 4, 1, 3, 2, 4, 1, 3, 4, 2, 2, 2, 4, 3, 4, 5$$

10.5.4 Wilcoxon Matched-Pairs Signed Rank Test

This test is similar to two sample paired t-test. Here one has matched groups, or the same group of people who are tested twice on a dependent measure, and we want to know whether the difference between the matched groups (or the change in scores for the same subjects) is significant. For example, we may like to find out if a specific type of exercise program had a positive effect on performance of the players in a certain sport. In such situations, the Mann–Whitney test described in previous section, can not be used.

This test is applicable to the problems with the following characteristics:

1. The problem objective is to compare two populations.
2. The data are numerical or are ranked and normality requirement necessary to perform a parametric test is not satisfied.
3. The samples are matched pairs.

The Wilcoxon matched-pair signed-ranks test is a powerful nonparametric test for comparing related samples. It is used for determining whether the population of differences is centered at zero when dealing with two dependent (paired) samples. This method determines the differences of the paired observations and then calculates a value of the test statistics using the ranks of the differences. Both small sample (using special tables) and large samples (using an approximate normal distribution) are used frequently in practice.

EXAMPLE 10.5.3 A college basketball coach has 15 players named A,B, ...O, and their statistics including the total number of points each player scored in six selected games. He designed an adventure program to raise their self-confidence and administrates that. Table below provides data for this group of players before and after an adventure activity program. His goal is to assess whether they gained confidence and measures that by increase in the number of points they scored against the same teams. The coach is using a one-tailed test because he is hypothesizing a positive change (increase in total number of points scored) and is not interested in any changes that might occur in the opposite direction. Help the coach to draw appropriate conclusion.

Players Name	Points Scored Before	Points Scored After	Difference	Rank	Signed Rank	
A	33	36	+3	6.5	+6.5	...
B	30	31	+1	2	+2	...
C	40	37	-3	6.5	-6.5	-6.5
D	27	36	+9	13	+13	...
E	18	24	+6	10	+10	...
F	26	25	-1	2	-2	-2
G	35	35	0	...	...	...
H	20	16	-4	8	-8	-8
I	38	33	-5	9	-9	-9
J	16	24	+8	12	+12	...
K	26	28	+2	4.5	+4.5	...
L	21	20	-1	2	-2	-2
M	18	20	+2	4.5	+4.5	...
N	24	24	0	...	...	...
O	11	18	+7	11	+11	...
						T=27.5

SOLUTION: Note that some differences are positive and some are negative. In two cases, no score changes occurred. Therefore, these two are dropped from further analysis because they would have no influence. We started with the lowest difference. Three players have a difference of one (either more or less). As before, because they occupy the first three ranks, each is given the average rank of 2. After ranking the differences, in the next column, ranks are given a negative or positive sign according to whether change was negative or positive, respectively. Here, there are eight plus ranks and five minus ranks, so all the minus ranks are entered to column 7. These ranks are summed, which is designated as T.

The coach had specified in advance that he would be using a one-tailed test with significance level of 0.01. With a sample size of greater than 10 (some authors recommend normal approximation for sample sizes greater than 25), we compute z and consult Table B3 for the critical value. For sample sizes less than 10, special tables are used.

$$z = \frac{T - n(n+1)/4}{\sqrt{n(n+1)(2n+1)/24}} = \frac{27.5 - 13(13+1)/4}{\sqrt{13(13+1)[2(13)+1]/24}}$$
$$= \frac{-18}{14.31} = -1.26.$$

Chapter 10: Nonparametric Statistics

The sign of the z does not matter. Using Table B3 the p-value is $p(z < -1.26) = 0.1038$. This is greater than 0.01 (not significant). So unlike what coach may have hoped for, no significant gains in self confidence is achieved after participation in the adventure program.

A PRACTICE PROBLEM Twelve sets of identical twins are given a test that measures their athletic ability in scale from 60–100. Do the data provide sufficient evidence to indicate that the firstborn of a pair of twins is more athletic than the second born? Use $\alpha = 0.05$ and show that the p-value is 0.2249.

Set Number	Firstborn	Second born
1	71	80
2	71	77
3	77	76
4	68	64
5	91	96
6	72	72
7	77	65
8	91	90
9	70	65
10	86	88
11	88	81
12	87	72

10.5.5 Paired-Sample Sign Test

In Section 10.5.1, we introduced and used sign test for a single sample. The sign test can also be used for comparison of two means with dependent samples, such as a before and an after test. Recall that when dependent samples are taken from normally distributed populations, the paired t-test is used. When the condition of normality is not met, the nonparametric sign test can be used. This test is employed in the following situations:

1. The problem objective is to compare two populations.
2. The data are ranked.
3. The experimental design is matched pairs.

EXAMPLE The coach of a water polo team believes that the number of ear infections can be reduced if players use a special kind of ear plugs. A sample of 10 players was selected, and the number of infections for a 4-month period was recorded. During the first 2 months, players did not use the ear plugs; but during the second 2 months, they did. At the beginning of the second 2-month period, each player was examined to make sure that no infections were present. At $\alpha = 0.05$, can the coach conclude that using ear plugs reduces the number of ear infections? Table below presents the number of ear infections for the first and the second periods.

Player	First Period	Second Period	Difference	Rank of Absolute Difference
A	3	2	1	3
B	0	1	-1	3
C	5	4	1	3
D	4	0	4	9
E	2	1	1	3
F	4	3	1	3
G	3	1	2	7
H	5	3	2	7
I	2	2	0	(Eliminated)
J	1	3	-2	7

SOLUTION 1: The hypothesis to be tested are;

H_0: The number of ear infections is not be reduced

H_1: The number of ear infections is reduced

To find the test quantity, we count the number of + and − signs found in differences and use the smaller value as the test quantity. There are 2 negative (−) signs, so the test quantity is 2.

From Table 10.5.1, with $n = 9$ (the total number of + and − signs; the 0 is not counted) and $\alpha = 0.05$ (one-tailed), at most 1 negative (−) sign is needed to reject the null hypothesis because 1 is the smallest entry in the $\alpha = 0.05$ column of Table 10.5.1.

There are 2 negative(−) signs. The decision is do not reject the null hypothesis. The reason is that $1 < 2$. There is not sufficient evidence to support the claim that the use of ear plugs reduces the number of ear infections, at 5% significant level.

Player	First Period	Second Period	Sign of Difference
A	3	2	+
B	0	1	−
C	5	4	+
D	4	0	+
E	2	1	+
F	4	3	+
G	3	1	+
H	5	3	+
I	2	2	0
J	1	3	−

SOLUTIONS 2: Here to find the test statistics, we first rank the differences and then calculate the rank sum of the negative (T_-) and positive (T_+) differences. The test statistics is the smaller of these

two numbers. In this example, rank sums are, respectively, 10 and 35 for negative and positive differences. Therefore, the test statistics is 10 (= T_-). Using Table B9, we have $T_0 = 8$ for a one-tailed test with $\alpha = 0.05$ and $n = 9$. Therefore, the test statistics and rejection region are, respectively,

$$\text{Test Statistics: } T = \text{The Negative Rank Sum}$$

$$\text{Rejection Region } T \leq 8$$

Since $T = 10$ exceeds the critical value $T_0 = 8$, we conclude that sample provides insufficient evidence at $\alpha = 0.05$ level to support the claim that the use of ear plugs reduces the number of ear infections.

PRACTICE PROBLEMS

(1) A coach must choose between two plans for an upcoming game. As an aid in reaching a decision, he asks 10 trainees to examine the plans and rate them on a scale from 1 to 10 with 10 being the highest. The coach will then adopt plan A unless results provide sufficient evidence that plan B is more effective. Do the data provide evidence at $\alpha = 0.05$ level that distribution of rating for plan B lies above that for plan A?

Trainer:	1	2	3	4	5	6	7	8	9	10
Plan A:	5	9	10	8	6	3	9	8	4	7
Plan B:	9	4	8	9	10	6	8	8	5	9

Show that the test statistics is $T = 15.5$.

(2) Suppose you want to compare two different treatments for a minor injury to be applied during the game. Your goal is to find out if responses to one treatment tend to be larger than those for the other. If there are 25 pairs, show that the rejection region for the large sample Wilcoxon signed rank test is $T > 1.645$. If $T_+ = 273$ show that the test statistics equals $T = 2.97$ and the p-value equals 0.0015.

10.5.6 Kruskal–Wallis and Friedman Tests

In Chapter 6 and in the previous sections, we briefly introduced nonparametric methods and identified the factors that dictate the use of the Wilcoxon rank sum tests. These methods are used to compare two populations. In this section, we introduced nonparametric version of the analysis of variance (ANOVA), which is applied to compare two or more populations. One of the requirements of the parametric ANOVA is that, the response variable is normally distributed. If the response variable is not normal or the data are ranked, we must use a nonparametric procedure. The Kruskal–Wallis test is employed when the samples are independent (completely randomized design). The Friedman test is applied when the samples are blocked (randomized block design).

The Kruskal–Wallis test is applicable in the following situation:

1. The problem objective is to compare two or more populations.
2. The data are either ranked or quantitative but nonnormal.
3. The samples are independent.

Kruskal–Wallis test is, in fact, an extension of the Mann–Whitney test. It is used to test if two or more populations differ in location when using independent samples. As in the Mann–Whitney test, the samples are pooled and ranked from smallest to largest. The resulting ranks are then used to define a test statistics which, as we shall see, has a chi-square distribution.

A nonparametric alternative to the randomized block design and analysis is the Friedman test. Unlike the randomized block test, the Friedman test does not require normal populations with equal variances. Values are ranked within each block and then summed for the various populations under consideration. In what follows these tests will be explained in more detail.

Kruskal–Wallis Test

The Kruskal–Wallis test is applied to problems that involves comparing two or more populations, data are ranked or are quantitative but nonnormal and experimental design is based on independent samples. To apply this test, first rank all the observations. As before, 1 is assigned to the smallest observation, and n is assigned to the largest observation, where $n = n_1 + n_2 + \ldots + n_K$. For ties, average the ranks. Then calculate the sum of the ranks of each sample, and call them $R_1, R_2, \ldots, R_K$ (K is the number of populations).

The test statistic is

$$T = H = [\frac{12}{n(n+1)} \sum \frac{R_i^2}{n_j}] - 3(n+1).$$

If the sample sizes are at least 5, then H is distributed approximately as chi-squared with $K - 1$ degrees of freedom.

EXAMPLE 10.5.4 A research was concerned with the effectiveness of running and swimming on loss of fat measured by the sum of skin folds. In the study other than treatment groups, a control group was also used in which the subjects did not engage in any exercise program but sum of skin folds were measured before and after the 10-week period. For this analysis, a small numbers of subjects was used in each group. Assume that the data were skewed, thus prompting the use of a nonparametric test. The table below shows the measured values for the three groups. They represent the loss of fat; thus, negative numbers are good, and positive numbers represent gains in fatness and are not considered good.

Running		Swimming		Control	
Score	Rank	Score	Rank	Score	Rank
-10	10	-25	15.5	+10	3
-29	19	-31	20	-1	7
-8	8	-19	12	+20	1
-32	21	0	6	-10	10
-22	14	-28	18	+9	4
-25	15.5	-27	17	+13	2
+2	5	-10	10		
-20	13				
$R_1 = 105.5$		$R_2 = 98.5$		$R_3 = 27$	

Chapter 10: Nonparametric Statistics

The averages are, respectively, 13.2, 14.1, and 4.5 for running, swimming, and control group. Apply the Kruskal–Wallis test and draw appropriate conclusions.

SOLUTIONS: The worst score (a gain of fatness of 20 mm in skinfold thickness) is given a rank of 1 and the best score (a loss of 32 mm) the highest rank of 21. As in previous examples, ties are given the average ranking. It should be obvious by now that if the null hypothesis of no difference was true, the average rank for each group would be the same. In other words, any difference in scores would be attributable to the sampling error. We see that the average ranks for the three groups are running: 13.2, swimming: 14.1, and control: 4.5.

To find test statistics H for each group, we need to square the sum of ranks and divide them by the number of scores in that group; we then sum for all three groups:

$$\frac{(105.5)^2}{8} + \frac{(98.5)^2}{7} + \frac{(27)^2}{6} = 2{,}898.8$$

This sum is multiplied by $\frac{12}{n(n+1)}$

$$\left(\frac{12}{21(21+1)}\right)(2{,}898.8) = (.26)(2{,}898.8) = 75.37$$

from which we subtract $3(n+1)$

$$75.37 - 3(21+1) = 75.37 - 66 = 9.37.$$

Thus, we have $T = H = 9.37$. When there are at least five subjects in each group, test H using a chi-square with $K-1$ degrees of freedom (where K is number of groups). If there are five or fewer subjects in each group, we should use a special table. Because we have sufficient numbers in each group, we consult Table B5 and find that $\chi^2_{0.95,(2)} = 5.99$. As 9.37 surpasses this, we conclude that there are significant differences among the groups. This means that at least one of these sports was effective for loosing fat.

EXAMPLE Young boys of age 7 are taught basic moves of soccer using one of the three different methods. They were randomly assigned to three groups. One group used programmed instruction, a second used standard techniques, and the third used an open approach that teaches the moves while playing the games. The improvements in implementing the moves correctly was graded from 0 to 2. The results are shown in the table below. The methods are named 1, 2, and 3, respectively. Do the data provide sufficient evidence to indicate that the probability distributions of improvements in implementation of moves differ for at least two of the methods? Use $\alpha = 0.05$.

Method 1	Method 2	Method 3
0.9	1.0	1.7
1.5	0.8	0.5
0.7	0.9	1.6
1.1	1.2	1.4
0.5	1.4	1.0

SOLUTION:

Method 1	Rank	Method 2	Rank	Method 3	Rank
0.9	5.5	1.0	7.5	1.7	15
1.5	13	0.8	4	0.5	1.5
0.7	3	0.9	5.5	1.6	14
1.1	9	1.2	10	1.4	11.5
0.5	1.5	1.4	11.5	1.0	7.5
	$R_1 = 32$		$R_2 = 38.5$		$R_3 = 49.5$

To determine whether the probability distribution of scores differ among the three methods, we test:

H_0: The probability distribution of scores are the same for the three groups

H_1: At least two of the three distributions differ in location

The test statistics is

$$T = H = \frac{12}{n(n+1)} \Sigma \frac{R_j^2}{n_j} - 3(n+1)$$

$$= \frac{12}{15(15+1)} \left[\frac{(32)^2}{5} + \frac{(38.5)^2}{5} + \frac{(49.5)^2}{5} \right] - 3(15+1)$$

$$= 49.565 - 48$$

$$= 1.565.$$

The reject region is $H > \chi^2_{0.95,(2)} = 5.99147$.

Since the observed value of the test statistic does not fall in the rejection region, do not reject H_0. There is insufficient evidence to indicate a difference in implementation among the three methods at $\alpha = 0.05$.

A PRACTICE PROBLEM The athletic director of small college is interested in knowing how students rate the quality of the programs in three major sports: basketball, soccer, and tennis. To find out, students are given the opportunity to fill out comment cards. The following data present response of 10 students on each sports selected randomly. Do these data provide sufficient evidence at a 5% significant level to indicate that students perceive the quality of programs to be different among three sports? Here 1, 2, 3, and 4 present, poor, fair, good, and excellent, respectively.

Basketball	Rank	Soccer	Rank	Tennis	Rank
4	27.0	3	16.5	3	16.5
4	27.0	4	27.0	1	2.0
3	16.5	2	6.5	3	16.5
4	27.0	2	6.5	2	6.5
3	16.5	3	16.5	1	2.0
3	16.5	4	27.0	3	16.5
3	16.5	3	16.5	4	27.0
3	16.5	3	16.5	2	6.5
2	6.5	2	6.5	4	27.0
3	16.5	3	16.5	1	2.0
	$R_1 = 186.5$		$R_2 = 156.0$		$R_3 = 122.5$

Friedman Test

The Friedman test is applied to problems that involve comparison of two or more populations, data are ranked or are quantitative but nonnormal and experimental design uses blocks. To apply the test, rank all observations within each block, where 1 is the smallest observation and n is the largest observation. Average the rank for the ties. Then compute the rank sums $R_1, R_2, \ldots, R_K$. The test statistic is

$$F_r = [\frac{12}{K(J)(J+1)} \sum R_j^2] - 3K(J+1).$$

The Friedman statistics measures the extent to which the J samples differ with respect to their relative ranks within the blocks. If either the number of populations (J) or the number of blocks is at least 5, F_r is distributed approximately as chi-squared with $J-1$ degrees of freedom.

The Friedman test is analogous to repeated measures ANOVA in parametric statistics. The scores are ranked, and chi-square is used to test the hypothesis. An example will explain the steps in the analysis.

EXAMPLE 10.5.5 Twelve people have been measured for percent fat by three different methods: hydrostatic weighing (H), total body water (W), and potassium-40 (P). The investigator wishes to find out whether there is any difference among the three methods. The measurements for methods H, W, and P for the 12 subjects are shown in table below. Use $\alpha = 0.05$.

| | Method | | | Ranks | | |
Subject	H	W	P	R(H)	R(W)	R(P)
1	16.0	15.7	15.2	3	2	1
2	11.6	10.9	11.5	3	1	2
3	18.1	18.0	17.6	3	2	1
4	16.3	16.8	17.0	1	2	3
5	12.0	13.1	12.8	3	1	2
6	12.5	12.2	12.3	3	1	2
7	9.3	8.8	9.0	3	1	2
8	18.8	17.5	18.0	3	1	2
9	19.2	19.7	19.7	1	2.5	2.5
10	22.3	22.8	21.6	2	3	1
11	20.7	20.3	19.6	3	2	1
12	24.1	23.7	24.4	2	1	3
				28	21.5	22.5

SOLUTION: First the three measurements for each individual are ranked from 1 to 3. Subject 1, for example, has percent fat values of 16.0, 15.7, and 15.2 for methods H, W, and P, respectively. Method P yields the lowest percent fat, so it is ranked 1, the next lowest for W is ranked 2, and method H is ranked 3. This procedure is followed for the rest of the subjects (ranks are also shown in the table). The familiar procedure of averaging tied ranks is followed. Under the null hypothesis, these sums should be about the same because the ranks should be evenly distributed over all three methods. On the other hand, if for example, one method shows consistently low percent fat values it would have more of rank 1, and the result would be a lower sum of ranks.

The Friedman two-way ANOVA by ranks yields a chi-square with degrees of freedom $J - 1$. The formula is as follows:

$$F_r = \frac{12}{KJ(J+1)} \sum R_j^2 - 3K(J+1)$$

where K = the number of rows, J = the number of columns, and $\sum R_j^2$ = the sum of the squared column totals. The computation for the example are as follows:

$$F_r = [\frac{12}{(12)(3)(3+1)}][(28)^2 + (21.5)^2 + (22.5)^2] - 3(12)(3+1)$$

$$= (12/144)(784 + 462.25 + 506.25) - 144$$

$$= \frac{1}{12}(1,752.5) - 144 = 2.04.$$

From Table B5, $\chi^2_{0.95,(2)} = 5.99$. Therefore, we conclude that there are no significance differences among the three methods of estimating percent body fat.

Chapter 10: Nonparametric Statistics

EXAMPLE An experiment involving players and their performance was conducted using a randomized block design involving for plans (treatments) using six players (blocks). The performances are ranked for each player. Use the Friedman test for a randomized block design to determine whether the data provide sufficient evidence to indicate that at least two of the plans probability distributions differ in location. Using $\alpha = 0.05$.

			Players			
Plans	1	2	3	4	5	6
A	3	3	2	3	2	3
B	1	1	1	2	1	1
C	4	4	3	4	4	4
D	2	2	4	1	3	2

SOLUTION:

$$R_1 = 16, \quad R_2 = 7, \quad R_3 = 23, \quad R_4 = 14$$

To determine whether at least two of the plan probability distributions differ in location, we test:

H_0: The probability distributions of the four plans are identical

H_1: At least two of the probability distribution differ in location

The test statistics is $F_r = \frac{12}{6(4)(4+1)}[16^2 + 7^2 + 23^2 + 14^2] - 3(6)(4+1) = 13$.

The rejection region is $F_r > \chi^2_{0.95,(4)} = 7.81473$.

Since the observed value of the test statistic falls in the rejection region, reject H_0. There is sufficient evidence to indicate a difference in the location for at least two of the probability distribution at $\alpha = 0.05$.

A PRACTICE PROBLEM A certain type of sports injury is treated with one of the three different drugs. Suppose that we want to compare these using a randomized block design by administrating them to the same players in different times. The order in which the drugs are administered is randomly determined for each subject. Results of one such experiment is given below that presents reaction time for these three drugs. What conclusions can be drawn from these data? Employ a 5% significance level.

Players	Drug 1	Rank	Drug 2	Rank	Drug 3	Rank
1	1.21	1	1.48	2	1.56	3
2	1.63	1	1.85	2	2.01	3
3	1.42	1	2.06	3	1.70	2
4	2.43	2	1.98	1	2.64	3
5	1.16	1	1.27	2	1.48	3
6	1.94	1	2.44	2	2.81	3

10.5.7 Spearman Rank Correlation

Suppose that an investigator wishes to determine the strength of the relationships between two or more variables when the assumptions of parametric statistics cannot be met or perhaps when precise measurements are not available. If the data can be converted to ranks, the Spearman rank difference method can be used to measure the correlation between two sets of ranks. It is a quick and simple process based on the following quantity:

$$r_s = 1 - 6\left(\sum D^2\right)/[n(n^2-1)] \quad (10.26)$$

where $\sum D^2$ = the sum of the squared differences between ranks, and n = the number of pairs of ranks.

An alternative method for calculation of r_s is to use the formula for ordinary correlation and replace ranks for the values of x and y respectively. That is

$$r_s = \frac{\sum R(x)R(y) - \left(\sum R(x)\right)\left(\sum R(y)\right)/n}{\sqrt{\sum R^2(x) - \left(\sum R(x)\right)^2/n}\sqrt{\sum R^2(y) - \left(\sum R(y)\right)^2/n}}.$$

EXAMPLE 10.5.6 To demonstrate the method, suppose a manager of sport organization has ranked 10 of his players and the coach has ranked the same 10 players. The research question is the following, are players perceived similarly by the manager and the coach? Table below contains the data.

	Players									
	A	B	C	D	E	F	G	H	I	J
Rank From Manager:	4	6	8	1	7	3	9	5	2	10
Rank From Coach:	1	9	8	6	2	4	3	10	5	7
Differences Squared: (D^2)	9	9	0	25	25	1	36	25	9	9
$\sum D^2 = 148$										

SOLUTION: To answer the research question we first calculate r_s

$$r_s = 1 - 6(148)/10(100-1) = 1 - 888/990 = 0.103$$

Note that using the alternative formula, we obtain

$$r_s = \frac{311 - (55)(55)/10}{\sqrt{385 - (55)^2/10}\sqrt{385 - (55)^2/10}} = 8.5/82.5 = 0.103$$

Table 10.5.2 due to Olds (1938, 1949) contains the values of r_s that are necessary for significance at 0.05 and 0.01 levels. For $n = 10$ and 0.05, the table value for r_s is 0.65 (Table 10.5.2). Because r_s is much lower than this value, we conclude that the correlation coefficient is not significantly different from zero. That is, the rankings of the manager and the coach are very different.

Note that it is possible to use normal approximation and set up a test regarding the correlation. For large n, we have $E(r_s) = \rho_s$ and $\text{Var}(r_s) = 1/(n-1)$ leading to $r_s \sim N(\rho_s, 1/\sqrt{n-1})$. Thus to test $H_0 : \rho_s = 0$, we use the test statistics $T = z = (r_s - 0)\sqrt{n-1} = (0.10 - 0)\sqrt{9} = 0.3$. The rejection region is $z < -1.96$ or $z > 1.96$. Since $0.3 < 1.96$ we cannot reject H_0.

Chapter 10: Nonparametric Statistics

TABLE 10.5.2 Critical Values for Spearman Rank Correlation for a Two-Tailed Test

Sample Size	Level of Significance 0.05	Level of Significance 0.01	Sample Size	Level of Significance 0.05	Level of Significance 0.01
6	0.886		19	0.462	0.608
7	0.786		20	0.450	0.591
8	0.738	0.881	21	0.438	0.576
9	0.683	0.833	22	0.428	0.562
10	0.648	0.818	23	0.418	0.549
11	0.623	0.794	24	0.409	0.537
12	0.591	0.780	25	0.400	0.526
13	0.566	0.745	26	0.392	0.515
14	0.545	0.716	27	0.385	0.505
15	0.525	0.689	28	0.377	0.496
16	0.507	0.666	29	0.370	0.487
17	0.490	0.645	30	0.364	0.478
18	0.476	0.625			

The Spearman r_s can also be computed for variables in which one or both of the values are interval measurements. The measurements simply must be converted to ranks. Remember that tied ranks are averaged. When D^2 is very large, the fraction to be subtracted from 1 is greater than 1, which results in a negative r_s.

The Spearman r_s and the Pearson r do not necessarily yield the same coefficient for the same data, especially if there are a number of tied ranks. Also, because r_s is based on ranks that are not continuous or normally distributed, the coefficient may differ from Pearson r. However, the Spearman r_s is a valuable tool for special cases in which r cannot or should not be used.

EXAMPLE 10.5.7 Six brands of tennis rackets are compared and ranked by two experienced tennis players

Brand	A	B	C	D	E	F
Player 1	6	5	1	3	2	4
Player 2	5	6	2	1	4	3

Do data indicate a positive correlation in the rankings of the two players?

SOLUTION:

$$H_0: \rho_s = 0 \quad vs \quad H_1: \rho_s > 0$$

The test statistics is

$$r_s = 1 - \frac{6\Sigma D^2}{n(n^2-1)} = 1 - \frac{12}{6(6^2-1)} = 0.657$$

Using Table 10.5.2 with $\alpha = 0.05$ and $n = 6$ the rejection region is $r_s > 0.829$.

Since $0.657 < 0.829$ we do not reject H_0. There is insufficient evidence to indicate a positive correlation in the rankings at 5%, significance level.

A PRACTICE PROBLEM Two college coaches have observed 10 soccer players trying for the team and each ranked them from 1 (best) to 10 (worst). Determine whether the coaches' ranks are related (are consistent) by calculating r_s. Use $\alpha = 0.05$ to test its significance.

	Ranking	
Player	Coach 1	Coach 2
1	4	5
2	1	2
3	9	10
4	5	6
5	2	1
6	10	9
7	7	7
8	3	3
9	6	4
10	8	8

A Research Problem At the end of year 2000, several organizations such as ESPN ranked the best athletes of the century. Find two of these rankings and select few of your favorite athletes. Calculate r_s to see if these rankings were consistent. You can do this for rankings for your favorite teams in a sport of your choice.

10.6 EXAMPLES OF APPLICATIONS

As pointed out earlier the nonparametric techniques for comparing two or more populations have wide applicability and are particularly useful when observations cannot be assigned specific values, but can be ranked. The techniques require fewer restrictive assumptions than their parametric counterparts, and allow for a comparison of the probability distributions, rather than specific parameters, of the populations of interest.

The Wilcoxon rank sum test may be used to compare two populations when data arise from an independent sampling design. When a paired difference design is employed, the Wilcoxon signed rank test is appropriate. The Kruskal–Wallis H test uses data from a completely randomized design to compare populations. The Friedman F_r test is appropriate for comparing J populations based on a randomized block design. This section reviews each of these using an example of each.

Chapter 10: Nonparametric Statistics

10.6.1 Wilcoxon Rank Sum Test for Comparing Two Populations, Independent Samples

It is generally known that private universities are more expensive that state universities and usually offer more scholarship. The amount of scholarship available to a member of varsity teams is often an important factor in the selection of a college or university for high school athletes. Nine student athletes from a state university and seven athletes from a private university in the same state were randomly selected and the amount of scholarship received during the last academic year was recorded for each. The results are shown below:

State University ($)	Private University ($)
1,400	1,900
1,500	2,500
1,200	3,000
1,100	4,600
2,000	2,800
1,600	2,200
1,400	1,800
1,700	
2,400	

We want to use the Wilcoxon rank sum test to see if there is a difference in the probability distributions of scholarship amounts at two schools. First, we note that the application of t-test for the difference between two population means requires the assumptions of independent samples selected from normal populations with equal variances. In this case, the assumption of normality of the population distributions may not be reasonable, since histograms of these data show that the distributions of scholarship amounts are very skewed. It may thus be advisable to perform the nonparametric counterpart of the t-test, namely the Wilcoxon rank sum test for independent samples. Here we have

H_0 : The probability distributions of scholarship amounts are identical

H_1 : The probability distributions of the scholarship amounts are different

At $\alpha = 0.05$, the null hypothesis will be rejected if

$$T \leq 41 \quad or \quad T \geq 78$$

where T is the rank sum of the amounts from the smaller sample (private university), and the critical values are obtained from Table B7.

To compute the value of the test statistic, T, we first rank all the sample observations as though they were selected from the same population:

State University		Private University	
Observation ($)	Rank	Observation ($)	Rank
1,400	3.5	1,900	9
1,500	5	2,500	13
1,200	2	3,000	15
2,000	10	2,800	14
1,600	6	2,200	11
1,400	3.5	1,800	8
1,700	7		
2,400	12		

Note that the two equal observations ($1,400) would have received ranks 3 and 4, thus, each is assigned their average rank of $(3+4)/2 = 3.5$.

Now, the rank sum corresponding to the smaller sample is

$$T_B = 9 + 13 + 15 + 16 + 14 + 11 + 8 = 86.$$

This value of the test statistic lies in the rejection region. Thus, we conclude that the distributions of scholarship amounts awarded at the two universities are different. This, as mentioned earlier, is expect.

10.6.2 Wilcoxon Signed Rank Test for Comparing Two Populations, Paired Difference Experiment

Many large sport equipment chains now produce their own goods for sale under house label. They advertise that, although the house brands are of the same quality as the national brands, they can be sold at lower prices because of a lower production costs. A comparison of the daily sales (number of units sold) of 11 products at a local store produced the data shown in the following table.

It is desired to compare the sales of the national brand and house brand products. We want to specify what type of experimental design is represented here and perform the Wilcoxon signed rank test for the hypothesis that the probability distributions of the sales of national and house brands are identical using $\alpha = 0.01$. The alternative hypothesis of interest is that, the sales of national brands tend to exceed those of house brands.

Chapter 10: Nonparametric Statistics

Product	National Brand	House Brand
Running shoes	303	237
Tank top	504	428
Tennis shoes	205	127
Hat	157	136
Volleyball	205	49
Tennis racket	273	302
Socks	394	147
Bags	93	248
Sneakers	188	188
Indoor soccer shoes	126	147
Baseball glove	303	29

First we note that, this is a paired difference experiment, and the analysis will be based on the differences between the pairs of measurements. The Wilcoxon signed rank test for the paired difference design provides a test for the following:

H_0: The probability distributions of the sales for national and house brand products are identical

H_1: The sales for national brands tend to exceed those for house brands

To compute the value of the test statistic, we first obtain the ranks of the absolute values of the differences between the measurements. The calculations are shown in the following table:

	Sales				
Product	National Brand	House Brand	Difference (National-House)	Absolute Value of Difference	Rank of Absolute Value
1	303	237	66	66	4
2	504	428	76	76	5
3	205	127	78	78	6
4	157	136	21	21	1.5
5	205	49	156	156	8
6	273	302	−29	29	3
7	394	147	247	247	9
8	93	248	−155	155	7
9	188	188	0	0	Drop
10	126	147	−21	21	1.5
11	303	29	274	274	1.0

Observe that differences of zero are eliminated, since they do not contribute to the rank sums. In addition, as before ties in absolute differences receive the average of the ranks they would be assigned if they were unequal but successive measurements. Thus, the absolute differences tied at 21, which would have received ranks 1 and 2, are each assigned the average rank of $(1+2)/2 = 1.5$.

For this one-sided test, the test statistic is the negative rank sum. This is because, if the alternative hypothesis is true, we expect most of the national minus house brand sale differences to be positive. Thus, we would expect the negative rank sum T to be small if the alternative hypothesis is true. The critical value will be based on $n = 10$ paired observations and is obtained from Table B9. Thus, at significance level $\alpha = 0.01$, the null hypothesis will be rejected if $T \leq 5$. We now compute the negative rank sum, the sum of the ranks of the negative differences. It is

$$T = 3 + 7 + 1.5 = 11.5$$

This value does not lie within the rejection region. There is insufficient evidence (at $\alpha = 0.01$) to indicate that the sales of national brands (significantly) exceed sales of the house bands.

10.6.3 Kruskal–Wallis H Test for a Completely Randomized Design

A major sport organization has initiated a campaign to make coming to the games a more pleasant experience for the fans. Previous surveys conducted by the organization have indicated that the aspect most in need of improvement is ticket sale (or ticket pick up). A consulting firm has provided the following data on time required (in minutes) to buy tickets and enter the facility, at four of the nation's large cities:

New York	Chicago	Washington	Boston
14.2	11.9	10.8	13.1
12.8	12.2	12.1	11.9
11.9	12.9	13.1	13.2
15.3	11.8	12.8	13.0
13.9	12.8	11.9	12.5
13.0	13.1		

We want to use a nonparametric procedure to test for a difference in the probability distributions of time required to obtain ticket and enter the facility in the four cities using $\alpha = 0.05$. We also want to answer the following questions: what type of experimental design is represented here? what assumptions are necessary for the analysis of these data using ANOVA.

First, we note that the data are from a completely randomized design, in which independent random samples were selected from each of the four populations of times to be compared. The ANOVA requires the assumption that the four distributions are approximately normal, with equal variances. If we do not want to make these restrictive assumptions, then a nonparametric procedure may be preferred. Here, we will perform a Kruskal–Wallis test of

H_0: The probability distributions of the populations of times required to buy ticket and enter the facility are identical for the four cities

H_1: At least two of the distributions differ in location (mean)

Chapter 10: Nonparametric Statistics

At significance level α, the null hypothesis will be rejected if the value of the test statistic, H, exceeds $\chi^2_{1-\alpha}$ with $(J-1)$ degrees of freedom, where J is the number of independent samples upon which the test is based. Thus, for $\alpha = 0.05$ and $J - 1 = 3$, H_0 will be rejected if $H > 7.815$. To compute the value of H, it is necessary to obtain the rank sum for each of the four samples, where the rank of each observation is computed according to its relative magnitude when all four samples are combined. The rankings are shown in the following table:

New York		Chicago		Washington		Boston	
Time	Rank	Time	Rank	Time	Rank	Time	Rank
14.2	21	11.9	4.5	10.8	1	13.1	17
12.8	11	12.2	8	12.1	7	11.9	4.5
11.9	4.5	12.9	13	13.1	17	13.2	19
15.3	22	11.8	2	12.8	11	13.0	14.5
13.9	20	12.8	11	11.9	4.5	12.5	9
13.0	14.5	13.1	17				
	$R_1 = 93.0$		$R_2 = 55.5$		$R_3 = 40.5$		$R_4 = 64.0$

Tied observations are handled in the usual manner, by assigning the average value of the ranks to each of the tied observations.

$$H = \frac{12}{n(n+1)} \Sigma \frac{R_i^2}{n_i} - 3(n+1).$$

For this example, $n_1 = 6, n_2 = 6, n_3 = 5, n_4 = 5, J = 4$, and $n = n_1 + n_2 + n_3 + n_4 = 22$. Substitution of these values and the rank sums computed in the table yields:

$$H = \frac{12}{22(23)} \left[\frac{(93.0)^2}{6} + \frac{(55.5)^2}{6} + \frac{(40.5)^2}{5} + \frac{(64.0)^2}{5} \right] - 3(23) = 73.57 - 69 = 4.57.$$

Since the computed value of H does not exceed the critical value of 7.815, there is insufficient evidence to support the alternative hypothesis that the probability distributions of service times differ in location for at least two of the cities.

10.6.4 The Friedman F_r Test for a Randomized Block Design

It is generally believed that most people are influenced by a product's price, label, and advertising. To test this, a recreational facility that offers ice cream in its stand invited college students to participate in a taste test. Each student was given three containers of ice cream, in a randomized order. Containers I and II contained the same, inexpensive local brand of ice cream, but students were told that ice cream I was very expensive and that ice cream II was not. Container III contained a very expensive brand but students were told nothing about it. Tasters were asked to rate each of the three ice cream on a scale from 1 to 20, with higher values indicating better taste. Ratings submitted by eight randomly selected students are shown below:

	Ratings		
Student	Ice Cream I	Ice Cream II	Ice Cream III
1	17	11	13
2	19	15	12
3	16	15	19
4	13	11	17
5	15	15	11
6	17	14	17
7	19	15	17
8	20	16	19

We want to use a nonparametric test procedure to determine whether there is a difference in the probability distributions of student ratings using $\alpha = 0.10$ and discuss the experimental design employed here.

First note that this represents a randomized block design with $K = 8$ blocks (students) and $J = 3$ treatments (types of ice cream). The J population means may be compared using the ANOVA techniques. However, these parametric methods required the assumptions that the populations of student ratings for the three ice creams are normally distributed and that their variance are all equal. Here, we will perform the nonparametric counterpart, which requires no distributional assumptions. The elements of the Friedman F_r test for this randomized block design are as follows:

H_0: The probability distributions of student ratings are identical for the three ice creams

H_1: At least two of the probability distributions differ in location

At significance level α, the null hypothesis is rejected if the test statistic, F_r, exceeds the critical value $\chi^2_{1-\alpha,(J-1)}$. For this example, we have $J - 1 = 3 - 1 = 2$ degrees of freedom and $\alpha = 0.10$. Thus, H_0 will be rejected if $F_r > 4.605$. In order to compute the value of F_r, it is first required to rank the observations within each block (student), and then obtain the rank sums for each of the three treatments (ice cream). The results are shown below:

	Ice Cream *I*		Ice Cream *II*		Ice Cream *III*	
Student	Rating	Rank	Rating	Rank	Rating	Rank
1	17	3	11	1	13	2
2	19	3	15	2	12	1
3	16	2	15	1	19	3
4	13	2	11	1	17	3
5	15	2.5	15	2.5	11	1
6	17	2.5	14	1	17	2.5
7	19	3	15	1	17	2
8	20	3	16	1	19	2
		$R_1 = 21.0$		$R_2 = 120.5$		$R_3 = 16.5$

Chapter 10: Nonparametric Statistics

Again the tied observations within blocks are assigned the average value of the ranks each tied observation would receive if they were unequal but successive measurements. The test statistic is computed as follows:

$$F_r = \frac{12}{KJ(J+1)} \Sigma R_j^2 - 3K(J+1)$$
$$= \frac{12}{8(3)(4)}[(21.0)^2 + (10.5)^2 + (16.5)^2] - 3(8)(4) = 102.9375 - 96 = 6.9375.$$

Since the calculated value of $F_r = 6.9375$ exceeds the critical value of 4.605, we conclude that the student rating distributions differ in location for at least two of the ice creams at 10% significance level.

10.7 USING R

In this section, we demonstrate how to perform mentioned nonparametrical analysis using R software. In fact, you will find some functions were used before.

10.7.1 Sign Test

We focus on small sample size cases only because it will be a normal test when the size is large enough. In order to run nonparametric sign test, we must install and load a package named *BSDA* first. This is because we have no built-in function to do so, however, the authors of package BSDA created a function for us. In R, one can install packages available on CRAN, the public R package repository, using **install.packages()** function. The first parameter of this function call must be a String value indicating the package name. So you can install BSDA package like this.

```
install.packages('BSDA')
```

That's it. Whenever you see success message in the R console, the package is ready to use.

Now, we need to load a package, thus the objects and functions inside the package will be available for us. There are two popular ways to load a package: **library()** and **require()**, for example,

```
library(BSDA)
```

Recall the example in Section 10.5.1, we would like to check whether the median sales would be 40. The function **SIGN.test()** in BSDA package will be used.

```
# In fact we only need load package ONCE.
library(BSDA)
sale = c(22, 16, 40, 43, 18, 30, 29, 32, 37, 36, 39, 34, 39, 45, 28,
         36, 40, 34, 39, 52)
SIGN.test(sale, md=40)
```

You should be able to get following results.

```
        One-sample Sign-Test

data: sale
s = 3, p-value = 0.007538
alternative hypothesis: true median is not equal to 40
95 percent confidence interval:
 30.23294 39.00000
sample estimates:
median of x
        36

Achieved and Interpolated Confidence Intervals:

                 Conf.Level   L.E.pt    U.E.pt
Lower Achieved CI    0.8847  32.0000        39
Interpolated   CI    0.9500  30.2329        39
Upper Achieved CI    0.9586  30.0000        39
```

The test statistics is 3 and *p*-value 0.007538 suggests rejection of null hypothesis.

10.7.2 Wilcoxon Test

Fortunately, there is a built-in function for Wilcoxon Rank Sum Test named **wilcox.test()**. You may provide two vectors of data for comparison. If the data are paired, you need to set the parameter **paired** to be TRUE. Recall the Example 10.5.3, we may solve it in this way.

```
before = c(33, 30, 40, 27, 18, 26, 35, 20, 38, 16, 26, 21, 18, 24, 11)
after  = c(36, 31, 37, 36, 24, 25, 35, 16, 33, 24, 28, 20, 20, 24, 18)
wilcox.test(before, after, paired = TRUE, exact=FALSE)


        Wilcoxon signed rank test with continuity correction

data: before and after
V = 27.5, p-value = 0.2205
alternative hypothesis: true location shift is not equal to 0
```

10.7.3 Friedman Test

A built-in function named **friedman.test()** can be used for this analysis. Recall Example 10.5.5, there were 12 subjects taking 3 different methods.

```
# Make a matrix
fat = matrix(c(16, 11.6, 18.1, 16.3, 12, 12.5,
9.3, 18.8, 19.2, 22.3, 20.7, 24.1,
15.7, 10.9, 18, 16.8, 13.1, 12.2,
8.8, 17.5, 19.7, 22.8, 20.3, 23.7,
15.2, 11.5, 17.6, 17, 12.8, 12.3,
9, 18, 19.7, 21.6, 19.6, 24.4), ncol = 3, byrow = FALSE)
friedman.test(fat)
```

    Friedman rank sum test

data: fat
Friedman chi-squared = 2.0851, df = 2, p-value = 0.3526

However, we may type in data in vectors, therefore, we need to provide a "formula" to the function, just like what we did in regression. Recall the example in Section 10.5.6: an experiment involving players and their performance was conducted using a randomized block design involving for plans (treatments) using six players (blocks). The performances are ranked for each player. Use the Friedman test for a randomized block design to determine whether the data provide sufficient evidence to indicate that at least two of the plans probability distributions differ in location. Using $\alpha = 0.05$.

	\multicolumn{6}{c}{Players}
Plans	1	2	3	4	5	6
A	3	3	2	3	2	3
B	1	1	1	2	1	1
C	4	4	3	4	4	4
D	2	2	4	1	3	2

```
d = c(3, 3, 2, 3, 2, 3,
1, 1, 1, 2, 1, 1,
4, 4, 3, 4, 4, 4,
2, 2, 4, 1, 3, 2)
plan = as.factor(c(rep('A', 6), rep('B', 6),
rep('C', 6), rep('D', 6)))
player = as.factor(rep(1:6, 4))
friedman.test(d~plan|player)
```

    Friedman rank sum test

data: d and plan and player
Friedman chi-squared = 13, df = 3, p-value = 0.004637

In above code, we used **rep()** function which simply repeat a symbol for given number of times. For example **rep('A', 6)** gave 'A', 'A', 'A', 'A', 'A', 'A', and **rep(1:6, 4)** repeats a sequence, 1 through 6, 4 times, i.e. **1, 2, 3, 4, 5, 6, 1, 2, 3, 4, 5, 6, 1, 2, 3, 4, 5, 6, 1, 2, 3, 4, 5, 6**.

10.7.4 Spearman's correlation

We keep using **cor.test()** function to calculate the correlation. However, we need to specify a parameter **method** with value of "spearman" in String. Let's solve the Example 10.5.6 below.

```
manager = c(4, 6, 8, 1, 7, 3, 9, 5, 2, 10)
coach = c(1, 9, 8, 6, 2, 4, 3, 10, 5, 7)
cor.test(manager, coach, method = 'spearman')
```

```
        Spearman's rank correlation rho

data: manager and coach
S = 148, p-value = 0.785
alternative hypothesis: true rho is not equal to 0
sample estimates:
rho
0.1030303
```

You will see the test statistic, $\sum D^2$, is 148, and *p*-value is 0.785. The correlation is 0.103.

A PRACTICE PROBLEM Try to solve Example 10.5.7 using R and interpret the output.

REFERENCES

Olds, E. G. 1938. "Distribution of Sums of Squares of Rank Differences for Small Numbers of Individuals." *Annals of Mathematical Statistics* 9:133–48.

Olds, E. G. 1949. "The 5 Percent Significance Levels of Sums of Squares of Rank Difference and Correction." *Annals of Mathematical Statistics* 20:117–8.

Table B1. Binomial Distribution

n	x	0.1	0.2	0.25	0.3	0.4	0.5	0.6	0.7	0.75	0.8	0.9
2	0	0.81	0.64	0.563	0.49	0.36	0.25	0.16	0.09	0.063	0.04	0.01
	1	0.18	0.32	0.375	0.42	0.48	0.5	0.48	0.42	0.375	0.32	0.18
	2	0.01	0.04	0.063	0.09	0.16	0.25	0.36	0.49	0.563	0.64	0.81
3	0	0.729	0.512	0.422	0.343	0.216	0.125	0.064	0.027	0.016	0.008	0.001
	1	0.243	0.384	0.422	0.441	0.432	0.375	0.288	0.189	0.141	0.096	0.027
	2	0.027	0.096	0.141	0.189	0.288	0.375	0.432	0.441	0.422	0.384	0.243
	3	0.001	0.008	0.016	0.027	0.064	0.125	0.216	0.343	0.422	0.512	0.729
4	0	0.656	0.41	0.316	0.24	0.13	0.063	0.026	0.008	0.004	0.002	0
	1	0.292	0.41	0.422	0.412	0.346	0.25	0.154	0.076	0.047	0.026	0.004
	2	0.049	0.154	0.211	0.265	0.346	0.375	0.346	0.265	0.211	0.154	0.049
	3	0.004	0.026	0.047	0.076	0.154	0.25	0.346	0.412	0.422	0.41	0.292
	4	0	0.002	0.004	0.008	0.026	0.063	0.13	0.24	0.316	0.41	0.656
5	0	0.59	0.328	0.237	0.168	0.078	0.031	0.01	0.002	0.001	0	0
	1	0.328	0.41	0.396	0.36	0.259	0.156	0.077	0.028	0.015	0.006	0
	2	0.073	0.205	0.264	0.309	0.346	0.313	0.23	0.132	0.088	0.051	0.008
	3	0.008	0.051	0.088	0.132	0.23	0.313	0.346	0.309	0.264	0.205	0.073
	4	0	0.006	0.015	0.028	0.077	0.156	0.259	0.36	0.396	0.41	0.328
	5	0	0	0.001	0.002	0.01	0.031	0.078	0.168	0.237	0.328	0.59
6	0	0.531	0.262	0.178	0.118	0.047	0.016	0.004	0.001	0	0	0
	1	0.354	0.393	0.356	0.303	0.187	0.094	0.037	0.01	0.004	0.002	0
	2	0.098	0.246	0.297	0.324	0.311	0.234	0.138	0.06	0.033	0.015	0.001
	3	0.015	0.082	0.132	0.185	0.276	0.313	0.276	0.185	0.132	0.082	0.015
	4	0.001	0.015	0.033	0.06	0.138	0.234	0.311	0.324	0.297	0.246	0.098
	5	0	0.002	0.004	0.01	0.037	0.094	0.187	0.303	0.356	0.393	0.354
	6	0	0	0	0.001	0.004	0.016	0.047	0.118	0.178	0.262	0.531
7	0	0.478	0.21	0.133	0.082	0.028	0.008	0.002	0	0	0	0
	1	0.372	0.367	0.311	0.247	0.131	0.055	0.017	0.004	0.001	0	0
	2	0.124	0.275	0.311	0.318	0.261	0.164	0.077	0.025	0.012	0.004	0
	3	0.023	0.115	0.173	0.227	0.29	0.273	0.194	0.097	0.058	0.029	0.003
	4	0.003	0.029	0.058	0.097	0.194	0.273	0.29	0.227	0.173	0.115	0.023
	5	0	0.004	0.012	0.025	0.077	0.164	0.261	0.318	0.311	0.275	0.124
	6	0	0	0.001	0.004	0.017	0.055	0.131	0.247	0.311	0.367	0.372
	7	0	0	0	0	0.002	0.008	0.028	0.082	0.133	0.21	0.478
8	0	0.43	0.168	0.1	0.058	0.017	0.004	0.001	0	0	0	0
	1	0.383	0.336	0.267	0.198	0.09	0.031	0.008	0.001	0	0	0
	2	0.149	0.294	0.311	0.296	0.209	0.109	0.041	0.01	0.004	0.001	0
	3	0.033	0.147	0.208	0.254	0.279	0.219	0.124	0.047	0.023	0.009	0
	4	0.005	0.046	0.087	0.136	0.232	0.273	0.232	0.136	0.087	0.046	0.005
	5	0	0.009	0.023	0.047	0.124	0.219	0.279	0.254	0.208	0.147	0.033
	6	0	0.001	0.004	0.01	0.041	0.109	0.209	0.296	0.311	0.294	0.149
	7	0	0	0	0.001	0.008	0.031	0.09	0.198	0.267	0.336	0.383
	8	0	0	0	0	0.001	0.004	0.017	0.058	0.1	0.168	0.43
9	0	0.387	0.134	0.075	0.04	0.01	0.002	0	0	0	0	0
	1	0.387	0.302	0.225	0.156	0.06	0.018	0.004	0	0	0	0
	2	0.172	0.302	0.3	0.267	0.161	0.07	0.021	0.004	0.001	0	0
	3	0.045	0.176	0.234	0.267	0.251	0.164	0.074	0.021	0.009	0.003	0

Table B1. Binomial Distribution

n	k											
	4	0.007	0.066	0.117	0.172	0.251	0.246	0.167	0.074	0.039	0.017	0.001
	5	0.001	0.017	0.039	0.074	0.167	0.246	0.251	0.172	0.117	0.066	0.007
	6	0	0.003	0.009	0.021	0.074	0.164	0.251	0.267	0.234	0.176	0.045
	7	0	0	0.001	0.004	0.021	0.07	0.161	0.267	0.3	0.302	0.172
	8	0	0	0	0	0.004	0.018	0.06	0.156	0.225	0.302	0.387
	9	0	0	0	0	0	0.002	0.01	0.04	0.075	0.134	0.387
10	0	0.349	0.107	0.056	0.028	0.006	0.001	0	0	0	0	0
	1	0.387	0.268	0.188	0.121	0.04	0.01	0.002	0	0	0	0
	2	0.194	0.302	0.282	0.233	0.121	0.044	0.011	0.001	0	0	0
	3	0.057	0.201	0.25	0.267	0.215	0.117	0.042	0.009	0.003	0.001	0
	4	0.011	0.088	0.146	0.2	0.251	0.205	0.111	0.037	0.016	0.006	0
	5	0.001	0.026	0.058	0.103	0.201	0.246	0.201	0.103	0.058	0.026	0.001
	6	0	0.006	0.016	0.037	0.111	0.205	0.251	0.2	0.146	0.088	0.011
	7	0	0.001	0.003	0.009	0.042	0.117	0.215	0.267	0.25	0.201	0.057
	8	0	0	0	0.001	0.011	0.044	0.121	0.233	0.282	0.302	0.194
	9	0	0	0	0	0.002	0.01	0.04	0.121	0.188	0.268	0.387
	10	0	0	0	0	0	0.001	0.006	0.028	0.056	0.107	0.349
11	0	0.314	0.086	0.042	0.02	0.004	0	0	0	0	0	0
	1	0.384	0.236	0.155	0.093	0.027	0.005	0.001	0	0	0	0
	2	0.213	0.295	0.258	0.2	0.089	0.027	0.005	0.001	0	0	0
	3	0.071	0.221	0.258	0.257	0.177	0.081	0.023	0.004	0.001	0	0
	4	0.016	0.111	0.172	0.22	0.236	0.161	0.07	0.017	0.006	0.002	0
	5	0.002	0.039	0.08	0.132	0.221	0.226	0.147	0.057	0.027	0.01	0
	6	0	0.01	0.027	0.057	0.147	0.226	0.221	0.132	0.08	0.039	0.002
	7	0	0.002	0.006	0.017	0.07	0.161	0.236	0.22	0.172	0.111	0.016
	8	0	0	0.001	0.004	0.023	0.081	0.177	0.257	0.258	0.221	0.071
	9	0	0	0	0.001	0.005	0.027	0.089	0.2	0.258	0.295	0.213
	10	0	0	0	0	0.001	0.005	0.027	0.093	0.155	0.236	0.384
	11	0	0	0	0	0	0	0.004	0.02	0.042	0.086	0.314
12	0	0.282	0.069	0.032	0.014	0.002	0	0	0	0	0	0
	1	0.377	0.206	0.127	0.071	0.017	0.003	0	0	0	0	0
	2	0.23	0.283	0.232	0.168	0.064	0.016	0.002	0	0	0	0
	3	0.085	0.236	0.258	0.24	0.142	0.054	0.012	0.001	0	0	0
	4	0.021	0.133	0.194	0.231	0.213	0.121	0.042	0.008	0.002	0.001	0
	5	0.004	0.053	0.103	0.158	0.227	0.193	0.101	0.029	0.011	0.003	0
	6	0	0.016	0.04	0.079	0.177	0.226	0.177	0.079	0.04	0.016	0
	7	0	0.003	0.011	0.029	0.101	0.193	0.227	0.158	0.103	0.053	0.004
	8	0	0.001	0.002	0.008	0.042	0.121	0.213	0.231	0.194	0.133	0.021
	9	0	0	0	0.001	0.012	0.054	0.142	0.24	0.258	0.236	0.085
	10	0	0	0	0	0.002	0.016	0.064	0.168	0.232	0.283	0.23
	11	0	0	0	0	0	0.003	0.017	0.071	0.127	0.206	0.377
	12	0	0	0	0	0	0	0.002	0.014	0.032	0.069	0.282
13	0	0.254	0.055	0.024	0.01	0.001	0	0	0	0	0	0
	1	0.367	0.179	0.103	0.054	0.011	0.002	0	0	0	0	0
	2	0.245	0.268	0.206	0.139	0.045	0.01	0.001	0	0	0	0
	3	0.1	0.246	0.252	0.218	0.111	0.035	0.006	0.001	0	0	0
	4	0.028	0.154	0.21	0.234	0.184	0.087	0.024	0.003	0.001	0	0

Table B1. Binomial Distribution

n	k											
	5	0.006	0.069	0.126	0.18	0.221	0.157	0.066	0.014	0.005	0.001	0
	6	0.001	0.023	0.056	0.103	0.197	0.209	0.131	0.044	0.019	0.006	0
	7	0	0.006	0.019	0.044	0.131	0.209	0.197	0.103	0.056	0.023	0.001
	8	0	0.001	0.005	0.014	0.066	0.157	0.221	0.18	0.126	0.069	0.006
	9	0	0	0.001	0.003	0.024	0.087	0.184	0.234	0.21	0.154	0.028
	10	0	0	0	0.001	0.006	0.035	0.111	0.218	0.252	0.246	0.1
	11	0	0	0	0	0.001	0.01	0.045	0.139	0.206	0.268	0.245
	12	0	0	0	0	0	0.002	0.011	0.054	0.103	0.179	0.367
	13	0	0	0	0	0	0	0.001	0.01	0.024	0.055	0.254
14	0	0.229	0.044	0.018	0.007	0.001	0	0	0	0	0	0
	1	0.356	0.154	0.083	0.041	0.007	0.001	0	0	0	0	0
	2	0.257	0.25	0.18	0.113	0.032	0.006	0.001	0	0	0	0
	3	0.114	0.25	0.24	0.194	0.085	0.022	0.003	0	0	0	0
	4	0.035	0.172	0.22	0.229	0.155	0.061	0.014	0.001	0	0	0
	5	0.008	0.086	0.147	0.196	0.207	0.122	0.041	0.007	0.002	0	0
	6	0.001	0.032	0.073	0.126	0.207	0.183	0.092	0.023	0.008	0.002	0
	7	0	0.009	0.028	0.062	0.157	0.209	0.157	0.062	0.028	0.009	0
	8	0	0.002	0.008	0.023	0.092	0.183	0.207	0.126	0.073	0.032	0.001
	9	0	0	0.002	0.007	0.041	0.122	0.207	0.196	0.147	0.086	0.008
	10	0	0	0	0.001	0.014	0.061	0.155	0.229	0.22	0.172	0.035
	11	0	0	0	0	0.003	0.022	0.085	0.194	0.24	0.25	0.114
	12	0	0	0	0	0.001	0.006	0.032	0.113	0.18	0.25	0.257
	13	0	0	0	0	0	0.001	0.007	0.041	0.083	0.154	0.356
	14	0	0	0	0	0	0	0.001	0.007	0.018	0.044	0.229
15	0	0.206	0.035	0.013	0.005	0	0	0	0	0	0	0
	1	0.343	0.132	0.067	0.031	0.005	0	0	0	0	0	0
	2	0.267	0.231	0.156	0.092	0.022	0.003	0	0	0	0	0
	3	0.129	0.25	0.225	0.17	0.063	0.014	0.002	0	0	0	0
	4	0.043	0.188	0.225	0.219	0.127	0.042	0.007	0.001	0	0	0
	5	0.01	0.103	0.165	0.206	0.186	0.092	0.024	0.003	0.001	0	0
	6	0.002	0.043	0.092	0.147	0.207	0.153	0.061	0.012	0.003	0.001	0
	7	0	0.014	0.039	0.081	0.177	0.196	0.118	0.035	0.013	0.003	0
	8	0	0.003	0.013	0.035	0.118	0.196	0.177	0.081	0.039	0.014	0
	9	0	0.001	0.003	0.012	0.061	0.153	0.207	0.147	0.092	0.043	0.002
	10	0	0	0.001	0.003	0.024	0.092	0.186	0.206	0.165	0.103	0.01
	11	0	0	0	0.001	0.007	0.042	0.127	0.219	0.225	0.188	0.043
	12	0	0	0	0	0.002	0.014	0.063	0.17	0.225	0.25	0.129
	13	0	0	0	0	0	0.003	0.022	0.092	0.156	0.231	0.267
	14	0	0	0	0	0	0	0.005	0.031	0.067	0.132	0.343
	15	0	0	0	0	0	0	0	0.005	0.013	0.035	0.206
16	0	0.185	0.028	0.01	0.003	0	0	0	0	0	0	0
	1	0.329	0.113	0.053	0.023	0.003	0	0	0	0	0	0
	2	0.275	0.211	0.134	0.073	0.015	0.002	0	0	0	0	0
	3	0.142	0.246	0.208	0.146	0.047	0.009	0.001	0	0	0	0
	4	0.051	0.2	0.225	0.204	0.101	0.028	0.004	0	0	0	0
	5	0.014	0.12	0.18	0.21	0.162	0.067	0.014	0.001	0	0	0
	6	0.003	0.055	0.11	0.165	0.198	0.122	0.039	0.006	0.001	0	0

Table B1. Binomial Distribution

n	k											
	7	0	0.02	0.052	0.101	0.189	0.175	0.084	0.019	0.006	0.001	0
	8	0	0.006	0.02	0.049	0.142	0.196	0.142	0.049	0.02	0.006	0
	9	0	0.001	0.006	0.019	0.084	0.175	0.189	0.101	0.052	0.02	0
	10	0	0	0.001	0.006	0.039	0.122	0.198	0.165	0.11	0.055	0.003
	11	0	0	0	0.001	0.014	0.067	0.162	0.21	0.18	0.12	0.014
	12	0	0	0	0	0.004	0.028	0.101	0.204	0.225	0.2	0.051
	13	0	0	0	0	0.001	0.009	0.047	0.146	0.208	0.246	0.142
	14	0	0	0	0	0	0.002	0.015	0.073	0.134	0.211	0.275
	15	0	0	0	0	0	0	0.003	0.023	0.053	0.113	0.329
	16	0	0	0	0	0	0	0	0.003	0.01	0.028	0.185
17	0	0.167	0.023	0.008	0.002	0	0	0	0	0	0	0
	1	0.315	0.096	0.043	0.017	0.002	0	0	0	0	0	0
	2	0.28	0.191	0.114	0.058	0.01	0.001	0	0	0	0	0
	3	0.156	0.239	0.189	0.125	0.034	0.005	0	0	0	0	0
	4	0.06	0.209	0.221	0.187	0.08	0.018	0.002	0	0	0	0
	5	0.017	0.136	0.191	0.208	0.138	0.047	0.008	0.001	0	0	0
	6	0.004	0.068	0.128	0.178	0.184	0.094	0.024	0.003	0.001	0	0
	7	0.001	0.027	0.067	0.12	0.193	0.148	0.057	0.009	0.002	0	0
	8	0	0.008	0.028	0.064	0.161	0.185	0.107	0.028	0.009	0.002	0
	9	0	0.002	0.009	0.028	0.107	0.185	0.161	0.064	0.028	0.008	0
	10	0	0	0.002	0.009	0.057	0.148	0.193	0.12	0.067	0.027	0.001
	11	0	0	0.001	0.003	0.024	0.094	0.184	0.178	0.128	0.068	0.004
	12	0	0	0	0.001	0.008	0.047	0.138	0.208	0.191	0.136	0.017
	13	0	0	0	0	0.002	0.018	0.08	0.187	0.221	0.209	0.06
	14	0	0	0	0	0	0.005	0.034	0.125	0.189	0.239	0.156
	15	0	0	0	0	0	0.001	0.01	0.058	0.114	0.191	0.28
	16	0	0	0	0	0	0	0.002	0.017	0.043	0.096	0.315
	17	0	0	0	0	0	0	0	0.002	0.008	0.023	0.167
18	0	0.15	0.018	0.006	0.002	0	0	0	0	0	0	0
	1	0.3	0.081	0.034	0.013	0.001	0	0	0	0	0	0
	2	0.284	0.172	0.096	0.046	0.007	0.001	0	0	0	0	0
	3	0.168	0.23	0.17	0.105	0.025	0.003	0	0	0	0	0
	4	0.07	0.215	0.213	0.168	0.061	0.012	0.001	0	0	0	0
	5	0.022	0.151	0.199	0.202	0.115	0.033	0.004	0	0	0	0
	6	0.005	0.082	0.144	0.187	0.166	0.071	0.015	0.001	0	0	0
	7	0.001	0.035	0.082	0.138	0.189	0.121	0.037	0.005	0.001	0	0
	8	0	0.012	0.038	0.081	0.173	0.167	0.077	0.015	0.004	0.001	0
	9	0	0.003	0.014	0.039	0.128	0.185	0.128	0.039	0.014	0.003	0
	10	0	0.001	0.004	0.015	0.077	0.167	0.173	0.081	0.038	0.012	0
	11	0	0	0.001	0.005	0.037	0.121	0.189	0.138	0.082	0.035	0.001
	12	0	0	0	0.001	0.015	0.071	0.166	0.187	0.144	0.082	0.005
	13	0	0	0	0	0.004	0.033	0.115	0.202	0.199	0.151	0.022
	14	0	0	0	0	0.001	0.012	0.061	0.168	0.213	0.215	0.07
	15	0	0	0	0	0	0.003	0.025	0.105	0.17	0.23	0.168
	16	0	0	0	0	0	0.001	0.007	0.046	0.096	0.172	0.284
	17	0	0	0	0	0	0	0.001	0.013	0.034	0.081	0.3
	18	0	0	0	0	0	0	0	0.002	0.006	0.018	0.15

Table B1. Binomial Distribution

n	k	0.05	0.10	0.15	0.20	0.30	0.40	0.50	0.60	0.70	0.80	0.90
19	0	0.135	0.014	0.004	0.001	0	0	0	0	0	0	0
	1	0.285	0.068	0.027	0.009	0.001	0	0	0	0	0	0
	2	0.285	0.154	0.08	0.036	0.005	0	0	0	0	0	0
	3	0.18	0.218	0.152	0.087	0.017	0.002	0	0	0	0	0
	4	0.08	0.218	0.202	0.149	0.047	0.007	0.001	0	0	0	0
	5	0.027	0.164	0.202	0.192	0.093	0.022	0.002	0	0	0	0
	6	0.007	0.095	0.157	0.192	0.145	0.052	0.008	0.001	0	0	0
	7	0.001	0.044	0.097	0.153	0.18	0.096	0.024	0.002	0	0	0
	8	0	0.017	0.049	0.098	0.18	0.144	0.053	0.008	0.002	0	0
	9	0	0.005	0.02	0.051	0.146	0.176	0.098	0.022	0.007	0.001	0
	10	0	0.001	0.007	0.022	0.098	0.176	0.146	0.051	0.02	0.005	0
	11	0	0	0.002	0.008	0.053	0.144	0.18	0.098	0.049	0.017	0
	12	0	0	0	0.002	0.024	0.096	0.18	0.153	0.097	0.044	0.001
	13	0	0	0	0.001	0.008	0.052	0.145	0.192	0.157	0.095	0.007
	14	0	0	0	0	0.002	0.022	0.093	0.192	0.202	0.164	0.027
	15	0	0	0	0	0.001	0.007	0.047	0.149	0.202	0.218	0.08
	16	0	0	0	0	0	0.002	0.017	0.087	0.152	0.218	0.18
	17	0	0	0	0	0	0	0.005	0.036	0.08	0.154	0.285
	18	0	0	0	0	0	0	0.001	0.009	0.027	0.068	0.285
	19	0	0	0	0	0	0	0	0.001	0.004	0.014	0.135
20	0	0.122	0.012	0.003	0.001	0	0	0	0	0	0	0
	1	0.27	0.058	0.021	0.007	0	0	0	0	0	0	0
	2	0.285	0.137	0.067	0.028	0.003	0	0	0	0	0	0
	3	0.19	0.205	0.134	0.072	0.012	0.001	0	0	0	0	0
	4	0.09	0.218	0.19	0.13	0.035	0.005	0	0	0	0	0
	5	0.032	0.175	0.202	0.179	0.075	0.015	0.001	0	0	0	0
	6	0.009	0.109	0.169	0.192	0.124	0.037	0.005	0	0	0	0
	7	0.002	0.055	0.112	0.164	0.166	0.074	0.015	0.001	0	0	0
	8	0	0.022	0.061	0.114	0.18	0.12	0.035	0.004	0.001	0	0
	9	0	0.007	0.027	0.065	0.16	0.16	0.071	0.012	0.003	0	0
	10	0	0.002	0.01	0.031	0.117	0.176	0.117	0.031	0.01	0.002	0
	11	0	0	0.003	0.012	0.071	0.16	0.16	0.065	0.027	0.007	0
	12	0	0	0.001	0.004	0.035	0.12	0.18	0.114	0.061	0.022	0
	13	0	0	0	0.001	0.015	0.074	0.166	0.164	0.112	0.055	0.002
	14	0	0	0	0	0.005	0.037	0.124	0.192	0.169	0.109	0.009
	15	0	0	0	0	0.001	0.015	0.075	0.179	0.202	0.175	0.032
	16	0	0	0	0	0	0.005	0.035	0.13	0.19	0.218	0.09
	17	0	0	0	0	0	0.001	0.012	0.072	0.134	0.205	0.19
	18	0	0	0	0	0	0	0.003	0.028	0.067	0.137	0.285
	19	0	0	0	0	0	0	0	0.007	0.021	0.058	0.27
	20	0	0	0	0	0	0	0	0.001	0.003	0.012	0.122

Table B2. Poisson probability

x	λ=0.1	0.2	0.3	0.4	0.5	0.6	0.7	0.8	0.9	1
0	0.9048	0.8187	0.7408	0.6703	0.6065	0.5488	0.4966	0.4493	0.4066	0.3679
1	0.0905	0.1637	0.2222	0.2681	0.3033	0.3293	0.3476	0.3595	0.3659	0.3679
2	0.0045	0.0164	0.0333	0.0536	0.0758	0.0988	0.1217	0.1438	0.1647	0.1839
3	0.0002	0.0011	0.0033	0.0072	0.0126	0.0198	0.0284	0.0383	0.0494	0.0613
4	0	0.0001	0.0003	0.0007	0.0016	0.003	0.005	0.0077	0.0111	0.0153
5	0	0	0	0.0001	0.0002	0.0004	0.0007	0.0012	0.002	0.0031
6	0	0	0	0	0	0	0.0001	0.0002	0.0003	0.0005
7	0	0	0	0	0	0	0	0	0	0.0001

x	λ=1.1	1.2	1.3	1.4	1.5	1.6	1.7	1.8	1.9	2
0	0.3329	0.3012	0.2725	0.2466	0.2231	0.2019	0.1827	0.1653	0.1496	0.1353
1	0.3662	0.3614	0.3543	0.3452	0.3347	0.323	0.3106	0.2975	0.2842	0.2707
2	0.2014	0.2169	0.2303	0.2417	0.251	0.2584	0.264	0.2678	0.27	0.2707
3	0.0738	0.0867	0.0998	0.1128	0.1255	0.1378	0.1496	0.1607	0.171	0.1804
4	0.0203	0.026	0.0324	0.0395	0.0471	0.0551	0.0636	0.0723	0.0812	0.0902
5	0.0045	0.0062	0.0084	0.0111	0.0141	0.0176	0.0216	0.026	0.0309	0.0361
6	0.0008	0.0012	0.0018	0.0026	0.0035	0.0047	0.0061	0.0078	0.0098	0.012
7	0.0001	0.0002	0.0003	0.0005	0.0008	0.0011	0.0015	0.002	0.0027	0.0034
8	0	0	0.0001	0.0001	0.0001	0.0002	0.0003	0.0005	0.0006	0.0009
9	0	0	0	0	0	0	0.0001	0.0001	0.0001	0.0002

x	λ=2.1	2.2	2.3	2.4	2.5	2.6	2.7	2.8	2.9	3
0	0.1225	0.1108	0.1003	0.0907	0.0821	0.0743	0.0672	0.0608	0.055	0.0498
1	0.2572	0.2438	0.2306	0.2177	0.2052	0.1931	0.1815	0.1703	0.1596	0.1494
2	0.27	0.2681	0.2652	0.2613	0.2565	0.251	0.245	0.2384	0.2314	0.224
3	0.189	0.1966	0.2033	0.209	0.2138	0.2176	0.2205	0.2225	0.2237	0.224
4	0.0992	0.1082	0.1169	0.1254	0.1336	0.1414	0.1488	0.1557	0.1622	0.168
5	0.0417	0.0476	0.0538	0.0602	0.0668	0.0735	0.0804	0.0872	0.094	0.1008
6	0.0146	0.0174	0.0206	0.0241	0.0278	0.0319	0.0362	0.0407	0.0455	0.0504
7	0.0044	0.0055	0.0068	0.0083	0.0099	0.0118	0.0139	0.0163	0.0188	0.0216
8	0.0011	0.0015	0.0019	0.0025	0.0031	0.0038	0.0047	0.0057	0.0068	0.0081
9	0.0003	0.0004	0.0005	0.0007	0.0009	0.0011	0.0014	0.0018	0.0022	0.0027
10	0.0001	0.0001	0.0001	0.0002	0.0002	0.0003	0.0004	0.0005	0.0006	0.0008
11	0	0	0	0	0	0.0001	0.0001	0.0001	0.0002	0.0002
12	0	0	0	0	0	0	0	0	0	0.0001

Table B3. Standard Normal Probability

z	0	0.01	0.02	0.03	0.04	0.05	0.06	0.07	0.08	0.09
-3.4	0.0003	0.0003	0.0003	0.0003	0.0003	0.0003	0.0003	0.0003	0.0003	0.0002
-3.3	0.0005	0.0005	0.0005	0.0004	0.0004	0.0004	0.0004	0.0004	0.0004	0.0003
-3.2	0.0007	0.0007	0.0006	0.0006	0.0006	0.0006	0.0006	0.0005	0.0005	0.0005
-3.1	0.001	0.0009	0.0009	0.0009	0.0008	0.0008	0.0008	0.0008	0.0007	0.0007
-3.0	0.0013	0.0013	0.0013	0.0012	0.0012	0.0011	0.0011	0.0011	0.001	0.001
-2.9	0.0019	0.0018	0.0018	0.0017	0.0016	0.0016	0.0015	0.0015	0.0014	0.0014
-2.8	0.0026	0.0025	0.0024	0.0023	0.0023	0.0022	0.0021	0.0021	0.002	0.0019
-2.7	0.0035	0.0034	0.0033	0.0032	0.0031	0.003	0.0029	0.0028	0.0027	0.0026
-2.6	0.0047	0.0045	0.0044	0.0043	0.0041	0.004	0.0039	0.0038	0.0037	0.0036
-2.5	0.0062	0.006	0.0059	0.0057	0.0055	0.0054	0.0052	0.0051	0.0049	0.0048
-2.4	0.0082	0.008	0.0078	0.0075	0.0073	0.0071	0.0069	0.0068	0.0066	0.0064
-2.3	0.0107	0.0104	0.0102	0.0099	0.0096	0.0094	0.0091	0.0089	0.0087	0.0084
-2.2	0.0139	0.0136	0.0132	0.0129	0.0125	0.0122	0.0119	0.0116	0.0113	0.011
-2.1	0.0179	0.0174	0.017	0.0166	0.0162	0.0158	0.0154	0.015	0.0146	0.0143
-2.0	0.0228	0.0222	0.0217	0.0212	0.0207	0.0202	0.0197	0.0192	0.0188	0.0183
-1.9	0.0287	0.0281	0.0274	0.0268	0.0262	0.0256	0.025	0.0244	0.0239	0.0233
-1.8	0.0359	0.0351	0.0344	0.0336	0.0329	0.0322	0.0314	0.0307	0.0301	0.0294
-1.7	0.0446	0.0436	0.0427	0.0418	0.0409	0.0401	0.0392	0.0384	0.0375	0.0367
-1.6	0.0548	0.0537	0.0526	0.0516	0.0505	0.0495	0.0485	0.0475	0.0465	0.0455
-1.5	0.0668	0.0655	0.0643	0.063	0.0618	0.0606	0.0594	0.0582	0.0571	0.0559
-1.4	0.0808	0.0793	0.0778	0.0764	0.0749	0.0735	0.0721	0.0708	0.0694	0.0681
-1.3	0.0968	0.0951	0.0934	0.0918	0.0901	0.0885	0.0869	0.0853	0.0838	0.0823
-1.2	0.1151	0.1131	0.1112	0.1093	0.1075	0.1056	0.1038	0.102	0.1003	0.0985
-1.1	0.1357	0.1335	0.1314	0.1292	0.1271	0.1251	0.123	0.121	0.119	0.117
-1.0	0.1587	0.1562	0.1539	0.1515	0.1492	0.1469	0.1446	0.1423	0.1401	0.1379
-0.9	0.1841	0.1814	0.1788	0.1762	0.1736	0.1711	0.1685	0.166	0.1635	0.1611
-0.8	0.2119	0.209	0.2061	0.2033	0.2005	0.1977	0.1949	0.1922	0.1894	0.1867
-0.7	0.242	0.2389	0.2358	0.2327	0.2296	0.2266	0.2236	0.2206	0.2177	0.2148
-0.6	0.2743	0.2709	0.2676	0.2643	0.2611	0.2578	0.2546	0.2514	0.2483	0.2451
-0.5	0.3085	0.305	0.3015	0.2981	0.2946	0.2912	0.2877	0.2843	0.281	0.2776
-0.4	0.3446	0.3409	0.3372	0.3336	0.33	0.3264	0.3228	0.3192	0.3156	0.3121
-0.3	0.3821	0.3783	0.3745	0.3707	0.3669	0.3632	0.3594	0.3557	0.352	0.3483
-0.2	0.4207	0.4168	0.4129	0.409	0.4052	0.4013	0.3974	0.3936	0.3897	0.3859
-0.1	0.4602	0.4562	0.4522	0.4483	0.4443	0.4404	0.4364	0.4325	0.4286	0.4247
-0.0	0.5	0.496	0.492	0.488	0.484	0.4801	0.4761	0.4721	0.4681	0.4641
0.0	0.5	0.504	0.508	0.512	0.516	0.5199	0.5239	0.5279	0.5319	0.5359
0.1	0.5398	0.5438	0.5478	0.5517	0.5557	0.5596	0.5636	0.5675	0.5714	0.5753
0.2	0.5793	0.5832	0.5871	0.591	0.5948	0.5987	0.6026	0.6064	0.6103	0.6141
0.3	0.6179	0.6217	0.6255	0.6293	0.6331	0.6368	0.6406	0.6443	0.648	0.6517
0.4	0.6554	0.6591	0.6628	0.6664	0.67	0.6736	0.6772	0.6808	0.6844	0.6879
0.5	0.6915	0.695	0.6985	0.7019	0.7054	0.7088	0.7123	0.7157	0.719	0.7224
0.6	0.7257	0.7291	0.7324	0.7357	0.7389	0.7422	0.7454	0.7486	0.7517	0.7549
0.7	0.758	0.7611	0.7642	0.7673	0.7704	0.7734	0.7764	0.7794	0.7823	0.7852
0.8	0.7881	0.791	0.7939	0.7967	0.7995	0.8023	0.8051	0.8078	0.8106	0.8133
0.9	0.8159	0.8186	0.8212	0.8238	0.8264	0.8289	0.8315	0.834	0.8365	0.8389
1.0	0.8413	0.8438	0.8461	0.8485	0.8508	0.8531	0.8554	0.8577	0.8599	0.8621

Table B3. Standard Normal Probability

1.1	0.8643	0.8665	0.8686	0.8708	0.8729	0.8749	0.877	0.879	0.881	0.883
1.2	0.8849	0.8869	0.8888	0.8907	0.8925	0.8944	0.8962	0.898	0.8997	0.9015
1.3	0.9032	0.9049	0.9066	0.9082	0.9099	0.9115	0.9131	0.9147	0.9162	0.9177
1.4	0.9192	0.9207	0.9222	0.9236	0.9251	0.9265	0.9279	0.9292	0.9306	0.9319
1.5	0.9332	0.9345	0.9357	0.937	0.9382	0.9394	0.9406	0.9418	0.9429	0.9441
1.6	0.9452	0.9463	0.9474	0.9484	0.9495	0.9505	0.9515	0.9525	0.9535	0.9545
1.7	0.9554	0.9564	0.9573	0.9582	0.9591	0.9599	0.9608	0.9616	0.9625	0.9633
1.8	0.9641	0.9649	0.9656	0.9664	0.9671	0.9678	0.9686	0.9693	0.9699	0.9706
1.9	0.9713	0.9719	0.9726	0.9732	0.9738	0.9744	0.975	0.9756	0.9761	0.9767
2.0	0.9772	0.9778	0.9783	0.9788	0.9793	0.9798	0.9803	0.9808	0.9812	0.9817
2.1	0.9821	0.9826	0.983	0.9834	0.9838	0.9842	0.9846	0.985	0.9854	0.9857
2.2	0.9861	0.9864	0.9868	0.9871	0.9875	0.9878	0.9881	0.9884	0.9887	0.989
2.3	0.9893	0.9896	0.9898	0.9901	0.9904	0.9906	0.9909	0.9911	0.9913	0.9916
2.4	0.9918	0.992	0.9922	0.9925	0.9927	0.9929	0.9931	0.9932	0.9934	0.9936
2.5	0.9938	0.994	0.9941	0.9943	0.9945	0.9946	0.9948	0.9949	0.9951	0.9952
2.6	0.9953	0.9955	0.9956	0.9957	0.9959	0.996	0.9961	0.9962	0.9963	0.9964
2.7	0.9965	0.9966	0.9967	0.9968	0.9969	0.997	0.9971	0.9972	0.9973	0.9974
2.8	0.9974	0.9975	0.9976	0.9977	0.9977	0.9978	0.9979	0.9979	0.998	0.9981
2.9	0.9981	0.9982	0.9982	0.9983	0.9984	0.9984	0.9985	0.9985	0.9986	0.9986
3.0	0.9987	0.9987	0.9987	0.9988	0.9988	0.9989	0.9989	0.9989	0.999	0.999
3.1	0.999	0.9991	0.9991	0.9991	0.9992	0.9992	0.9992	0.9992	0.9993	0.9993
3.2	0.9993	0.9993	0.9994	0.9994	0.9994	0.9994	0.9994	0.9995	0.9995	0.9995
3.3	0.9995	0.9995	0.9995	0.9996	0.9996	0.9996	0.9996	0.9996	0.9996	0.9997
3.4	0.9997	0.9997	0.9997	0.9997	0.9997	0.9997	0.9997	0.9997	0.9997	0.9998

Table B4. Critical values of Student's T distribution

Degrees of Freedom	Right tail probability							
	0.0005	0.001	0.005	0.01	0.025	0.05	0.1	0.2
1	636.6192	318.3088	63.6567	31.8205	12.7062	6.3138	3.0777	1.3764
2	31.5991	22.3271	9.9248	6.9646	4.3027	2.92	1.8856	1.0607
3	12.924	10.2145	5.8409	4.5407	3.1824	2.3534	1.6377	0.9785
4	8.6103	7.1732	4.6041	3.7469	2.7764	2.1318	1.5332	0.941
5	6.8688	5.8934	4.0321	3.3649	2.5706	2.015	1.4759	0.9195
6	5.9588	5.2076	3.7074	3.1427	2.4469	1.9432	1.4398	0.9057
7	5.4079	4.7853	3.4995	2.998	2.3646	1.8946	1.4149	0.896
8	5.0413	4.5008	3.3554	2.8965	2.306	1.8595	1.3968	0.8889
9	4.7809	4.2968	3.2498	2.8214	2.2622	1.8331	1.383	0.8834
10	4.5869	4.1437	3.1693	2.7638	2.2281	1.8125	1.3722	0.8791
11	4.437	4.0247	3.1058	2.7181	2.201	1.7959	1.3634	0.8755
12	4.3178	3.9296	3.0545	2.681	2.1788	1.7823	1.3562	0.8726
13	4.2208	3.852	3.0123	2.6503	2.1604	1.7709	1.3502	0.8702
14	4.1405	3.7874	2.9768	2.6245	2.1448	1.7613	1.345	0.8681
15	4.0728	3.7328	2.9467	2.6025	2.1314	1.7531	1.3406	0.8662
16	4.015	3.6862	2.9208	2.5835	2.1199	1.7459	1.3368	0.8647
17	3.9651	3.6458	2.8982	2.5669	2.1098	1.7396	1.3334	0.8633
18	3.9216	3.6105	2.8784	2.5524	2.1009	1.7341	1.3304	0.862
19	3.8834	3.5794	2.8609	2.5395	2.093	1.7291	1.3277	0.861
20	3.8495	3.5518	2.8453	2.528	2.086	1.7247	1.3253	0.86
21	3.8193	3.5272	2.8314	2.5176	2.0796	1.7207	1.3232	0.8591
22	3.7921	3.505	2.8188	2.5083	2.0739	1.7171	1.3212	0.8583
23	3.7676	3.485	2.8073	2.4999	2.0687	1.7139	1.3195	0.8575
24	3.7454	3.4668	2.7969	2.4922	2.0639	1.7109	1.3178	0.8569
25	3.7251	3.4502	2.7874	2.4851	2.0595	1.7081	1.3163	0.8562
26	3.7066	3.435	2.7787	2.4786	2.0555	1.7056	1.315	0.8557
27	3.6896	3.421	2.7707	2.4727	2.0518	1.7033	1.3137	0.8551
28	3.6739	3.4082	2.7633	2.4671	2.0484	1.7011	1.3125	0.8546
29	3.6594	3.3962	2.7564	2.462	2.0452	1.6991	1.3114	0.8542
30	3.646	3.3852	2.75	2.4573	2.0423	1.6973	1.3104	0.8538
40	3.551	3.3069	2.7045	2.4233	2.0211	1.6839	1.3031	0.8507
50	3.496	3.2614	2.6778	2.4033	2.0086	1.6759	1.2987	0.8489
60	3.4602	3.2317	2.6603	2.3901	2.0003	1.6706	1.2958	0.8477
70	3.435	3.2108	2.6479	2.3808	1.9944	1.6669	1.2938	0.8468
80	3.4163	3.1953	2.6387	2.3739	1.9901	1.6641	1.2922	0.8461
90	3.4019	3.1833	2.6316	2.3685	1.9867	1.662	1.291	0.8456
100	3.3905	3.1737	2.6259	2.3642	1.984	1.6602	1.2901	0.8452
120	3.3735	3.1595	2.6174	2.3578	1.9799	1.6577	1.2886	0.8446

Table B5. Critical values of Chi-square distribution

Degrees of Freedom	Right tail probability									
	0.005	0.01	0.025	0.05	0.1	0.9	0.95	0.975	0.99	0.995
1	7.8794	6.6349	5.0239	3.8415	2.7055	0.0158	0.0039	0.001	0.0002	0
2	10.5966	9.2103	7.3778	5.9915	4.6052	0.2107	0.1026	0.0506	0.0201	0.01
3	12.8382	11.3449	9.3484	7.8147	6.2514	0.5844	0.3518	0.2158	0.1148	0.0717
4	14.8603	13.2767	11.1433	9.4877	7.7794	1.0636	0.7107	0.4844	0.2971	0.207
5	16.7496	15.0863	12.8325	11.0705	9.2364	1.6103	1.1455	0.8312	0.5543	0.4117
6	18.5476	16.8119	14.4494	12.5916	10.6446	2.2041	1.6354	1.2373	0.8721	0.6757
7	20.2777	18.4753	16.0128	14.0671	12.017	2.8331	2.1673	1.6899	1.239	0.9893
8	21.955	20.0902	17.5345	15.5073	13.3616	3.4895	2.7326	2.1797	1.6465	1.3444
9	23.5894	21.666	19.0228	16.919	14.6837	4.1682	3.3251	2.7004	2.0879	1.7349
10	25.1882	23.2093	20.4832	18.307	15.9872	4.8652	3.9403	3.247	2.5582	2.1559
11	26.7568	24.725	21.92	19.6751	17.275	5.5778	4.5748	3.8157	3.0535	2.6032
12	28.2995	26.217	23.3367	21.0261	18.5493	6.3038	5.226	4.4038	3.5706	3.0738
13	29.8195	27.6882	24.7356	22.362	19.8119	7.0415	5.8919	5.0088	4.1069	3.565
14	31.3193	29.1412	26.1189	23.6848	21.0641	7.7895	6.5706	5.6287	4.6604	4.0747
15	32.8013	30.5779	27.4884	24.9958	22.3071	8.5468	7.2609	6.2621	5.2293	4.6009
16	34.2672	31.9999	28.8454	26.2962	23.5418	9.3122	7.9616	6.9077	5.8122	5.1422
17	35.7185	33.4087	30.191	27.5871	24.769	10.0852	8.6718	7.5642	6.4078	5.6972
18	37.1565	34.8053	31.5264	28.8693	25.9894	10.8649	9.3905	8.2307	7.0149	6.2648
19	38.5823	36.1909	32.8523	30.1435	27.2036	11.6509	10.117	8.9065	7.6327	6.844
20	39.9968	37.5662	34.1696	31.4104	28.412	12.4426	10.8508	9.5908	8.2604	7.4338
21	41.4011	38.9322	35.4789	32.6706	29.6151	13.2396	11.5913	10.2829	8.8972	8.0337
22	42.7957	40.2894	36.7807	33.9244	30.8133	14.0415	12.338	10.9823	9.5425	8.6427
23	44.1813	41.6384	38.0756	35.1725	32.0069	14.848	13.0905	11.6886	10.1957	9.2604
24	45.5585	42.9798	39.3641	36.415	33.1962	15.6587	13.8484	12.4012	10.8564	9.8862
25	46.9279	44.3141	40.6465	37.6525	34.3816	16.4734	14.6114	13.1197	11.524	10.5197
26	48.2899	45.6417	41.9232	38.8851	35.5632	17.2919	15.3792	13.8439	12.1981	11.1602
27	49.6449	46.9629	43.1945	40.1133	36.7412	18.1139	16.1514	14.5734	12.8785	11.8076
28	50.9934	48.2782	44.4608	41.3371	37.9159	18.9392	16.9279	15.3079	13.5647	12.4613
29	52.3356	49.5879	45.7223	42.557	39.0875	19.7677	17.7084	16.0471	14.2565	13.1211
30	53.672	50.8922	46.9792	43.773	40.256	20.5992	18.4927	16.7908	14.9535	13.7867
40	66.766	63.6907	59.3417	55.7585	51.8051	29.0505	26.5093	24.433	22.1643	20.7065
50	79.49	76.1539	71.4202	67.5048	63.1671	37.6886	34.7643	32.3574	29.7067	27.9907
60	91.9517	88.3794	83.2977	79.0819	74.397	46.4589	43.188	40.4817	37.4849	35.5345
70	104.215	100.425	95.0232	90.5312	85.527	55.3289	51.7393	48.7576	45.4417	43.2752
80	116.321	112.329	106.629	101.88	96.5782	64.2778	60.3915	57.1532	53.5401	51.1719
90	128.299	124.116	118.136	113.145	107.565	73.2911	69.126	65.6466	61.7541	59.1963
100	140.17	135.807	129.561	124.342	118.498	82.3581	77.9295	74.2219	70.0649	67.3276
110	151.949	147.414	140.917	135.48	129.385	91.471	86.7916	82.8671	78.4583	75.55
120	163.648	158.95	152.211	146.567	140.233	100.624	95.7046	91.5726	86.9233	83.8516

Table B6. Critical values of F distribution

DF1	alpha	DF2, Degrees of Freedom in the denominator							
		1	2	3	4	5	6	7	8
1	0.1	39.8635	8.5263	5.5383	4.5448	4.0604	3.7759	3.5894	3.4579
	0.05	161.4476	18.5128	10.128	7.7086	6.6079	5.9874	5.5914	5.3177
	0.025	647.789	38.5063	17.4434	12.2179	10.007	8.8131	8.0727	7.5709
	0.01	4052.181	98.5025	34.1162	21.1977	16.2582	13.745	12.2464	11.2586
	0.001	405284.1	998.5003	167.0292	74.1373	47.1808	35.5075	29.2452	25.4148
2	0.1	49.5	9	5.4624	4.3246	3.7797	3.4633	3.2574	3.1131
	0.05	199.5	19	9.5521	6.9443	5.7861	5.1433	4.7374	4.459
	0.025	799.5	39	16.0441	10.6491	8.4336	7.2599	6.5415	6.0595
	0.01	4999.5	99	30.8165	18	13.2739	10.9248	9.5466	8.6491
	0.001	499999.5	999	148.5	61.2456	37.1223	27	21.689	18.4937
3	0.1	53.5932	9.1618	5.3908	4.1909	3.6195	3.2888	3.0741	2.9238
	0.05	215.7073	19.1643	9.2766	6.5914	5.4095	4.7571	4.3468	4.0662
	0.025	864.163	39.1655	15.4392	9.9792	7.7636	6.5988	5.8898	5.416
	0.01	5403.352	99.1662	29.4567	16.6944	12.06	9.7795	8.4513	7.591
	0.001	540379.2	999.1666	141.1085	56.1772	33.2025	23.7033	18.7723	15.8295
4	0.1	55.833	9.2434	5.3426	4.1072	3.5202	3.1808	2.9605	2.8064
	0.05	224.5832	19.2468	9.1172	6.3882	5.1922	4.5337	4.1203	3.8379
	0.025	899.5833	39.2484	15.101	9.6045	7.3879	6.2272	5.5226	5.0526
	0.01	5624.583	99.2494	28.7099	15.977	11.3919	9.1483	7.8466	7.0061
	0.001	562499.6	999.2499	137.1004	53.4358	31.085	21.9235	17.198	14.3916
5	0.1	57.2401	9.2926	5.3092	4.0506	3.453	3.1075	2.8833	2.7264
	0.05	230.1619	19.2964	9.0135	6.2561	5.0503	4.3874	3.9715	3.6875
	0.025	921.8479	39.2982	14.8848	9.3645	7.1464	5.9876	5.2852	4.8173
	0.01	5763.65	99.2993	28.2371	15.5219	10.967	8.7459	7.4604	6.6318
	0.001	576404.6	999.2999	134.58	51.7116	29.7524	20.8027	16.2058	13.4847
6	0.1	58.2044	9.3255	5.2847	4.0097	3.4045	3.0546	2.8274	2.6683
	0.05	233.986	19.3295	8.9406	6.1631	4.9503	4.2839	3.866	3.5806
	0.025	937.1111	39.3315	14.7347	9.1973	6.9777	5.8198	5.1186	4.6517
	0.01	5858.986	99.3326	27.9107	15.2069	10.6723	8.4661	7.1914	6.3707
	0.001	585937.1	999.3333	132.8475	50.525	28.8344	20.0297	15.5208	12.858
7	0.1	58.906	9.3491	5.2662	3.979	3.3679	3.0145	2.7849	2.6241
	0.05	236.7684	19.3532	8.8867	6.0942	4.8759	4.2067	3.787	3.5005
	0.025	948.2169	39.3552	14.6244	9.0741	6.8531	5.6955	4.9949	4.5286
	0.01	5928.356	99.3564	27.6717	14.9758	10.4555	8.26	6.9928	6.1776
	0.001	592873.3	999.3571	131.5829	49.6579	28.1626	19.4634	15.0186	12.398
8	0.1	59.439	9.3668	5.2517	3.9549	3.3393	2.983	2.7516	2.5893
	0.05	238.8827	19.371	8.8452	6.041	4.8183	4.1468	3.7257	3.4381
	0.025	956.6562	39.373	14.5399	8.9796	6.7572	5.5996	4.8993	4.4333
	0.01	5981.07	99.3742	27.4892	14.7989	10.2893	8.1017	6.84	6.0289
	0.001	598144.2	999.3749	130.619	48.9962	27.6495	19.0303	14.634	12.0455
9	0.1	59.8576	9.3805	5.24	3.9357	3.3163	2.9577	2.7247	2.5612
	0.05	240.5433	19.3848	8.8123	5.9988	4.7725	4.099	3.6767	3.3881
	0.025	963.2846	39.3869	14.4731	8.9047	6.6811	5.5234	4.8232	4.3572
	0.01	6022.473	99.3881	27.3452	14.6591	10.1578	7.9761	6.7188	5.9106
	0.001	602284	999.3888	129.86	48.4745	27.2445	18.6882	14.3299	11.7665

Table B6. Critical values of F distribution

df2	α	1	2	3	4	5	6	7	8
10	0.1	60.195	9.3916	5.2304	3.9199	3.2974	2.9369	2.7025	2.538
	0.05	241.8817	19.3959	8.7855	5.9644	4.7351	4.06	3.6365	3.3472
	0.025	968.6274	39.398	14.4189	8.8439	6.6192	5.4613	4.7611	4.2951
	0.01	6055.847	99.3992	27.2287	14.5459	10.051	7.8741	6.6201	5.8143
	0.001	605621	999.3999	129.2467	48.0526	26.9166	18.4109	14.0833	11.5401
11	0.1	60.4727	9.4006	5.2224	3.9067	3.2816	2.9195	2.6839	2.5186
	0.05	242.9835	19.405	8.7633	5.9358	4.704	4.0274	3.603	3.313
	0.025	973.0252	39.4071	14.3742	8.7935	6.5678	5.4098	4.7095	4.2434
	0.01	6083.317	99.4083	27.1326	14.4523	9.9626	7.7896	6.5382	5.7343
	0.001	608367.7	999.409	128.7408	47.7043	26.6456	18.1816	13.8791	11.3525
12	0.1	60.7052	9.4081	5.2156	3.8955	3.2682	2.9047	2.6681	2.502
	0.05	243.906	19.4125	8.7446	5.9117	4.6777	3.9999	3.5747	3.2839
	0.025	976.7079	39.4146	14.3366	8.7512	6.5245	5.3662	4.6658	4.1997
	0.01	6106.321	99.4159	27.0518	14.3736	9.8883	7.7183	6.4691	5.6667
	0.001	610667.8	999.4166	128.3165	47.4118	26.418	17.9888	13.7073	11.1945
15	0.1	61.2203	9.4247	5.2003	3.8704	3.238	2.8712	2.6322	2.4642
	0.05	245.9499	19.4291	8.7029	5.8578	4.6188	3.9381	3.5107	3.2184
	0.025	984.8668	39.4313	14.2527	8.6565	6.4277	5.2687	4.5678	4.1012
	0.01	6157.285	99.4325	26.8722	14.1982	9.7222	7.559	6.3143	5.5151
	0.001	615763.7	999.4333	127.3736	46.7612	25.9108	17.5587	13.3237	10.8413
18	0.1	61.5664	9.4358	5.1898	3.8531	3.2172	2.8481	2.6074	2.438
	0.05	247.3232	19.4402	8.6745	5.8211	4.5785	3.8957	3.4669	3.1733
	0.025	990.349	39.4424	14.196	8.5924	6.3619	5.2021	4.5008	4.0338
	0.01	6191.529	99.4436	26.7509	14.0795	9.6096	7.4507	6.2089	5.4116
	0.001	619187.7	999.4444	126.7378	46.3217	25.5677	17.2673	13.0633	10.6012
20	0.1	61.7403	9.4413	5.1845	3.8443	3.2067	2.8363	2.5947	2.4246
	0.05	248.0131	19.4458	8.6602	5.8025	4.5581	3.8742	3.4445	3.1503
	0.025	993.1028	39.4479	14.1674	8.5599	6.3286	5.1684	4.4667	3.9995
	0.01	6208.73	99.4492	26.6898	14.0196	9.5526	7.3958	6.1554	5.3591
	0.001	620907.7	999.4499	126.4178	46.1003	25.3946	17.1201	12.9316	10.4797
25	0.1	62.0545	9.4513	5.1747	3.8283	3.1873	2.8147	2.5714	2.3999
	0.05	249.2601	19.4558	8.6341	5.7687	4.5209	3.8348	3.4036	3.1081
	0.025	998.0808	39.4579	14.1155	8.501	6.2679	5.1069	4.4045	3.9367
	0.01	6239.825	99.4592	26.579	13.9109	9.4491	7.296	6.058	5.2631
	0.001	624016.8	999.4599	125.8379	45.6986	25.0803	16.8525	12.692	10.2583
30	0.1	62.265	9.4579	5.1681	3.8174	3.1741	2.8	2.5555	2.383
	0.05	250.0951	19.4624	8.6166	5.7459	4.4957	3.8082	3.3758	3.0794
	0.025	1001.414	39.4646	14.0805	8.4613	6.2269	5.0652	4.3624	3.894
	0.01	6260.649	99.4658	26.5045	13.8377	9.3793	7.2285	5.992	5.1981
	0.001	626099	999.4666	125.4486	45.4286	24.8688	16.6722	12.5304	10.1087
40	0.1	62.5291	9.4662	5.1597	3.8036	3.1573	2.7812	2.5351	2.3614
	0.05	251.1432	19.4707	8.5944	5.717	4.4638	3.7743	3.3404	3.0428
	0.025	1005.598	39.4729	14.0365	8.4111	6.175	5.0125	4.3089	3.8398
	0.01	6286.782	99.4742	26.4108	13.7454	9.2912	7.1432	5.9084	5.1156
	0.001	628712	999.4749	124.959	45.0886	24.602	16.4445	12.326	9.9194
50	0.1	62.6881	9.4712	5.1546	3.7952	3.1471	2.7697	2.5226	2.3481
	0.05	251.7742	19.4757	8.581	5.6995	4.4444	3.7537	3.3189	3.0204

Table B6. Critical values of F distribution

	0.025	1008.117	39.4779	14.0099	8.3808	6.1436	4.9804	4.2763	3.8067
	0.01	6302.517	99.4792	26.3542	13.6896	9.2378	7.0915	5.8577	5.0654
	0.001	630285.4	999.4799	124.6635	44.8832	24.4407	16.3067	12.202	9.8044
60	0.1	62.7943	9.4746	5.1512	3.7896	3.1402	2.762	2.5142	2.3391
	0.05	252.1957	19.4791	8.572	5.6877	4.4314	3.7398	3.3043	3.0053
	0.025	1009.8	39.4812	13.9921	8.3604	6.1225	4.9589	4.2544	3.7844
	0.01	6313.03	99.4825	26.3164	13.6522	9.202	7.0567	5.8236	5.0316
	0.001	631336.6	999.4833	124.4658	44.7457	24.3326	16.2143	12.1189	9.7272
80	0.1	62.9273	9.4787	5.1469	3.7825	3.1316	2.7522	2.5036	2.3277
	0.05	252.7237	19.4832	8.5607	5.673	4.415	3.7223	3.286	2.9862
	0.025	1011.908	39.4854	13.9697	8.3349	6.096	4.9318	4.2268	3.7563
	0.01	6326.197	99.4867	26.2688	13.6053	9.157	7.013	5.7806	4.989
	0.001	632653.1	999.4874	124.218	44.5731	24.1969	16.0981	12.0143	9.63
100	0.1	63.0073	9.4812	5.1443	3.7782	3.1263	2.7463	2.4971	2.3208
	0.05	253.0411	19.4857	8.5539	5.6641	4.4051	3.7117	3.2749	2.9747
	0.025	1013.175	39.4879	13.9563	8.3195	6.08	4.9154	4.2101	3.7393
	0.01	6334.11	99.4892	26.2402	13.577	9.1299	6.9867	5.7547	4.9633
	0.001	633444.3	999.4899	124.0688	44.4692	24.1151	16.028	11.9512	9.5714

DF1	alpha	DF2							
		9	10	11	12	15	18	20	25
1	0.1	3.3603	3.285	3.2252	3.1765	3.0732	3.007	2.9747	2.9177
	0.05	5.1174	4.9646	4.8443	4.7472	4.5431	4.4139	4.3512	4.2417
	0.025	7.2093	6.9367	6.7241	6.5538	6.1995	5.9781	5.8715	5.6864
	0.01	10.5614	10.0443	9.646	9.3302	8.6831	8.2854	8.096	7.7698
	0.001	22.8571	21.0396	19.6868	18.6433	16.5874	15.3793	14.8188	13.8767
2	0.1	3.0065	2.9245	2.8595	2.8068	2.6952	2.6239	2.5893	2.5283
	0.05	4.2565	4.1028	3.9823	3.8853	3.6823	3.5546	3.4928	3.3852
	0.025	5.7147	5.4564	5.2559	5.0959	4.765	4.5597	4.4613	4.2909
	0.01	8.0215	7.5594	7.2057	6.9266	6.3589	6.0129	5.8489	5.568
	0.001	16.3871	14.9054	13.8116	12.9737	11.3391	10.3899	9.9526	9.2225
3	0.1	2.8129	2.7277	2.6602	2.6055	2.4898	2.416	2.3801	2.317
	0.05	3.8625	3.7083	3.5874	3.4903	3.2874	3.1599	3.0984	2.9912
	0.025	5.0781	4.8256	4.63	4.4742	4.1528	3.9539	3.8587	3.6943
	0.01	6.9919	6.5523	6.2167	5.9525	5.417	5.0919	4.9382	4.6755
	0.001	13.9018	12.5527	11.5611	10.8042	9.3353	8.4875	8.0984	7.4511
4	0.1	2.6927	2.6053	2.5362	2.4801	2.3614	2.2858	2.2489	2.1842
	0.05	3.6331	3.478	3.3567	3.2592	3.0556	2.9277	2.8661	2.7587
	0.025	4.7181	4.4683	4.2751	4.1212	3.8043	3.6083	3.5147	3.353
	0.01	6.4221	5.9943	5.6683	5.412	4.8932	4.579	4.4307	4.1774
	0.001	12.5603	11.2828	10.3461	9.6327	8.2527	7.4593	7.096	6.4931
5	0.1	2.6106	2.5216	2.4512	2.394	2.273	2.1958	2.1582	2.0922
	0.05	3.4817	3.3258	3.2039	3.1059	2.9013	2.7729	2.7109	2.603
	0.025	4.4844	4.2361	4.044	3.8911	3.5764	3.382	3.2891	3.1287
	0.01	6.0569	5.6363	5.316	5.0643	4.5556	4.2479	4.1027	3.855
	0.001	11.7137	10.4807	9.5784	8.8921	7.5674	6.8078	6.4606	5.8851
6	0.1	2.5509	2.4606	2.3891	2.331	2.2081	2.1296	2.0913	2.0241
	0.05	3.3738	3.2172	3.0946	2.9961	2.7905	2.6613	2.599	2.4904

Table B6. Critical values of F distribution

	α								
	0.025	4.3197	4.0721	3.8807	3.7283	3.4147	3.2209	3.1283	2.9685
	0.01	5.8018	5.3858	5.0692	4.8206	4.3183	4.0146	3.8714	3.6272
	0.001	11.1281	9.9256	9.0466	8.3788	7.0917	6.355	6.0186	5.4617
7	0.1	2.5053	2.414	2.3416	2.2828	2.1582	2.0785	2.0397	1.9714
	0.05	3.2927	3.1355	3.0123	2.9134	2.7066	2.5767	2.514	2.4047
	0.025	4.197	3.9498	3.7586	3.6065	3.2934	3.0999	3.0074	2.8478
	0.01	5.6129	5.2001	4.8861	4.6395	4.1415	3.8406	3.6987	3.4568
	0.001	10.6979	9.5175	8.6553	8.0009	6.7408	6.0206	5.692	5.1484
8	0.1	2.4694	2.3772	2.304	2.2446	2.1185	2.0379	1.9985	1.9292
	0.05	3.2296	3.0717	2.948	2.8486	2.6408	2.5102	2.4471	2.3371
	0.025	4.102	3.8549	3.6638	3.5118	3.1987	3.0053	2.9128	2.7531
	0.01	5.4671	5.0567	4.7445	4.4994	4.0045	3.7054	3.5644	3.3239
	0.001	10.368	9.2041	8.3548	7.7104	6.4707	5.7628	5.44	4.9063
9	0.1	2.4403	2.3473	2.2735	2.2135	2.0862	2.0047	1.9649	1.8947
	0.05	3.1789	3.0204	2.8962	2.7964	2.5876	2.4563	2.3928	2.2821
	0.025	4.026	3.779	3.5879	3.4358	3.1227	2.9291	2.8365	2.6766
	0.01	5.3511	4.9424	4.6315	4.3875	3.8948	3.5971	3.4567	3.2172
	0.001	10.1066	8.9558	8.1163	7.4797	6.2559	5.5575	5.2392	4.7131
10	0.1	2.4163	2.3226	2.2482	2.1878	2.0593	1.977	1.9367	1.8658
	0.05	3.1373	2.9782	2.8536	2.7534	2.5437	2.4117	2.3479	2.2365
	0.025	3.9639	3.7168	3.5257	3.3736	3.0602	2.8664	2.7737	2.6135
	0.01	5.2565	4.8491	4.5393	4.2961	3.8049	3.5082	3.3682	3.1294
	0.001	9.8943	8.7539	7.9224	7.292	6.0808	5.39	5.0752	4.5551
11	0.1	2.3961	2.3018	2.2269	2.166	2.0366	1.9535	1.9129	1.8412
	0.05	3.1025	2.943	2.8179	2.7173	2.5068	2.3742	2.31	2.1979
	0.025	3.9121	3.6649	3.4737	3.3215	3.0078	2.8137	2.7209	2.5603
	0.01	5.1779	4.7715	4.4624	4.2198	3.7299	3.4338	3.2941	3.0558
	0.001	9.7183	8.5864	7.7614	7.1362	5.9352	5.2505	4.9386	4.4233
12	0.1	2.3789	2.2841	2.2087	2.1474	2.0171	1.9333	1.8924	1.82
	0.05	3.0729	2.913	2.7876	2.6866	2.4753	2.3421	2.2776	2.1649
	0.025	3.8682	3.6209	3.4296	3.2773	2.9633	2.7689	2.6758	2.5149
	0.01	5.1114	4.7059	4.3974	4.1553	3.6662	3.3706	3.2311	2.9931
	0.001	9.57	8.4452	7.6256	7.0046	5.8121	5.1324	4.8229	4.3116
15	0.1	2.3396	2.2435	2.1671	2.1049	1.9722	1.8868	1.8449	1.7708
	0.05	3.0061	2.845	2.7186	2.6169	2.4034	2.2686	2.2033	2.0889
	0.025	3.7694	3.5217	3.3299	3.1772	2.8621	2.6667	2.5731	2.411
	0.01	4.9621	4.5581	4.2509	4.0096	3.5222	3.2273	3.088	2.8502
	0.001	9.2381	8.1288	7.321	6.7092	5.5351	4.8663	4.5618	4.0587
18	0.1	2.3123	2.2153	2.138	2.075	1.9407	1.8539	1.8113	1.7358
	0.05	2.96	2.798	2.6709	2.5684	2.3533	2.2172	2.1511	2.0353
	0.025	3.7015	3.4534	3.2612	3.1081	2.7919	2.5956	2.5014	2.3381
	0.01	4.8599	4.4569	4.1503	3.9095	3.4228	3.128	2.9887	2.7506
	0.001	9.0121	7.9131	7.1131	6.5074	5.3452	4.6833	4.3819	3.8839
20	0.1	2.2983	2.2007	2.123	2.0597	1.9243	1.8368	1.7938	1.7175
	0.05	2.9365	2.774	2.6464	2.5436	2.3275	2.1906	2.1242	2.0075
	0.025	3.6669	3.4185	3.2261	3.0728	2.7559	2.559	2.4645	2.3005
	0.01	4.808	4.4054	4.099	3.8584	3.3719	3.0771	2.9377	2.6993

Table B6. Critical values of F distribution

DF1	alpha				DF2				
					30	40	50	60	80	100
25	0.001	8.8976	7.8037	7.0076	6.4048	5.2484	4.5899	4.29	3.7944
	0.1	2.2725	2.1739	2.0953	2.0312	1.8939	1.8049	1.7611	1.6831
	0.05	2.8932	2.7298	2.6014	2.4977	2.2797	2.1413	2.0739	1.9554
	0.025	3.6035	3.3546	3.1616	3.0077	2.6894	2.4912	2.3959	2.2303
	0.01	4.713	4.3111	4.0051	3.7647	3.2782	2.9831	2.8434	2.6041
30	0.001	8.6888	7.6041	6.8147	6.2172	5.071	4.4182	4.1208	3.6291
	0.1	2.2547	2.1554	2.0762	2.0115	1.8728	1.7827	1.7382	1.6589
	0.05	2.8637	2.6996	2.5705	2.4663	2.2468	2.1071	2.0391	1.9192
	0.025	3.5604	3.311	3.1176	2.9633	2.6437	2.4445	2.3486	2.1816
	0.01	4.6486	4.2469	3.9411	3.7008	3.2141	2.9185	2.7785	2.5383
40	0.001	8.5476	7.4688	6.6839	6.0898	4.9502	4.3009	4.005	3.5155
	0.1	2.232	2.1317	2.0516	1.9861	1.8454	1.7537	1.7083	1.6272
	0.05	2.8259	2.6609	2.5309	2.4259	2.2043	2.0629	1.9938	1.8718
	0.025	3.5055	3.2554	3.0613	2.9063	2.585	2.3842	2.2873	2.1183
	0.01	4.5666	4.1653	3.8596	3.6192	3.1319	2.8354	2.6947	2.453
50	0.001	8.3685	7.2971	6.5178	5.9278	4.7959	4.1507	3.8564	3.3692
	0.1	2.218	2.1171	2.0364	1.9704	1.8284	1.7356	1.6896	1.6072
	0.05	2.8028	2.6371	2.5066	2.401	2.178	2.0354	1.9656	1.8421
	0.025	3.4719	3.2214	3.0268	2.8714	2.5488	2.3468	2.2493	2.0787
	0.01	4.5167	4.1155	3.8097	3.5692	3.0814	2.7841	2.643	2.3999
60	0.001	8.2597	7.1927	6.4165	5.829	4.7015	4.0584	3.765	3.2787
	0.1	2.2085	2.1072	2.0261	1.9597	1.8168	1.7232	1.6768	1.5934
	0.05	2.7872	2.6211	2.4901	2.3842	2.1601	2.0166	1.9464	1.8217
	0.025	3.4493	3.1984	3.0035	2.8478	2.5242	2.3214	2.2234	2.0516
	0.01	4.4831	4.0819	3.7761	3.5355	3.0471	2.7493	2.6077	2.3637
80	0.001	8.1865	7.1224	6.3483	5.7623	4.6377	3.996	3.703	3.2171
	0.1	2.1965	2.0946	2.013	1.9461	1.8019	1.7073	1.6603	1.5755
	0.05	2.7675	2.6008	2.4692	2.3628	2.1373	1.9927	1.9217	1.7955
	0.025	3.4207	3.1694	2.974	2.8178	2.493	2.289	2.1902	2.0169
	0.01	4.4407	4.0394	3.7335	3.4928	3.0037	2.705	2.5628	2.3173
100	0.001	8.0944	7.0338	6.2623	5.6782	4.5569	3.9167	3.6242	3.1386
	0.1	2.1892	2.0869	2.005	1.9379	1.7929	1.6976	1.6501	1.5645
	0.05	2.7556	2.5884	2.4566	2.3498	2.1234	1.978	1.9066	1.7794
	0.025	3.4034	3.1517	2.9561	2.7996	2.4739	2.2692	2.1699	1.9955
	0.01	4.415	4.0137	3.7077	3.4668	2.9772	2.6779	2.5353	2.2888
	0.001	8.0387	6.9802	6.2102	5.6272	4.5079	3.8685	3.5762	3.0905

DF1	alpha	30	40	50	60	80	100
1	0.1	2.8807	2.8354	2.8087	2.7911	2.7693	2.7564
	0.05	4.1709	4.0847	4.0343	4.0012	3.9604	3.9361
	0.025	5.5675	5.4239	5.3403	5.2856	5.2184	5.1786
	0.01	7.5625	7.3141	7.1706	7.0771	6.9627	6.8953
	0.001	13.293	12.6094	12.2221	11.973	11.6714	11.4954
2	0.1	2.4887	2.4404	2.412	2.3933	2.3701	2.3564
	0.05	3.3158	3.2317	3.1826	3.1504	3.1108	3.0873
	0.025	4.1821	4.051	3.9749	3.9253	3.8643	3.8284
	0.01	5.3903	5.1785	5.0566	4.9774	4.8807	4.8239

Table B6. Critical values of F distribution

3	0.001	8.7734	8.2508	7.9564	7.7678	7.5401	7.4077
	0.1	2.2761	2.2261	2.1967	2.1774	2.1535	2.1394
	0.05	2.9223	2.8387	2.79	2.7581	2.7188	2.6955
	0.025	3.5894	3.4633	3.3902	3.3425	3.2841	3.2496
	0.01	4.5097	4.3126	4.1993	4.1259	4.0363	3.9837
4	0.001	7.0545	6.5945	6.3364	6.1712	5.9723	5.8568
	0.1	2.1422	2.0909	2.0608	2.041	2.0165	2.0019
	0.05	2.6896	2.606	2.5572	2.5252	2.4859	2.4626
	0.025	3.2499	3.1261	3.0544	3.0077	2.9504	2.9166
	0.01	4.0179	3.8283	3.7195	3.649	3.5631	3.5127
5	0.001	6.1245	5.6981	5.4593	5.3067	5.1231	5.0167
	0.1	2.0492	1.9968	1.966	1.9457	1.9206	1.9057
	0.05	2.5336	2.4495	2.4004	2.3683	2.3287	2.3053
	0.025	3.0265	2.9037	2.8327	2.7863	2.7295	2.6961
	0.01	3.699	3.5138	3.4077	3.3389	3.255	3.2059
6	0.001	5.5339	5.1283	4.9013	4.7565	4.5824	4.4815
	0.1	1.9803	1.9269	1.8954	1.8747	1.8491	1.8339
	0.05	2.4205	2.3359	2.2864	2.2541	2.2142	2.1906
	0.025	2.8667	2.7444	2.6736	2.6274	2.5708	2.5374
	0.01	3.4735	3.291	3.1864	3.1187	3.0361	2.9877
7	0.001	5.1223	4.7306	4.5117	4.3721	4.2043	4.1071
	0.1	1.9269	1.8725	1.8405	1.8194	1.7933	1.7778
	0.05	2.3343	2.249	2.1992	2.1665	2.1263	2.1025
	0.025	2.746	2.6238	2.553	2.5068	2.4502	2.4168
	0.01	3.3045	3.1238	3.0202	2.953	2.8713	2.8233
8	0.001	4.8173	4.4355	4.2224	4.0864	3.9232	3.8286
	0.1	1.8841	1.8289	1.7963	1.7748	1.7483	1.7324
	0.05	2.2662	2.1802	2.1299	2.097	2.0564	2.0323
	0.025	2.6513	2.5289	2.4579	2.4117	2.3549	2.3215
	0.01	3.1726	2.993	2.89	2.8233	2.742	2.6943
9	0.001	4.5814	4.207	3.998	3.8648	3.7049	3.6123
	0.1	1.849	1.7929	1.7598	1.738	1.711	1.6949
	0.05	2.2107	2.124	2.0734	2.0401	1.9991	1.9748
	0.025	2.5746	2.4519	2.3808	2.3344	2.2775	2.2439
	0.01	3.0665	2.8876	2.785	2.7185	2.6374	2.5898
10	0.001	4.393	4.0243	3.8185	3.6873	3.5298	3.4387
	0.1	1.8195	1.7627	1.7291	1.707	1.6796	1.6632
	0.05	2.1646	2.0772	2.0261	1.9926	1.9512	1.9267
	0.025	2.5112	2.3882	2.3168	2.2702	2.213	2.1793
	0.01	2.9791	2.8005	2.6981	2.6318	2.5508	2.5033
11	0.001	4.2388	3.8744	3.6711	3.5415	3.3859	3.2959
	0.1	1.7944	1.7369	1.7029	1.6805	1.6526	1.636
	0.05	2.1256	2.0376	1.9861	1.9522	1.9105	1.8857
	0.025	2.4577	2.3343	2.2627	2.2159	2.1584	2.1245
	0.01	2.9057	2.7274	2.625	2.5587	2.4777	2.4302
12	0.001	4.11	3.749	3.5476	3.4193	3.2652	3.176
	0.1	1.7727	1.7146	1.6802	1.6574	1.6292	1.6124

Table B6. Critical values of F distribution

	α						
	0.05	2.0921	2.0035	1.9515	1.9174	1.8753	1.8503
	0.025	2.412	2.2882	2.2162	2.1692	2.1115	2.0773
	0.01	2.8431	2.6648	2.5625	2.4961	2.4151	2.3676
	0.001	4.0006	3.6425	3.4426	3.3153	3.1624	3.0739
15	0.1	1.7223	1.6624	1.6269	1.6034	1.5741	1.5566
	0.05	2.0148	1.9245	1.8714	1.8364	1.7932	1.7675
	0.025	2.3072	2.1819	2.109	2.0613	2.0026	1.9679
	0.01	2.7002	2.5216	2.419	2.3523	2.2709	2.223
	0.001	3.7527	3.4003	3.2035	3.0781	2.9274	2.8402
18	0.1	1.6862	1.6249	1.5884	1.5642	1.534	1.516
	0.05	1.9601	1.8682	1.8141	1.7784	1.7342	1.7079
	0.025	2.2334	2.1068	2.033	1.9846	1.925	1.8897
	0.01	2.6003	2.421	2.3178	2.2507	2.1686	2.1203
	0.001	3.581	3.2318	3.0367	2.9123	2.7627	2.6761
20	0.1	1.6673	1.6052	1.5681	1.5435	1.5128	1.4943
	0.05	1.9317	1.8389	1.7841	1.748	1.7032	1.6764
	0.025	2.1952	2.0677	1.9933	1.9445	1.8843	1.8486
	0.01	2.5487	2.3689	2.2652	2.1978	2.1153	2.0666
	0.001	3.4928	3.145	2.9506	2.8266	2.6774	2.5909
25	0.1	1.6316	1.5677	1.5294	1.5039	1.472	1.4528
	0.05	1.8782	1.7835	1.7273	1.6902	1.644	1.6163
	0.025	2.1237	1.9943	1.9186	1.8687	1.8071	1.7705
	0.01	2.4526	2.2714	2.1667	2.0984	2.0146	1.9652
	0.001	3.3296	2.9838	2.7902	2.6665	2.5176	2.4311
30	0.1	1.6065	1.5411	1.5018	1.4755	1.4426	1.4227
	0.05	1.8409	1.7444	1.6872	1.6491	1.6017	1.5733
	0.025	2.0739	1.9429	1.8659	1.8152	1.7523	1.7148
	0.01	2.386	2.2034	2.0976	2.0285	1.9435	1.8933
	0.001	3.2171	2.8721	2.6787	2.5549	2.4057	2.3189
40	0.1	1.5732	1.5056	1.4648	1.4373	1.4027	1.3817
	0.05	1.7918	1.6928	1.6337	1.5943	1.5449	1.5151
	0.025	2.0089	1.8752	1.7963	1.744	1.679	1.6401
	0.01	2.2992	2.1142	2.0066	1.936	1.8489	1.7972
	0.001	3.0716	2.7268	2.5329	2.4086	2.2582	2.1704
50	0.1	1.5522	1.483	1.4409	1.4126	1.3767	1.3548
	0.05	1.7609	1.66	1.5995	1.559	1.5081	1.4772
	0.025	1.9681	1.8324	1.752	1.6985	1.6318	1.5917
	0.01	2.245	2.0581	1.949	1.8772	1.7883	1.7353
	0.001	2.9813	2.636	2.4413	2.3162	2.1644	2.0756
60	0.1	1.5376	1.4672	1.4242	1.3952	1.3583	1.3356
	0.05	1.7396	1.6373	1.5757	1.5343	1.4821	1.4504
	0.025	1.94	1.8028	1.7211	1.6668	1.5987	1.5575
	0.01	2.2079	2.0194	1.909	1.8363	1.7459	1.6918
	0.001	2.9196	2.5737	2.3782	2.2523	2.0992	2.0094
80	0.1	1.5187	1.4465	1.4023	1.3722	1.3337	1.31
	0.05	1.7121	1.6077	1.5445	1.5019	1.4477	1.4146
	0.025	1.9039	1.7644	1.681	1.6252	1.5549	1.5122

Table B6. Critical values of F distribution

	0.01	2.1601	1.9694	1.8571	1.7828	1.6901	1.6342
	0.001	2.8407	2.4934	2.2965	2.1693	2.0139	1.9224
100	0.1	1.5069	1.4336	1.3885	1.3576	1.318	1.2934
	0.05	1.695	1.5892	1.5249	1.4814	1.4259	1.3917
	0.025	1.8816	1.7405	1.6558	1.599	1.5271	1.4833
	0.01	2.1307	1.9383	1.8248	1.7493	1.6548	1.5977
	0.001	2.7923	2.4439	2.2458	2.1175	1.9604	1.8674

Table B7. Wilcoxon Rank-Sum Test (Mann-Whitney Test) critical values

1-tail	α = 0.025			α = 0.05			1-tail	α = 0.025			α = 0.05		
2-tail	α = 0.05			α = 0.10			2-tail	α = 0.05			α = 0.10		
m	n	W	d	P	W	d	P	m	n	W	d	P	W	d	P
---	---	---	---	---	---	---	---	---	---	---	---	---	---	---	---
3	3				6 15	1	.0500	5	10	23 57	9	.0200	26 54	12	.0496
3	4				6 18	1	.0286	5	11	24 61	10	.0190	27 58	13	.0449
3	5	6 21	1	.0179	7 20	2	.0357	5	12	26 64	12	.0242	28 62	14	.0409
3	6	7 23	2	.0238	8 22	3	.0476	5	13	27 68	13	.0230	30 65	16	.0473
3	7	7 26	2	.0167	8 25	3	.0333	5	14	28 72	14	.0218	31 69	17	.0435
3	8	8 28	3	.0242	9 27	4	.0424	5	15	29 76	15	.0209	33 72	19	.0491
3	9	8 31	3	.0182	10 29	5	.0500	5	16	30 80	16	.0201	34 76	20	.0455
3	10	9 33	4	.0245	10 32	5	.0385	5	17	32 83	18	.0238	35 80	21	.0425
3	11	9 36	4	.0192	11 34	6	.0440	5	18	33 87	19	.0229	37 83	23	.0472
3	12	10 38	5	.0242	11 37	6	.0352	5	19	34 91	20	.0220	38 87	24	.0442
3	13	10 41	5	.0196	12 39	7	.0411	5	20	35 95	21	.0212	40 90	26	.0485
3	14	11 43	6	.0235	13 41	8	.0456	5	21	37 98	23	.0243	41 94	27	.0457
3	15	11 46	6	.0196	13 44	8	.0380	5	22	38 102	24	.0234	43 97	29	.0496
3	16	12 48	7	.0237	14 46	9	.0423	5	23	39 106	25	.0226	44 101	30	.0469
3	17	12 51	7	.0202	15 48	10	.0465	5	24	40 110	26	.0219	45 105	31	.0445
3	18	13 53	8	.0233	15 51	10	.0398	5	25	42 113	28	.0246	47 108	33	.0480
3	19	13 56	8	.0201	16 53	11	.0435	6	6	26 52	6	.0206	28 50	8	.0465
3	20	14 58	9	.0232	17 55	12	.0469	6	7	27 57	7	.0175	29 55	9	.0367
3	21	14 61	9	.0203	17 58	12	.0410	6	8	29 61	9	.0213	31 59	11	.0406
3	22	15 63	10	.0230	18 60	13	.0443	6	9	31 65	11	.0248	33 63	13	.0440
3	23	15 66	10	.0204	19 62	14	.0473	6	10	32 70	12	.0210	35 67	15	.0467
3	24	16 68	11	.0229	19 65	14	.0421	6	11	34 74	14	.0238	37 71	17	.0491
3	25	16 71	11	.0205	20 67	15	.0449	6	12	35 79	15	.0207	38 76	18	.0415
4	4	10 26	1	.0143	11 25	2	.0286	6	13	37 83	17	.0231	40 80	20	.0437
4	5	11 29	2	.0159	12 28	3	.0317	6	14	38 88	18	.0204	42 84	22	.0457
4	6	12 32	3	.0190	13 31	4	.0333	6	15	40 92	20	.0224	44 88	24	.0474
4	7	13 35	4	.0212	14 34	5	.0364	6	16	42 96	22	.0244	46 92	26	.0490
4	8	14 38	5	.0242	15 37	6	.0364	6	17	43 101	23	.0219	47 97	27	.0433
4	9	14 42	5	.0168	16 40	7	.0378	6	18	45 105	25	.0236	49 101	29	.0448
4	10	15 45	6	.0180	17 43	8	.0380	6	19	46 110	26	.0214	51 105	31	.0462
4	11	16 48	7	.0198	18 46	9	.0388	6	20	48 114	28	.0229	53 109	33	.0475
4	12	17 51	8	.0209	19 49	10	.0390	6	21	50 118	30	.0244	55 113	35	.0487
4	13	18 54	9	.0223	20 52	11	.0395	6	22	51 123	31	.0224	57 117	37	.0498
4	14	19 57	10	.0232	21 55	12	.0395	6	23	53 127	33	.0237	58 122	38	.0452
4	15	20 60	11	.0243	22 58	13	.0400	6	24	54 132	34	.0219	60 126	40	.0463
4	16	21 63	12	.0250	24 60	15	.0497	6	25	56 136	36	.0231	62 130	42	.0473
4	17	21 67	12	.0202	25 63	16	.0493	7	7	36 69	9	.0189	39 66	12	.0487
4	18	22 70	13	.0212	26 66	17	.0491	7	8	38 74	11	.0200	41 71	14	.0469
4	19	23 73	14	.0219	27 69	18	.0487	7	9	40 79	13	.0209	43 76	16	.0454
4	20	24 76	15	.0227	28 72	19	.0485	7	10	42 84	15	.0215	45 81	18	.0439
4	21	25 79	16	.0233	29 75	20	.0481	7	11	44 89	17	.0221	47 86	20	.0427
4	22	26 82	17	.0240	30 78	21	.0480	7	12	46 94	19	.0225	49 91	22	.0416
4	23	27 85	18	.0246	31 81	22	.0477	7	13	48 99	21	.0228	52 95	25	.0484
4	24	27 89	18	.0211	32 84	23	.0475	7	14	50 104	23	.0230	54 100	27	.0469
4	25	28 92	19	.0217	33 87	24	.0473	7	15	52 109	25	.0233	56 105	29	.0455
5	5	17 38	3	.0159	19 36	5	.0476	7	16	54 114	27	.0234	58 110	31	.0443
5	6	18 42	4	.0152	20 40	6	.0411	7	17	56 119	29	.0236	61 114	34	.0497
5	7	20 45	6	.0240	21 44	7	.0366	7	18	58 124	31	.0237	63 119	36	.0484
5	8	21 49	7	.0225	23 47	9	.0466	7	19	60 129	33	.0238	65 124	38	.0471
5	9	22 53	8	.0210	24 51	10	.0415	7	20	62 134	35	.0239	67 129	40	.0460

	1-tail 2-tail	$\alpha = 0.025$ $\alpha = 0.05$			$\alpha = 0.05$ $\alpha = 0.10$			1-tail 2-tail	$\alpha = 0.025$ $\alpha = 0.05$			$\alpha = 0.05$ $\alpha = 0.10$		
m	n	W	d	P	W	d	P	m	n	W	d	P	W	d	P
7	21	64 139	37	.0240	69 134	42	.0449	10	20	110 200	56	.0245	117 193	62	.0498
7	22	66 144	39	.0240	72 138	45	.0492	10	21	113 207	59	.0241	120 200	65	.0478
7	23	68 149	41	.0241	74 143	47	.0481	10	22	116 214	62	.0237	123 207	68	.0459
7	24	70 154	43	.0241	76 148	49	.0470	10	23	119 221	65	.0233	127 213	72	.0482
7	25	72 159	45	.0242	78 153	51	.0461	10	24	122 228	68	.0230	130 220	75	.0465
8	8	49 87	14	.0249	51 85	16	.0415	10	25	126 234	72	.0248	134 226	79	.0486
8	9	51 93	16	.0232	54 90	19	.0464	11	11	96 157	31	.0237	100 153	34	.0440
8	10	53 99	18	.0217	56 96	21	.0416	11	12	99 165	34	.0219	104 160	38	.0454
8	11	55 105	20	.0204	59 101	24	.0454	11	13	103 172	38	.0237	108 167	42	.0467
8	12	58 110	23	.0237	62 106	27	.0489	11	14	106 180	41	.0221	112 174	46	.0477
8	13	60 116	25	.0223	64 112	29	.0445	11	15	110 187	45	.0236	116 181	50	.0486
8	14	62 122	27	.0211	67 117	32	.0475	11	16	113 195	48	.0221	120 188	54	.0494
8	15	65 127	30	.0237	69 123	34	.0437	11	17	117 202	52	.0235	123 196	57	.0453
8	16	67 133	32	.0224	72 128	37	.0463	11	18	121 209	56	.0247	127 203	61	.0461
8	17	70 138	35	.0247	75 133	40	.0487	11	19	124 217	59	.0233	131 210	65	.0468
8	18	72 144	37	.0235	77 139	42	.0452	11	20	128 224	63	.0244	135 217	69	.0474
8	19	74 150	39	.0224	80 144	45	.0475	11	21	131 232	66	.0230	139 224	73	.0480
8	20	77 155	42	.0244	83 149	48	.0495	11	22	135 239	70	.0240	143 231	77	.0486
8	21	79 161	44	.0233	85 155	50	.0464	11	23	139 246	74	.0250	147 238	81	.0490
8	22	81 167	46	.0223	88 160	53	.0483	11	24	142 254	77	.0237	151 245	85	.0495
8	23	84 172	49	.0240	90 166	55	.0454	11	25	146 261	81	.0246	155 252	89	.0499
8	24	86 178	51	.0231	93 171	58	.0472	12	12	115 185	38	.0225	120 180	42	.0444
8	25	89 183	54	.0247	96 176	61	.0488	12	13	119 193	42	.0229	125 187	47	.0488
9	9	62 109	18	.0200	66 105	22	.0470	12	14	123 201	46	.0232	129 195	51	.0475
9	10	65 115	21	.0217	69 111	25	.0474	12	15	127 209	50	.0234	133 203	55	.0463
9	11	68 121	24	.0232	72 117	28	.0476	12	16	131 217	54	.0236	138 210	60	.0500
9	12	71 127	27	.0245	75 123	31	.0477	12	17	135 225	58	.0238	142 218	64	.0486
9	13	73 134	29	.0217	78 129	34	.0478	12	18	139 233	62	.0239	146 226	68	.0474
9	14	76 140	32	.0228	81 135	37	.0478	12	19	143 241	66	.0240	150 234	72	.0463
9	15	79 146	35	.0238	84 141	40	.0478	12	20	147 249	70	.0241	155 241	77	.0493
9	16	82 152	38	.0247	87 147	43	.0477	12	21	151 257	74	.0242	159 249	81	.0481
9	17	84 159	40	.0223	90 153	46	.0476	12	22	155 265	78	.0242	163 257	85	.0471
9	18	87 165	43	.0231	93 159	49	.0475	12	23	159 273	82	.0243	168 264	90	.0496
9	19	90 171	46	.0239	96 165	52	.0474	12	24	163 281	86	.0243	172 272	94	.0486
9	20	93 177	49	.0245	99 171	55	.0473	12	25	167 289	90	.0243	176 280	98	.0475
9	21	95 184	51	.0225	102 177	58	.0472	13	13	136 215	46	.0221	142 209	51	.0454
9	22	98 190	54	.0231	105 183	61	.0471	13	14	141 223	51	.0241	147 217	56	.0472
9	23	101 196	57	.0237	108 189	64	.0470	13	15	145 232	55	.0232	152 225	61	.0489
9	24	104 202	60	.0243	111 195	67	.0469	13	16	150 240	60	.0250	156 234	65	.0458
9	25	107 208	63	.0249	114 201	70	.0468	13	17	154 249	64	.0240	161 242	70	.0472
10	10	78 132	24	.0216	82 128	28	.0446	13	18	158 258	68	.0232	166 250	75	.0485
10	11	81 139	27	.0215	86 134	32	.0493	13	19	163 266	73	.0247	171 258	80	.0497
10	12	84 146	30	.0213	89 141	35	.0465	13	20	167 275	77	.0238	175 267	84	.0470
10	13	88 152	34	.0247	92 148	38	.0441	13	21	171 284	81	.0231	180 275	89	.0481
10	14	91 159	37	.0242	96 154	42	.0478	13	22	176 292	86	.0243	185 283	94	.0491
10	15	94 166	40	.0238	99 161	45	.0455	13	23	180 301	90	.0236	189 292	98	.0467
10	16	97 173	43	.0234	103 167	49	.0487	13	24	185 309	95	.0247	194 300	103	.0476
10	17	100 180	46	.0230	106 174	52	.0465	13	25	189 318	99	.0240	199 308	108	.0485
10	18	103 187	49	.0226	110 180	56	.0493	14	14	160 246	56	.0249	166 240	61	.0469
10	19	107 193	53	.0250	113 187	59	.0472	14	15	164 256	60	.0229	171 249	66	.0466

| 1-tail | | α = 0.025 | | | α = 0.05 | | | 1-tail | | α = 0.025 | | | α = 0.05 | | |
| 2-tail | | α = 0.05 | | | α = 0.10 | | | 2-tail | | α = 0.05 | | | α = 0.10 | | |
m	n	W		d	P	W		d	P	m	n	W		d	P	W		d	P
14	16	169	265	65	.0236	176	258	72	.0463	17	24	282	432	130	.0239	294	420	141	.0492
14	17	174	274	70	.0242	182	266	78	.0500	17	25	288	443	136	.0238	300	431	147	.0480
14	18	179	283	75	.0247	187	275	83	.0495	18	18	270	396	100	.0235	280	386	109	.0485
14	19	183	293	79	.0230	192	284	88	.0489	18	19	277	407	107	.0246	287	397	116	.0490
14	20	188	302	84	.0235	197	293	93	.0484	18	20	283	419	113	.0238	294	408	123	.0495
14	21	193	311	89	.0239	202	302	98	.0480	18	21	290	430	120	.0247	301	419	130	.0499
14	22	198	320	94	.0243	207	311	103	.0475	18	22	296	442	126	.0240	307	431	136	.0474
14	23	203	329	99	.0247	212	320	108	.0471	18	23	303	453	133	.0248	314	442	143	.0478
14	24	207	339	103	.0233	218	328	114	.0498	18	24	309	465	139	.0240	321	453	150	.0481
14	25	212	348	108	.0236	223	337	119	.0492	18	25	316	476	146	.0248	328	464	157	.0484
15	15	184	281	65	.0227	192	273	73	.0488	19	19	303	438	114	.0248	313	428	123	.0482
15	16	190	290	71	.0247	197	283	78	.0466	19	20	309	451	120	.0234	320	440	130	.0474
15	17	195	300	76	.0243	203	292	84	.0485	19	21	316	463	127	.0236	328	451	138	.0494
15	18	200	310	81	.0239	208	302	89	.0465	19	22	323	475	134	.0238	335	463	145	.0486
15	19	205	320	86	.0235	214	311	95	.0482	19	23	330	487	141	.0240	342	475	152	.0478
15	20	210	330	91	.0232	220	320	101	.0497	19	24	337	499	148	.0241	350	486	160	.0496
15	21	216	339	97	.0247	225	330	106	.0478	19	25	344	511	155	.0243	357	498	167	.0488
15	22	221	349	102	.0243	231	339	112	.0492	20	20	337	483	128	.0245	348	472	138	.0482
15	23	226	359	107	.0239	236	349	117	.0474	20	21	344	496	135	.0241	356	484	146	.0490
15	24	231	369	112	.0235	242	358	123	.0486	20	22	351	509	142	.0236	364	496	154	.0497
15	25	237	378	118	.0248	248	367	129	.0499	20	23	359	521	150	.0246	371	509	161	.0478
16	16	211	317	76	.0234	219	309	84	.0469	20	24	366	534	157	.0242	379	521	169	.0484
16	17	217	327	82	.0243	225	319	90	.0471	20	25	373	547	164	.0237	387	533	177	.0490
16	18	222	338	87	.0231	231	329	96	.0473	21	21	373	530	143	.0245	385	518	154	.0486
16	19	228	348	93	.0239	237	339	102	.0474	21	22	381	543	151	.0249	393	531	162	.0482
16	20	234	358	99	.0247	243	349	108	.0475	21	23	388	557	158	.0238	401	544	170	.0478
16	21	239	369	104	.0235	249	359	114	.0475	21	24	396	570	166	.0242	410	556	179	.0497
16	22	245	379	110	.0242	255	369	120	.0476	21	25	404	583	174	.0245	418	569	187	.0492
16	23	251	389	116	.0248	261	379	126	.0476	22	22	411	579	159	.0247	424	566	171	.0491
16	24	256	400	121	.0238	267	389	132	.0476	22	23	419	593	167	.0244	432	580	179	.0477
16	25	262	410	127	.0243	273	399	138	.0476	22	24	427	607	175	.0242	441	593	188	.0486
17	17	240	355	88	.0243	249	346	97	.0493	22	25	435	621	183	.0240	450	606	197	.0494
17	18	246	366	94	.0243	255	357	103	.0479	23	23	451	630	176	.0249	465	616	189	.0499
17	19	252	377	100	.0243	262	367	110	.0499	23	24	459	645	184	.0242	474	630	198	.0497
17	20	258	388	106	.0242	268	378	116	.0485	23	25	468	659	193	.0246	483	644	207	.0495
17	21	264	399	112	.0242	274	389	122	.0473	24	24	492	684	193	.0241	507	669	207	.0486
17	22	270	410	118	.0241	281	399	129	.0490	24	25	501	699	202	.0241	517	683	217	.0496
17	23	276	421	124	.0240	287	410	135	.0477	25	25	536	739	212	.0247	552	723	227	.0497

Table B8. Wilcoxon Signed Ranks Test

n	Two Tails Test		One Tail Test	
	α = .05	α = .01	α = .05	α = .01
5	--	--	0	--
6	0	--	2	--
7	2	--	3	0
8	3	0	5	1
9	5	1	8	3
10	8	3	10	5
11	10	5	13	7
12	13	7	17	9
13	17	9	21	12
14	21	12	25	15
15	25	15	30	19
16	29	19	35	23
17	34	23	41	27
18	40	27	47	32
19	46	32	53	37
20	52	37	60	43
21	58	42	67	49
22	65	48	75	55
23	73	54	83	62
24	81	61	91	69
25	89	68	100	76
26	98	75	110	84
27	107	83	119	92
28	116	91	130	101
29	126	100	140	110
30	137	109	151	120

Table B9. Critical Values for Spearman Rank Correlation (two-tailed)

Sample Size	Level of Significance .05	Level of Significance .01	Sample Size	Level of Significance .05	Level of Significance .01
6	.886		19	.462	.608
7	.786		20	.450	.591
8	.738	.881	21	.438	576
9	.683	.833	22	.428	.562
10	.648	.818	23	.418	.549
11	.623	.794	24	.409	.537
12	.591	.780	25	.400	.526
13	.566	.745	26	.392	.515
14	.545	.716	27	.385	.505
15	.525	.689	28	.377	.496
16	.507	.666	29	.370	.487
17	.490	.645	30	.364	.478
18	.476	.625			

CPSIA information can be obtained
at www.ICGtesting.com
Printed in the USA
BVHW052154210120
570123BV00004B/8